On the origin of Mind

Life's need to re-represent itself

Part I

Martin Wurzinger

10th anniversary print edition

Original digital edition

Website
http://www.otoom.net

Contact
http://www.otoom.net/contact.htm

Printed by CreateSpace

Available from Amazon.com and other retail outlets

ISBN 978-0-9944053-0-2

Cover:
The image on the front cover is a Mandelbrot set, especially configured to represent the fractal nature of complex dynamic systems.
The large image on the back cover is part of that Mandelbrot set, magnified almost 1 million times.

Contents

To Galileo Galilei and his kind, who have to steer their ship of reason amongst the rocks of superstition.

1. The start of the journey

It is an old cliché that everything which is defined by humans has a beginning and an end. Our mind is responsible for that—it's the way it has been taught. *Why* and *how* such an arrangement has come to be (and many other phenomena besides) is the subject matter of this book.

We begin with some of the major philosophies that have come to us through the ages. They are useful for two reasons. They present explicitly described systems of ideas, formed and connected according to rules which are interesting in themselves because they tell us something about the assumed values and priorities of their people. When it comes to the European intellectual development they also provide us with the conceptual building blocks out of which have emerged the responses, attacks, and substitutes we in the West are living with today. By identifying these streams and their effects we have the opportunity to learn about the mechanics behind them, just as one would reverse-engineer a system that obviously operates, but whose internal functionalities are currently hidden. It requires a top-down approach, where identifiable phenomena are described through meaningful labels, which then in turn become the subsystems that may suggest further deconstruction and more detailed naming. We will gain a reasonable understanding about the mind's broad cognitive dynamics by tracing such thoughts, comparing their influences upon each other and what mental images result from these processes. This is not about testing the belief systems for their ultimate truth value; we are interested in the way the mind works, not so much in the data themselves. Having written and detailed records of human musings allows us to do just that. We gradually build up our insights into thought structures, their dynamics and their organisation. Although later philosophies and/or religions tend to be more complex than earlier ones (and we will see why) using some of their elements for the purpose of explaining a certain dynamic in terms of thought structures does not imply that these elements are a comprehensive representation of this or that philosophy. This should be born in mind, especially in the beginning when the emerging picture is still relatively simple.

In Part II we continue by investigating how a bunch of slightly different cells in an organism could have grown into a comprehensive system that *thinks*. The knowledge derived from Part I will have made it easier to envisage its very first beginnings, to see how it is possible that a potentially active state of latent dynamics (*whatever* its material nature) can turn into a multitudinous representation of sensory input, at once proactive and reactive.

Then we will look at the things that bunch of neurons does. By then we will have acquired useful interpretative tools which help us observe the principle goings-on in what can be called the *emerging mind*, struggling as it does at this stage to come to terms with the multi-faceted environment of its surroundings. We will follow the thoughts from their simple beginnings to ever more complex structures, until we come to understand not only their interweaving patterns but also their abstractions, which in turn form intrinsic mosaics, connected with each other but also with their lower level origins and their higher level progeny. This is the powerhouse which produces scientific theories, epics, and symphonies—but also superstitious terror. In contrast to Part I we now employ a bottom-up approach, where one function-detail after another is grouped together to create larger and larger concepts until the entire system has been covered.

And finally, having observed and described that inner space of ours, can we but muse about the purposes we put it to, our grace as we stock it with material, and about the selectivity we apply when it comes to the travels we conduct amongst its vast and awesome landscape. To gaze into this majestic mirror will be the prize and the duty we have to pay for having come into the extended possession of such a precious instrument.

But first there are a few points which have to be addressed. When it comes to the mind we are not dealing with just any object; this is a special case, and so the mental preparations will have to be diligent and particular.

Since we like having things organised in the form of start, middle, and end, it makes sense to build our initial investigation of the mind around what is familiar, and therefore comfortable. This is not merely another way of saying "let's take it easy". Just as a tool gives better service if it is used the way it has been designed, so will any system shine if it is allowed to perform in harmony with its own self. The mind is no exception, although we have to be careful with our analogies. To call something a *tool* implies it being an object which is handled by someone, and under a higher agenda at that. To call the mind a tool would mean to say that here is a system which is used by some agent in order to perform a certain service. There is no doubt the mind performs a service, but who is its master?

If we say "us" then we are forced to ascribe the control we are supposed to exercise over our minds to yet another agent; but isn't the act of control precisely what our minds are meant to do? Continuing with our argument will very quickly lead into deep philosophical and psychological waters, and at this point we are not yet equipped for such an adventure. So let us leave this question for later, except for the realisation that words are very powerful tools in themselves, and it is important to take great care in how we apply them. This advice is doubly significant for the present topic.

The reason for the significance lies in a fundamental fact: the topic of mind is the only one in our entire existence where the object and subject are one and the same. Nothing more than the mind investigating the whole of itself—now there is a challenge!

One has to be very careful therefore how one proceeds. A wrong step (or word in this case) can evoke a context which possesses an undercurrent of meaning too subtle to be understood immediately; yet a however slight prodding towards this or that direction will sooner or later result in a significant course difference, and there is not even a guarantee the difference will be understood in time. If there is an error of definition in any other topic the feedback process, carried as it is by perceptions from the outside, facilitates a correction somewhere along the way. For example, if somebody describes to me how to fix something on a car, then in the case of my misinterpretation of the advice the problem will become obvious rather quickly once I start turning the wrong screw or whatever. But if I—that is, my mind—carries a certain thought the meaning of which is not quite right, who is there to look over my shoulder and point this out? Moreover, if such a thought forms part of a larger train then a minute error can very quickly turn into a major problem if the mistake is unwittingly propagated down the line. Alfred Korzybski, in his book *Science and Sanity* [1], warns,

> We do not realise what tremendous power the structure of habitual language has. It is not an exaggeration to say that it enslaves us through the mechanism of semantic reactions and that the structure which a language exhibits, and impresses upon us subconsciously, is 'automatically projected' upon the world around us.

I hope I do not seem to be labouring the point. Yet the problem of a system analysing its own self does produce many dangers, and throughout human history the effects of these dangers have made themselves felt.

One more point about words. Thinkers from many different areas have grappled with the analysis, the understanding, even the modeling of the mind; philosophy, psychology, cognitive science, social and political science, artificial intelligence, to name some. Since the focus of these individuals had often been a differing one, their manners of description and therefore their vocabulary tended to reflect their varying perspectives and even their priorities. It would be quite confusing to mix their languages indiscriminately, but it would be equally inappropriate to centre on one set of expressions and disregard the rest. My solution to the problem could be described as an egocentric compromise: since the following text proceeds in a sequential manner, words are used that come 'naturally' in the course of the explanation (mostly, these would be my own). If, on the other hand, a certain aspect features so prominently in one particular area that its description is really the only one which seems fitting, then its vernacular is used. In any case, the bibliography provides the reader with the references to texts where the authors' original expressions can be readily accessed.

Here are two examples right away, *thought structures* and *thought patterns*. If we disregard for a moment the question of what constitutes thought itself, and how an idea can be represented amongst the brain's neurons, we can settle on an image for thoughts as being a construct of whatever it is that thoughts are composed of. For the purposes of dealing with thought dynamics in a functional-semantic sense such an imagery serves us quite well. If one conjures up a picture of some structure which consists of words, of sentences, of qualities and characteristics and whatever else one might include, and comes up with a kind of lattice work of imaginary struts and planks, held together by the relationships of words, sentences, etc, in the form of nuts and bolts, then what we have is certainly not a true replica of what is in our heads; it is however a useful model to visualise the connectivities, the extent, and the mutual relationships that exist within it. For that reason we are going to use the term thought structures to refer to not just a single thought, but an idea, something which is composed of many thoughts. It allows us to play around with its contents through the use of descriptives which do not imply any other unwanted interpretations.

Thought patterns are the next step up from thought structures. They are like buildings made out of the rods-and-beams blocks mentioned before, where these form identifiable but separate units which, when put together, give the whole an overall character. Various words have been used by others to describe what are called thought structures and patterns here; for our investigations the terms include a sufficient degree of evoked functionality to make talking about such things easier, but at the same time do not create too much complexity in the reader's mind to permit interpretations which were not on the author's agenda.

The title of the book is an allusion to Darwin's *On the Origin of Species*, because both undertakings contain certain similarities. In Darwin's case the work provided a new and challenging explanation for the existence of species in nature, their evolutionary patterns and dynamics. Through this panorama our outlook widened by engaging a more encompassing perspective, and it allowed us to view nature in its various manifestations without the parochial tendencies that arise out of centric attitudes. It enabled humans to be seen as another element of evolution, part and parcel of a world full of variety, yet all derived from common building blocks. As so many other human endeavours it was an important step in the quest for perceiving reality. And, again like so many other endeavours, it antagonised those groups and individuals who use their minds not to understand but to prescribe, not to search but to hide.

This book seeks to open our understanding to yet another part of existence. Apart from deep space it is the least known entity so far, its ultimate complexity and its dynamics only now just being touched upon. It could even be argued that it is more awesome than deep space itself, at least from the human perspective. After all, for anything to be apparent to us it has to exist in our mind. It is possible for a human to imagine planets, suns, galaxies—and still there is room for more. Ultimately, we do not really live and move anywhere; it is the image our mind presents us which we hold as arbiter, and nothing is thought, or done, or acted upon except that which exists in some distributed form amongst the millions of neurons in our brain. And yet it is also sometimes true that what appears to be simple for one person is incomprehensible to the next—an unbreachable gulf which separates cultures, societies, and indeed individuals.

Such arbitrary difference between what seems to be a reality common to us all and the imagery of the inner space has led to the assumption that there is a profound dichotomy dividing body and mind. It is a theme returned to again and again by philosophers from many ages. This idea of body vs mind affects the image we have of our own selves, our position in the greater scheme of things and how we ought to react to the environmental forces perceived from day to day. Just as in pre-Darwinian times our humano-centric view directed us to assume a particular uniqueness amongst nature, so does the mind-body differentiation provide us with the justification to imbue the phenomenon of mind with a licence to control, yet at the same time impose a set of restrictions that drive many, if not most, of our plans and endeavours. We will see how false such an image really is, and how illusory therefore our presumptuous authority can be; and how, by freeing ourselves from the restrictions derived from those boundaries, the mind can be utilised much better.

There is a further parallel to Darwin's work. Much of the material for its insights came from an extensive journey around the globe, observing and analysing species in many lands. Only through such a wide-ranging research could a comprehensive picture emerge, because in the end it takes the experience of reality at large to see and therefore to understand.

The observations presented in this book also come from many parts of the world (with much of it visited under sail as well, and all of it as a traveller—not tourist). People and individuals in forty different countries provided the background through their speech, their mannerism, their cultural behaviour. It does not follow however that theoretical thinking and controlled experiments have no place in an undertaking such as this. On the contrary, if a particular activity can be reproduced within a specifically created environment and still come up with results that confirm the overall model, so much the better. On the other hand, the very act of setting up an experiment implies the special focus the experimenter is taking. In other words, there is a certain expectation which one sets out to prove. For the purpose of surveying an unfamiliar landscape where its features are largely unknown, such an approach is rather unproductive. Therefore, if anything-can-happen is the name of the game, it is best to go 'out there' and observe whatever presents itself. At the same time, in the course of our investigations we will visit many individuals who have through their own research, sometimes coming from quite surprising directions, shed light upon the intrinsic dynamics of the mind.

Such findings are interesting for two reasons. Firstly, they provide a kind of 'inside-out' confirmation to the wider reference which is employed in this book. A result from a relatively narrow focus which also links positively with the greater picture demonstrates a two-way validity that is certainly satisfying. If a description of an object concurs with the description of one of its parts, then the degree of certainty of both views is being raised. We could term this the unified field phenomenon, and throughout our discourse we will experience it many times.

Secondly, given that we investigate the mind itself and its associated activities, it is always fascinating to observe the process through which it works itself from a question (which in itself is significant) to the intermediate stages of musings to an eventual result. The very nature of the result becomes then a matter of interest. It is the old 'black-box' phenomenon all over again, where there is a system whose technical innards are unknown to us, but whose input and output we are able to identify, at least to a certain degree. Comparing what has entered the system with that what eventually emerges, we can make a pretty good guess as to the nature of the processes at work here. In the case of the mind (our black box in this case) we know there are neurons that are responsible for processing information, but we do not know their functional properties to the extent that they can explain what actually happens to the data which enter the brain via its nerve strands. We do however have the reaction of the person to go by, and through a comparison of the input (visual signals, spoken words, justifiably assumed experiences and ideas), with the output (spoken sentences, expressions via gestures, drawings, actions and initiatives taken) we can reason about the connections between the two.

Such a procedure is not as simple as it sounds. Anyone who has ever traveled a foreign country will know of encounters where the behaviour of a local had to be carefully interpreted because the cultural differences made a direct referencing impossible. In essence this problem exists all the time. Every perception we receive from another human being relies ultimately on its interpretation within our own mind, not the other's. Only common experience allows a degree of certainty, and most cultures have developed ways and means to steer around potential problem situations due to misunderstandings—they're called manners. Without these social graces our lives would be unnerving at best, if not downright dangerous.

In the context of this book the interpretation of human behaviour has been done through a process of detailed identification of the person's condition, and not the observer's. While the identification is *done* by the observer, nonetheless the result has not been accepted unless and until the condition of the observed had been satisfactorily identified in terms of the phenomenon in question. If that sounds somewhat obtuse I only ask to bear with me until we get to the first examples, when the method will become clear.

There is one detail that should be dealt with right here and now, and it concerns the investigative method per se. Consider the task of teaching a course for prospective car mechanics. You would have to assume that your students know next to nothing about engines etc, not because this may or may not be true but because the course ought to guide the newcomer through the various stages in a systematic, comprehensive way. In order to be systematic you need to know what the students know, and the only way to be sure about that is to provide the knowledge yourself. This means that you, the teacher, introduce the various elements which gradually come together to form the whole picture. It has to be done sensibly and logically for a student to be able to understand what you are teaching.

Now consider explaining the mind. Immediately there are two fundamental problems: one, the very thing that is doing the learning is the one which is being learnt about (I briefly touched on that just before) and two, everyone has a mind to begin with. It is like a car teaching cars about cars! If I wanted to explain the concept of a piston moving inside a cylinder, I could use a temporary example of a tube and a plug, demonstrating the changing volume while talking about the forces at work. In other words, I take recourse to another object, while drawing similarities between both concepts to illustrate the unfamiliar through the familiar. The more familiar a topic becomes, the easier it will be to build upon existing knowledge and so the learning curve will steepen as time goes on.

This approach does not work well when dealing with the mind. For one thing, there is no similarly functioning object I could use for comparison, no matter how far I am prepared to stretch the point. We are stuck inside our minds and that's that. Moreover, the only way I can access anyone's brain is through words or signs—but since it is the functionality we are concerned with here I have to rely on the recipient's own functionality to explain how a word works, even if what I talk about is meant to be taken differently from what the very act of talking induces in the other person. To return to the analogy with cars again, it is like a teacher-car holding up an axle before a class of student-cars and explaining an entirely new idea about the axle—but the only thing the students have access to is their own axles and how *they* have used them all along. I strongly suspect these are the underlying circumstances which have prevented us from fully understanding ourselves despite centuries-long efforts; and yet in a technical sense it is hardly touched upon in the relevant literature.

So, since this book is being written not merely for my own amusement (for want of a better word!) but is supposed to help its readers in understanding our condition, certain measures have to be taken to overcome the problem as best as possible. Firstly, I endeavoured to make my analogies as simple and as straightforward as possible (a point I alluded to already when I mentioned thought structures and patterns). In this case analogies can only work if their conceptual domain is quite limited, so that the reader cannot 'think into' a comparison something which should not be there. Secondly, since it is impossible to identify any actual spot inside the brain which contains say, a 'house' or a 'butterfly', let alone a wider-spanning idea, we need to corral the focus of our attention by using the forthcoming assortment of reactions that have been triggered by our object of interest. A brief example: suppose I wanted to describe the existence of 'honour' in a group of people, and what role such a concept plays in their affairs. I would illustrate a number of circumstances (represented by defined situations or expressions the people gave) which would point to something like 'honour', each in their own particular way. While one single incident does not explicitly mark the presence of the idea, taken in their sum-total however they do circumscribe the occurrence of *something*, and the more obvious that *something* is made to become through all these incidents the more easily it can be identified for what it is. While such an approach may sound cumbersome at first, it has the advantage of portraying the subtle nuances through which a concept may exist and—especially useful in terms of the above mentioned problem—it prepares the reader for the general context in which the idea should be placed. There is also a cost: the disassociation of my own moral judgment throughout the process, as well as the trust placed in the reader to attempt to do likewise. Alas, I have no control over the latter.

It may seem that these procedures are not as clear-cut and unassailable as an objectively measured, controlled experiment. In terms of discrete value units, of repeatable environments, and neutral observation, they certainly do come off second-best. For that reason I would never argue against controlled experiments in any way or form (indeed, on many occasions we will draw on their results). At the same time, a preconfigured environment is exactly that: preconfigured. Apart from shutting out any chance events as I mentioned earlier, there is another drawback. Because the quality of thought structures and their dynamics will always reflect the quality of their surroundings, nothing dramatically new can ever be gained by an *a priori* allocation of values under which the subject is expected to perform (unless of course there is a definite incentive for the subject to behave in a certain manner, but that's another story). While the reader will not have access to statistical sheets, print-outs of results, etc, the comprehensive description they will encounter will nevertheless provide the opportunity to test the situation for themselves. This, I hope, will compensate for the lack of any third-party material.

One more point. Nothing is created by anyone without an underlying reason, a trigger which provided the impetus to begin and to continue. In this case it was the stark realisation of how seemingly arbitrarily our composed values differ from one place or time to the next. Is it not a case for wonderment (and yes, sometimes despair) how a system of values appears so absolute, so indifferent to one's doubts, and yet is nothing more than a confluence of ideas, cemented into a structure, from thereon to hold sway over our thinking processes and so over our very lives? Consider for instance the care, the diligence, the sheer love and respect afforded to a people's effort such as the making of champagne in France—and how, somewhere else in another camp of an Islamic culture, the drinking of alcohol is judged harshly and draconian. Or, how a man's love for a boy has been elevated to a society's pantheon in ancient Greece, to be symbolised by the relationship between none other than the god Zeus himself and his lover Ganimede—and how today, at least in a Judaeo-Christian environment, the very same sentiments are relegated to the realm of perversity and abhorrence (note to the reader: observe your own emotions as you read this). What fundamental thoughts and conclusions are honed through such directives, and what subtle sensing, what secret niches are closed and hidden from the inner view by these acts of war!

To consider our human condition from that angle requires more than idle curiosity. It draws us to the foundations upon which our mental lives per se are performed. Is it surprising that throughout history the most illustrious civilisations have been those whose citizens enjoyed the greatest freedom of expression, the wildest frames of reference?

This then ought to be the true starting point: the awareness that arises once the deposits of coercion, of power, of belligerence have been slowly washed away. It is a point only in the semantic sense. For it is the entry to the vastness of what is, rather than what is allowed to be. This, in a very real sense, represents the true dimension of our universe. Thus the present book does not deny the existence of reality; it merely asserts its mental counterpart as ultimately the only one we have meaningful access to.

I have taken some time to outline the intellectual requirements necessary for our journey. They are like the provisions of a traveller embarking upon a long and adventuresome quest, placed in his luggage by a well-meaning inn-keeper. It behooves the traveller to take care of these provisions and use them with diligence, for the inn-keeper knows well what awaits the voyager. How valuable the journey turns out to be in the end will depend greatly upon how you, the reader, have made use of your supplies. This in turn affects the accounts you will give to others, and whether ultimately the places visited by yourself are of value to them.

It is time to take the road.

Part I:
The Stated

2. The mind-body myth

Socrates - Plato - Aristotle

The main philosophical framework of our Western culture incorporates the notion that there is a fundamental difference between the realm of ideas, thoughts, and feelings, and the physical world manifest by our bodies and all the other objects—animate and inanimate—we identify around us. We will see why such a distinction has provided the intellectual comfort for us to assert our stature in this form. While it does not assist in a fundamental interpretation of any objective reality, it nevertheless enables a human being to assign very specific qualities to mental manifestations and also assign equally specific but different qualities to matter.

This is important: we speak *to* an entity which we regard as being of mind, whereas we speak *of* another that we see as an object, a thing. By *speaking to* we allow the possibility of response which entails a certain readiness to readjust the current thought patterns, while the act of *speaking of* concentrates its energy on the descriptive processes alone. It is not as trivial as it sounds. By making allowances for being spoken to in return, the cognitive part of ourselves which is concerned with the ongoing organisation of thinking includes the thought structures which are concerned with the self, and so prepares the speaker for a certain positioning given the eventual response. This positioning involves the measure of the other person in relation to oneself, the stated and unstated aim of the encounter, a consideration as to the aftermath of that moment, and of course all those ideas which have to do with what a speaker feels the other person thinks about them.

Since most—if not all—these thoughts are by no means transparent, allowance has to be made for their dynamic re-interpretation as the situation proceeds. And because the next instant of that ongoing mental process is not much more predictable than its predecessors, the flexibility that has gone into the formulation of the relevant thought structures is being maintained. This happens on both sides. Both persons will employ their minds to a similar degree, and whether we are analytically aware of the processes or not, we still assume—however subconsciously—a similar *modus operandi* every time we communicate with an entity we consider having a *mind*.

By contrast, the mental operation with regards to an object contains none of that complexity. Yet because it lacks the inherent adjustment protocol so necessary to deal with someone that can 'think back', our thoughts are now relatively free to engage in a structure creation of a different kind. If I talk about a cup for example I may use words such as 'big' or 'small', but I can also cite measurements which show by their figures exactly what I mean. We can allow ourselves to be 'objective', that is to engage outside arbiters as an added evaluating device, knowing fully well that once a particular arbiter is chosen, its function will be employed to its own specific extent and nothing more. It is when I talk about such figures *to* someone that a whole range of indeterminables immediately enters the picture, and what had appeared to be some simple, unassailable numbers a moment ago are now quite possibly turned into political weapons.

An interesting demonstration of just that phenomenon can be found in Plato's *The Last Days of Socrates (Euthyphro)* [1]. We know the ancient Greek philosopher's ideas mostly through the writings of his disciple Plato, who generally rendered them in the form of discourses between his master and others. In terms of our present purposes such a format is particularly significant: following the repartees we are able to compare the semantic contents of statements which are designed as a response to another speaker with those that are descriptions of a fact. The trial of Socrates and his resultant, self-induced death in 399 BCE stands somewhat characteristically amongst the grandeur of a culture, where the sheer force of so many intellects still defines European thinking today but where also a mindless fear of the noble gathered its sycophants to destroy. Mindful of Plato's authorship we shall use 'Socrates' essentially as a descriptor.

In this section of the book the scene is outside the courthouse where Socrates is shortly to stand trial, and he and Euthyphro, a seer and expert on religion, are engaged in a discussion on what constitutes piety. Euthyphro, being the expert, is asked about his subject, so that Socrates might come to some definite conclusions about what to think and what to believe. When Euthyphro is about to tell him what he knows of the gods, Socrates interrupts him politely because he is after one simple answer only, namely the definition of piety, and Euthyphro with all the facts at his disposal surely should be able to supply it. According to the 'expert' then, piety is what is agreeable to the gods, whereas what is disagreeable to them is impious, one being the opposite to the other. Socrates then proceeds to take that answer apart, by asking Euthyphro whether in a dispute about the relative size of two numbers a settlement can be reached very quickly by taking recourse to arithmetic; to which Euthyphro readily agrees. Similarly, a disagreement about the different size of two objects can equally be solved by measuring both, and once more Euthyphro concedes that this is so. In these examples the opportunity to become hostile with one another hardly arises but on the other hand, says Socrates, there are many questions about which the two, and indeed the rest of humanity, could very well become hostile; questions relating to such subjects as right and wrong, honour and dishonour, and so on. Socrates now clinches the debate by suggesting that in the case of a concept such as honour we tend to look to the gods but, since even they can be in violent disagreement with each other, the final opinion a person holds depends very much on which god had been summonsed to provide guidance. The end result is that an idea can be both right and wrong or, in terms of the original question, pious and impious at the same time.

The debate then continues with Socrates trying to establish the futility of absolutist decisions about what is pious or impious with regard to moral behaviour, by showing Euthyphro that his preference for divine guidance is quite misplaced because it is illogical. The turns the debate is taking are a delight to follow, but for our purposes let us concentrate on the part I have just described. We can examine the sentiments expressed from the view of thought structures, and how the position taken—especially by someone like Euthyphro—is very much due to the internal organisation of the mind. As such it is not anything Socrates or Plato for that matter may or may not have implied, and I do not want to appear to be ascribing cognitive concepts to the translated scripts of someone who has lived over two thousand years ago (unless that someone would have explicitly stated as much and this is not the case here). Keeping this in mind, let us see what such a perspective may yield.

Socrates uses the example of number size to illustrate the ascertainable quality in something for which there exists a defining device accessible to us all in equal measure. A scale is a scale, and everyone can see its pointer easily enough. The image therefore which is represented in someone's mind can be justifiably assumed to be the same in any other's mind, and since that is a fact accepted by just about everyone the opportunity for disagreement would hardly arise. Contrast this with the idea of 'honour' however: where in one's surroundings is there such a thing which could be looked at and read in the way a scale is read for the observer to declare, "this thing is 2.5 honours"? And yet we all have to some degree a concept of what honour means, but our inner picture of it has not been derived from an objective measuring device.

Instead our picture of honour is a composite of experiences which have been identified as being pertinent because in some way or other they allowed us to make a connection between already existing thought structures and the additional concept, the newcomer having a perceived affinity with the former. Since the 'honour' patterns which had been established previously derived exclusively from their owner's experience through life they are uniquely personal, and therefore it is impossible to reference any part of them by means of objective measuring devices. Notice how Euthyphro had no objection calling upon such devices when the topic was a number or a mere object, but how quickly he turned to the realm of gods when the validity of piety for instance was to be established. One could see that as a simple means of escape to a higher authority, and in a sense of course it is just that; yet such a tactic has deeper roots still.

Any dispute is being conducted with the view towards settlement. In the case of a number that prospect is a relatively immediate one because it is possible to access the objective arbiter without much fuss. When it comes to such things as piety or honour however the argument could go on forever if a similarly deciding instrument cannot be found. In the Athens of antiquity a common cultural platform had been the Greek pantheon of gods, a scenario that was familiar to any citizen. Hence the reference to the gods in evoking a standard by which an abstract concept could be defined (*abstract* as in that which concentrates in itself the essential qualities of anything more extensive or more general; the definition of *essential* depends on the current abstraction process). We will return to the actual dynamics of the creation of links between thought structures and their further influence on the cognitive development, but right now we are mainly concerned with the fact that for a human being to be able to express anything, there has to have been a particular influence upon the mind that settled itself within the corresponding ideas. The newly formed thought pattern is now charged with presenting itself in the shape of a response, be that verbal or otherwise, and in order to do this effectively the target of its response needs to be clearly established. More importantly still, abstract concepts need to be assigned a realm which is made quite distinguishable from those areas that are concerned with the material world. Hence the development towards distinctly mind-related images to be contrasted with representations for the rest.

There is another passage in *The Last Days*, this time in the chapter *Phaedo*, where a conversation between Socrates and fellow philosophers is described, this time about the subject of 'death', in particular exploring the concept of 'noble death' [2]. Socrates guides his visitors (he is now in prison) through the intricacies of logical argument to show how one could reconcile the status of 'philosopher' with the more general demands of the body, such as eating and drinking, and sexual pleasure. He invites the forthcoming opinion from his company that a true philosopher, concerned as he is with intellectual matters, needs to disassociate himself from material concerns. The purpose of a philosopher is the "acquisition of knowledge", and because it is undoubtedly the mind which is the salient instrument in this enterprise there needs to be a distinction made between intellectual endeavours and such things as eating. Again the mind is faced with the necessity to organise itself in a certain manner. For the argument to distill the element of nobility out of the whole of a person's life, particularly within the context of a philosopher's existence, it could not proceed without the placing of 'soul' on one side, and 'body' on the other. If knowledge resides within the operation of human mental faculties, and its essence is sufficiently significant to warrant exclusive treatment in comparison with any other function a human being might be capable of, then the only way to preserve such exclusivity is to ensure that a summary description of the former can never be diffused by elements of the latter.

During the course of the conversation Socrates alludes to the idea of perfection—"absolute uprightness" and "absolute beauty" are mentioned—but modifies its nature immediately by declaring that within the reality of our bodily senses there can be no such thing, since the act of sensing itself dilutes their nature. It is the "real nature" of things which Socrates means when he talks about the unattainable, at least as far as humans are concerned. He then goes on to insist that the more "unaided" one's intellect is, the more unencumbered by the influence of the senses in other words, the closer one would get to the awareness of a thing's real nature. According to him the "pure and unadulterated thought to the pure and unadulterated object", is the only way to understanding.

Notice the choice of words here. In his endeavour to stress the difference between material matters and those of the soul (the word *soul* in ancient Greece denoted what we today understand by the word *mind*, and didn't necessarily have the religious overtones attached to it in later centuries), Socrates does two things: he apportions mutually distinct qualities to either, so as to make it quite difficult to mistake the two; and he couches his descriptions of the differences in phrases which leave no doubt about his own personal preferences (combining "soul" with "purity" and "truth", and "body" with "fancies" and "impediment", etc). We may assume that for an intellect such as his the choice of words and indeed the whole line of argument are conscious ones. Even less rational and more instinctive statements however feature the same favouritism when it comes to the expression of sentiments, and very rarely are any of them not tinged with one's personal taste.

From a wider perspective (and looking inwards) one could argue therefore that ancient Greece's culture, Socrates' fate, together with his understandable attempts to justify his position against the background of his own life and the atmosphere of the debates in question, prompted him (actually, his mind) to come up with a thought structure which fulfilled all the resultant conditions. And, being capable, mentally agile, and disciplined, could do so in a manner that is convincing. What is the by-product?

What we are left with is a philosophical argument which uses logic either seamlessly or, should the information become more rationally tenuous, informs us about it. In taking us through its pathways the sentences evoke pictures that are complementary to the stated ideas. By absorbing the sentences with their created attachments we are prompted to create similar thought structures so that *our* representations of the issues closely reflect Socrates' version. This need not be taken literally—one does not even have to agree with his conclusions. But if one does go along with what he says, the relative qualities of the issues involved will be found not just in him, but also in his audience.

The by-product then consists of a particular thought pattern arrangement where such things as purity, truth, nobility and so on, become firmly associated with the soul, spirit, mind, call it whatever you will. The impure in our world however, what can be described as adulterated and given all those epithets that fall into the category 'opposite' in relation to what is taken to be good, they are delegated to all that which is associated with body, and therefore matter. Out of personal necessity arises a judgment of the cosmos!

A less sophisticated set of neurons, one which is incapable of forming abstractions, could not have done it. It could not have taken an object, considered its qualities and then, like Socrates', distilled one salient feature and re-represented it as an ideal. It is true that the initial thrust of the argument was about *perception* of things, and the intrusion of physical distractions into that activity. To a true philosopher the soul cannot perceive reality in its actuality because the senses do not allow a clear reproduction of the outside. But Socrates also speaks about the bodily influences upon the *mind*, and how these prevent us from a proper understanding even if the input could have been improved. The fact that we need to eat and drink, acquire wealth and influence, wage war, and so on, all contribute to the deterioration of the soul, and hence the deterioration of pure knowledge. He can ask, "Do we recognise such a thing as absolute uprightness?", and then, "And absolute beauty and goodness too?", because his mind has abstracted a particular feature and expressed it as an ideal. The abstraction process is a functionality the mind possesses and is therefore not limited to particular thought structures only, just as a hammer does not loose its functionality because it is made to hit something other than a nail. Socrates of course appreciates what idealisation entails, because immediately he continues, "Have you ever seen any of these things with your eyes?" and "have you ever apprehended them with any other bodily sense?"—a question which, naturally enough, is being answered in the negative.

Notice again how a concept (eg beauty, goodness) is quite openly removed from the realm of matter, firstly through the application of abstraction, and then in the service of the desire to create a 'super concept', which at this point does not relate to anything that you or I have ever seen but which nevertheless is used as a guiding template for lower-level things. Such lower-level objects are being observed and handled in everyday life, they may contain particles which attract the description of 'beauty', but 'Beauty' they are not and never will be.

So, if our mind conceives of an ideal which is "pure", but our senses do not because they are "contaminated" and further, feed such 'dirt' into the mind which in turn becomes sullied with impurities, how else can the concept of any ideal be sustained unless we make an *a priori* distinction between mind and body?

It does not matter that the difference is one of perception only; that is to say is a construct of a system that tries to come to grips with the necessity to produce a cogent set of ideas without incriminating itself in the process. We, the recipients of the mind's output, cannot but accept what is offered because what is offered is *us*! It has often been argued by contemporary thinkers, especially in the field of artificial intelligence and robotics, that one major difference between a human and the machine is the human's situatedness in the world, whereas the robot knows only its immediate surroundings. In other words, we are able to bring together a whole set of data and derive from them the few salient elements that we need for the moment—the machine on the other hand does not relate to anything else bar its own narrow domain. But it could also be said that the world is situated in our minds, and the only reason we relate to something on the outside is due to its representation inside us. Therefore the act of perception—to go back to the beginning of the paragraph—may *seem* to offer a choice through substituting it with reality. This however is impossible for us; the perception is all we have. Now that the concept of an ideal has been mastered by the mind, the very process entailed the creation of a schism which is empowered to cast the world (that world we know of via the act of representation) into the categories of spirit and matter. From that moment onwards the thought processes which are concerned with a member of this or that category are going to follow very definite dynamics, depending on their auspices. The thought structures which these processes form will enable us to deal with something congruent to its respective category, but will also prevent us from stepping over the boundaries while still applying similar principles.

In Plato's *Phaedo* we find just such an example. It is doubly interesting, because not only does it give us an insight into Socrates' use of logical argument, but it shows how the act of idealisation virtually *forces* the mind to engage in a kind of deduction which otherwise would not have occurred.

Socrates and his visitors Cebes and Simmias continue the discussion about death. Having established that there is such a thing as a noble death (particularly for philosophers who strive for wisdom and the understanding of ideals), the question arises whether one ought to be afraid of being annihilated by it. It is the fear of nothingness after one dies which prompts many people to regard death with trepidation. Socrates proceeds to show how one can prove that just as death follows life, so is life the next step after death; therefore one does not need to be afraid of something which is only intermediate. How does he arrive at this conclusion?

He begins by juxtaposing two related characteristics, such as ‘slow’ and ‘faster’, or ‘big’ and ‘smaller’. ‘Slow’ being the opposite to ‘fast’, a thing must have been slower before it became faster, otherwise there would not be any distinction. Therefore, Socrates explains, the characteristic of something derives from its opposite. The same goes for ‘dead’ and ‘living’, since once again they are two opposites, and as one springs from its counterpart then ‘death’ must come from ‘life’, which can be readily observed. *Ergo*, ‘life’ also comes from ‘death’.

He supports that argument further by pointing to the act of recollection as being made possible because, although we may have forgotten a certain thing, we can draw it from our memory once again through something else. Let’s say I have forgotten about person A, but am reminded of him because I happen to meet person B, and she and A used to be friends. Socrates uses examples such as this one to illustrate the fact that we can only recollect something if we have had some sort of prior knowledge, and this implies of course that there must have been an opportunity where such knowledge could have been acquired.

He now refers to his previous assertion that there is such a thing as absolute beauty, or equality, and so on. To look at two items and declare them being equal can only ever be an incomplete exercise since the perception of these items occurs through our senses and they are not be trusted, as Socrates had pointed out earlier. So, if the act of perception per se does not give us sufficient information to call two things equal, then the nevertheless conscious notion of equality must have eventuated somewhere, it must have been “recollected”. It could not have come via the senses while being alive, since firstly they are unable to ascertain something of an absolute quality and secondly they are the vehicle we have had only since our birth. Therefore such prior knowledge must have come from a time *before* our birth, ie death, and this once again proves that life does indeed come from death.

Socrates makes an additional point about the soul. Just as the body which is composed of matter enters the process of dissipation once it has died, so would we have to regard the soul as being unchanging because it does not die with the body. The divine is immaterial, “glorious, pure, and invisible”, hence invariable. Now that we have classified our spirit as belonging to the divine, various inferences become possible. One is given directly in the current dialogue: “When soul and body are both in the same place”, Socrates says, “nature teaches the one to serve and be subject, the other to rule and to govern”. When he asks the company which of the two—body or soul—was likely to govern and which would be the governed, the answer is predictable. How can something which is divine *not* be the one who governs, and the one that is impure *not* be the one that needs to be governed?

Our purpose in examining the steps in the foregoing arguments is not to split hairs against the background of another two millennia of philosophical debate. It is rather to make use of the convenient form of discourse presented here so that we can identify and trace the underlying structures of the thoughts before us. An analogy with railway tracks suggests itself. One could view a conclusion as a switch, since the nature of the conclusion points henceforth to a particular direction and not another. It acts like a siding—now that we have branched onto a different track we are bound to follow the new route and encounter whatever it is able to offer. There are offerings waiting on the other track too, but they have been left behind. Conversely, one might backtrack a route taken and observe at which stage the branch-off occurred.

The conclusions Socrates had reached (the switches, the sidings along the route) brought us to a very particular area. It is a landscape which features a categorical differentiation between what we can see and what we cannot; where the visible things belong to matter, and those invisible to the spirit; where material objects have certain qualities which can be identified, but where such qualities are by no means perfect; where perfection, on the other hand, is an example of the pure, unseen, and therefore spiritual; and lastly, where the perfect, being preferable to the imperfect, is given reign and authority over the latter.

If we trace the thoughts in those arguments further and further back we eventually arrive at the emergence of the abstract (see the definition of *abstract* above). Without deducing 'Beauty' from something that is merely 'beautiful', we would forever be left without the quintessential quality against which anything else can be measured. Without 'Beauty' to act as a fundamental standard, a thing can either be more beautiful or less so, but whatever the degree can never be appraised to the fullest. In order for this to happen we need the notion of an Absolute, and only a neural system which is capable of a corresponding sophistication can come up with such an idea.

The identification of a relationship between two objects is comparatively simple. Socrates has used it on several occasions, such as in his assertion that relativities can easily be measured by using the correct device, or later in his description of incomplete sensory inputs which lack the comprehensiveness of the ideal. The thinking processes of animals include the act of comparison, but for us humans this is merely the first step: due to our mind's ability to abstract we can now proceed to build further thought structures which are quite artificial in nature. They do not have a concrete equivalent but offer us a substitute by developing scenarios in which the abstracted idea not only becomes feasible but is a necessity.

Consider the absolutist quality of 'Beauty'. For us to understand the abstract we needed to have the experience of something being beautiful, as Socrates had reminded his visitors. The act of observation also entails the awareness of other characteristics which diminish the concept of 'beauty', so we are satisfied with the idea of imperfection in anything we ascertain. Alongside the mental processes which give us 'beautiful' there are also those which give us the corresponding ideal, namely 'Beauty' in this case. Neither process is identifiable in itself, that is in neither case are we able to follow the detailed construction of a thought structure which causes 'beautiful' or 'Beauty' to pop up into our consciousness. But because the mind seeks linkages between anything it produces there is also a context brought into existence which allows the notion of 'a beautiful something', just as there has to be another context that permits us to declare an ideal. One example of such a process of construction is found in Socrates' arguments leading to the acceptance by his visitors of an entity such as the soul which is pure, immortal, and above all, capable of understanding the absolute version of an impure, material reality.

Notice how the emphasis is on agreeable things and how, through the act of synchronising their inherent understanding with the idea of similarly noble concepts such as purity, immortality, and spirit, a very specific characterisation is given to that which is made to represent such concepts, ie the soul. Quite logically then the elements pointing to the opposite, ie matter, body etc, would be associated with negatives. One could quite as easily provide 'Hate', 'Ugliness", 'Fear' and so on for a similar exercise, but it is significant that it hasn't been done. It is not difficult to see why: if there are two desires, one to ascertain an identity in us, and the other to align oneself with something 'good', then it is only natural to combine a positive form of idealisation with that part of ourselves which lends itself to this kind of treatment.

And yet, the neurological processes responsible for either type of thought structure are the same. While our level of understanding does not allow us yet to describe each functional detail of say, the thought ‘this is a beautiful rose’, it would nevertheless be reasonable to assume that the fundamental dynamics are the same between it and ‘ahh, such Beauty!’ for instance. Therefore the degree of validity between the description of a flower and that of some general state will have to be the same also (leaving aside the quality of information leading to this or that statement).

We have just made a very significant switch. It was possible due to a particular set of ideational preconditions. Firstly, we explicitly state that whatever thought structures we possess, they have been constructed in and are maintained by, our mind. Secondly, these thought structures are being modified internally, either through sensory input from the outside or by other thought structures on the inside. Thirdly, we cannot ascertain the constructive processes of such structures but are only cognisant of their results. None of these items is in any way controversial or revolutionary; they are merely reasonable assumptions for to claim otherwise would immediately invite a whole host of considerations that would lead us into the realms of fantasy and conjecture (as to (1)—if they are not inside us, where else are they? As to (2)—how else do we acquire new ideas, from some mysterious source? Regarding (3)—if we knew the actual, step-by-step process of thinking we would have described it by now). Given the three preconditions therefore we can follow the argument up to the point where Socrates posits the existence of an ideal—not because, let us remember, anyone has ever seen one, but because we can *imagine* the very idea of it.

For Socrates the act of having an idea is synonymous with the existence of that which the idea stands for. Granted, there are “impurities” induced by the senses, but thinking of a rose for example asserts implicitly that there is a such a thing as a rose. Indeed, thinking of an ideal such as ‘Beauty’ forces one to conclude that since we cannot view such a thing while we are alive (due to imperfect senses etc) we therefore must have ‘seen’ it while dead. At that moment we would find ourselves at a certain illogicality, were it not for the possibility (construed by further arguments) of something of us surviving death and so enabling us to ‘see’ after all, and with perfect vision at that. From now on the possibilities arising from the existence of a soul vs a body, the duties associated with the characteristics of either, and what all this means in our daily lives, can be seen and made use of as one travels along the track initially chosen.

But not if we had made that switch, mentioned above. Since the concrete and the abstract are different products but coming from the same source, their respective validity in terms of how truthful they can ever be is the same as well. While there may be a concept of ‘Beauty’ in this case, its degree of absolutism is dependent upon the assertion by the person declaring it—hence there cannot be anything ‘ideally beautiful’ because the reference to the absolute does not exist. If there is no reference point to be sought then there is no need for a special vehicle to look for it either, and the need to construe something for such a purpose never arises. This does not mean that no-one ever conjectures about such things as soul, spirit, life or death. However, since the urgency to idealise is gone there is no predilection for abstract ideas to be imbued with the halos of perfection (certainly not for those ideas which are meant to represent us), and with the absence of perfection also comes a more liberal interpretation of values. To “rule and to govern” may come naturally to the pure and immortal, with the role of subject assigned to those who come from dirt and are diluted; it matters whether humans regard themselves as possessing some of that governing quality. In any case, a thought about something concrete is, neurologically speaking, no different from a thought about an abstract, and so the perceptive quality leading us to consider a flower is no more or less enlightening than thoughts about beauty as such. Both originate in that same mixture of nerve cells and chemicals, dirty and imperfect all, where the only thing ennobling is the aspect we invent for ourselves. The moment an idea emerges from the subconscious is the moment we are aware of it; no sooner and no later. Even the expressiveness of the idea is tenuous: were it not for the myriad of references we were able to build gradually since our birth and through which we define and circumscribe that what wells up inside us, thoughts are merely chaotic sets of mental sensations without any dimensions of space, time, or values (more of this in Part II).

Furthermore, once the connection between the ideal (the absolute), its vehicle (the soul, spirit), and those who are in its service has been made, the ground is ready for the blossoming of an authority that sees itself outside the reach of mortals—by definition.

Plato, as mentioned before, was a student of Socrates and was responsible for setting down the discourses of his teacher. Plato's disciple in turn was Aristotle, and it is interesting to follow that man's arguments as they construct the relationships and significances of human values. In *Metaphysics* he explains what seem to him the fundamental aspects of such values and he weaves them into a logical pattern that seeks to demonstrate the purpose and therefore usefulness of what we ought to hold important. One can argue about their ultimate content, and indeed this has been done by several critics, but here we are concerned with their initial placement and definition.

To follow Aristotle's reasoning we have to get acquainted with a number of concepts he uses to build the structure [3]. A thing can be described by naming its characteristics such that we can know its essential nature. While some elements which make up a thing can change over its lifetime, there are others which are well entrenched and do not change. That which describes it fully, without which the thing would not be what it is, Aristotle terms a *hexis*. As he puts it, "Hexeis, either bodily or physical, are not alterations. For some hexeis are excellences and others are defects, and neither excellence nor defect is an alteration. Excellence, rather is a sort of perfection ... and defect is a perishing of and aberration from perfection." And further, "excellence is a sort of perfection, for everything is perfect and every substance is perfect when, in respect of the form of the appropriate excellence, it lacks no portion of its natural greatness." These definitions are important in themselves, since they are being used to augment several arguments that are interesting from a philosophical and ethical point of view. For our purposes their significance lies in the semantic hierarchy in which they have been placed, because it is these kind of relativities which hint at the role any pertinent thought structures have played in their usage.

Therefore we have hexeis as some sort of quintessential natures of a thing; they can be either well-disposed or ill-disposed, the former giving rise to perfection and the latter to imperfection. To be perfect the thing has to be in a state of excellence, provided none of its natural constituents have been compromised. In addition, only when it is most natural can it be perfect. There are no alternatives to those states—there either is perfection or there isn't. The idea that only a fully constituted *natural* state can lead to perfection invites the conclusion that for something to be perfect it has to be most natural.

Consider now where the notion of "most naturalness" comes from; we can only ascertain the nature of an object, an organism, indeed a human being, through observation within the context of the existing situation. Ultimately then it has to be us who can provide the label 'perfection' to anything we are in the process of considering. Hutchinson, the author of the text referenced under [3] above, points to just that problem by mentioning another school of thought, namely scientific objectivity, which disagrees with this kind of perspective. There concepts such as 'excellence' and 'perfection' are classified as anthropocentric categories, in other words descriptives that rely more on our personal interpretation than on any objective assessment.

Let us return to our current interest however. Rightly or wrongly, the thought structures which provide the notion of excellence under the guise of a particular viewpoint allow us nevertheless to build upon such basis further—an option which Aristotle had employed to construct a logical argument for defining, amongst other things, what makes a *good* life. As informative as those structures are, they are taking us outside our stated focus. What is more interesting for us is how Aristotle sees the relationship between a thing's nature and the quality of excellence. I offer the same passage as quoted already in the above mentioned text [4], for in the light of what we have found so far it is highly significant:

> The living creature consists primarily of soul and body, of which the one is naturally ruler, and the other naturally ruled. We must look for the natural things that are very much in a natural condition, and not in corrupted things. And so we must consider the man who is best disposed in respect of body and soul, in which this is evident. In wretched people, or in those who are in a wretched condition, it might seem that the body often rules the soul, because of [their] being in a bad and unnatural condition.

In defining the basic elements of which a living creature consists, Aristotle provides us with the premise that there is a body and a soul, and it remains to decide which of the two is dominant over the other for a qualitative assessment to be made. We have already seen such a perspective in Socrates, and its ramifications are being applied once again to lead to the same result: if the soul rules over the body, it is *good*; otherwise, it is *bad*. The argument, as far as Aristotle is concerned, then leads to the eventual definition of a good and/or bad life through a process of deductions and inductions which need us not concern here [5].

For us the important point is the starting position. Aristotle does not offer any explanations as to what the soul is or how precisely its influence could be identified amongst the machinations of the body, just as Socrates did not provide any kind of like definition. What Socrates did do however was to construct the semiotic (ie symbolic) need for a soul to be in existence, for otherwise the abstractions of goodness, purity, etc would not hold. Aristotle took it further by building upon such existing structures and using them as a mechanism to judge a person's life. The effect is important indeed.

Remember Socrates' debate with Euthyphro, mentioned earlier. In it he claimed—quite rightly—that to argue about qualities which cannot be measured objectively will never lead to a satisfactory outcome since there is no unassailable measuring tool by which they can be described, once and for all. So, a number can be properly defined, but something like 'honour' cannot. The same can be said about 'excellence' and 'perfection', since neither is dependent upon a conclusive representation of themselves but rely on their very existence on our willingness to declare them as such. And what does that willingness depend on? None other than our interpretation of the stated relationship between body and soul.

Let us consider the implications. Suppose you were asked to judge somebody's conduct within the context of the afore said. You would begin by listing samples out of that person's life in order to demonstrate a certain interplay between their body and soul. It may be relatively simple to observe physical actions, much harder to demonstrate any role the soul could have played. Were the actions determined by the person's higher self? If so, to what extent? Did the body follow the soul's instructions to the letter, or was there any advice which had not been heeded, or if at all then only incompletely? And in the case of some action, who is to say that the soul did not act upon knowledge and insight which a lesser human being cannot understand anyway and therefore is unable to judge fully (assuming those acts were implemented)?

We have the following situation:

- a life is perfect if it is most natural; what is natural?
- it is most natural if it is pure; when is it pure?
- it can only be pure if it is conducted under the auspices of the soul; what is the soul?
- the soul is the ruling agency, having access to knowledge which the rest of the body does not have; what is that knowledge?
- we cannot know properly since our physical senses are impure.

And yet we are asked, first by Socrates and more explicitly by Aristotle, to give judgment in matters of purity, perfection and excellence by appealing to the soul. In the end we have arrived at a point which by its very nature (defined as it was through the foregoing argument) is removed from any kind of assessments we are possibly capable of, since they all involve to some degree the senses, and they cannot be relied upon. So, in the final analysis the ultimate judgment (which one could say is the most important, the all-defining one) is that which is removed from analysis and reason, and is placed firmly into the hands of faith. A precarious outcome, to say the least!

There is another effect, one which we will have cause to re-visit later. A thought structure has been developed that gives rise to certain assumptions. They may or may not hold, but in any case are followed immediately by further declarations that are put up for debate. Notice it is those later declarations which are being offered for discussion, not the initial ones. This causes the focus to turn onto them (the later declarations) without their predecessors been given any thought at all. Therefore the debate may rage this way or that, but the underlying reasons are never being addressed; after all, at the moment of their presentation the discussion had not been formally 'declared open', so the audience merely waited for what was to follow. Then, with the new statements brought to the fore, everyone's attention is directed towards the current issues and the rest is forgotten. It is an old confusing trick: say something which cannot be argued one way or another because not enough information is at hand; then bring in further points, possibly controversial ones, and the debate now centres on those, to be prolonged at will. There aren't too many individuals who at that stage would step back, consider how it all started and question the very first opening statements. It brings us back to the beginning of this chapter where two mental functionalities were introduced, one when talking *about* something, and the other when talking *to* someone. When talking *about* a thing our mental processes are not invited to a repartee, but in the latter case we mentally gird ourselves to respond, question, criticise or attack. Referring to our example of judging someone's life, this means that we could well argue about the person's integrity, but under the circumstances we would never get to question the authority of the soul itself.

Before we leave Aristotle it may be worthwhile to take into account his general acceptance that there are certain things which cannot be demonstrated, the *first principles*, as Cooper has pointed out in his text on the philosopher [6]. Although there are procedures to explain why the sum of a triangle is 180 degrees for example and, continuing from there, to provide us with further insights and explanations, it is equally true that some principles have to be accepted as undemonstrated before embarking on any proofs whatsoever. These first principles, according to Aristotle, are known by intuition, and are applicable not only in scientific reasoning (eg in geometry), but also in the area of practical thinking. For humans therefore, debating what is right and what is wrong, may lead a person from one situation to the next along the path towards some justifiable end, but there has to be some intellectual awareness regarding the initial decision, the fundamental values as it were, and those can only be accessed through an intuitive understanding of what such values entail.

Given his position on the absolutist quality of the soul then it can be seen that any point of view about such ultimate starting points will have to accept the dominance of an abstract entity (ascribed perfect vision, awareness and understanding) and establish through such a construct their unassailability. This certainly reinforces the Aristotelian value system, and in turn provides further reasons for emphasising the dichotomy between body and soul.

By way of summary—and keeping in mind that we are not engaged in any critique of philosophies, rather an examination in terms of the salient points we may derive within the context of the functionality of thought structures—it is now possible to lay down the first beginnings of a systematic picture.

There are experiences that humans have, and first and foremost they are the product of sensory input. These experiences produce internal representations which lead to abstractions. We are capable of creating abstractions because our mind has evolved to a degree where such structures are possible. While it may seem to be a moot point whether the abstractions are formed intentionally or by default, further examinations of other philosophies suggest that the underlying similarities of these thought structures hint at a pre-ordained need for such images to be created (we will have ample opportunity to confirm that in the following chapters).

Therefore, the common experience of material objects (touching, handling things; observing their composition; etc) triggers the abstract idea of *matter*. Accompanied by the observation of impurities contained in the objects there also eventuates the abstract image of *impurity*. At the same time, a thinking being is confronted by the fact that there has to be a process which does the thinking, hence the abstract *mind*. Coupled with the awareness of cause and effect in this world the conclusion offers itself that our thinking processes are the result of the mind, and the data it deals with are imported via our senses.

Combine the abstract *mind* with the functional equivalent of *impurity*, posited on one hand against the identified non-existence of the pure version of an object and on the other the medium of our senses which allows the state of impurity to be ascertained, and we have the literal need for another abstract concept, namely that of an *ideal X* or *ideal Y*, provided that *ideal X,Y* has not been sullied by sensory degradation.

Why would the realisation of something being impure cause the aspect of negativity to emerge? Consider the state of craftsmanship, of technology. Any constructive endeavour by humans has to deal with the problem of corruption—whether in the form of an embedded branch inside a piece of wood, the appearance of rock particles in clay, even dirt in milk. Thus it is natural for humans to view any kind of material with mixed feelings. Depending on the ultimate purpose it is put to, the added effort of ridding it from its impurities is essential to the success of the operation. It is also reasonable to assume that people who have not developed the necessary degree of processing would not be driven to these kind of thoughts.

The stage is set for a two-fold mental universe: direct experiences through our senses produce an imagery of objects (constantly referenced by outside observations), and the abstraction process creates idealistic counterparts, including the mind itself. 'Good' things are those which have been purified (homogenous timber, soft clay, clean milk), and so the purified abstracts must also be better. Scenarios are being evoked to explain the status of 'goodness' in relation to the abstracts so formed and we can see how a combination of matter/impurity/bad on the one side and non-matter/purity/good on the other is quite inevitable. If the functional properties of an object are viewed under the auspices of matter the resultant connotations are in the negative, whereas if they have been placed within the arena of the non-material only they are deemed to be positive. Mind, being another object, is treated in the same way. If one's mental operations occur in conjunction with the senses they fall under the material category and are suspect, but for them to be positive they have to be removed from that sphere, thereby excluding the observational aspects and permitting internal processes only: meditation, pure knowledge, the spiritual.

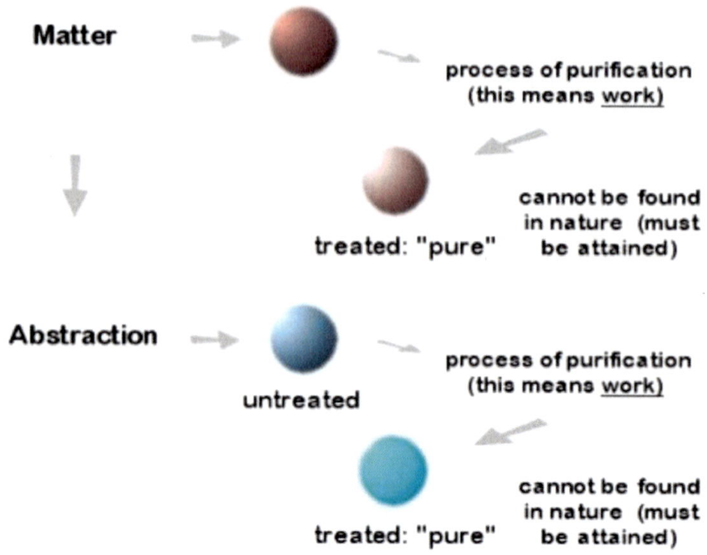

Fig. 2-1: Schematic diagram of the abstraction process. An object is first identified, then treated in the real world for the purpose of improvement. Its abstract equivalent is subjected to a similar process due to the perceived similarity.

The simple flow chart in Fig. 2-1 will illustrate the dynamics. Note that while Aristotle for example claims that the most pure is also the most natural, it nevertheless is not something which can be identified as such by our material senses, and so the ultimate arbiter has to be the materially unencumbered mind. Therefore the idea of "most natural" should be understood as another abstraction in itself and with this in mind the diagram makes sense again. This particular philosophical position by the way creates a problem when trying to define what is "most natural" by having to refer to everyday occurrences and their imaginary abstractions at one and the same time. Because "most natural" has not been recognised as an abstraction, its conceptual usage is likely to interfere with representations from the real world further down the line. That does not concern us at the moment however.

We have gained the first glimpse of the general dynamics of the human mind. While the picture is still far from completion, a major component of the mind's functionality has been identified. The example of Socrates' discourses and Aristotle's further delineations gave us the opportunity to observe the formation of thought structures as a process of mutual creation and feedback, against the background of a continuous need to explain and justify to ourselves.

In order to explore those processes more fully we turn our attention to a much more comprehensive cosmic model, one which predates Athens by at least 500 years: the metaphysics of the Jainas in ancient southern India.

3. The universe as a fractal

Jainism

Picture this:
You have emerged from a grove of trees and are now standing amidst a clearing on top of a hill. The night sky is cloudless and the stars stretch from one end of the horizon to the other. Through your sandal-clad feet you can feel the roughness of the ground, and blades of grass tickle your calves. A soft warm wind breathes over your face and the night calls of animals reach you from the surrounding shadows. As you gaze into the depths above, you gently move the weight of your body forwards and backwards, listening to the small sounds your feet make. You do not have words for most of the stars, indeed you are mystified by the brilliance of name-less lights that seem to move as a cohesive whole, with set direction day by day and yet you know of subtle changes across the seasons. You also know of animals and have compared them with each other but do not have a name for their inherent secrets. You know of timber you have hewn but do not understand the structure it seems to suggest. You know of life and death, of movement and stillness, of large and small, yet the meaning of it all eludes you. And you wonder.

The time is 3000 years ago, the location somewhere in the southern half of the Indian subcontinent. The people whose members looked at the world around them and who arrived at conclusions which eventually culminated in an entire philosophy at that time are known to us as the Jainas; their descendants still exist today. Through observation, but mostly through contemplation, they managed to create a comprehensive model of the cosmos that in its complexity is still likely to confuse some observers. As with any philosophy one can always argue about the ultimate truth of it all, yet there is no denying that their system is a finely crafted one, and its parts are put together with harmonious diligence.

Put into text by Sri Kundakunda-Acarya as the *Pañcastikayasara* [1] around the beginning of the 1st century CE, it offers a detailed conceptualisation of the world—from the large to the small, the 'real' to the abstract. For someone who seeks to follow the intellectual steps from sensory input to the mental representation of an idea it is a fascinating source of material.

Let us start with the big picture. Our imagined observer would be familiar with space. He would understand that there are dimensions in this world, that each thing has a length, a width and a depth. He also is able to form the abstraction of space, so that regardless of the inherent nature of an object it would have the general quality of Space. Just as a room contains many objects, and therefore being an entity which possesses the general quality of 'room' combined with the objects it contains (a room which cannot contain anything would not be a *room*), there surely then must also be a general *Whole* which contains all that which presents itself to the senses: people, animals, plants, mountains, rivers, and so on. He might well call it *Reality*.

Reality is first and foremost about space, but our observer is intelligent enough to consider certain attributes related to space which in themselves are too limiting to account for everything Reality contains. His linguistic dexterity enables him to speak of the Real which has extensive magnitude, and in this extensive magnitude there reside further concepts, represented by what we experience in daily life. One observes material things, and so that Whole which is Reality houses the principle of Matter. Material things do move, and therefore Reality also contains the principle of Motion. There is also the principle of Rest, likewise observable. When an object—having dimension—moves away from one place and comes to rest at another it has traversed space; its notion is added as a particular element alongside the others. Then there is the observer himself: he consists of matter, he moves, comes to rest, occupies space, but furthermore is endowed with a mind that thinks. That feature seems unique to humans, and surely as one can point to the general idea of Space, Motion, Rest, and Matter, one can also address the principle of Soul—again as being part of the Real.

The Real, that which has extensive magnitude, therefore unfolds into five primary characteristics (Soul, Matter, Space, Motion, Rest) without which it would not exist. Soul operates in Matter, Matter extends in Space, to which Motion and Rest are indirectly related. So far the abstractions form a subset that seems complete were it not for our experience of time. The sensations of passing moments relate to the mind, their conclusions can be applied to objects moving through space where its dimensions are obviously measurable through their numbers. Thus Time can be seen as standing apart, the ultimate calibre of an existence which is fluid and changing and yet provides continuity through creation and decay as well. All that can suggest itself to someone who moves from an experiential recognition of his surroundings to the formulation of abstracts, and considers carefully the individual roles each principled notion ought to fulfill.

Furthermore, such a person would have been fully aware of the abstracts as *abstracts*, because Reality holding Time together with Soul, Matter, Space, Motion and Rest have been recognised as empty ideas in Jainism—essential, but empty nevertheless. In order to manifest, in other words, to move from the Real to the real, these descriptive shells need the attributes of appearance, disappearance, and permanence. A thing comes into being, takes its place within the framework of spatial dimensions, rests and moves in instants of time, can be endowed with a soul—and disappears. That rhythm itself shows permanence, a fundamental, overriding corollary to the cycle of birth and death.

The act of realisation performed by the mind which enables it to classify a mental thought process as an abstraction of a perceived object must rank as one of the most significant steps in human evolution. Without the active awareness that my memory of a table for instance spawns a thought that encompasses the general principle of Table as a platform to put things on, I would not be able to surmise about platforms since I am unable to distill the pertinent features of any table and apply them *in my mind* without further recourse to any other sensory input; to look at any more tables is not necessary. The degree to which such features are recognised, their variety and how often they have to be repeated, determines a person's level of intelligence (amongst other things). Our modern abstract reasoning tests are designed to identify the subject's competence along those very lines. An example is shown in Fig. 3-1.

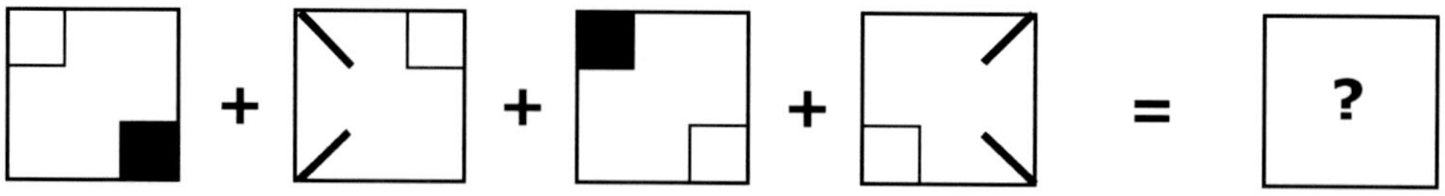

Fig. 3-1: An abstract reasoning test where four images are sequenced according to certain rules. The subject has to determine how the fifth image would have to look like if the rules are applied consistently (in this case they consist of the rotation of the white and the black squares by 180^0 in the 1st and 3rd image, and a similar movement of the white square and two black lines in image 2 and 4—that's one solution).

Once the relevant attributes of an object have been identified the same characteristics can be applied to its abstract. These qualities, although similar, are now abstract qualities. Let us call the qualities of the perceived object Q_{po} and their abstract counterparts Q_{ao} (the qualities of the abstract object). Q_{po} may have been identified through observation alone, or they may have been discovered through handling the perceived object. Note that the act of handling does not necessarily have to be a comprehensive process, that is our perception of how an object handles is based on what *we* have done with it, not what possibly could have been done. Thus Q_{po} reflect our personal understanding, and not the actual nature of the object itself. Therefore Q_{ao} are a mirror image of our subjective thinking. Once transferred into the realm of the abstract Q_{ao} fulfill the same functions as their perceived origin, *provided the idea is placed within the same functional context.* If it is not, we are faced with what could be termed an *incongruent functionality*, ie a manner of behaviour in the derived concept which does not reflect what its original would have done under similar circumstances in the real world.

For example, suppose I take an enclosed space, say a box, and proceed to fill it with various objects. There will come a point beyond which it is impossible to add anything else without bursting the walls of the box. I observe all this and come to the conclusion that the act of cramming things into a confined space will destroy the space (the Q_{po}). Suppose now I move that insight into the abstract and, applying the same functionality, I decide to view the human memory (an abstract) as more or less another kind of box (the Q_{ao}). The idea suggests itself then that if I would keep on memorising more and more data there will come a moment where my memory would be damaged as well, just as the box had been damaged before. This is clearly not so, and therefore the overall functionality of Q_{ao} is incongruent in relation to that of Q_{po}. Depending on how concerned I am about the possible consequences regarding Q_{ao} I may or may not put my concept to the test, and in the case of my own mind I might well be so afraid of my brain 'bursting' that I keep my personal learning curves to a minimum. Some people do just that!

In the case of the Jaina cosmic model it becomes quite apparent that its developers had spent a good deal of thought to ensure that the Q_{po}-Q_{ao} relationships are as congruent as they were able to make them. One implicit recognition of the $Q_{po} \rightarrow Q_{ao}$ transference is the remark,

> The soul [ie mind] is not distinct from its attribute of knowledge, and because the ways of knowledge are diverse the world of reality or the universe is also said to be multiverse by the wise. [2]

Or, as Upadhye put it, "There can be no knowledge apart from the knower". If knowledge cannot be separated from the knower, and knowledge is information structured by the Self, then it is the nature of the Self which ultimately determines our knowledge. Such insight is by no means trivial. Once a philosophy contains a structure such that its elements can only ever be traced back to the mind exercising that philosophy and no further, any mechanism which allows a concept to be asserted as the ultimate truth has been rendered useless.

A more direct focus is presented in the next verse [3] which discusses the relationship between a substance and its qualities. It is pointed out there that if a substance is distinct from its qualities, then "it [the substance] may change into other infinite substances". Likewise, "if the qualities can exist separate from their substance, there will be no necessity for the substance at all". And, it is added, "in either way the result would be absurd". From our perspective this represents an indirect admission that without a continual process of referencing, the respective functionalities belonging to Q_{po} and Q_{ao} will diverge and thereby lead to an incongruence.

The problem is the inexact definition of "quality". A thing is perceived and knowledge about it derived. Taking the external view, qualities are the originators of effects on an object's surroundings, but they also constitute what we perceive about an object. In this case they are not directly related to the object, but are a product of our thought processes and therefore thought structures. Given this approach any arguments regarding the difference between substance and quality dissolve. While their mental representations allow us to distinguish between the two (this is object X *and* it has such and such qualities), both (substance and quality) ultimately depend on the perceptive acumen of the observer. What keeps them apart intellectually are the representational processes on one hand and their abstractive interpretation on the other, yet the two need to go together in order for us to identify either the substance or its quality. Change the quality and you have another substance, change the substance and you are faced with different characteristics. The process of re-evaluation ensures that such a one-to-one relationship retains its integrity. To drive the point home even further, we are warned against accepting any kind of assessment as a revelation from a great spiritual being; rather, our insights ought to come from an observation of Reality [4].

The absolute need for thought structures based upon perception leads to the delineation of abstracts which can only be explained in terms of the dynamics of the everyday world. Keeping in mind our imperfect means of identification via our senses (this of course is relative—whether observing with the naked eye or using an electron microscope) we cannot do otherwise, but it is interesting to note how such a fact is openly admitted.

Consider the question of Karmas, held to be causal conditions which influence the soul in a particular form. One can see the justification for such a concept in the observed reality of external influences upon a thing, whatever that might be. The by now familiar phenomenon of a $Q_{po} \rightarrow Q_{ao}$ transference (let's call it the perception-abstraction transference—PAT for short) has taken place. Just as there are influences which act say, upon a jar of milk if left standing in the sun, the assumption is that there must be similar influences acting upon one's soul. Jaina thinkers are left with the task of how to deal with such interfering entities (the karmas). Given the notion of a general, underlying principle (in this case the Soul) paired with a momentary appearance (one's own soul), they needed an analogy which provided them with an equivalent scenario. In this particular case the search yielded the picture of an ocean, with the ever-changing wind creating waves upon its surface while leaving the actual body of water unaffected. The analogy seems to fit: there is the principle of Soul which we all have as part of the greater Whole, yet our current station in life subjects us to influences that identify *us* through a very specific manifestation of appearances.

Such a process does not only serve as explanation, it also provides confirmation through its very source. Since one can observe such a thing as waves on water in nature and, nature being an inherent aspect of the Real, incorporates the principle qualities just as any other aspect would, applying the PAT process to another aspect of the Real must per force contain the same truth value. The procedure can be formally stated as follows:

1. there is a set S of objects;
2. S features basic qualities Q;
3. S can be abstracted as an idea, S_A;
4. the same can be done with its basic qualities, yielding Q_A;
5. there is an identified subset s of S;
6. its abstract equivalent is s_A; what are its qualities?
7. find another s in S;
8. identify *its* qualities, q;
9. apply q to s_A, turning them into q_A; do they hold?
10. observe and re-confirm s and thereby q, observe and re-confirm s_A and thereby q_A; do they still hold?

11. $s_A + q_A$ fulfill the conditions dependent upon their allotted place in the scheme of things; therefore accept $s_A + q_A$.

Notice the repetition of subprocesses relating to perceived things and their abstracts, plus the ongoing references to the real world. As such the foregoing reflects what has been postulated throughout Jainism: a principle must not be self-contradictory, substance and quality go together, and the ultimate guidance comes from the Real.

Step 9 above requires the application of inference; if the sample observation being used (ie the s-q relationship) can be readily confirmed and s_A-q_A seems to accommodate the same principle, the Jaina thinker feels justified in accepting the same relationship for s_A-q_A. This is particularly significant when dealing with imperceptible objects, such as Karmic matter and its influence upon the soul. If world space is taken to include all the possible material bodies (observable as well as unobservable ones) and Karmic matter should fall into the second category, it nevertheless has to abide by the general principles applicable anywhere else. So firstly, Karmic particles can interact with the soul—a dynamic which can be perceived in concrete things. Secondly, there are interactions which can be directly attributed to the inherent and *observable* qualities of the agents involved, and there are others where the reasons for interactions are less clear, yet their result can be seen too. This creates two types of abstractions: one, where the mutual attraction is positive and directly causal; the other where the attraction is due to local (ie external) conditions, that is to say the interaction comes into being through both agents being subjected to the same kind of influences and not because of their inherent qualities per se.

Modern science may lead to a way of thinking in which the second option is not realistic. After all, if there is some evident action which cannot be understood right now then past experience would suggest that another, better, approach is needed and the causal relationships will become clear after all. Humans living 3000 years ago would not have had the opportunity to draw on such equivalent experiences, and so they would have been faced with situations that forced them to accept the possibility of interactions outside some direct causality. To differentiate between the two options, in Jainism the first is referred to as a causal relationship, the second as a contiguous one ('contiguous' as in: being in close proximity without actually touching). The question is: regarding the soul which case is chosen as the correct one?

Step 10 is applied: if option 1 is appropriate, then Karmic matter would have a direct causal effect upon the soul; this contradicts the general approach taken in Jaina philosophy however. Therefore option 2 is considered, but is there an equivalent in nature? Yes, according to the author of the Pañcastikayasara. The necessary example can be found in the observation that a casket of collyrium powder (a substance used for eye wash) becomes black by mere contact [5].

Since the detailed chemical composition of collyrium powder and that of the casket had not been known at the time, the observed abstraction was formed that two materials can co-exist and one influences the other without apparent cause—step 11 in the above algorithm has been fulfilled.

The foregoing would seem to imply that the thought processes move from the large picture to the small, that is from overall considerations about Soul for instance to what the existences of individual souls entail. It may well be presented as such, but consider the elementary ideas necessary for such thinking: without the memory of experience and its resultant abstractions derived therefrom no interpretations can be made. The awareness of an appropriate analogy that had been observed elsewhere gives the thinker the authority to declare his own musings as truthful. It is a dynamic we constantly use in daily life. In many cases our conclusions are not tested in terms of how the object fares now that a particular analogy has been applied to it, but how correct the original sample seems to us. Another PAT, this time with reference to truth values! Anticipating the possible problems a misapplied analogy can cause, we are therefore told by the Jainas to constantly check our references with the real world, just in case the similarity in principles should have been broken.

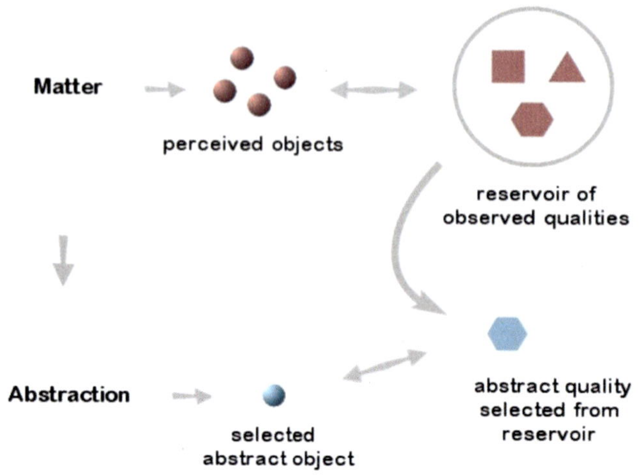

Fig. 3-2: The abstraction process including the transference of qualities from the perceived object to its abstracted equivalent (PAT). Given a choice of concrete qualities the most suitable for the current abstract is selected. Note that the ultimate truth value is dependent upon perception.

Our schematic representation of the process can now be enhanced by the inclusion of PAT (Fig. 3-2).

Two features need to be emphasised. Firstly, the validity of the resultant relationship between the abstract substance and its qualities ultimately depends on the astuteness of the accompanying observations. The fact that a correspondence has been found and the transference been applied does not in itself prove that there is such a quality which belongs to the abstract object. Under the Jaina auspices the constant reference to the real world would keep the plausibility of total failure to a minimum. Secondly, we have so far been dealing with observable things. The abstractions, although imaginary themselves, have their direct counterparts in the real world. They are understood to be a concept only, a general principle which has to be kept apart intellectually from their manifestations in the Real.

However, the PAT phenomenon can also be made to loop back towards the concrete. Its functionality still holds, although at first glance it may not seem that way. Rather than evoking an abstract s_A and placing it inside the world of ideas, s_A can be created for the purpose of gaining a real world item simply because there *ought* to exist one. If one's experience suggests a certain state of affairs, its abstractive counterpart can be used to form the idea of what is then taken to exist in the world as well.

Let us say we observe the grinding of rock into dust. No matter how much we grind and how fine the dust becomes, it is still identifiable as dust. We could conclude that even the most finely ground dust maintains the qualities of matter, and since matter is made up of particles they would have to become smaller and smaller until a certain limit is reached. If there was no limit the dust would give up its material quality and disappear; it does not disappear therefore there must be a minimum size of any particle. A new quality of Matter has been found: finite smallness, the indestructible basis for the real world. It also forms part of the Jaina cosmic model, described as the "primary atom" [6].

And yet, the descriptive form of "atom" must fit into the wider framework made up of the concrete (ie directly observable) characteristics as well as its abstract counterpart which include the Q_A. It becomes a measure of such a framework's overall integrity how plausible the mapping process turns out to be. The move from the set of possible Q_P towards those Q_A which seem suitable is shown in the further explanation about atoms. It is summarised by Upadhye in the following manner:

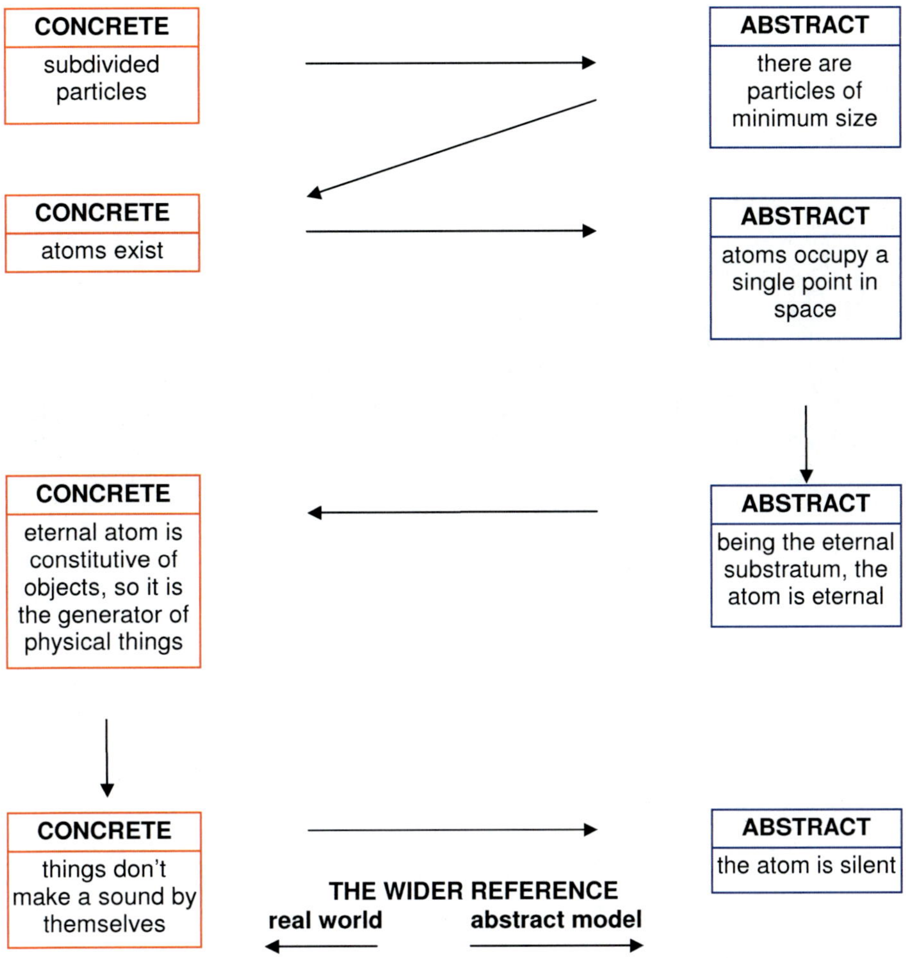

Fig. 3-3: A semantic flow chart of postulates regarding the nature of the atom. Moving from the concrete to the abstract the inherent truth value of the inference is jeopardised; even more so if moving from the abstract to the concrete. Although the truth value is high if remaining on the same level (ie concrete or abstract), after a jump its entire basis has been altered.

> The limiting unit of the process of division of molecules is ... the primary atom. Since it cannot be further subdivided it is called atom. Since it occupies a single point of space, it is one. Since it is the eternal and inalienable substratum of corporeal things, it is eternal. Since it is the constitutive basis of physical objects, it is the generator of physical things. Since it cannot by itself bring about sound vibrations, it is non-sounding or silent. These are the characteristics of the primary atom. [7]

The foregoing may seem to be a straightforward exercise in deduction since the descriptive nature of the atom is allowed to unfold in a logical manner from one feature to another. A closer examination of the inherent nature of those characteristics however reveals a number of significant jumps from the concrete level to the abstract, and back again. The problem is to identify the truth value of an inference that straddles both levels, from the concrete to the abstract and, even more problematic, from the abstract to the concrete. Therefore it becomes a matter of knowing the source of a particular declaration which may not always be openly accessible.

The above quote can be depicted in the form of a flow chart, indicating the semantic directions of its statements. Those areas requiring care in accepting their conclusions become clear.

It can be seen how moving into the abstract the deductive process becomes more tenuous. At the very least, it could be argued, the underlying observation has been sound, made as it were in the real world. But idealised qualities exist in their own world—it is their nature. A further delineation can only be as good as the previous step, especially when it is done without any available recourse to the real. Going from the abstract to the concrete makes it only worse, the absence of any definitive reference adding to the difficulty.

Consider what would happen if the analogy is not transferred into some idealised version of the overall scheme residing in the thinker's mind, but is being tested through reference with observable facts. Then the result becomes the next step in the cognitive process with a relatively high truth value, having bypassed a jump across to the abstract. The important word here is "relatively", since the ultimate reference is not so much to the actual object at hand, but the *perception* of it.

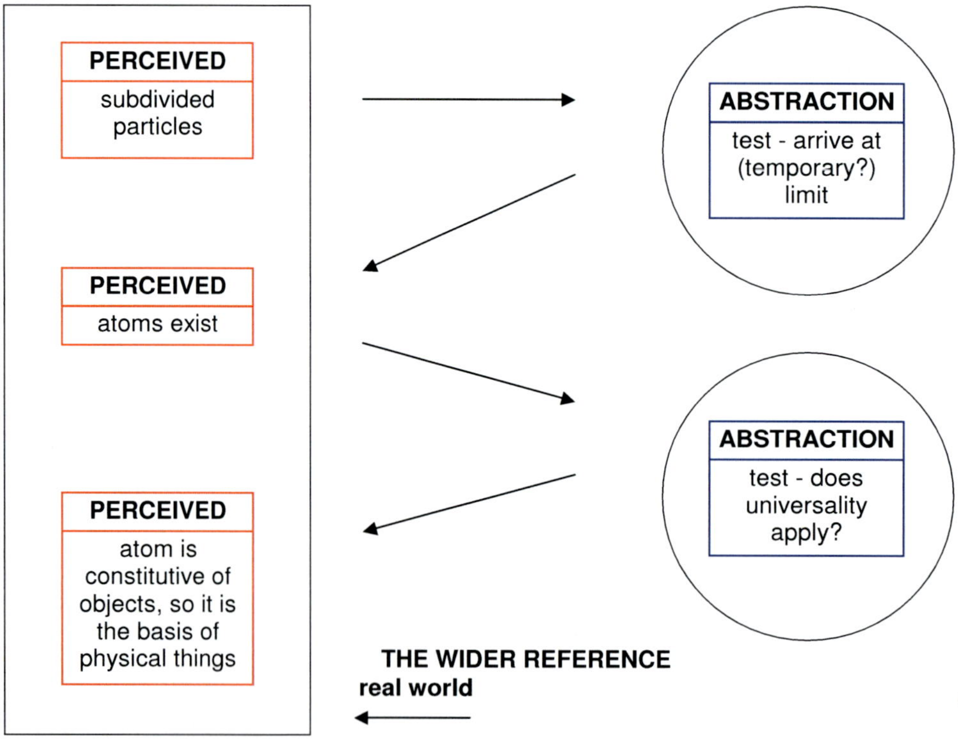

Fig. 3-4: Flow chart of postulates where the confirmation is with reference to the real world only. PAT is operating by abstracting perceived qualities, testing for them, and applying the result to the real world once again to test their validity. The truth value is determined by the perceived objects only. Note that subtle changes in some of the perceived qualities are now possible. Without sophisticated technology such a process would not be possible.

We revisit our previous flow chart leaving aside now the transference from the concrete to the abstract or vise versa. In addition we identify the concrete object as a perceived one, because that is what it is. PAT is still operating but is being constantly directed back towards the realm of the real where the confirmation is done at each step.

In the current scenario the abstractions are still being performed, but immediately tested against observations in the real world, thereby increasing the truth value of the results. Speaking from the here and now with a history of science behind us, we would be more inclined to view the latter version as more scientific than the former, yet as far as the mental dynamics are concerned the difference does not exist. Unless there is some overriding concern (eg working towards an ultimate outcome which has to be absolutely functional) the mind is just as happy delving into idealised concepts as it is taking the cues from the world around us—after all, some sort of abstractions are being applied in any case. To satisfy the mind it is important how well any intermediate model fits into the overall picture; fundamentally speaking therefore the entire process is a subjective one.

Thus one needs to be careful how the success of this operation is defined. Under the scientific view an outcome would only then deemed to be successful if it can be ascertained further. That however is a requirement which has only been added given the understood value of experimentation. Without that rider a thinking process will satisfy the owner if convenient analogies can be drawn, regardless of their sources' basic validity. It is a by-product of science which often remains neglected although an analysis of comparable thought structures reveal such differences clearly enough.

It is a reasonable assumption that people in antiquity did not have the means to view the atom in any way or form. In fact, the conventional test of that statement would be to perform a historical analysis to determine the state of technology in southern India around 1000BCE. Using thought structures this is no longer necessary. The reason the Jainas were able to talk about atoms lies in their ability to evoke analogies, as the process of PAT has demonstrated. Had it been their technology, the descriptive format of "atom" would have been different.

Further proof that it is not so much the objects themselves but their perceived quality which determine the direction of thought processes can be found in the Jaina explanation of Rest, that is its basic concept. The analogy used here is the earth, taken to be the absolute place of rest for all things [8]. Once again, in the absence of a wider reference (such as would be gained through astronomical observations, evidently unavailable at the time) the perceived quality of the earth has been found suitable for such a comparison. Without labouring the point, it needs to be emphasised that although going from one concluded step to the next seems obvious enough, such a path is in fact laid down by the *source* of data used at a particular point. If the source is not taken into consideration it is impossible to fully describe the entire process in terms of its ultimate validity. No matter how trivial that seems right now, a more complex scenario of cognitive dynamics will prove inaccessible to the analyst unless the focus is widened sufficiently to include the relevant thought structures surrounding the ideas.

Looking at the world around them the Jaina thinkers would have been struck by the inescapable fact that for every principle (ie every abstraction) there can be found its opposite (again an abstraction of course). So pervasive is this conclusion that the concept of opposites found its way into their philosophy, to the extent that its own abstraction (ie for every *is* there is an *is not*) forms part of the basic structure of their belief system. Take for instance their approach to dealing with the concepts of rest and motion. Not only can both be observed in daily life, but Jainas hold that in terms of their abstractions both versions are part of the same object. Thus although matter may be moving at the moment, it nevertheless incorporates the principle of rest at the same time. It is an interesting by-product of PAT: there is such a thing as opposites within a given area, say q_{P-1} and q_{P-2}; the q_{P-1}-q_{P-2} combination is another q_P; q_{P-1}

can be found in a given s_P; therefore the combination holds for that s_P too. A current observation does not necessarily need to confirm this.

Upadhye hints at such a state of affairs when discussing Rest and Motion (Adharma and Dharma). He says the following (my italics):

> This is something like Newton's gravitation; but Adharma is slightly different. Its main function is to arrest things. But then there is this difficulty. If there were Adharma alone how could there be motion in all things? There would be an eternal paralysis of Reality. *To remove this difficulty Dharma had to be posited.* The function of Dharma is to guarantee motion within the limits imposed by Adharma. [9]

In fact, the foregoing quote is part of a repudiation the author felt obliged to include, given the opinion of one interpreter of Jainism that the Rest/Motion question is really a moral one. There may well be reasons for entertaining such a possibility for the purpose of playing with a certain hypothesis, even for eventually deriving some moral construct further down the line. Upadhye makes use of philosophical points to show how, fundamentally speaking at least, such a notion is misplaced. There is another argument supporting Upadhye however, utilising the functionality of PAT. The creation of concepts, leading to the formulation of an opposites-abstract, is bound to evolve into the idea of a non-exclusive co-existence of abstract qualities, attached to any object found to be relevant. At that stage, morals do not come into it (which is not to say that some moral lesson might not be developed later on).

When we considered the procedural path taken by the transference from the perceived to the abstracted object, we used the atom as an example (see Fig 3-3). As it happens the outcome is not uncomfortable, even for modern minds. We have been fortunate that the deductive steps were relatively straightforward, and even a jump across to the abstract area did not evoke conclusions which were too alien when seen from the perspective of the real world. In other words, the possible options had been rather limited with no problematic results in the offing. This is not always the case however, since the overall schema does not discriminate in terms of truth values but in terms of relevant analogies; PAT ensures functional contiguousness amongst the Q_P/Q_A sets, not a functional model of the real world.

The greater the number of s_P and therefore s_A, the greater the reservoir of q_P and q_A as well. With that comes a greater choice for attaching any particular q_A to an object, whether that be purely in the abstract or also in the concrete realm. Should one abstraction refer to the greater wholeness of things, implying an inclusive status of several qualities, then that same status will also be transferred into the concrete, with sometimes confusing results in relation to the real (the *perceived* real, of course).

Consider for example the Jaina notion that there are five physical objects, namely earth, water, fire, air, and plants. These are not particularly evolved, having only one kind of feeling (that of contact). We are already familiar with the abstract concept of Soul, delineating into the many souls which constitute part of the Real together with other things. According to Jainism there are certain kinds of souls which through negative influences become associated with the above physical objects, thereby demonstrating their—undesirable—affinity with organic matter. As pointed out by Upadhye in his commentary [10] there is no problem with plants; their organic nature is understood well enough. Less clear is the connection with the other four, giving rise to some opinions that here is an indication of the inherent lack of sophistication in ancient thinking. There is no need to be so drastic. From the perspective of thought structures it can be argued that the schematic format of the entire system cannot help but come up with such attributes, given certain fundamental decisions, as will be shown.

We can apply the PAT system once again, with firstly identifying the Jainas' idea of an all-encompassing Whole. Its abstraction is endowed with certain qualities, with the understanding that these qualities, being distributed across the Whole, would permeate through the real world and manifest within so many perceived objects. The completeness must be maintained, so the notion of duality (*is* and *is not*), the affinity-seeking properties of matter, the causal and contiguous relationships, and the hierarchy of values are reflected in every segment of the Real, as indeed they are in terms of their conceptual dynamics. Souls, being objects again, attract the same attributes, spread across several kinds. And then there is the identification of five organic things—earth, water, fire, etc. What else can be done but apply the same schema once more, in order to uphold the consistency of the entire system? The problem is then not so much the integrity of a cosmic model, but the mental organisation in the observer's mind which tries to understand the underlying logic ("logic" as in: the principles governing correct inferences; "correct" as in: non-contradictory within itself). As mentioned already, great care had been taken by the Jaina philosophers to prevent disharmony, but in this case at least a certain price had to be paid. On one side are the organic objects, which at the point of their identification may not attract undue criticism. On the other are the attributes which are applied to the features of the system, including those five objects. Imbuing the objects with the attributes may be problematic, especially for an outside observer, but for a Jaina thinker a far bigger problem would be the modification of the attributes or the re-classification of objects. To change the attributes requires an adjustment right across the entire cosmic model, particularly given the repetitive nature of quality attribution to its various features. As for choosing different objects, one can only surmise about the specific world picture people in southern India 3000 years ago carried within them and what to them would be suggestive of viable, broad categories. In any case, one must not forget that in Jainism the absence of contradiction is far more important than revelation, as Upadhye reminds us on several occasions throughout the text.

The fundamental problem with the creation of such a model is the shifting between its concrete and abstract areas, where conclusions reached in one sector are applied as a consequence in the other. For instance, if the categorisation of objects (made in the concrete) was subjected to a critical evaluation through observations derived from the concrete as well, the result would be on firmer grounds, some possible side effects notwithstanding. One could also say the same thing for the other side: if a decison was made based on abstract reasoning alone, it would be prudent *not* to expose it to the concrete lest its shortcomings become apparent—a familiar scenario throughout the ages. Whatever option is taken, it can be seen that the mind accepts a conclusion as valid if its elements have been accepted as relevant to the issue. The relevance of its elements takes precedence over the effects the conclusion may have. So in the present example it was more important to maintain the collective nature of the attribute set, created as it had been in conjunction with the rest of the model, than to worry about what a fire organism might actually be like.

In the end a belief system is not about describing nature; it is an exhortation for humans to exercise their will in line with assumed truths. For the most part therefore philosophies spend their time explaining these truths so the rest of us may notice them as well. The processes of PAT play a significant part in the creation of ideals for which to strive, since only an idealised version of something can be made attractive enough. If the description of an everyday object (such as some piece of matter) invites a certain degree of confirmation from the real world, then when it comes to morals the need to identify a principle tends to promote some aloofness from the tangible.

In Jainism a person who wants to evolve spiritually is advised to believe in the true nature of reality. Of course, in order to do that one must know what the true nature is, and the list of what is correct and what is not demonstrates the now familiar creation of idealisations from the observable [11]. The quest for enlightenment is basically a struggle between life and matter. Thus matter—the diluted, the dirty, the indiscriminate mixture—has to give way to life—the pure, unencumbered, unsoiled and true reflection of the cosmos. In contrast to the previous process where the reference had been to perceived objects, we are now dealing with abstract objects per se; but this is an apparent contrast only. In terms of the perception/abstraction transference the dynamics are the same, that is to say they represent a derivative from an abstracted observation and are then moved into pure idealisation. Thus the "right path" involves three "jewels", ie belief in the true nature of reality (already mentioned), the understanding or knowledge of the constitutive categories of reality, and the right conduct based on the former two. Note that as far as thought structures are concerned, those three are perceptions which are being processed further.

In addition there are main conditions attached to the path, their conceptual nature being similar to the "jewels". They are pure abstractions: the co-operation of right belief and right knowledge; right conduct; the absence of desire and aversion; the path thus formed which leads to liberation; the taking of the right and reliable path; taking the path which is available only to matured souls; the matured souls having attained certain qualifications; and the need for the matured souls to have eradicated any gross emotions.

The reference system in this case evolves from a knowledge of oneself. Since the source of any delineation is the perceived notion of one's own person ("I am such and such and therefore I have to do that and that to attain enlightenment") the ultimate measure is the person's mind. While Jaina philosophers stress the importance of staying connected to the real world, it is not the objective assessment of a real-world performance which acts as the guiding principle but what representation thereof resides amongst the thinker's thought structures. It is significant in this context that of the two abstractive versions of souls it is the organic ones, ie those imbued with Life, which have the potential to become perfect. Inorganic souls do not have the capacity for spiritual growth. Describing that feature Upadhye comments, "This is one of the Jaina religious dogmas whose metaphysical basis is not quite clear" [12]. Metaphysically speaking they may well be obscure, but considering the effects of PAT upon a self-repeating system of well-defined classes of abstracts there is hardly any choice. The following schematic diagram in two parts will help make the situation clear.

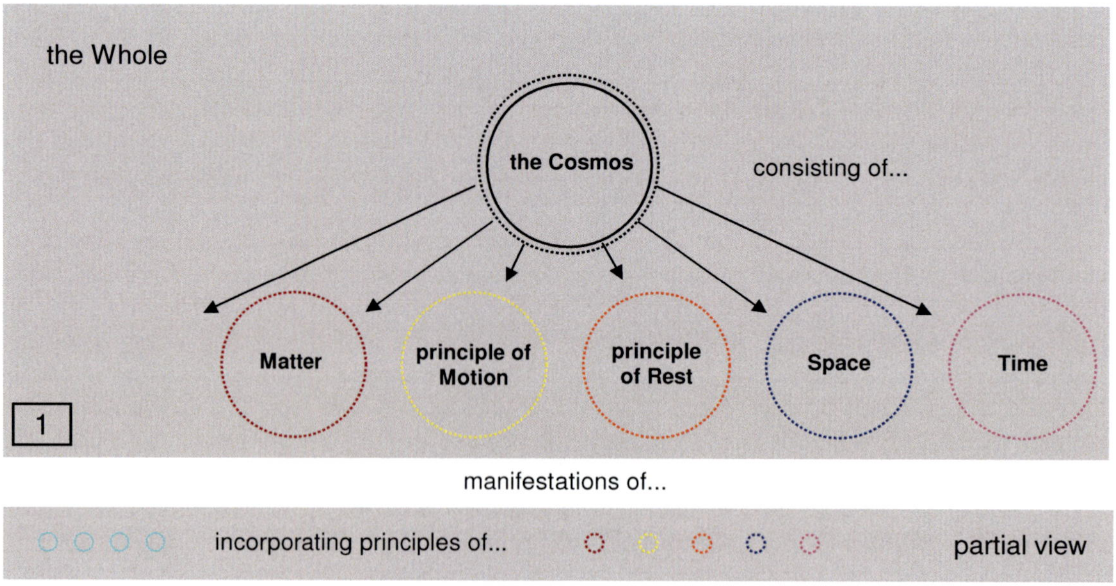

Once again we start with the big picture. The Cosmos itself is defined by the six main principles of Soul, Matter, the principle of Motion, of Rest, and of Time. Soul manifests in the many souls existing in the realm of the Real. With respect to these we again have the holistic view where we find the concept of souls in their pure state, that of souls enmeshed within the material world, and each soul's intrinsic value. The latter is further delineated into one understanding in which its use is sanctified by convention but without its intrinsic value, ie a figure of speech; and another version which depends upon its own intrinsic merit without any metaphorical meaning. But there is also a partial view, concentrating on a single soul only. The breakdown is similar to the wider view, with the individual soul being mixed with material qualities (these would be alien to the soul's true nature) and featuring the same aspects as before, namely a figurative and an actual one; and the intrinsic nature, again having the same aspects as the previous interpretation.

But souls are associated with the organic and, keeping with the general principle that each thing has to be capable of having the same predicates no matter where in the entire system one applies the focus, the evolutionary aspect of Life becomes linked with "soul". If there are two types of "soul", one attached to the organic and the other to the inorganic, it would have to be the former which is able to improve itself, not the latter. The process of conceptualisation is aided by the Jaina dictum that for any entity to fully develop there have to be two opposing and at the same time complimentary states. Hence the inclusion of the organic as well as the inorganic version. (We find the same idea applied to the Whole, where consciousness and the material are both embedded—one in Soul, the other in Space, Time, Motion and Rest which are held to have no consciousness [13].)

Such a repetitive breakdown into similar qualities suggests a fractal nature of the model. While not expounded in the mathematical sense (for this we have to wait until the present), it is nevertheless interesting to see a system of this nature, resulting from an attempt to explain one's world with the express intent not to run into contradictions. The perceptive representation of something allows recourse to observation in the real world, as stated before. In order to advance towards the ideal one needs an abstraction of the object's intrinsic nature, since in the Real that version cannot be found. Thus righteousness is an abstract, stripped of any diminishing ingredients that may have been accumulated while still placed within the concrete. One could say that the process of abstraction fundamentally serves this purpose alone. An implicit confirmation can be found in the advice that belief, knowledge, and conduct ought to be based upon the inherent nature of the self—a pure abstract. If they are not, that is if the realisation of belief etc. are gained from external things (perceptions, yes, but derived from the concrete), the result will be less substantial. The latter is classified as the "relative jewel", whereas the former constitutes the "real jewel" [14]. It is the "relative jewel" that serves to attain the real one: concrete sources of information are turned into representations, which are elevated into their respective ideals and so enable the mind to hold a picture of pure essence.

So far we have examined the descriptive elements of the Jaina model and the mental dynamics behind them. Yet the Pañcastikayasara goes further. We are also given a system of logic, a mechanism designed to organise the observations, abstractions, and conclusions processed by the mind. It is arguable whether a formal system of logic would watch over each and every act committed by any person as some sentinel of conscience. Most likely that won't be the case. On the other hand, members of a society do tend to be aware of an overriding law, be that in politics or commerce, in morals or just plain thinking. While not always acknowledged, in times of stress such rules are often used to underpin the validity of one's case, to the extent that the situation itself is being moulded in order to make it fit the rule. Indeed so common is that scenario that the degree to which people invoke logic to defend their argument is a measure of their general state of learning; a fact that can be observed anywhere in the world.

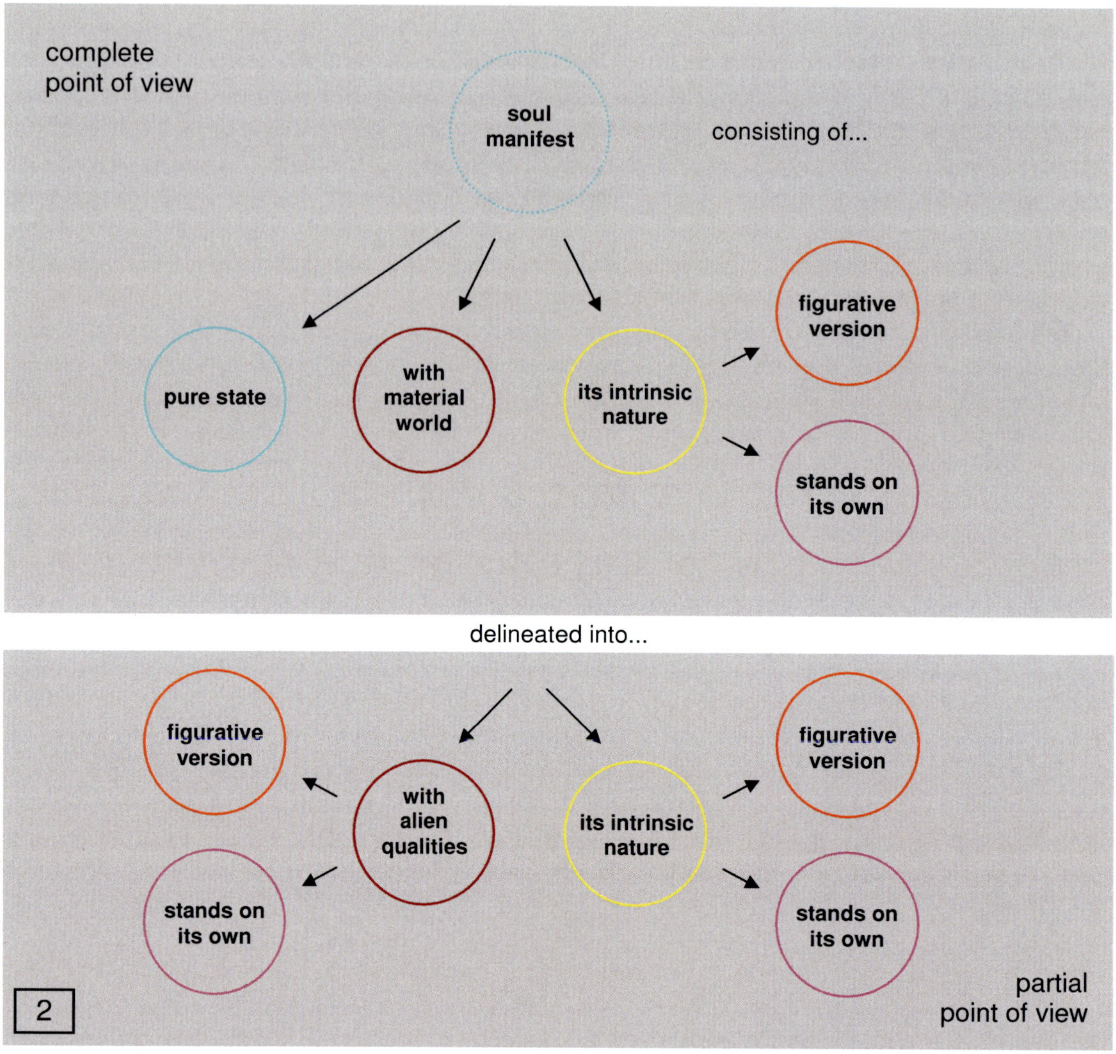

Fig. 3-5: The delineation of abstracted features in the cosmic model. (1- on previous page) represents the big picture, from the whole to its subsets. (2) shows the categorisation of Soul, again from the whole to the detail. For explanation see the text. NOTE: the colours are for identification only—they do not serve as analogies.

A brief listing of the Saptabhangī, or the Seven Modes of Predication, will illustrate their affinity with the conceptual relationships described so far [15].

Perhaps X is. Note the tentative format of this assertion. It is in line with the underlying concept that any conclusions reached must be ultimately referenced to observations of the real world and also founded within the consciousness of the thinker, progressing (and hopefully improving) as he is along the path through life. As such there can be no definite certainty.

Perhaps X is not.

Perhaps X is and is not. Considering the Whole of anything, the observer finds two aspects, one ascertaining the object's nature, the other positing its potential—but currently non-existing—otherness. Both are needed to define the object as a way of circumscription for which the first two predications alone are insufficient.

Perhaps X is indescribable. If predicate (3) encompasses the true and not-true aspects of a thing, then the significance of such a total may well require a language that reaches beyond the descriptive immediacy of common words. Hence a predicate which allows for such a circumstance.

Perhaps X is, although it is indescribable. Following from the above case the first three predicates could fall into the same general category.

Perhaps X is not, although it is indescribable.

Perhaps X is and is not, although it is indescribable.

How would those seven statements influence our thought structures?

As stated under (1) and earlier, the philosophy requires an ongoing development on behalf of the user. The thought structures created along the way, leading from the initial representations of the observed to their abstractions and eventually focusing on the purest idealisation of a given concept, will have established a hierarchy of values in due course. For example, given that a soul is being declared as having a pure aspect but also an impure one due to its involvement with Karmic bodies, the abstraction 'pure part' holds a greater degree of intrinsic grace than that which has been identified as being in a state of suffering. Therefore the qualities of the former will always override those of the latter in terms of their judgmental power. At the same time a person is also taught that the fundamental reference resides in one's mind. Suppose that Jaina logic committed the adherent to declare in absolutist fashion that something either 'is' or 'is not', at whatever stage of wisdom the person may be. The abstraction process itself does not guarantee any particular truth value within some Q, whether this be a quality held at the perception or at a later abstractive stage. The truth—such as it is—needs to be ascertained through reference to some outside input. A format of 'X is...' (or 'X is not...') would not leave the door open for further evaluation where the previous conclusion may well be updated to a more valid one, leading to a possible contradiction. This however is not allowed in Jainism. Thus declaring '*Perhaps* X is...' etc prevents the system from becoming rigid.

Jainism also incorporates the notion that within any particular concept there exists its assertive aspect in conjunction with the idea of what it is not. The two go together—for example, a jar is a *jar*, but equally the object defines its significance by what it is *not*. As such its abstractive, negating space is proportionally shared with similarly defined other representations. These create a common reservoir of negative assertions, where in an ever-changing existence (as far as material manifestations are concerned) a thing's uniqueness is merely a momentary occurrence, its characteristics drawn from a common pool whose own nature holds the sum-total of possible aspects. Therefore a change in an object's nature also affects the whole, not in its overall constituency (which is permanent), but in terms of its assertive potential. This may seem rather trivial, were it not for the resulting recognition that in such a case everything is part of everything else, and no single item can be viewed in total isolation from any other. An object's intrinsic nature then can intellectually only be dealt with via its abstraction, removing it from the perception process which had access to all the possible modifications. Once the thought structure relating to a particular thing is placed firmly on the abstract plane it can be upheld as an ideal, a never-changing image acting as a representative *par excellence*. Note the path thus established: firstly, from a perception to its abstraction and secondly, in case of any change, again through a process of confirming perceptions to another abstraction. The abstraction itself therefore cannot be altered ad hoc, lending a considerable degree of stability to the realm of ideals.

That pool of negative commonalities I mentioned in the last paragraph still has to be understood as the sum-total of discrete elements. A single element thereof however cannot be described through its set of characteristics, positive as well as negative. The function of human language is to express an affirmation, even if such an affirmation should have the purpose of negating. Our words apply a focus to the scenario of existence, giving our thoughts accessible contours to differentiate one from the other. Against the background of an all-encompassing mist of shared non-casts language becomes powerless. Thus the single element features—but cannot be described. The Jaina thinker saw fit to embed this state in a predication.

Compared to the first three statements we have a somewhat different process leading to the eventual abstract. Since all the 'is-not's are an abstraction in themselves, including their totality, that which cannot be expressed singly is an abstract also. Therefore altering a single constituent affects an abstract, without having taken the previous path through further perceptions. Where does that leave the system's stability? As it happens, leaving the abstract undefined in explicit terms does not really change anything about the item's nature, since its nature has not been described in the first place.

Keeping with the seemingly fractal nature of the entire model it can be expected that the non-describable aspects of our world would be repeated in terms of the first three predicates. If something *is* or *is not*, even if it contains both options, then such a positive and/or negative nature would also have to apply to the indescribable. We are now firmly in the abstract, but as explained above this does not constitute a problem for the overall integrity of the system.

The logic represented by Saptabhaṅgī has given rise to various arguments as to the correct position any one of the predicates holds in the light of objective assessment. Particularly items (3) and (7), because of their dualistic nature, have been linked with relevant theories by Menrong and Russel in the discussion on the topic [16]. From our point of view the question is not so much about the degree to which a thing's "objective fact" can be aligned with a given proposition; rather, to what extent does the abstractive quality of a certain perception lend itself to its idealisation. Even an "objective fact" about a thing is an abstraction in itself in terms of thought structures. Whether the character expressed through it is capable of calibrating our understanding of the thing depends not so much on the item itself but on the inherent truth potential residing in its abstraction. In other words, it is the abstracted delineation which is under scrutiny, with the accompanying proposition yet another abstract being paired with the former.

The example given in the text, "Socrates is living" (S_T) and "Socrates is not living" (S_F), highlights the issue. Right now S_F is true, although it ought to have a corresponding fact confirming just that. Yet such a fact does not exist, therefore S_F cannot be true—a state of affairs which is clearly wrong. Now consider the problem from the perspective of thought structures. S_T and S_F are both an idea although derived from different sources (say, a contemporary witness in ancient Athens; a visitor to a modern library reading about Greek history). The fact that both thought structures exist means that either one is a possible, and above all useable, result out of the transference from the perception to the abstraction. PAT has created S_T and it also was responsible for S_F. As abstracts they cannot be judged true or false in terms of "objective facts" if those facts are manifestations in the real world, like a rock somewhere on Mars would be an objective fact regardless of anyone knowing about it. To arrive at any conclusive evidence about Socrates requires the analysis of the original perceptive process, just as the assertion about a rock on Mars would demand some sort of proof that such a rock has actually been identified. Identification however is one thing; the resulting image quite another. The steps one needs to take to agree with an identification process are not the same as those which allow us to hypothesise about an idea.

Suppose someone today declared, “Socrates is alive and well”. Assuming I had never heard of the man before, I can nevertheless hear the words and imagine a picture of a live Socrates. Place the same scenario 2000 years back in the past and the mental picture could be exactly the same. Not only that, both abstractions can be used to form other thoughts, such as Socrates striding through a street, holding a speech before the senate, or indeed discussing his philosophical ideas. Let’s call the Socrates in my mind S_{PAT}, since he is the product of a transference. S_{PAT} is no more or less true because I can invoke whatever scenes with him, therefore the truth value of these scenes is not dependent upon my S_{PAT} but on the kind of connections one can make between *them* and the outside world; that is to say, how valid the scenes are as potential outcomes of some other PAT.

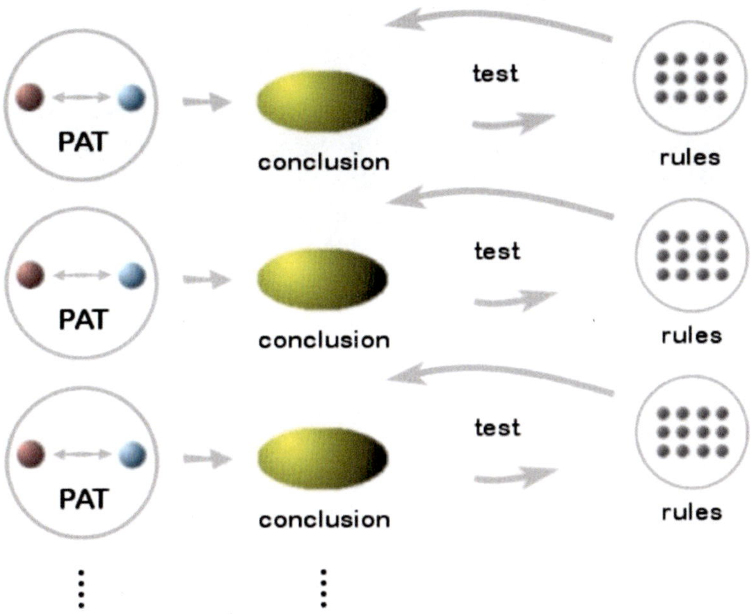

Fig. 3-6: An entire imaginary system is created by repeated processes of perception/ abstraction transferences (PATs), subject to certain rules. The rules are applied to each PAT and its conclusion, which itself is used as a reference to the rules. Because they have been declared paramount they are active all the time, regardless of any results from PATs or their conclusions.

We will have cause to visit similar arguments later on, simply because the concept of reality is often combined with reality per se, and the functionality of either is used haphazardly without considering their respective domains. We can summarise the insights we have gained through our examination of Jainism and present an update on the previous schematic diagram of thought structure dynamics (see Fig. 3-2 in this chapter).

There are two main significant features; firstly, the process of PAT which creates an ideation through outside input which is imbued with certain characteristics. Secondly, the type of characteristics are dependent upon another PAT, again derived from some input which gets its qualities either from the outside or from a proposition that has its origin amongst the thought structures already formed. For instance, the various aspects of ‘soul’ come from ideations themselves, whereas the rule that the ultimate reference has to be the real world is an outside input (note however that the other required reference, one’s own consciousness, is another internal construct). Because these rules are a dominant feature, any conclusions reached will have to be subordinate and therefore any contradiction demands a further analysis of the original data, rather than some modification of the result itself. It is an ongoing process, working its way from the initial concepts down to more and more detailed delineations, employing the given rules to monitor the temporary outcomes.

The diagram is not concerned with the actual content of observations, conclusions, or rules. Our focus is on the dynamics of idea-creation, using the examples from Jainism to follow the processes. It is not a complete picture of our mind yet, since we have dealt with a philosophy, that is an already established system of principles and ideas which its adherent is meant to follow. As such the contents (ie the data) will vary from person to person, but the procedure (ie the algorithm) remains the same. What the diagram does not show are the processes of creating the rules, the hierarchies resulting from differing inputs (external and internal), and how the hierarchies themselves can change and re-arrange the results. In daily life, without the guiding force that a belief system provides, the dynamics are similar but the direction of its processes would be more at random.

But above all, the people who created their picture of the cosmos were human. Their minds dealt with the fundamentals of life, and it was the daily pleasures, pains, fears, and desires which influenced their thoughts. The protagonist in *Foucault's Pendulum* has it explained to him by his lover [17]:

"Pow, archetypes don't exist; the body exists. The belly inside is beautiful, because the baby grows there, because your sweet cock, all bright and jolly, thrusts there, and good, tasty food descends there, and for this reason the cavern, the grotto, the tunnel are beautiful and important, and the labyrinth, too, which is made in the image of our wonderful intestines. When somebody wants to invent something beautiful and important, it has to come from there, because you also came from there the day you were born, because fertility always comes from inside a cavity, where first something rots and then, lo and behold, there's a little man, a date, a baobab.

"And high is better than low, because if you have your head down, the blood goes to your brain, because feet stink and hair doesn't stink as much, because it's better to climb a tree and pick fruit than end up underground, food for worms—you really have to be in an attic—while you often hurt yourself falling. That's why up is angelic and down devilish." ...

"The sun is good because it does the body good, and because it has the sense to reappear every day; therefore, whatever returns is good, not what passes and is done with. The easiest way to return from where you've been without retracing your steps is to walk in a circle. The animal that coils in a circle is the serpent; that's why so many cults and myths of the serpent exist, because it's hard to represent the return of the sun by the coiling of a hippopotamus. Furthermore, if you have to make a ceremony to invoke the sun, it's best to move in a circle, because if you go in a straight line, you move away from home, which means the ceremony will have to be kept short. The circle is the most convenient arrangement for any rite, even the fire-eaters in the marketplace know this, because in a circle everybody can see the one who's in the center, whereas if a whole tribe formed a straight line, like a squad of soldiers, the people at the ends wouldn't see. And that's why the circle and rotary motion and cyclic return are fundamental to every cult and every rite."

Unfortunately, life tends to become more complicated by the addition of meanings and mental connections. To continue our exploration of the mind we go to the beginnings of Christianity, because it was there that many of the fundamental building blocks of the current European perspective have been put together. Our first port of call is Persia.

4. Building Europe's mind: the foundations

Manichaeism - St Augustine

Whenever an issue is raised in today's Western world, chances are that somewhere amongst its elements live the moral and social values of Christianity. Sometimes their presence is obvious, often less so. Sometimes their influence is hidden altogether, working through indirect channels that had grown slowly over the centuries, their origins buried under layers and layers of sentiments, interpretations, and surreptitious measures. At other times they are being proclaimed loudly, inviting people to rally openly to their cause. After all, a desire can be expressed through full voice and big gesture, but it also can be conveyed through a wink of an eye, or the subtle movement of a limb. Even more delicately, it can be transmitted by the omission of words, an emphasis on a pause, the weighty silence.

The variety of means demonstrates the role of content paired with its delivery, because it is not only the idea itself which is of interest, but the meaning of the concept behind its presentation before another mind. To express one's agreement with mere words without any accompanying body movement casts doubt over the words themselves, indeed may render them utterly useless if not counterproductive. On the other hand, gestures alone can carry an impact provided they are being understood in the appropriate manner. Thus there are hierarchies, precisely defined values, and priorities that translate into mannerisms and etiquette. The ranking of values, that is to say the relative bias the effect of a manifested idea is capable of inducing in the recipient's mind, is dependent upon their creation, the modes of introduction, and their affinity with already established concepts; in other words, their history.

Not only individuals relate to such circumstances; society at large does. The minds of a million people can turn on a single point of faith, and the lives of millions can be crushed by the incongruous weight of one idea. For sheer survival value, it pays to understand the ideational atmosphere one breathes.

Europeans had by no means been a tabula rasa before the advent of Christianity. Through centuries of domination however any lingering traces of other interpretations of one's reality had been wiped out with varying degrees of brutality. Those elements which are surviving, such as Christmas trees or solstice fires, are thoroughly stripped of their original meanings to the extent that a response to their re-enactment would cause consternation or even legal proceedings (which is not to deny the existence of groups who seek to preserve certain traditions—always with a careful eye towards the local authorities). For example, as of to date 'witchcraft' is still illegal in Queensland, Australia, a law which a local mayor tried to invoke in order to ban a conference on paganism (unsuccessfully, as it turned out) [1]. It is of course significant *which* aspects of an old custom have been preserved and which have been replaced.

To understand the conceptual layout of the Western mind it is therefore necessary to place together the pieces as they have been introduced, in terms of their content as well as their relative values. We need to identify the mutual influence of one idea against another, because when it comes to playing out a given mentality the process is not random. Once again, this text is not a critique of philosophies or religions as such; but recognising the role a thought structure plays requires knowledge about its nature and its relative placement amongst neighbouring concepts. Just as individuals acquire their attitudes as they move through life, so do societies at large.

Where to start? If the task was to map out the philosophies which contributed towards the Western mind set one could choose Judaism as an entry point, given that it had been the precursor to Christianity. At the same time it could be argued that Judaism itself drew its sources from various belief systems, and one could not understand its structure unless one were familiar with the ancient Middle East in general. Since we are interested in the abstract equivalents of perceptions, precepts, and subsequent actions, it is not necessary to examine *all* streams which one way or another made up a given period in history for the purpose of understanding the inherent functionality of those abstracts. Provided there is the opportunity to observe their existence and the ramifications resulting from their connectivities any interested readers will hopefully be sufficiently equipped to strike out on their own to wherever their fancy may take them. The choice of subjects therefore is a personal one; the overriding concern being the transparency of such phenomena, their historical chronology coming a close second (for reasons already mentioned), and the specific nature of this or that belief, third.

One of the belief systems which played a part in providing Christianity in its early days with certain ingredients was developed by the Persian Mani. It is difficult to attach a comprehensive and accurate description to his persona using the vocabulary of the 20th century. Every age has its own categories, a combination of factors which is unique and cannot be simply mapped from one era to another. One could call him a prophet, a preacher, an artist (he was famous for his skills in painting and calligraphy). Observing his life and how he single-handedly created a religion which was to survive until virtually the present in some parts of the world, notably China, he must also have possessed the talents of a politician and a diplomat. Last but not least, he must have been a teacher and orator of considerable power. Establishing a significant religion amidst the political and spiritual cauldron of Mesopotamia in the third century after Christ's birth is surely a major achievement. Alas, all that eventually became his downfall and led to his death during imprisonment in 276 or 277CE—the victim of a religious competitor.

Yet he also was a child of his time; his mind, being unusually gifted (or cursed?) with the craving for conceptual matter with which to fill the spaces of its inner universe, sought and found its pieces amongst the Old Testament of the Jews, the new messages coming from the Christians, the Syrian-Babylonian religions. Not all were to his liking, but even in his rejection one can find the effect of this or that cult. Let us then reconstruct the world according to Mani and see how the thought structures fit together. As always, the description is in line with our particular perspective, treating any concept as a thought structure or pattern, placed within its basin according to its required significance.

At the heart of Manichaeism lies the notion of a dualistic universe, but with certain reservations [2]. There are two primary elements which are being qualified through the names given, automatically imbuing them with definite values: one is "Good", the other "Evil". In this case the idea originated with the Iranian primary principles of Ahura Mazda and Ahriman respectively, although there both were descendants of the primordial godhead, Zervan. If both, Good and Evil, are divine brothers, then the conceptual weight of either would have to be equal. The consequences of such a picture are not necessarily straightforward. One can argue that Evil, being of divine origin, would attract sterner measures from the—human—defenders of "Good" when charged with the task of confronting it in the here and now. It is also true that it can gain greater status, a preference which should translate into some reverence directed towards a person subscribing to its influence. In such a case room is made in society for the worship of appropriate beings and symbols, where their specific qualities are woven into the general fabric of life-values, to be evoked under special circumstances. One example is the representation of Kali, a Hindu goddess personifying the dark forces of nature, in tantric belief systems [3]. To remove "Evil" from the divine will certainly diminish its stature and serve to relegate the follower to nether-regions. It also enhances the god-figure since *its* characteristics do not even touch the negative—not in the interpreted physical, the spiritual, nor moral sense. Consequently, a human who finds himself outside the circle of grace will be mortified through the absence of any divine influence, a disadvantage for him yet a good fortune for his adversaries.

In any case, for Mani the idea of Good and Evil being equal brothers was unacceptable. God was given the epithet "Father of the blessed light", or "Father of greatness", as Mani called him. At the same time this realm of light was also synonymous with God, and both had been in existence for all eternity. To the all-constituting, permanent godhead were ascribed certain delineating features, which are worthwhile subjects for students of religion but do not need concern us here. For instance, Mani decided on five "dwellings", attributed to God and representing the intellect, namely sense, reason, thought, deliberation, and intention. Then there is the idea of God's four-sidedness, a certain geometric perspective used to circumscribe divine aspects; and so on.

If a detailed examination of Mani's psyche were possible, combined with an equally thorough understanding of the social psychology in those days, one could venture into the layered significance of religious strata. As it is, such an undertaking would not be productive for several reasons. First and foremost, the sheer distance in time prevents us from empathising with a people who lived under totally different conditions—social, political, spiritual—from those of today, anywhere. Secondly, the material available to us is by no means complete. There are enough texts and other artifacts to deduce the broad picture, but while we may know the hymns sung we cannot know the inner thoughts held at the same time. A particular text may give us certain hints about what it meant to be pious, but further deliberations should be done with care. For example, referring to the prohibition of blasphemous thoughts, the uprooting of plants, killing of animals, and sexual lust as far as the "elect" class was concerned, we find the following in the *Acta Archelai*, chapter 10, about an adherent ready to partake of a meal [4]:

> "I did not mow thee, did not grind thee, nor knead,
> Nor lay thee in the oven,
> But another did do this and bring thee to me.
> I eat thee without sin."

According to Widengren, that "elect is more or less a liar". This may well be so, but there is also another explanation.

Let us remember that religious ideals are thought structures and as such they represent abstracts. Abstracts are embedded ideations, cross-linked with relevant patterns according to their perceived values and significances. In Part II we will follow the creation of these constructs from their very beginnings, but for the moment it suffices to say that a concept, by virtue of its acceptance, is part and parcel of the person's inner landscape. Whether the relationship is a positive or negative one, that is whether the individual has identified with its nature or with its counter-nature, that concept in a very real sense *is* the person. Furthermore, just as religious ideas form a contextual part of one's psyche, so do other characteristics. One's identity therefore is not only what one perceives as being uniquely own, but also extends to other, collective features, that is the common denominators of social manifestations. Thus a Manichaean elect can see himself as a religious disciple, but also as an Iranian, a subject of King Shapur. He can also see himself in terms of an elect with all the relevant attributes and duties, *and* in contrast to somebody else to whom such duties do not apply, regardless of that person's standing in any other context. The question is, what comes first?

The answer lies in the analysis of the thought structures which have been formed on the basis of whatever input we can find. In this case it is the texts and religious dogmas as stated by Mani, as well as illustrations of behaviour. What ideations they conjure and their relative weights and biases help to build a reasonable mental panorama. There is a difference then between such information as, 'The god-concept manifests as an apprehensible tetrad including, God, light, force, wisdom' [5], and what we can glean from a biography of the Bishop Porphyrius of Gaza by Marcus Diaconus, himself a Christian chronicler, in which a scene is described where a woman is led before the bishop so she could denounce her Manichaean faith [6]. While not directly contributing to an understanding of the Mani world-view, it is nevertheless instructive to observe the sub-text of such a testimony because it gives us some idea of the intensity and sheer immediacy of people's convictions. For this reason a brief diversion is worthwhile. The event occurred in Palestine around 375 CE.

The quote begins with, "About this date a woman of Antiochia named Julia, who belonged to the abominable heresy of the so-called Manichees, came to live in the city." It goes on to say how she converted some of the city's residents and in doing so attracted the attention of Porphyrius, who subsequently had her brought before him—"But after some days the holy Porphyrius, being informed by certain believers, sent for and questioned her who she was and whence and what her faith". Julia told him she was Manichaean and was ready to explain her faith, when the bishop interrupts and tells her to come back tomorrow.

We are told, "The blessed man, having fasted and prayed much onto Christ that he might shame the devil, prepared himself for the following day and summoned some from among the pious, priests as well as laymen, to listen to the disputation between him and the woman." Thus fortified he received the visitor once more, but now, "Upon Julia the years had gathered. And their behaviour [Julia was accompanied by four others] was modest and their air was meek, but—as already said—outwardly they were sheep and inwardly ravenous wolves and venomous beasts, for hypocrisy marks all their words and deeds. And the holy man, holding the sacred writings in his hands and after he had struck the cross upon his mouth, challenged them to declare their faith, and they began to speak."

What the woman said has not been recorded, most likely out of fear that a reader might be attracted to these heresies after all, as Widengren quite rightly surmises. In any case, after about four hours of "blasphemies" Porphyrius "was inflamed by divine passion" and pronounced his punishment, effectively cursing her,

> "God, who has created all things,
> Who alone is eternal,
> Who has neither beginning nor end,
> Who exalted in trinity,
> He thy tongue will strike,
> And thy mouth will close,
> That thou evil things mayst not speak!"

After which Julia "began to tremble", the features of her face changed and she "was bereft of speech and movement, her eyes open and fast upon the most holy bishop". Eventually she "gave up the spirit" and was given a "befitting burial".

Several things are worth noting. The time is towards the end of the 4th century, when Christianity had been officially accepted not long before under the Roman Emperor Constantine. Although their newly found status freed them from the brutal persecutions suffered previously, for Roman citizens in general life became tedious. Taxes were assessed year by year, and peasants, city dwellers, craftsmen, up to the town's nobles were obliged to contribute to the state's coffers. The position of nobles was particularly precarious, since they were held personally responsible for reaching the set quota, and shortfalls were punished harshly, often with death. The political situation was unstable, and it can be well imagined how a provincial atmosphere could contribute to an already charged state of affairs, given inadequate logistics, rumours, and small-town allegiances [7]. Considering the social respect an office in a religion can confer, coupled with the opportunity to remove oneself from the pressing demands of secular life by adopting the ascetic aspects of the faith, plus the need to demonstrate one's righteousness in such a case, and we have a heady mix of psychological pressure just waiting for the right moment to show itself to the public; a public which is living under exactly the same conditions.

We can expect therefore to have individuals who would be more than happy to report anything untoward to their superiors, and what better superior than a representative of divinity? In times of stress, especially if a good portion of that stress comes from one's authorities, a certain alertness develops amongst those affected which enables them to find faults away from themselves. In addition, although Christians were free to follow their religion, the faith had to compete with innumerable other cults and sects, some more established than others. So a bishop, needing to affirm his stature as a religious figure of respect and also remoteness from mundane pressures, would be well disposed towards anyone coming to him with reports of matters relevant to just such needs. The actions serve the informers as well as the bishop.

We do not know whether Porphyrius did or did not have any prior knowledge of Mani's teachings. In either case his office would demand that due ceremony should be observed when presented with the woman, if only for diplomatic reasons: give the accused enough rope to hang herself. Why the interruption and postponement? Provinces are small places, and even a bishop would have only a narrow circle of officials, who in turn are acquainted with few citizens across town. Directly or indirectly everyone would know everyone else, and what the bishop did during any 24-hour period could not have been much of a secret. To insert a dramatic pause therefore, dedicated to fasting and religious observation, ostentatiously apportioning to oneself divine guidance, could only be in his favour. If nothing else, it was bound to give his subsequent actions more credibility.

Next day various people are summoned to act as witnesses. Again, a politically wise move, repeated so many times hence by astute performers when faced with a potentially delicate situation. The description offered by Marcus Diaconus is typical of someone who is at great pains to ingratiate himself with the authorities. No epithet is spared, no adjective missed to make the writer's own integrity clear. Thus Julia has markedly aged (although the treatment by her captors may well have contributed), and Marcus takes it upon himself to provide the reader with an insight into the minds of heretics— outwardly they are sheep and inwardly ravenous wolves and venomous beasts. One should not underestimate the effect such insinuations have, and even modern societies can be as much the victim of public slander than those of old (in this context victimhood should be understood as being subject to whipped-up emotions, leading to actions undertaken on behalf of the agitator while taking the attention away from more realistic issues).

No doubt the delivery of whatever Julia had to say had already been compromised by the presumptuous attitude of her audience—the chronicler's remarks are sufficient confirmation. *What* actually had been said there we do not know, but for a pious reader it is enough to learn that her words caused "divine passion" to inflame a holy man's mind and with that reject their contents utterly. Indeed, as we will see later in Part II, the linking of some information to an emotional abstract as its representative is far more effective in categorising said data than taking them aboard per se. We can be sure that by the end of that hearing the attendant officials had settled upon an appropriate demeanour of righteous outrage, suitably prompted by the bishop's exclamations. Whether the woman's response to Porphyrius' condemnation was purely a psychosomatic reaction, brought on by her already weakened state (physically and mentally), or simply dramatic licence on behalf of Diaconus to demonstrate the power of God, is difficult to determine. Most likely it would have been a combination of the two. There is a third possibility, based on the construction of mutually supportive abstracts which then in turn evoke other, congruent ideations; more of this in Part II.

The current point of our argument then is this: a religion, being a main part of a culture, positions certain concepts within the inner realm of humans such that at any given time the behaviour of the person will be configured according to the functional weight of those concepts. It is up to the observer to identify them in terms of their influence, their priorities, and their effect upon one's behaviour. The previous diversion showed how informative apparent trivialities can be.

Let us return to Mani. Central to his world view is the struggle between good and evil, not a unique scenario in itself but nevertheless showing certain significant features. Firstly, God is seen as pure and ideal, not concerned with the affairs of the world and as such with the playing out of material forces. Therefore, in order to enable the interaction between the positive and the negative, a conduit needed to be created for the relevant imagery to take place. In Manichaeism it is the "Mother of life", interestingly not "created" by the god, but "called" [8]. The choice of words is significant, already implying—later confirmed—a certain antipathy towards sexual connotations which otherwise would be invoked. The "Mother of life" in turn "called" into existence Primaeval Man, who now goes forth to battle the forces of darkness. Although he was initially defeated the foray had been voluntary, thereby indicating that Primaeval Man took it upon himself to descend into the lower regions and become corrupted by inferior elements. Due to his ancestry however there remained within him the essential power to overcome his downfall and eventually re-affirm his connection with the divine. The process is not a straightforward one, emphasising the tortuous nature of self-improvement. Upon his return from victory the mother embraces and kisses her son, thought to have gone forever.

To Mani himself as well as to his followers such a deep-reaching story must have been a heady psychological and emotional mix. They could see themselves as pale representations of Primaeval Man, having coalesced in an asexual environment from the highest divinity and, after spending some time downtrodden and beset with worldly grime, being able to long for the ascendancy towards the everlasting light, amplified through the loving welcome by the prototypical mother. There would have been little need to illustrate the suffering of daily life—death and disease were constantly at one's side, accompanied by pain the likes of which would hardly be understood by a modern citizen. Yet a person could also relate to a state of comfort, warmth, and cleanliness, at least sufficiently so to idealise such experiences into a picture of other-worldly bliss. To link these visions with a denial of sexuality creates a powerful combination, enough to counteract the instinctual and sensory attractiveness of eroticism in a very fundamental way.

The foregoing should not be taken as suggesting that Mani construed his religion with a degree of well-developed psychological insight, pitching his ideas towards the purposefully analysed psyche of his people. That he was a dedicated and shrewd operator can hardly be in question. It equally is apparent however that Manichaeism contains many elements of other faiths, and the juxtaposition of good vs evil, played out against the background of symbolised human characteristics and affairs, can be seen in various forms elsewhere. For example, the notion of Man redeeming himself through battling the forces of darkness while at the same time being redeemed by higher echelons still, is a feature of Zoroastrianism and so of old Iranian origin; yet the emotional element included in the Manichaean version seems to have its roots in the myth of Tammuz, a Babylonian and Assyrian god [9]. Interestingly, Tammuz is held to descend to the nether regions for part of the year from where he is rescued by his sister, the heaven goddess Ishtar.

The significant aspect in terms of thought structures is the placement of a particular notion within the greater framework of the entire system, the reasons for which can be divided between the sheer existence of this or that idea on one side and Mani's personal inclination on the other. As mentioned before, while scholars can point us to the former we have no historically accessible grounds on which to formulate the latter. We can however construct a relatively loose network of thought patterns based on their resultant effects upon the attitudes and behaviours of their owners. One principled demonstration of that approach was given above using the testimony of Marcus Diaconus.

When Primaeval Man was rescued, there were still particles of light that were caught up in darkness. The role of their recovery fell to the "Living Spirit", another derivative from the Iranian tradition. There he was called "the god Mithra", the creator of the visible universe who overcame the demons of darkness, punished them and used their skins to make the Sky, their bones to erect the mountains, and with their excrement formed the earth [10]. Thus the overall mythical picture is brought into heightened relief. The divine, placed on high and being the residence of perfect goodness and bliss, untainted by sexuality, yet harbouring the potential for falling into corruption which itself is a definite region and is found downwards; there the existence of evil and darkness, where even the features of its landscape are the deathly leftovers and scatological products of demons.

It stands to reason that any interaction involving Man and higher forces in the worldly realm would manifest in activities which are first and foremost identifiable as material, incorporating all the notions we have touched upon already. Therefore the process of recovering the light particles from their intermingling with evil entailed the use of a cosmic wheel that drew the particles towards the moon and from there to the sun, their respective passages coinciding with the waxing and waning of the moon. The spreading of the light elements on earth was caused by our "Third Messenger" seducing the male and female Archons (ie demons) by showing himself alternatively as a naked virgin and as a naked youth, both suitably attractive. In response the male demons discharged their sperm which fell upon the Earth, and the already pregnant females bore their offspring prematurely in the form of monsters which in turn ate the tree-buds and so the light particles within. Both scenes, the column of light raising upwards from earth as well as the seduction of the demons, are derived from Iranian sources.

Widengren reminds us that the circumstances surrounding the spreading of the light particles had been particularly repulsive to the Christian Church. The reasons seem fairly obvious. Given Mani's general antipathy towards sex their inclusion in his religion is already somewhat curious and is only partly explained by his contemporary situatedness. Again there are similarities with Middle Iranian texts; furthermore, the Greek schools of medicine at the time taught about sperm in terms of a fiery flow emanating from the spinal chord, including the belief that fire was in fact the highest element within the human body. So far the situation is fairly straightforward. In the context of our perspective we should at least posit an additional option derived from his persona. Since the scenario involves the negative part of existence (darkness, earth, demons, condemned light), it invites a continuing interpretation along very particular lines. As such any occurrence of sexual matters would not merely be a matter of a simple listing, but the opportunity for stark portrayal which is not going to be missed. To describe how certain representations of the divine must languish amidst corruption is one thing; ascribing their fate to passionate displays of female and male sexual desire surely elevates them to still higher levels of impiety. One could of course muse as to the allegorical realism of such stories, considering their historic remoteness. What modern pregnant woman would give immediate birth upon the sight of a—admittedly very beautiful—male youth, driven by erotic passion only? Even if one allows for the element of exaggeration in order to drive the point home, then the choice itself is indicative of a certain mindset which, by today's standards, would seem somewhat alien. And yet, it is exactly through such imagery that the underlying meanings have been transported through the ages and their abstractions, no matter how removed from their ancient source they may be, form the kernels of present-day, living and acting memes.

The battle between light and dark is not over however. In order to keep those light particles still not ascended to heaven, personified Matter decided to enlist the services of a male and a female demon. The former devoured all the male monsters, the latter all the female ones. Afterwards they both had intercourse which resulted in the birth of Adam, then Eve, the first human beings. From his demonic heritage Adam possessed animal instincts, but he was also holder of the remaining light and therefore became the first candidate for redemption. His redeemer is a manifestation of the Third Messenger, called the Son of God, or "brilliant Jesus", who awakens the languishing Adam by freeing him from the influence of the demons and instructing him about his history and linkage to God.

There follow a number of notions regarding the end of the world. The struggle of moving the light particles from earth to heaven is a long and arduous one, eventually resulting in Jesus returning to earth for a short reign which terminates with the annihilation of earth. At that point the material world, together with all its demons and the damned, will be formed into a gigantic mass and sunk into a moat of cosmic proportions, to be secured by an immense rock [11].

The ideas contain several significant features. The emergence of the human race is couched in terms of excess and perversity. Even allowing for different moral values the connection between sex and demons, especially if preceded by an orgy of slaughter and gorging, surely must cast definite shadows upon eroticism in general. Thus to make humans fundamentally interdependent with evil, to the extent where our very existence is deemed to be obscene, creates powerful basics (ie thought patterns) across the realms of mental imagery. Note that while humans have a certain potential to escape their state of degeneration, it is only through the intervention of the god, and then only via knowledge about the religion that they gain the opportunity for salvation. Needless to say, it has to be the *right* religion which makes it possible.

All this of course exists in the form of abstracts. There is a thought structure within the mind of a devout Manichaean that holds a picture of Man as mired in nastiness, quite possibly without even conjuring any concrete examples of what such a morass might entail. "It's too terrible to even think about" is a functionality of human minds which saves some people from worrying too much about what it actually *is* that is so terrible, but nevertheless supplying them with the appropriate amount of moral indignation to send them on their way with amplified emotions. Because the concepts are abstracts they relate to their specific domain only, a region which is defined by the meaning and its semantic extent. So the idea of a crushed and destroyed world sits within its own basin of moral righteousness, protected by emotional and egotistical factors from the outside (ie other abstracts). It allows another abstract, this time featuring a cataclysmic aftermath, to exist unperturbed. If cause and effect, matter and existence and similar basic themes were placed within an overall framework of connecting patterns, the thought of annihilated matter would not be able to coexist with another which carries that same world having been sunk into a moat and covered with a rock, as materially as ever. By erecting semantic fences around particular concepts the mind is forced to move along the directions so imposed, while at the same time being diverted away from unwanted views. Compare the current construct with the Jaina system in the previous chapter, where one fundamental aspect had been the constant referral to the real coupled with the need for consistency. Under those auspices the Manichaean scenario would not hold.

Our examination of Mani's belief system enables us to recognise a certain pattern. We have the awareness of a greater whole, a general state of existence per se. Since it cannot be articulated through lack of understanding of its underlying dynamics, it has to be retained as a pure abstract. By "pure" we mean an idea which in itself cannot be delineated into further elements that could be described in any way or form. In order to derive any 'lessons' from such a recognition, whatever is deemed to serve along those lines must be personified so that the notion becomes addressable. Therefore, something like a greater whole, or an all-encompassing godhead, could not demonstrably create the first human beings—but a personified derivative, ie a god manifest, can do so. As soon as a god figure has been cast into a human-like form, an interdependent link is formed which allows certain characteristics to be shifted between the god-abstract and the human-abstract, keeping in mind of course that the latter always represents a selective image of Man. (Not every belief system employs such a two-tiered hierarchy. Simpler versions leave out the higher level, so that the god-human interaction can still be maintained via an acting entity.)

The meaning of the word 'fundamental' is well understood, even 2000 years ago. It implies the formulation of a particular character at the very first time some act of creativity takes place. Whether it is the making of food, the construction of a house, or the setting-up of an organisation, the importance of the initial steps is very much appreciated. To incorporate the creation of humans in a belief system is also a fundamental act which determines the basic character of humans from then on. The argument about the wilfulness of such an inclusion is really besides the point. Regardless of any psychological insight that may or may not have been present at that time, the functional relationship between a creationist belief coupled with its cognitive ramifications and its real-world equivalent cannot be overlooked, given that a creationist belief exists. To ascribe the emergence of the human race to something nauseating must therefore evoke sentiments which can and do influence the attitude the adherents have towards themselves and their fellow citizens, no matter what qualities a contemporary sample may yield. In terms of default perceptions there is a strong similarity between the current theme and the law: the difference between the presumption of guilt and the idea of 'innocent until proven guilty' is well appreciated in enlightened circles. It is significant that in Europe the principle of presumed guilt has been mostly adhered to until the influence of the Christian Church became limited in an effective way; this shows how influential thought structures can be if their pre-established nature is allowed to trickle down into other areas.

It is too early in our investigation to fully delve into the status of sex. From now on we will collect circumstantial evidence of its position within our minds, followed in Part II by a detailed description. Suffice to say that it constitutes one of the most basic building blocs of our self image. It matters a great deal therefore how erotic behaviour is portrayed and what connotations one attaches to its role in everyday life. To include a sexual component within the story of human creation seems only natural, given that its worldly aspects can be observed everywhere. To place something which in itself is only natural amongst an antagonistic setting however causes definite results. As the dynamics of thought structures in general show, the hierarchy of established notions is important. Although one's sexuality is at the core of one's being, if its abstract becomes governed by an even more authoritative idea then its own influence will be diminished. How natural or how logical such an arrangement may be is neither here nor there—what ultimately matters is the *perception* of things and not their actual nature. So, to define oneself in whatever sexual terms is one thing, but if there is an overriding notion which the mind says comes from God, the situation changes dramatically. The effect can also be gauged by the theme's chronology: the thought patterns feature God first, then the creation of humans within a certain—negative—context, and only then somewhere down the line the various frameworks dealing with the issues of the day. Note that the creation of such patterns either occurs very early in life or, if introduced later, is made possible through a dramatic ceremony (ie initiation) which in today's terms would be described as a form of brain washing.

A construct of this nature ensures the dominance of selected attitudes and the dependency of certain others, regardless of the circumstances. The result is that a person's behaviour is first and foremost sinful, and only the following of a religious practice can assuage the severity of inherent damnation. If such a practice is not forthcoming, the mind of the person will invoke a state akin to paralysis, that is to say any potentially ameliorating thought structures are deactivated by the overriding control of the initial determinants. For example, the following hypothetical line of thinking is *not* possible for a devout Manichaean:

> I have experienced lust;
> my faith tells me I have sinned;
> it is bad, so I have to do something about it;
> I should repent, but isn't my faith the very reason I feel to be in a state of sin?
> therefore I will re-examine the faith to see how bad my act really has been;
> if I can find flaws in my faith then I have done nothing wrong, and so there is no need to repent.

Step 4 is only an option if the mind permits one to adopt a higher perspective for the purpose of examining the situation. Yet the dominant position of the thought structure pertaining to the role of humans relative to the role of God will keep this avenue closed; one's faith is never questioned because the pattern does not create a need to question.

In addition to the concepts about the beginning of existence as an authoritative cognitive sentinel we have a similar role relating to the end of the world. Beginning and End are another set of concrete experiences that have their abstracts imbedded very early in life, and therefore every belief system developed by thinking beings will contain these notions. To connect the End with the concept of 'effect' is also quite automatic, since any observed final stage can be identified as the effect of some cause. The cognitive function which has developed along these lines assures the very same principle being thought into the big picture coming from a religion. Since according to Mani humanity's state of grace is precarious to say the least, the end of its existence will thus be construed as a consequence of its piety or otherwise. Again we have the End as an abstract which exerts its semantic power not only to what can be matched with concrete experience (ie one's own life), but also stretches into infinity by virtue of being mapped onto Existence as such. At that point its status becomes influential indeed. "Infinity? Infinity is B-I-G," a character in a comic-strip says. No need to be explicit, because the mind is satisfied with the abstraction of "B-I-G" as a representation of immensity. Or, consider a slightly different version in which the concept is presented via a gradual loosing of focus; in other words, despite each item being identifiable enough, their arraignment eventually dissolves the detail and leaves the idea itself for the recipient to contemplate. The following example comes from an old children's story, its source unknown to the author:

> In a land far, far away, there is a huge mountain. Once every hundred years, a little bird comes and sharpens its beak. When the bird has worn down the entire mountain, one second of eternity has passed.

In such cases, words like 'end' and 'eternity' are not defined in a strictly concrete sense. Rather, they represent open-ended concepts which become linked with whatever the mind finds appropriate. The process of establishing the appropriateness of some thought structure can thus be manipulated.

Our next visit brings us closer to Europe: St. Augustine. For nine years he had been a member of the Manichaeans, and after his conversion to Christianity (referred to as the Catholic Church in his writings) he became a fervent follower of his newly found calling. The two works we will draw on mostly have been written in 388 CE and revised in 390 [12]. The first, *The Way of Life of the Catholic Church*, could arguably be called a rationalisation for the need to have a religion and the Christian variety at that. His second, *The Way of Life of the Manichaeans*, is a criticism of Mani's doctrine and therefore particularly interesting in the light of our own discourse. Both serve us well in introducing us to the intellectual constellations of a religious thinker, especially one living so close in mind and body to a relatively new religion when it had still been free from many subsequent layers of political and spiritual reconstructions.

The core question he seeks to answer for his audience deals with the need for humans to be happy. Somewhat echoing Aristotle (see chapter 2) he looks for a schema which not only allows him to ascertain the nature of happiness but also ensures the correctness of the identification process. The former, according to him, is found through diligence, ie knowledge; the latter through piety, since only someone pious "is worthy to know" [13]. Immediately we are given a set of priorities in principle, established even before coming to the actual content. That tier is reinforced by the rhetorical question of whether he should begin "with authority, or with reason" [14]. He needed not have answered it: to declare these two concepts in terms of a contrast is sufficient to arouse a sentiment which effectively differentiates between information that is force-fed and another that is allowed to grow on its own accord. Being authoritative himself however he does not withhold his position and says, "The order of nature is such that, when we learn anything, authority precedes reason; for reason may seem weak when, having stated its argument, it turns to authority for support". The next sentence is significant too: "And because the minds of men are obscured by the habitual darkness of sin and evil which enshrouds them and, as a consequence, lack the clarity for perception proper to reason, it has been beneficially provided that the dazzled eye be led into the light of truth beneath the boughs of authority."

By the way, Augustine wrote in Latin, a language for which the skills of the translator are particularly pertinent, harbouring as it does in its grammatical logic many potential traps that might lead to misinterpretations of the original. We are assured by the publishers via their *nihil obstat* and *imprimatur* that the translation is in accordance with Church teaching and as such is "free of doctrinal or moral error" [15]. One can assume therefore that the message from the book finds approval at least up until the time it was published, which was the year 1966.

The above statements regarding authority and reason give us the opportunity to introduce the *functional abstract*. So far we have dealt with items which represented a general object, such as the cosmos, the world, divine figures, but also concepts such as Good and Evil and so on. Whether their real-world equivalents are actual things or ideas, they nevertheless have one quality in common: they are perceptions of something that *is*. A functional abstract on the other hand is the cognitive representation of something that *does*. The other we can call an *object abstract*. Note that both are names for a particular, localised, mental state, applied for the purpose of describing a certain occurrence—in other words, a metaphor. In Part II we will represent a more technical description of these phenomena, but for now the use of such metaphors is sufficient.

The manner in which "authority" and "reason" are presented create a functional abstract because there is now a rule in the mind which says, "treat anything under the heading of 'authority' in one specific way, and treat anything which comes under 'reason' in another specific way". *How* this is meant to be done depends on the rule itself, and becomes clear for the person upon further explanation. Functional abstracts can be invoked in various ways, often by way of sentence construction. In this case it is the implied difference established between two words through use of the comma, followed by the conjunct 'or'. The difference can either be outlined in due course, or the recipient could be aware of it immediately because the associated contexts are known a priori. Suppose a guest is welcomed at a party with the following words, "Would you like a glass of wine, or are you driving?" No further hints are necessary, since the mind knows that drinking wine occurs under one set of rules, but being the driver of a car suggests another. They are of course abstracts, because 'drinking' (the object abstract) is associated with the functional abstract representative of say, relaxation, not needing to be careful, and so on. On the other hand, the object abstract of 'driving' is connected with the functional abstract of behaving in a disciplined, responsible manner, and whatever else that may imply during the course of the party.

Both types of abstract are the product of a confluence of thought structures which have been able to form the ones in question. As such they can be compared to features of a landscape which are dependent upon the underlying contours for their existence. Their cognitive effect is relative, that is to say their influence draws on the relationship they have with their immediate surroundings, just like a mountain river is directly determined by its shores, but where those shores themselves are the elevations created by the terrain contrasting with its encircling plains. From the perspective of the river, its waters erode the flanking cliffs; but from the perspective of the plains, a canyon comes into being. It is a matter of reference, and the greater the radius, the more distant the horizon, and therefore the more truthful the resultant picture will be. What determines the horizon of our inner landscape is our existence's potential for expansion. What nurtures the potential is a culture's ability to absorb, willingly as well as unwillingly. The process is performed by one, yet carried by many.

So Augustine uses his introduction to set down some ground rules. While human reason will enable the person to understand what he, Augustine, has to say, unless there exists the overriding authority within that self-same person's mind which points to Church doctrine as fundamental reference, any subsequent considerations will turn out to be invalid. From that basis he proceeds to explain the steps leading to the answer to the question of what constitutes happiness. He builds his argument upon the premise that for man to be happy there has to be a focus on some form of achievement, in other words a striving towards something which can be identified as an improvement. Thus anything inferior to the current status is dismissed outright. What remains then are two possibilities: either something is being achieved which is similar to man, or there are things which are superior. Provided man is perfect, the first option holds. Augustine disputes this however and guides the reader towards an understanding of what is better still.

There appears an inconsistency depending on what perspective one takes. If Augustine saw his exemplary man as just one example out of general humanity, then some individual status may well be compared to somebody else's and found wanting—in which case the move towards the better one is certainly valid. Under that view we would populate our value categories with human-related elements, to be associated with certain people under certain circumstances, and evaluate our proposition according to desirable or undesirable distributions of the same. But Augustine was not a humanist; for him the status of humans may be better or worse, but never Good (remember the "habitual darkness of sin and evil"). The existence of something better than what can be found in man is thus a foregone conclusion and what is left is to identify what that can be. Note the progress of the argument: once again we have an a priori assertion in order to direct our thinking processes away from any possible appeasement with ourselves.

Having declared that man consists of body and soul (more shades of classical Greece here) he refuses to delve into an exact definition of both features in terms of the whole human being. Consider the two opposing analogies used. One that of a lamp which is called such although it features not only the vessel itself but also the flame, and so could serve as a simile if one were to focus on the body alone, although there is the soul too. The other a horseman, by which we mean the man although the word encompasses the rider as well as the horse, since it is he who guides the horse and so that simile would fit the soul alone, governing the body. Quite wisely Augustine rejects both, no doubt being aware of the difficulties which arise very quickly if the constitutive parts of those analogies were pursued to any great detail. Instead he opts for a more inclusive alternative where body *and* soul, or the soul alone, are the deciding references against which the superior good should be judged. In terms of logical frameworks his solution may well be found wanting, but given the cultural background of his era the decision is quite a good one. Note that we now deal with a schema in which the argument about body/soul is avoided altogether (at least as far as the current problem is concerned), while at the same time the reader's mind is prodded back towards the centre of the issue [16].

Since the soul is held to be the governing element in man, whatever is better must therefore improve upon the soul. We have come to another example of an a priori approach, this time related to the process of value assignment. One could draw a comparison with a clerk in an office who deals with in- and out-trays. A file is placed in the out-tray once it has been dealt with so that the out-tray signifies the file's status *after* its due process. But in a satirical twist the situation is turned around: the file is placed in the out-tray *in order to* make it finished no matter what the file contains. Augustine is doing the same by declaring "virtue" as the governing element of the soul first, and so is free to define the character of virtue itself afterwards without compromising his erstwhile definition. Given his fundamental opinion about humanity in general, if virtue is to be imbued with anything better at all than such a thing has to come from somewhere. The stage is set for revealing what that would be.

There is the choice between the advice coming from a wise man or from God. We ought to prefer God, since a wise man can be taken from us or we ourselves could choose to reject such advice. With God these possibilities don't exist. Note again that the initiative is taken out of our hands. There is a thought structure therefore which, firstly, ensures the dominance of a named concept (in this case "virtue") over any other patterns that may govern our perceptions. Secondly, such concept is later defined as being associated with the thought patterns pertaining to the divine, with all its ramifications.

The next step is particularly interesting in light of our perspective. Consider the following passage, justifying the superposition of the divine over—human—reason [17]:

> We must have recourse, then, to the teachings of those who were in all probability wise. This is as far as reason can take us. For in human things, reason, although lacking the certitude of truth, is thought secure. But when it

> approaches divine things, it turns away unable to behold; it trembles, pants, and burns with love, and, driven back from the light of truth, returns, not from choice but from exhaustion, to its familiar darkness. What a dreadful thing it would be if the weary soul, seeking repose after such an experience, should suffer greater weakness! Ineffable Wisdom has so arranged it, therefore, that when we seek to retreat into the darkness, *authority comes to our aid* and appeals to us with the wonderful deeds and utterances of its sacred books, which like shadows make attractive the brightness of truth. [My italics]

Having installed the thought structure 'virtue' in conjunction with the associated patterns of the divine, the influence this arrangement exercises is taken as an outside entity that operates on the dynamics of the rest of the mind. Augustine speaks from his own experience; accompanying his conversion to Christianity the effects of these dynamics would be familiar to him. Yet the result is a perception which enables the carrier of such thoughts to describe the process as the manifestation of an outside force, disassociated from the self and acting according to rules not adjustable by the individual. This is the same phenomenon which makes some people declare, "The devil made me do it", or, closer to home, "See what you made me do now". It is a direct outcome from the overall arrangement: the imposition of a thought pattern which is ascertained as being non-self, subsequently imbued with certain influential characteristics, and being used from then on as the final arbiter of the person's actions within the context of that pattern's semantics. So "see what you made me do now" usually refers to a minor situation only; "the devil made me do it" will encompass a series of scenarios deemed to be open to abuse; and "it's God's will" covers anything a human life can harbour.

For Augustine that authoritative entity is filled with notions about Jesus Christ and what can be found about him in the Bible. With the status of the entity assured anything can be of significance. When we are given Christ's quote from the Bible, "I am the truth", then these words have an automatic authority which they would otherwise lack. Compare that assertion with a spiel by someone who ends by saying, "Trust me". Immediately we become cynical because there is no preparatory thought pattern in our mind which forces us to accept the person's declaration by default. Devout people on the other hand carry in their heads the semantic rule set which ensures compliance with their religion's precepts.

Another element of control is love. As an idea in the general sense it would be well understood by humans everywhere, although there exists no accurate definition even today. Any concept which is commonly accepted but which at the same time remains outside the realm of objectively identifiable items is a prime candidate for re-representation under a certain guise. If love therefore (which everybody 'understands', ie 'knows' about) can be harnessed to a purpose other than the usual scenario while still maintaining its accepted meaning, it has the potential to turn into a powerful instrument of coercion. Once installed in a mind it becomes virtually impossible to switch off without doing some kind of damage to all the associated thought structures and patterns. Naturally enough, the consequences would be serious.

'Love' is another special abstract since it stems from the mind itself, similar to 'hunger', 'thirst', or 'joy', 'sadness'. Because it is not derived from an outside experience per se (which after all could be arranged or modified) but represents an inherent default state it cannot be switched on and off at will by an outside control. It can however be apportioned to a created container (such as 'virtue'), where it fulfills a function which is aligned with the container. Random reactions by the individual are largely prevented and control is exercised via the container. Let us call them *default abstracts*. Augustine spells it out [18]:

> If, then, we ask what it means to live rightly, that is, to strive for happiness by an upright life, it will most certainly mean to love virtue and wisdom and truth—to love with our whole heart and with our whole soul and with our whole mind the virtue which is inviolate and invincible, the wisdom which never gives way to folly, and the truth which is not altered but remains ever the same.

Note that "virtue", "wisdom", and "truth" are not actual objects themselves but abstracts, derived through a process which results in ascribing a particular thought structure or even pattern to their own semantics. Therefore, while we all have similar ideas of what 'truth' or 'wisdom' is, the things they actually stand for within the context of some scenario are open to manipulation. If their content is under attack, and if handled superficially (which is mostly the case), then for the owner of such patterns the attack is not directed towards the contents but the idea itself and furthermore, since the meaning of say, 'truth' is one's own, any polemic is perceived to be a criticism of one's own person. The text goes on to say,

> It is by this that we come to see the Father Himself, for it has been said: 'No one comes to the Father but through me'. It is to this we adhere by sanctification for, when sanctified, we are inflamed with that full and perfect love which prevents us from turning away from Him and causes us to be conformed to Him rather than to the world. 'He has predestined us,' as the Apostle says, 'to become conformed to the image of His Son.'

Compare these mental constellations with the situation relating to a police force. We have its members wearing uniform and therefore being identifiable first and foremost as 'police', only later being followed by what a member might actually do. If the action had been reprehensible the public is far more concerned than if the same had been done by a general citizen; not so much because of the action itself, but because the police uniform projects a certain, overriding image first without giving direct access to its contents. We would like to be certain that behind the uniform's facade is something we can approve of in any case, without having to worry about the details. Under Augustine's schema such certainty is by no means assured. One effective way to prevent the subterfuge is the separation of powers in a modern democracy, but as in the case of the notion 'innocent until proven guilty' (see earlier in the chapter) the significance and indeed necessity of such a measure could only be understood after Church thinking had been rolled back across an entire culture.

Through this self-confirming argument (posit 'virtue' which is 'good' by default, imbue it with certain characteristics, and then declare such characteristics as being 'good' because they are now 'virtuous') Augustine has arrived at the conclusion that 'happiness' is the "perfect love of God" [19].

Another example of subsuming an abstract to a previously established thought structure is Augustine's definition of sobriety. Basically, 'sobriety' means temperance, or solemnity. In his text however it is linked to the "knowledge of truth", which further delineates to knowledge about God, especially since he explains 'temperance' as promising a "certain integrity and incorruptibility in the love that unites us to God" [20]. As such it may still be connected with the common interpretation were it not for the possibility, if not requirement, for the individual to be "inflamed with that full and perfect love" which comes with serving one's god. Acts committed when inflamed with an emotion can hardly be described as 'temperate', nor are their outcomes usually 'solemn'. Under the regime before us these considerations do not arise however.

At this stage in our journey we are already familiar with the abstract as being a mental representation derived from some source. Initially the abstracts are formed from outside inputs (later in the process they may well draw their contents from other abstracts). So a rose can turn into a Rose, and something beautiful can serve to create the idea of Beauty. Consider how Augustine deals with it (in general, since obviously he did not write his tracts in terms of thought structures etc) [21]. He quotes the New Testament:

> 'While we look not at the things that are seen, but at the things that are not seen. For the things that are seen are temporal, but the things that are not seen are eternal.'

"The things that are not seen" stand for all the imagery created in our minds by religious teaching. The others are cognitive derivatives from objects in the real world. In chapter 2 we have seen how an abstract such as Beauty can be construed as being the ideal, and therefore everlasting, template of something which is merely 'beautiful', simply because it can be made to reach across any other perceived instances. That principle is recognised in the above passage provided we allow a certain re-interpretation of the meaning of words due to our differing perspective. Augustine continues:

> We can understand from this how far from being Christians those individuals are who believe that the sun and moon ought not only to be loved but also worshipped. For what do we see if not the sun and moon? But we are forbidden to look to the things that are seen.

Keeping in mind the dynamics of thought structures we can understand sun-worship for example as the result of a particular perception-abstraction transference involving the sun as an object and its idealised stature turned into a sun-god. On the other hand, the Christian religion takes something like 'virtue'—already a result of a PAT—and shapes that into another thought pattern: a PAT riding on top of a PAT (denoted by PAT^2). What is the difference?

In the first case we have a concept which relies in its ultimate interpretation upon the initially perceived characteristics of its real-world equivalent. If the sun should change one day (which of course it will), or if its overall image changes because its nature becomes better understood, the abstract will be affected too. Naturally, this has ramifications for the entire belief system. If the object of worship however is the product of a PAT^2 then any changes in the real world will not necessarily affect whatever hangs from PAT^2, buffered as it is by its—abstract—predecessor. Compared to say, ancient Egyptian religion, Christianity is therefore more sophisticated; the small detail of its teachings are more amenable to re-interpretation, the doctrines are more easily transferred to different cultures, while at the same time making use of more robust mental structures to protect itself. The above admonition then not to look "to the things that are seen" can be understood as a preventative measure to safeguard against the deconstruction of any PAT^2 or higher.

One of the most far-reaching effects the appropriation of certain ideas under a greater umbrella has is shown by the use of 'love' in the services of the Church. Under the topic of neighbourly love in particular, Augustine affirms the desirability of being well disposed towards one's fellow man. He explains it in the following manner [22]:

> Now you love yourself in a salutary manner only if you love God more than yourself. Therefore, what you do in your own behalf, you must do also for your

> neighbour, so that he, too, may love God with perfect love. But you do not love him as yourself unless you endeavour to lead him to the good which you yourself are pursuing. For this is the one good which is not diminished by the multitude striving to reach it with you.

Had the love for one's fellow man not been acceptable, the conclusion could not be reached; had there been no concept for love per se, neighbourly love could not be dealt with; and finally, had 'love' not been directed towards the divine there would be no connection between it and neighbourly love. The end result: a devout follower can persuade himself that the act of preaching, coercion, even force, in relation to a non-believer is done out of an act of love and, provided the consequence is the non-believer's conversion in any way or form, "God's will" has been done. (It is sometimes pointed out that Christians, compared to other religions, are most ready to lend assistance in a given situation. Considering the foregoing the act of helping is not so much a sign of altruism but one of self-gratification. One may argue that it is the result that counts; unfortunately, the result is not always considered, whether it is the inquisitors treating heretics, or psychiatrists treating homosexuals, etc etc.)

In his *The Way of the Life of the Manichaeans* Augustine sets out to show the errors committed by his former brethren and points to the Catholic alternative [23]. To list them all here would be rather unproductive, but a couple deserve our attention.

His first argument is about the nature of evil, and it is because of the basic layout of mental configurations that the debate takes certain turns. If God is the quintessence of perfection then whatever he has created has to be at least 'good'. In Augustine's words, "it would be madness to demand that the works be equal to the workman, the creatures equal to the Creator" [24]. However, if everything made is initially good, then something else which is identifiably evil must have been corrupted at some stage. Since it cannot have been God who did the corrupting it must have been evil and furthermore, the thing's nature is now changed as well. In fact, a thing's nature has been bestowed by God and therefore the act of corruption has turned nature to non-nature. Ergo, evil is not a nature, and to be affected by evil means changing from being to not being. This way of thinking is accepted by Augustine as the basis of any other considerations.

Let us illustrate the argument in terms of thought structures. There is a flow of perceptions, resulting in a generated abstract imagery about the world through the processes of PAT. Further abstractions present the idealisations such as Virtue, but also what is understood to be the quintessential fruit, house, man, or woman. Since they are perceptions and as such are situated within their own basin of associated—and associable!—patterns, their format depends on external circumstances, the history of their representations, and cultural values and perceptions. There is also the inculcated notion of God being the creator of everything that is good. The teaching of such a doctrine creates its own functional abstracts which subordinate any further input according to their rules. One of the rules says that God's work is good, another that anything tainted by evil is contrary to its nature and thus by generalisation against nature itself. Therefore, a PAT product which conforms with accepted expectations will be deemed to be 'good' and as such linked to God. Should the PAT product not agree with accepted expectations its source will be regarded as evil and even refused its own nature to the point where its very existence becomes deniable. Obviously, a thing which occupies a rightful place in the world will be safer than another which does not even properly exist anyway. Note that the ultimate determinants in these thought processes are not the actual characteristics of the object under discussion, but the contents of the governing structure which in this case defines what it means to be God. Schematically the process looks like in Fig. 4-1.

Any time a philosophy features a configuration such as the one illustrated above, that is where value judgments are arrived at via the application of some rule set which in itself draws its validation from another idealisation, that philosophy will spend a great deal of time explaining its values under varying circumstances. The argument about 'evil' is a case in point. Under the present auspices it is not good enough to match an abstract with its direct object-equivalent; rather, the following derivative needs to be justified through its status and the applicability between what a PAT^2 actually means and the derivative will have to be affirmed as well. Compare two examples, one a direct-case argument, the other what Augustine had to come up with in order to clarify the Church's position on the influence of evil.

Suppose workplace regulations stipulate that a main electricity switch be made inactive by attaching a warning label to it. Anyone with any understanding of electricity will recognise the usefulness of that rule. In case of doubt the confirmation can be had very easily, either through direct observation or through recourse to information which makes the potential danger quite clear. A fanatical leftist may link the rule to authoritarian standover policies by management, but for anyone else the suggestion would be ludicrous. In other words, only extreme emotional pressures will prevent the mind from coming to an appropriate conclusion.

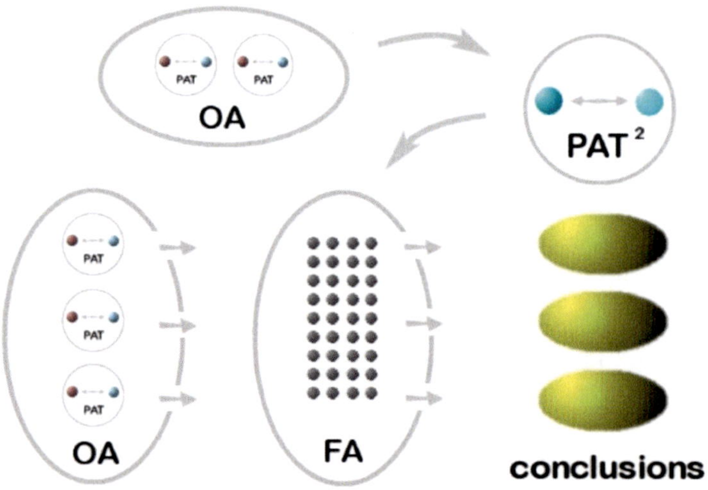

Fig. 4-1: Some object abstracts (OA) are delineated into further abstracts (PAT^2) which set up rules as functional abstracts (FA). Other OA are subjected to these rules during the process of forming conclusions.

Now consider Augustine. The problem starts with the definition of evil, and a circuitous route via the debate about a thing's nature has to be taken to get any description at all. Then we need to become assured about what 'good' is, and having arrived at that point the influence in question needs to examined. In the present text it happens to be 'corruption', and so a logical argument develops about how one may view a corrupting influence in terms of acting on some good vs acting on something evil. Through religious precepts the connection between 'good' and the 'supreme good' is cemented into place and only then can the semantics pertaining to 'corruption' be exposed to what is held to be 'good' together with a thing's nature. Similarly, a definition of 'evil' in terms of the divine order (previously established) needs to be supplied and another correlation between it and the corrupting influence outlined. In the text the eventual conclusion is that no harm can befall what is the supreme good, nor can corruption defile what is evil because it contains nothing 'good' which could be damaged in any way [25]. Since all the definitions which matter are found in the products of some PAT^2 their actual content is constantly left open to interpretation. It is not without reason of course that the argument about good vs evil has still not been resolved, certainly not under the semantic framework of religious doctrine.

The second example demonstrated the alternative to and the potential flaws of the first. Augustine berates the Manichaeans for seeing part of God in such things as grain or flowers, while denying his existence in others. Whereas in Christianity the idealisation of a PAT was linked with divinity, that is to say the end result of a process which exposed a PAT to a rule set, for Mani it was enough to go from say, grain, to God directly. While escaping a tortuous debate about abstract detail, the shortcomings of such a system become obvious with equal ease. Augustine does not hesitate to labour the point [26]:

> Why do you enjoy in the violet the same colour you despise in victims of cholera, in those suffering from jaundice, and in the excrement of infants? Why do you maintain that the brightness and sheen of oil bespeak a plentiful admixture of goodness, and seek to purify it by taking it into your throat and stomach, while at the same time you are afraid to let a drop of meat grease touch your lips although it has the same shining quality as oil? Why do you look upon the golden melon as one of God's treasures and not the rancid fat of bacon or the yolk of an egg? Why does the whiteness of lettuce speak to you of God while the whiteness of milk does not? [And so on]

Note that the emphasis is on the interpreted meaning of an object in terms of its relevance to God, not the nature of the thing per se. For this reason his way of thinking can be described as being more sophisticated, because it does not allow a relationship to exist by virtue of its nature only. By way of illustration we are told of a man who lives frugally in order to be disciplined but who ingests some bacon during the course of the day. We are told of another who indulges but abstains from meat as the rules demand. According to Mani the latter is more pious than the first, a position Augustine vehemently attacks. The interposition of a further abstract between two others opens the mind to arrive at a more logical and also just resolution, although for Augustine the reason for its inclusion is not found in logic but in the need for religious precepts. As the indirect system provides greater flexibility there is a price to be paid—its interpretative quality becomes more unstable.

Now that we have identified the dynamics behind PAT^2 we can draw a rough outline of the process as made use of by Augustine.

A human being is subjected to a constant flow of information from the outside as well as perceptions welling up from within, such as the feelings of hunger, love, sadness, and so on. These cause abstracts to be formed which are defaults because of their constant presence in one's life. Inculcated notions about God set up their own abstracts and are accompanied by rules which influence the processing of incoming information plus the representations coming from within. Since the default abstracts are being supplied throughout life, their regulation is an ongoing task under the auspices of a system which imposes different value judgments. For example, to be able to project one's love toward God and continue to do so undistracted from the more natural alternatives, requires a diligent form of supervision reaching well beyond a person's formative years. Note that each process is only subject to so much influence from some functional abstracts as these are being perceived to be relevant to that process. Such perceptions can be modified by the rules themselves, or they possess a more inherent affinity or otherwise due to their nature. It is therefore possible for a Manichaean about to eat to see himself as pious because it was someone else who collected the food; we can also be quite certain that Augustine would be more willing to accept advice from a fellow Christian than from someone who described himself as a Manichaean.

While functional abstracts are to be found in each one of the main groups shown above, the most important rule sets are being generated by those thought patterns which deal with the ideas about the divine. How would the relevant abstracts have to be organised amongst themselves in order to satisfactorily justify the existence of God for someone like Augustine? One of the more elaborate proofs can be found in his *De Libero Arbitrio* [27]. Let us follow his way of reasoning and see what his mind finds acceptable.

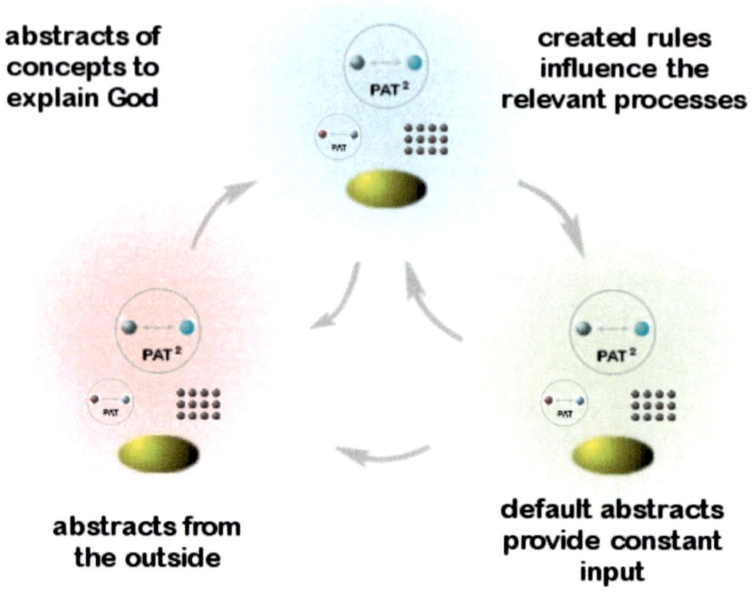

Fig. 4-2: The interplay of three types of perception-abstraction processes: perceptions sourced from the general experience with the outside, concepts derived from religious teachings, and perceptions formed by internal processes. Their coloured halos suggest a distributed influence decreasing away from their centres.

The tract is in the form of a dialogue not unlike those composed by Plato to illustrate Socrates' thoughts; here it is between Augustine and a student, Evodius. The discussion centres around the question of what can be considered good or bad behaviour, given that either one is conducted under the auspices of free will and so made available by God. If whatever God provides is inherently good, the problem is how to approach bad behaviour (if indeed there is such a thing) under the mantle of voluntary choice performed by Man.

Earlier in the piece we (ie Evodius) are advised to "lean upon piety and follow the paths of reason", so that whatever we may think about the whole issue it ultimately follows the guidelines of the Church [28]. Since reason is not being rejected however it becomes a matter of drawing a line somewhere beyond which reason stays mute. Following that basic rule there is another: the "eternal and immutable law" must be above the "temporal law" [29]. For Augustine a temporal law is one which can be changed in the course of time because circumstances demand it. Both versions, that is the one before and after the change, may be perfectly just but cannot co-exist at one and the same time. He uses the example of a state in which the citizenry had eventually become depraved (under the old law) and there now arises the need for someone to take over and put things right (and creating a new set of laws for that purpose). These temporal laws do not appear at random, says Augustine, but are derived from a higher authority manifest in fundamental criteria according to which we judge right and wrong, and these are immutable. He calls them the "highest reason". Hence for him, evil men are wretched and the good happy, and a "well-ordered and serious nation" ought to govern itself but a "wicked nation" should not have that freedom. It follows then that whatever is desirable under a temporal legislature must have been derived from its immutable counterpart; or, to turn this around, whatever has been taken from the highest law for the purpose of inclusion under the lower must perforce be just. 'Highest reason' is another type of container, similar to 'virtue' (see our previous discussion about love in this chapter). It holds all those characteristics which are deemed to be god-derived and are thereby removed from any direct assessment involving current considerations. We may argue about the finer points of a given law provided we are able (allowed?) to identify it as 'temporal'; we are prohibited from doing so if it had been declared 'divine' from the outset. Such restrictions are by no means insignificant. Once a notion has been inculcated in one's mind as unassailable by having it completely removed from any potentially critical thought processes, that notion will be held, declared, and defended—sometimes to the death. As Augustine asks his student, "Therefore, although there is one law according to which all the temporal laws for governing men are changed, the eternal law itself cannot be changed, can it?" To which Evodius replies, "I understand it cannot be changed. No force, no chance, no misfortune can ever bring it about that it is not just for everything to be ordered in the highest degree." [30]

Note the exact wording here, "..to be ordered in the highest degree". The creation of an abstract starts with the observation from the outside, and in this case there would be any number of PATs which came from the realisation that there is such a thing as a hierarchical order to be observed everywhere. Thus the notion of an "order in the highest degree" is a PAT which would not be disputed. Augustine uses that insight about order in general (knowing it would not be questioned per se) to re-represent the idea he introduced in terms of the relationship between the two kinds of laws. This is already an application of using containers so that a particular thought might be accepted. People make use of it all the time. For example, let's say a city council is faced with a widely-publicised mismanagement of some sorts. The next day a councillor wants to bring attention to yet another problem by concluding a speech with "Isn't that yet another one of those stuff-ups?!" Everyone knows what is meant by "stuff-ups" because the details have been done over again and again. It is not necessary to imbue the new story with emotional superlatives; all that needs to be done is to drag the container "stuff-ups" across to the current image and watch the faces go red.

Gradually the conversation turns to the inherent ability of men to act wisely. If Man has never been wise (his entire status being defined by God in the first place), how can he be punished justly for any misdeeds? Augustine responds by saying, "You say this as though you had clearly proved that we were never wise. You are thinking only of the time since we were born into this life." At this point he briefly brings in the mind/body dichotomy to show that there must have been a time when the spirit, before its union with the body, availed itself of the wisdom which is to be found in regions beyond the material (not being directly related to the present problem he moves on however). Recall a similar theme in chapter 2 where Socrates seeks to justify his belief in an afterlife by questioning how one single life time could possibly contain all the opportunities for one's current state of insight. That particular line of thinking had been prompted by the need to somehow substantiate the existence of 'soul' as a definite counterpart to the concept of 'body'. Although Socrates developed *his* argument for different reasons, the initial problem of how to enable the mind to hold the ideation of 'soul' productively is the same as Augustine faces. We know that Augustine availed himself of Roman and Greek sources, but it is nevertheless interesting to observe a familiar trait within certain thought structures that one way or another have the task to prepare the individual mind for positioning itself amongst the wider context of life.

Another differentiation is made between the deeds being made and the things used to do them. He arrives there because he wants to make it clear that there are goods which can be taken away from one's person, and there are others that belong to us forever once we have acquired them. In the case of the former, so he argues, it is the love men have for them which in turn prompts the resentment in case of theft which in turn necessitates laws and punishment to address the resentment. So, if men did not love transitory things there would be no punishment; hence the comparatively greater value of permanent goods, such as virtue, wisdom, and so on.

One could argue with such a perspective on misdeeds in general within the greater scheme of a social fabric, but Augustine needed a precept of this kind to assert his next point, namely the possibility that some members of society make evil use of things and as such cling to them with a misplaced love; misplaced because through this love the person becomes in thrall to the goods and therefore tends to serve them, rather than the goods serving the person. In other words, goods do not make the person better—if anything they are improved by someone's handling of them.

Or, differently still, the initiative resides with humans and not with the goods: a functional abstract which prevents the viewpoint that "gold and silver should be blamed because of greedy men, or food and wine because of gluttons and drunkards, or womanly beauty because of adulterers and fornicators" [31]. A perspective which has remained relevant right up to the present moment and which is often used in a highly selective manner to suit one's purposes.

We are now getting to the heart of the matter, namely the existence of God as such. Having established that men do evil things ('evil' as in 'contrary to what highest reason would dictate') it follows that it must be free will which enables them to act this way. Since all things come from God, including free will itself, in the face of the seeming moral contradiction the question remains whether it is indeed from God that all these things—including ourselves—come. The answer is rather circulatory, relying on precepts. One, the immutable laws established by God, is already familiar, with the corollary that God is just. Delineated from that, one needs to know that God rewards acts according to these laws whereas transgressions are being punished. Next, the more worldly notion that justice is concerned with honouring foreigners rather than punishing them for any possible misdeeds, but such largesse does not apply to society's own members. Therefore, since God is not only concerned with our good deeds but disciplines us for bad behaviour as well we must be part of God's realm.

The custom of allowing strangers a certain leeway in their behaviour is common right around the world, even today. Indeed, it is a mark of a society's degree of uniqueness to allow outsiders their freedom, with a corresponding level of intolerance towards their own. It is like saying to a visitor, "Look, we know we have our own special ways and means, and we don't expect you to have any knowledge of them." The less exotic the culture, the less pronounced this difference in response will be too. Thus there are functional abstracts related to concepts about the divine with the exception of hospitality, which is a PAT derived from daily life. At the beginning of our discussion of Augustine's proof I mentioned a line which separates presumed ideas from those that are accessible to argumentation. Let's call it the *logic boundary*, if by logic we mean a system that is not contradictory within the realm for which it is designed and so allowing discourse on that basis. Presently we have an example which runs between the God-associated thought patterns on one side and local custom-related ones on the other.

Back to the argument. Now that we can accept divine punishment (and, being divine, it has to be just) we can also accept the obligation to do good which accompanies free will. If we did not have free will then being chastised for behaving badly would be unjust, as would be any reward for acting properly. But the punishment is not unjust, as the previous constellation of thought structures has made us accept; therefore free will comes from God and is morally necessary to ensure justice [32].

Up to that point it has been established that Man has a choice, and rightly so. This only works however if we believe in the source itself, namely God. How can we prove his existence? Augustine starts by asking whether Evodius is sure he himself exists at all. If

Evodius did not exist, he would not only be afraid of being deceived by Augustine's questioning, he could not be deceived at all—hence he must exist. Evodius agrees, and this would not be possible unless he was alive. Does Evodius understand the reasoning? Yes he does, and therefore (having the ability to understand) it follows that he (1) exists, (2) is alive, and (3) understands. Augustine asks his student to compare these three points in order to rank them by degree of importance. As such, something which understands must be alive and must also exist; and something that lives must exist. Something that only exists does not necessarily live and if that is the case it does not understand either. So, a stone exists but does not live; a beast exists and is alive, but does not understand; only Man comprises all three qualities, having the power to understand. Therefore understanding is the most excellent of the three.

Augustine now draws on an analogy involving the five senses (seeing, hearing, smelling, touching, and tasting) to show that there is another, higher, sense which perceives that which the senses present. The eye sees, but does not 'know' that it cannot hear, smell, etc. Likewise the ear hears, but does not 'know' that it cannot see, smell, etc. And so on. It is our reasoning faculty which enables us to distinguish between the senses and contemplate the presence or otherwise of one particular sensory input rather than another [33].

It is an interesting development. Augustine posits the inescapable fact that although we have senses which allow us to perceive the world around us, there is nevertheless something else which enables us to draw conclusions about what it all *means*. The senses alone cannot do this; a higher faculty is responsible for the eventual interpretation of what we perceive. He expresses it thus [34],

> Then, when we perceive colour, we do not likewise perceive that we perceive it by the sense of sight itself; when we hear a sound, we do not hear our own hearing; when we smell a rose, something has fragrance for us, but not our sense of smell. When we taste something, the sense of taste itself does not taste in our mouth. We touch something, but we cannot touch the sense of touch itself. Since this is so, it is clear that these five senses cannot perceive themselves, although all corporeal objects may be deceived by them.

An astute observation, giving us the first inkling of what today we would term cognitive dynamics. He goes further: not only does reason perceive the input from the senses, it gives us the awareness of perception itself. As such reason is to be placed above the five senses, as well as the inner sense which controls them. Considering the hierarchy represented by non-existing things, existing things, what is alive, the inner sense, what understands, and reason, the next step must be God unless it can be shown that there is something between human reason (which is mutable) and the highest (God, which is immutable). To do this one cannot draw on anything which is supplied by a faculty inferior to reason; the logic drawn so far does not allow it. If, on the other hand, reason discovers something which is eternal and immutable (and anything eternal and immutable must perforce belong to God) then the existence of God has been proven.

Again we observe a logic boundary. The definition of the hierarchical functionalities, laying down their particular order but also the step-by-step sequential process of listing, implies a similar, continuing, arrangement. On the other side of the line are the characteristics pertaining to the divine, such as eternal existence and immutability. The latter come from a thought pattern which contains the functional abstracts and with them the rule sets of Church doctrine. Like in the previous step of the argument which concerned itself with Man's inclusion in the realm of God, the debate covered first worldly aspects until it could go no further; at which point recourse was taken to other, unquestionable principles.

The same process is repeated in the next step. Augustine distinguishes objects which can be perceived with the senses although the objects themselves remain unchanged, and others that via the act of perception become part of us. The sun and the moon would be representatives of the former, the food we ingest of the latter. But there are also objects which are perceived by the faculty of reason and yet stay unaltered, namely mathematical numbers. A number, its truth, its significance, is unaffected regardless of whether it is understood, used in a calculation, or not grasped at all. There follows an example of integer series to show that no matter how far one is prepared to go, the relationship between a number and its double, the ranges in between, and their ratios exist in a very particular order. We can deduce all this through our reasoning ability. We are then asked whether the same can be true for wisdom, so that whatever goods can be ascribed to wisdom are the same for all of us. Augustine answers this in the affirmative, because there are notions (eg equals should be compared to equals; the worse should be subordinate to the better; to each should be given his own) that can be shared by everyone. Such truths, therefore, do not belong to single individuals, indeed cannot be appropriated by anyone for himself alone—"Truth is one and common to all" [35]. Moreover, since truth is unchanging but the mind is not, it is superior to the mind and "more excellent".

Note *what* is included under the title 'truth': they are all abstracts, the results of some PAT which somewhere along the line have been derived from outside observations. The senses perceive an object, however distinctly, but objects can change. The abstraction therefrom however does not change, especially if it has to undergo an evaluation according to set standards. In other words, the text tacitly acknowledges that whereas perceptions (ie first-order PATs) can differ widely from person to person, their abstractions in turn (ie the second-order PATs) are common to all of us; the data may vary, but the process is the same *provided* it is performed under the same rules. Compare the present approach to Socrates' discourse in chapter 2 when he talked about the difficulty of ascertaining an idea such as 'honour' (a PAT^2), and the relative ease by which we identify an object through its description (an ordinary PAT). In Augustine's doctrinaire world of religious principles the situation is exactly the reverse. As he puts it, "But no one makes judgment about the rules themselves." [36]

The problem had been to find something which is better than reason. If reason deals with the mutable and truth contains the immutable, then truth is better than reason. If there is something higher still, it must be God. Since God is acknowledged as being the highest it becomes rather immaterial whether the truth or God are one and the same. There *is* something more excellent than the human mind and therefore God exists [37].

Further in the text Augustine says something very revealing, at least from our perspective [38]:

> Therefore, Evodius, if you look at something mutable, you cannot grasp it either with the bodily senses or the consideration of the mind, unless it possesses some numerical form. If this form is removed, the mutables resolve into nothing; do not, then, doubt that there is some eternal and immutable Form which prevents mutable objects from being destroyed and allows them to complete their temporal course, as it were, by measured movements and in a distinct variety of forms.

Recall the simile regarding numbers in order to demonstrate the existence of abstractions (which Augustine calls "immutables", or "truth"). He now asserts not only do these abstractions exist, without them we would not have any perceptions at all. His is an oblique way of saying that for cognition proper to function the mind must have the ability to turn these representations into their abstracted form. At the same time, such ideations are not seen as mental products per se but are taken to belong to some outside entity, in this case God. We are witnessing the phenomenon of projection, whereby an internal imagery is not referred to its actual owner but ascribed to another representation which in turn is taken for what it is meant to stand for—not an unusual occurrence. Consider the person who procrastinates and declares, "I would like to do such and such but X is holding me back." It is not *really* X who is the culprit here, rather the imagined version of X, and so it is the person himself who is ultimately responsible; nevertheless, poor X is suffering the consequences to his reputation. We have a word for the Xs of this world: scapegoats.

Towards the end of the text concerning the proof of God's existence Augustine reiterates once more his attitude regarding the significance of various types of abstracts. He refers to them by "goods", saying [39],

> Therefore the virtues, by which men live rightly, are great goods, while all kinds of physical beauty, without which men can live rightly, are the lowest goods. The powers of the spirit, without no one can live rightly, are the intermediate goods between these two.

In terms of their religious connotations the hierarchy ultimately relies upon the goods' source. The virtues, held to come directly from God, rank the highest; the faculty of reason, being the highest in Man, ranks second; and beauty, with its material basis, comes last. In terms of thought structures the ranking concurs with the degree of abstraction, in other words, their method of creation. As such "virtues" are the result of a second order PAT (PAT^2) *and* their placement under certain auspices, in this case the thought pattern concerned with God. This makes them at least a type PAT^3. The "powers of the spirit", that is human reason which ascertains the validity of this or that sensory input, deal with the further abstracted form of some abstract and therefore a PAT^2. And finally, the abstracts of direct sensory input are ordinary PATs. To put it into a concrete example: I observe a group of women and I am aware of their beauty—PAT^1; watching their activity I realise they are engaged in a religious service—PAT^2; I acknowledge this to be virtuous since I have learned the importance of accepting God's will and understand the ways through which this should be done—PAT^3.

We can also see why their respective significances differ from each other. A PAT^1 is unique with respect to myself; the perception in my mind which prompts me to say "beautiful!" is mine and mine only and cannot be shared with anyone else. PAT^{2}s can be shared provided the basic criteria according to which judgment is made can be agreed upon. So, while two people may agree that racing cars are beautiful (and still disagree over a Lotus vs a Ferrari), to an aviation enthusiast the cars may not mean much at all. In other words, the concept of Beauty is more difficult to override than one example of it, and therefore it carries more weight in our thought processes. And further, if Beauty becomes linked to the philosophical agenda of what constitutes value in general, then to dismantle *that* concept is harder still. For that reason any religion, any doctrine, contains a functional mechanism which enables mere experiences to be elevated to an overriding paradigm. It also features defence mechanisms which become more and more belligerent the further up in the hierarchy the things are they are meant to defend.

Augustine asks why someone would—or even could—move his will from an immutable good to a changeable one and thereby commit an evil act. He admits he doesn't know the answer [40], advising us only to be steadfast in our beliefs. Considering the schema underlying the PAT processes it becomes clear how such a movement could come about. If a PAT^3 gets deconstructed so that one of its parts emigrates from its basin, that part has lost some of its significance and with that its inherent weight. As a PAT^2 it is now easily redirected. An example would be the idea of virginity. A religious person accepts it as something God-given and therefore to be defended until the point where another religious imperative takes over (ie marriage). Should religion be less important the decisions rely more upon the personal situation but still retain a degree of significance (as in, "I want to save myself for my future wife/husband"). If however the concept (a PAT^2) itself gets questioned so that virginity per se becomes irrelevant then sexual behaviour will be determined by the demands of the moment (which is not to say of course that there can't be other PAT^3s which lay down some rules).

Fig. 4-3 shows the basic schema of the proof, expressed in terms of the roles the PAT processes play.

The daily experience of life would have generated perceptions and their abstracts dealing with situations which produced such ideas as highest, virtue, hospitality etc as PAT^2s, developed from their concrete sources. Augustine's learning and exposure to the Christian faith would also have instilled notions of 'the highest', and idealised versions of PAT^2 derivatives, such as God's virtue, justice, and so on.

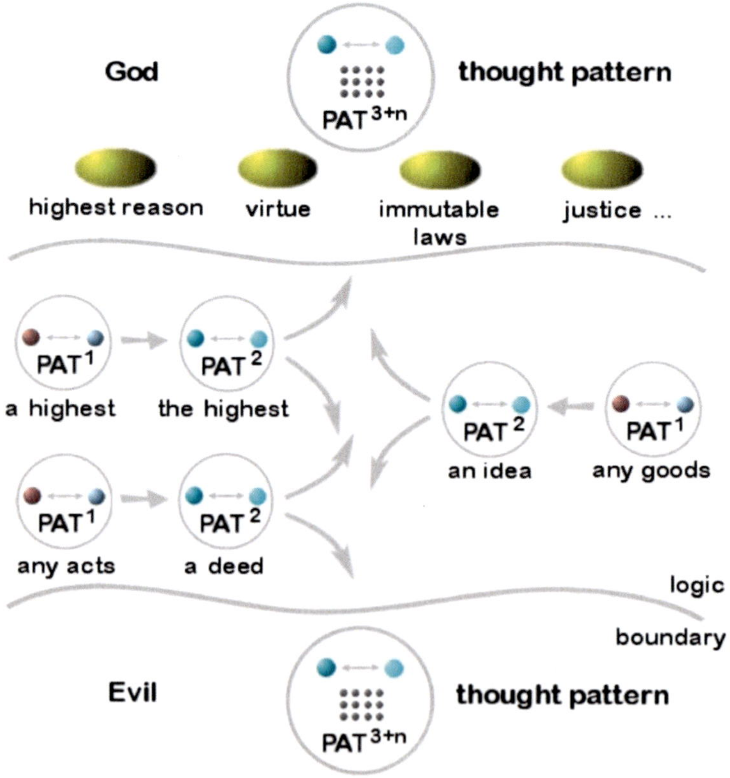

Fig. 4-3: The dynamics of thought structures as they occur in Augustine's proof of the existence of God. God and Evil are thought patterns which hold third- or higher generation PATs within their domains. The logic boundary keeps these apart from second- and lower generation PATs which are ultimately derived from outside observations. A PAT^2 becomes a PAT^3 etc by being delegated to God or Evil according to the rules situated in the realm of the higher PATs. Only some thought structures are shown.

The rules in which they were embedded influenced his mindset to the extent that any PAT^2 could be assessed in terms of its value defined by these rules, with the result that their content would be relegated to either God's domain if found positive, or the category of Evil if held to be negative. Although the PAT^2s, dependent as they are on outside experiences, can be evaluated by reason, that is to say are accessible to individual contemplation for which the principles have been distilled from a juxtaposition of other PAT^2s, their ultimate destination (ie the basins containing the PAT^3s) are shielded from the personal by the logic boundary. Beyond that line the quality of any PAT^3 is determined by the rule set specific to that basin. This works both ways: a PAT^2 will not be admitted unless it accords with the basin's rule set; once let in however it is taken as a rightful subset of that particular PAT^3 realm and as such it becomes imbued with an authority any PAT^2 simply does not have. To use an extreme example, the act of killing may be rejected if subject to the reasoning process enabled by PAT^2; on the other hand, if there should be a PAT^3 which under the auspices of a 'higher law' sanctions such behaviour, then reason will be overruled with obvious consequences. It should be kept in mind that the creation of a PAT^3 territory is done in a relativistic manner, in other words a weighty indoctrination process is not necessary for it to exist.What we term 'free will' operates between the logic boundaries. Once they have been transgressed the person acts the way he or she 'has to act', even if that should cause a feeling of unease or worse. In art the concept of the classical tragedy occurs in virtually all cultures, in which humans are the actors in a play that inexorably moves towards its sinister outcome, predetermined as it is by the rules of social convention and morals. Nothing can change fate: "it is life".

In the next chapter we will explore one of these dramas, played out in the 12/13th century against the Cathars (also known as the Albigenses), a Christian sect which was deemed to be heretical by the establishment and inspired a most cruel campaign of suppression and conquest. Both sides were locked into their respective paradigms, an unyielding atmosphere constructed by PAT^2s and PAT^3s that only utter annihilation could resolve.

5. Rights and wrongs

St Thomas Aquinas - Cathars - etc!

We have reached a stage in our exploratory journey where some changes in our approach become necessary. So far the focus has been on one particular philosophy, one particular people, in order to point out the significant features of their mental imagery that were suitable for explaining the dynamics of thought structures and patterns at that point in our progress. As mentioned earlier, such features should not be taken as representing the sum total of this or that mind set—we merely picked out some highlights to illustrate the operations of thought structures. Nevertheless, due to historical distance (where episodes concatenate and differences tend to shrink), but also because of relative isolation, it was possible to consider Manichaeism for instance and, having dealt with that religion, jump across to some other in order to continue the explanation of mind dynamics. Having arrived at the Europe of St Augustine (in terms of our theme that is) we need to follow the conceptual evolution in the context of character as well as outlining the dynamics themselves, rather than portraying our examples as seemingly isolated models of human thought. From now on we will use the philosophies not only to build our understanding of thought structures, but also to demonstrate the process which went into building our contemporary mind set as such. But first a brief summary of what we have discovered so far may be in order.

Input from outside reaches the mind via the senses, where by 'outside' we mean anything beyond that conceptual realm we choose to call the 'mind'. This can mean a mountain in the distance as well as the rumbling of our stomach. Once the signal has entered the mind it becomes a representation carried by the neurons, which we term an *abstract*. (Note that words like 'abstract' and similar descriptors yet to come are metaphors, that is symbols which attach a 'handle' to a concept without having to go into the detail of the thing for which the symbol stands. In some cases this would be impossible for lack of knowledge, in others counterproductive given the current stage of explanation.) We have called the process leading to an abstract the *perception-abstraction transference*, or *PAT* for short. This indicates a 'meaning' being transferred from the fullness of the signal itself into a cluster of excited neurons so that the resultant representation can be described in terms of a specific relationship between the basic signal data and the abstract. *What* exactly is being transferred depends on the type of transmission, ie the senses and therefore the type of data being transferable in the first place. That first representation, coming directly after the input, we have designated as a PAT^1. For convenience of description we call the results of sensory input which do not relate to some automatic process (such as the production of gastric juices) as *object abstracts*, and those that do, *default abstracts*. In addition there are *functional abstracts*, which are related to object abstracts with the proviso that they represent some type of activity, or movement. The stream of perception input occurs throughout one's life, and series of PAT^1s eventually form patterns because of their inherent similarities. When enough PAT^1s have accumulated to form a meaningful complex, we can call them a *thought structure*, and many—again meaningfully related—thought structures in turn form *thought patterns*. What constitutes a 'meaningful complex' we will leave for later when we consider the development from the very beginning. Right now the usage of this phrase serves us a purpose, because it allows us to identify certain structures which *do* belong to each other and, more importantly, which are evaluated by the mind as a unique but possibly relevant idea that is available for further processing. Since there is a difference between the real object and its mental representation, we distinguish between qualities of the perceived objects and those of the abstracted objects (Q_{po} and Q_{ao}). That further processing is another abstraction: just as the senses carried a raw signal to be turned into a basin of particularly organised neurons, so does the thought structure itself propagate towards its surroundings. There it results in another PAT, which we designate as a *PAT*^{2}. In our everyday experience a PAT^1 would be a given table for example; the PAT^2 emerges once we have considered that table as a particular kind of platform, to be used in a wider sense. This means we can say things like, "be honest and lay it on the table!", or "tabling a report". PAT^{2}s enable us to use commonly perceived things (including actions) and turn them into intellectually wider-reaching concepts which, for all intents and purposes, acquire a unique role for themselves. We can discuss 'developments', argue about 'safety' 'issues', teach about 'materials', and so on; all these words being processed PAT^{1}s and put to a wider use. It doesn't stop there: PAT^{2}s influence their surroundings and create PAT^{3}s, to enable even wider-ranging applications. One can turn the accumulated concepts of a specific class of PAT^{2}s into a type of super-idea, which is then used as a *container* for any other suitable ingredients. 'God' is such a container, its functional type being a PAT^3 (usually), to be populated with concepts appropriate to a certain religion. Higher-level PATs come at a cost however: the connection between one and the original PAT^{1}s has become tenuous, if not non-existent. Whereas in the development of a PAT from a lower to a higher level the Q_{ao} is largely maintained, once the higher level PATs can be shifted across to other PAT regions such links can be broken. It can happen that the underlying qualities of the moved PAT do not match the qualities of the lower-level PAT with which it is meant to be connected. In that case there exists an *incongruent functionality* between two PATs, a situation only identifiable by tracing the origins of the respective functionalities involved. Functional abstracts, representing movements, provide the rules according to which other objects are being placed into context. Initially such rules are perceived in reality, but after becoming representational abstracts they can be applied anywhere amongst thought structures, provided the mind comes up with a reason to do so (other functional PATs are responsible for that). Once the rules have been so far removed from continuous outside input or some discriminatory inside input for them to be shielded from such influence, they now exist in a zone where whatever can be termed 'logic' does not apply; we can say they exist on the other side of the *logic boundary*, and from then on become inaccessible to the rule sets governing the conceptualisations within the boundaries. Now back to the chapter...

St Thomas Aquinas (1225/6 - 1274) has been and still is one of the most influential philosophers within the Catholic church. Since the Christian faith, to a large extent carried by Catholicism in Europe and indeed the rest of the world, has become a standard against which many other ways of thinking needed to measure themselves, the words of Aquinas are worthwhile studying. The material that follows has been taken from Herman Reith's book, *The Metaphysics of St. Thomas Aquinas* [1]. It provides the reader not only with excerpts from Aquinas' writings, but also includes a commentary with a wider reference to works by others as well as to contemporary thought. Because the philosopher's text is still adhered to by the Catholic church and Reith—himself a Catholic—argues within that framework, I have considered the thoughts of both men to be relevant to our investigation. Throughout the next paragraphs, while dealing with the topic 'St Thomas Aquinas', we will therefore regard these ideas generally as one source without a particular differentiation as to one or the other persona. This is justifiable since we are not really concerned with Aquinas or Reith *per se*, but the belief system outlined before us. That belief system is shared by both men.

Keeping in mind our agenda, we will not follow the writings in a chronological order; rather, we are going to examine Aquinas' ideas—and those of Reith of course—within the context of our perspective. This means we will range across his works in order to combine and relate the concepts as they pertain to their own significance, and not to their sequential presentation in Aquinas' works in general.

In his foreword Reith points out via a quote from Father Chenu, O.P., that the "medieval commentator is plainly concerned not only with the letter of the text but also with the 'intention' of the author..." [2]. We can take this to mean that, reading Aquinas, we are presented with the author's position in terms of his own opinions as well as his attitude towards any material he chose to refer to (for example, Aristotle's writings). Such a stance is only advantageous to us: even when referring to others we can partake of the full flavour of the theologian, while at the same time not disparaging Aquinas' scholarship. It also presents us with the opportunity to include his comments as a representation of personal thought structures and their respective relationships, since they are obviously embedded within his wider patterns. In other words, we know we are dealing with Aquinas himself all the time. This is in contrast to the historian's purpose, who would have to apply a certain amount of discrimination when it comes to any referenced work by a third party.

Thomas Aquinas was as much a teacher as he was a philosopher and theologian. He therefore approached his subject with a clear method, namely that of introducing his thoughts and conclusions not merely to present their contents per se, but also mindful of an hitherto uninformed audience. To this end we often find substantial explanations of words and their origins, so that the exact, intended, meaning might be conveyed. At the very outset Aquinas therefore feels the need to identify the role of a philosopher; not as a "wise man", that is a holder of knowledge, but as a "lover of wisdom", ie one who is in the process of seeking knowledge. The ultimate aim of knowledge is truth, and it is towards this end that any considerations are employed. "What is truth" is a question asked thousands of times by many, but it is in its definition that we find crucial differences. To concentrate on the contents of what we label thus is, in my opinion, secondary to what the understanding of the epithet itself implies.

'Truth' being an insight, an idea, is not corporeal; this means it is not transferred from one person to another like a piece of material would be—now it is here, then in someone else's possession. While it can be shared it cannot be owned, if by "owned" we mean an exclusivity compared to those who do not "own". In addition, if an idea is 'true' then this is in contrast to any other position which is 'false', and naturally there can be many of the latter while we can have only one justifiably belonging to the former. As such it represents a relatively simplistic interpretation, but one which nevertheless would suggest itself when dealing with straightforward observations as would have been the case in Aquinas' times. His contemporaries could surmise about the quality of milk for instance: is it stale, fresh, sour, sweet, clean, contaminated, hot, cold, etc etc. There are many possibilities, but as far as one particular jug of milk is concerned, only one case out of many other related ones can be the true one at any one time. It is an observation which, when abstracted into a principle, produces the definition of 'truth' as just mentioned. Today we can be more sophisticated. We can say that according to one system a certain state can be true, but in another system that same state can be false—all without changing the appearance of the objects in question. For example, in the decimal system we can say "1 + 1 = 2", which is true. In the binary system on the other hand that same statement is false, since '1 + 1' produces '10' ('one-zero'). It becomes obvious that our previous definition, relying implicitly on the axiom that a thing can only be at one place at the same time, needs further fine-tuning. In days gone by such a refinement did not exist, nor was there the recognised need for one. So the functional abstract associated with 'truth', ie 'there is only one truth otherwise something would have to be false', constituted the guideline which accompanied any thought structures dealing with "truth". But 'truth' is also associated with 'usefulness', because working with items that had been identified truthfully would obviously produce better results than some process which was based on falsehoods. Therefore ideas such as 'usefulness', a 'well-done act', and 'success' go hand in hand with 'truth', so that eventually the abstract 'Truth' embodies all those qualities—if not explicitly then certainly implicitly.

Another functional abstract is derived from the recognition that for something to be classified as 'true' it must have been the result of a process which entailed work, study, insight; in other words, understanding. If one does not understand something, one is not qualified to judge truthfully, and so being able to do so after all incorporates the ability to understand—an intellectual qualifier. If all the ultimate wisdom resides in God, so must 'truth'; and God equates Truth. Observe the steps in terms of PATs: first there is the functional abstract laying down the rules about how to treat something that is true, a PAT^2. The notion of 'God' had at that stage of our history become a PAT^3 (see chapter 4, regarding St Augustine), placed on the other side of the logic boundary where direct references with reality no longer hold. The result of all this is that Reith can declare, "Truth is by its nature independent of men's minds, for it is divine, as even the pagan Greek philosophers knew" [3]. Under the current scenario the 'truth'-building process itself is therefore critical, since it determines what exactly ends up under the auspices of Truth.

That very process is presented to us by Aquinas, putting it within the concept of metaphysics. Again, using an exemplary observation from nature, he compares the acquisition of knowledge to seeds which fall on some ground, maybe fertile, possibly barren. The ground however stands for 'soul', so that the seeds of knowledge, that is the "first concepts of understanding" [4], are taken to lodge there in order to—hopefully—germinate. The functional abstract relating to the seed-observation delineates the qualitative environment (fertile, barren), which has definite ramifications towards the understanding of 'soul'. The functional abstracts surrounding 'God' lay down the interpreted manner in which knowledge is gained as well as its ultimate goal, namely fusion with the divine. Similar to Reith above, Aquinas is therefore prompted to write,

> In these general principles, however, all the consequences are included as in certain seminal principles. When, therefore, the mind is led from these general notions to actual knowledge of the particular things, which it knew previously in general and, as it were, potentially, then one is said to acquire knowledge. [5]

Note the sentence, "In these general principles, ...". From a pedagogical point of view the procurement of a certain concept, derived from a set of general principles and then further outlined by more and more specific explanations, is indeed valuable. Yet in modern educational systems the objects being taught are not mysteries, waiting to be explored; on the contrary, they represent subject matter which has already been implemented, tested, probably modified and fine-tuned, and now forms part of our usable environs. The ultimate topic of Thomas Aquinas, on the other hand, consists of philosophical conjecture, historical backgrounds, and emotional and intellectual attachment to certain ideas. There are no precise developmental sequences which guide the cognitive mind from one principle to the next, and to impose upon oneself the prerequisite that there must be some pre-ordained, uninterrupted line of thought stretching from some axiom to the final (and already accepted) conclusion is to constrain whatever logic may be inherent in the themes under discussion. Yet this is exactly what Aquinas is demanding, as shown in the quote above. For us it becomes interesting because a functional abstract of that nature is bound to influence any future contemplations and, at the very least, will have the effect of corralling their potential outcomes into pre-set categories (eg acceptable or unacceptable). At worst, it will prevent the emergence of a particular thought outright.

In his attempt to demonstrate the justification of moving from the observable to those realms which cannot be sourced similarly, Aquinas takes recourse to a type of categorisation which allows the differentiation between the material and the spiritual. It is more than a mere play with words. Let us not forget that concepts such as 'philosophy', 'nature', 'spirit', the 'divine', and therefore 'metaphysics' have not been a contemporary invention, but were descriptions of containers (to use our present nomenclature) which proved useful for what they seemed to encompass—which was of course the very reason for their creation. When we are told that philosophy means the pursuit of knowledge, then word combinations such as "philosophy of nature" has the potential to create confusion in a modern mind. The reflective, meditative, and ultimately self-referencing nature of a traditional philosopher must be contrasted with the observing, descriptive, analysing, and ultimately exterior-referencing mode of a more modern investigator. Consider the following quote by Aquinas [6]:

> If there were no other substance beyond those in nature of which physics treats, physics would be first science. But if there is some other immobile substance, that will be prior to natural substance, and consequently the philosophy treating of this kind of substance will be first philosophy. And because it is first, it will be universal and its function will be to investigate being as such.

The perception of the world as being divided into the material, ie nature, and something non-material calls for the introduction of a two-faceted intellectual approach; we have seen the presence of a similar need in chapter 2 when we followed the development of the mind/body dichotomy. In both cases the subject matter required a specific format of thinking—not because of some cognitive discord that had been discovered during the process of evaluation, but because of an a priori position that there are two aspects of the cosmos and each had to be treated separately. The thought pattern pertaining to nature (what can be classified as material), the essence of things (what can be identified as a thing's inherent nature), truth (what can be known as true to the exclusion of any perturbations to such a quality), cause and effect (what can be regarded as a source of something), and the ideal (what can be termed the extreme form of a singular manifestation of something); the combination of these aspects, all gleaned from daily and straightforward observations, can and will lead to singular interpretations once the afore-mentioned dichotomy is made use of.

To accept that life in general functions under divine auspices leads almost automatically from the prosaic to the elevated, where forces and entities beyond one's comprehension perform tasks which can nevertheless be described in a conventional format. As Reith says, "Consequently, the philosophy of nature must establish that there are beings that are independent of the laws of matter in their operation and existence" [7]. For a present-day thinker the above sentence would call for further explanation. For example, what exactly is meant by "nature" (chemical elements, organisms, forms of energy, principles of mathematics...); what exactly is meant by "beings" (inanimate or animate things); and what is meant by the "laws of matter" (relating to chemical compounds, geometric structures, organic cell structures...)?

The function of a container is to hold one or several ideas within a basin where certain rules are made to apply. The integrity of the contents is not significant for the container's existence. When Reith therefore continues, "..the philosophy of inanimate nature shows that there must be an immobile mover to cause motion, and the philosophy of animate nature demonstrates the existence of a substantial, spiritual soul" then such a statement is not the product of analytical interpretation, but illustrates the assumed purposes of Reith's and Aquinas' mental imagery. If, whatever causes us to conjecture the existence of a soul for instance, is already placed into a different intellectual framework from what would be used for say, thinking about the building of houses, then it becomes possible to take such a concept as 'mover' and apply its functionality to whatever is found in the alternative. And furthermore, if one's overall model *demands* the existence of certain phenomena, it is not difficult to reconcile oneself to the notion of separate realities on one hand, but still obeying similar metaphorical principles on the other. Aquinas writes,

> Because it is joined to the body, the human intellect has as its proper object the essence or nature that exists in corporeal matter; and from the natures of visible things one can rise to some knowledge of invisible things. [8]

Although we have seen the emergence of complex thought structures in terms of ongoing abstractions when it comes to the conceptualising of what cannot be seen with one's senses (the use of PAT^2s and their deployment into PAT^3s), the reference to sensory experiences as a basis for deeper insights still shows a certain simplicity. We are being presented with a thought pattern which derives its validity from the view that the power of human sensory organs define the extent of natural manifestations. At the same time Aquinas is sufficiently astute to recognise the significance of analogies. He speaks of two kinds: a "proportion" which indicates a certain ratio between the qualities of two things (eg one being twice as big as another, or one being influenced in a similar way to another), and "proportionality", where the relationship between two proportions is identified. In our terminology the former would be a PAT^2, whereas the latter would constitute a PAT^3. However, PAT^3s do not link directly to the specific mechanism which produced a PAT^2, and so there is the risk that the idea of proportionality is lent to concepts which are only artificially brought together. Aquinas demonstrates this by saying,

> When, however, things are said to be proportionate by way of proportionality, their relation to each other is not considered. All that is considered is the similarity of the relationship of two things to two other things. *Thus nothing prevents* an infinite from being proportionate to an infinite; for just as a particular finite is equal to a certain finite, so an infinite is equal to another infinite. [9] [My italics]

A readily acceptable use of the term proportionality is given in the same passage:

> Again, the agreement is occasionally noted not between two things which have a proportion between them, but rather between two related proportions, for example, six has something in common with four because six is two times three, just as four is two times two. The first type of agreement is one of proportion; the second, of proportionality.

By contrast, the following is somewhat more controversial:

> Sometimes, however, a thing is predicated analogously according to the second type of agreement [ie proportionality], as when *sight* is predicated of bodily sight and of the intellect because understanding is in the mind as sight is

> in the eye. ..But in the other type of analogy [ie proportionality], no definite relation is involved between the things which have something in common analogously, so there is no reason why some name cannot be predicated analogously of God and creature in this manner.

In the two statements given, the similarity between the product of two numbers would be a PAT^3 well inside the logic boundary; the parallel between a feature belonging to an organism and its transposed equivalent to God would be found outside it. Yet in both the veracity of the underlying principles *per se* can be demonstrated.

What we have said about the two types of relationships applies not only to objects or things in general, but also to principles, or rules. Just as object abstracts lend themselves to a transposition from the observable realm to the metaphysical, so can an observed principle, that is a functional abstract, be mapped onto a concept which has been created in one's mind only. Aquinas expresses it as follows,

> Now the proportion of the created intellect to the understanding of God is not, in fact, based on a commensuration of an existing proportion, but on the fact that proportion means any relation of one thing to another, as of matter to form, or of cause to effect. In this sense, then, *nothing prevents* there being a proportion of creature to God on the basis of a relation of one who understands to the thing understood, just as on the basis of the relation of effect to cause. [10] [My italics]

Once the relationship of proportionality has been established, the association between any PAT^2s on one hand and the relevant connection between certain PAT^2s and one or more PAT^3s has become a more satisfying one to the mind in question. As a consequence its acceptance value has increased and it will be made use of again from here onwards.

Notwithstanding the above (or maybe because of it!) the point is made about an inherent reciprocity not being transferable to its equivalent when that other example resides within a PAT^3 domain—in other words, where proportionality applies rather than a mere proportion. For instance, one may know about an object, and therefore there exists a relationship between the knower and the focus of his attention; but the object does not depend upon someone knowing about it. As Aquinas puts it, "The relation by which knowledge is referred to its object is accordingly real, but the relation by which the object is referred to its knowledge is only conceptual" [11]. Thus, in terms of our nomenclature, while the former referred to in this quote is a straightforward PAT^2, in order to deal consciously with the latter a PAT^3 is needed. One could call it a higher form of understanding and as such an advance upon a more simplistic view, but that comes at a cost: the possibility to accept the underlying principle as a given in both cases, because the veracity of the simpler version can be established and therefore there is the temptation to regard the reciprocal version as equally valid.

The problem is highlighted by Reith's comment (unwittingly, I presume),

> "..in the order of learning we first name a thing from what we are familiar with. We name the divine attributes and think about them according to the ways that we understand. But this does not mean that the divine attributes are dependent on those of creatures. Being, truth, and goodness in God cannot be measured by what is found in creatures because 'it is not necessary that the capacity of the passive potency be commensurate with the active power of the agent'. The kind of proportion between God and creatures is a relationship of

> order and does not imply that the effect is commensurate with the cause. Between the finite and the infinite there is only a similitude of proportions". [12]

It is certainly correct to identify a fundamental difference between 'real' relationships and those which are conceptualised within the mind only; the problem lies in transferring any logical relationships regardless. Considering that the mind can only ever understand what it is capable of understanding at any given time (subject to its knowledge base and its functional limitations), the derivative of any subsequent conclusions can only be as good as their initial source. To assume, as Aquinas suggests, that an identified principle can safely be applied to a higher concept because of some found proportionality is neglecting the inherent—and differing—complexity of either idea. This is radically different from the scientific approach, where higher-level concepts can certainly be entertained, but where any assumed, common principle requires to be tested in terms of that higher-level concept. For that reason alone the gulf between science and religion is unbreachable. Should a religious thinker decide to proceed by measuring a given concept according to its inherent status, metaphysics as demonstrated here falls by the wayside; conversely, should a scientist examine a hypothesis by applying the principles of a different-level concept, the outcome would not inspire confidence in its ultimate truth value. It needs to be kept in mind however that either approach is defined beforehand, that is to say there exists a set of functional abstracts which determine the path taken by their owner. Or, to put it yet another way, there is no inherent, automatic necessity in either *modus operandi* which is capable of suggesting one before the other. Rather, the first option (Aquinas' way) represents what could be called the 'raw' version inasmuch as it manifests a certain mind dynamic without recourse to a critical analysis of what the outcome would actually be like. The scientific paradigm, on the other hand, is more sophisticated because through trial and error (*and* the application of intellect) the greater exactitude and therefore reliability to be gained there has been shown. Such an insight however was not accessible to the 13th century mind.

For Thomas Aquinas then the consideration as to whether the "human soul of Christ could know in the Word of God all that the Word knows" [13] does not present a problem as far as the statement itself is concerned. His mind, having positioned the God-concept (a PAT^3) beyond the logic boundary, having identified the principles involved in a viewer-viewed relationship derived from observations within his environment (PAT^2s), and feeling comfortable with the applicability of such principles across any idea inside as well as beyond the logic boundary, his mind has no reason to object to such a question on the grounds of a possible functional incongruity. Functional incongruities, let us remember, are a concept created through analysis and posited as such; they are not flaws in a system like a leaky valve would be a flaw in an engine. While the engine functions insufficiently with a leaky valve, the mind suffers no impediment while performing a functional incongruity. It is only the outcome of actions resulting from such a way of thinking which are hurtful—the mind responsible for that need not be concerned with it at all.

In fact, as has been hinted at in an earlier paragraph, once a higher-level concept has been imbued with the appropriateness of a certain proportionality, it has been sanctified as it were (no pun intended!) by the original analogy and its previously confirmed truth value. Consequently it is possible for Reith to declare,

> The common being of metaphysics, as the subject of the science, is the analogical bond that brings together all created being; although it does not give direct knowledge of infinite being, it opens the way to the possibility of a science about God based on human reason. [14]

Would a constant reference to the real world prevent such thought processes from going off the rails? In chapter 3 we learned from Jainism that any conclusion drawn from abstractions should still be tested against one's experiences within the Real, even if this means re-examining a divine revelation. In a sense Aquinas is advising the same, holding as he does that first and foremost the human mind is dependent upon the senses, and from those experiences is any knowledge of the transfinite derived. He does not however caution against disregarding the context levels of ideas (nor does the Pañcastikayasara for that matter), but prescribes the power of reason to solve any question regardless of what domains such a question may straddle. For Man to "rise through his power of reason to a knowledge of the transfinite world that is the efficient and final cause of the finite world" is, in Reith's and so Aquinas' words, a possible and indeed desirable undertaking; having pre-defined the nature of the ultimate goal ensures that the thought processes run along a controlled and constraining path.

Yet even the reference to the real world is fraught with danger. Consider the issue of substance vs accident, ie an object and its characteristics. We will return to this theme several times as many philosophers have dealt with the significance of one against the other. For Aquinas it is a matter of defining the treatment of the concept 'substance' such that it allows an efficient treatment of the concept 'accident', all that within the context of a religious world view as outlined by him. Let us see what thought structures there are and how they are being used.

As Reith explains, "Substance is hidden from our senses as the foundation of a building is hidden from our eyes" [15]. What he means is that accidents, ie the attributes of a certain object, are the things we can determine through our senses, but that the object itself exists underneath such a facade and therefore cannot be identified as readily. It is only through the application of our intellect that we come to any conception at all of what the characteristics represent. Reith mentions in passing some strands of philosophy where the existence of substance is denied outright because it is invisible to our senses, while at the same time admitting to its appearance since *it* can be ascertained. A curious proposition, and one we will deal with in due course although it is not as simple as Reith makes it out to be.

Nevertheless, through an examination of the thought structures involved we can see how such positions can be arrived at. Let's suppose I look at a stone and through its colour, texture, and shape I determine that it is indeed a piece of rock. But I *really* don't know it is a rock, since I have made the conclusion based on the object's appearance. A skillful modeller could fabricate a pseudo-stone which would fool most people (the movie industry for instance depends on such skills). Therefore, the reason why I would categorise the thing as a stone is because I can draw on my experience of similar events where I had been able to identify the underlying substance and now I am applying the analogy to the current situation. To translate this into our terminology I can say that, based on particular PAT^1s (the result of the stone's appearance) and certain PAT^2s (the representation in memory of similar circumstances), I have now in my mind the idea of 'stone', which is a PAT^2. The same process applies if I were to undertake a chemical analysis, since the result is once again interpreted at the PAT^2 level.

So far so good. Note however that my PAT^2-stone is neither a real-this nor real-that but in actual terms is wholly imaginary—unlike its colour, shape etc it has never been ascertained in the real world but acquired its existence through my personal thought processes. We can see how philosophers of a certain intensity could exaggerate their predicament by denying the stone's substance altogether, but it should likewise be obvious that regardless of my internal activities that stone (or whatever it is) does exist in the real world. In case of any doubt I could raise the object and drop it onto my head—a simple yet effective proof.

At this point, for a theologian like Aquinas the work is not done yet. It is not enough to declare substances and their appearances according to our perceptive capabilities; the human intellect can also make analogies, draw conclusions, identify principles. Our intellectual lives would be poor indeed if we could only ever conceptualise the world via attributes. By the same token, to invoke a proportion through the use of some substance (with or without accompanying characteristics—eg "hard as a rock") necessitates the abstraction of that substance. What suggests itself here is a kind of intellectual common denominator which allows the thinker to posit 'substance' and 'accident' alongside each other without having to emphasise their conceptual differences to the detriment of the desired analogy. In other words, if "hard as a rock" is the analogy and/or proportion to be applied to a particular focus, then in terms of shedding light on the latter, both, substance ("rock") and accident ("hard"), will have to be useful conceptual entities in relation to the focus. Aquinas recognises this:

> Hence the supervening accident, by union with its subject, does not cause the act of existing in which the thing subsists, rendering the thing a substantial being. Rather, it causes a certain *secondary act of existing*, without which the subsistent thing can be understood, as what is first can be understood without what is second. [16]

Reith describes the same relationship in the following words, preceding the above quote,

> The mode of existence of substance is unlike that of accidents, and that is why accidents are not simply said to *exist* as substances are; rather accidents *exist in* a substance; they modify the being of substance and are said to be *beings of a being*.

While 'substance' and 'accident' have to be kept apart in terms of their actual meaning (maintaining the difference between a thing and a characteristic), when it comes to the conceptual usage of either they carry the same weight.

We have come to the heart of the matter. The intellectual use of substances is important because without them the significance of analogies would be dramatically diminished. Thomas Aquinas:

> The metaphysician does not investigate sensible substances precisely because they are sensible but rather because they are substances or because they are beings, or finally, because through them we are led to the knowledge of immaterial substances. [17]

How is their use made possible? Well, we have seen that in order for an ideation to be carried across analogously to another scenario it has to be abstracted and, since 'substance' is already a PAT^2, it will need to be turned into a PAT^3. There is now the possibility of a conceptual mismatch between the PAT^2-type carrier of a characteristic and the PAT^3-type carrier of an otherwise relevant object, ie an incongruent functionality. The problem is exemplified by Aquinas' intent to move the derived meanings of a substance with its accompanying attributes towards the metaphysical concept of divinity, which in itself requires a very particular set of attributes being attached to a very particular type of substance, and all of this together being incorporated within the domain of yet another PAT^3, namely God. Thus the qualities of the substances under discussion, that is the Q_{ao} of substances, need to be aligned with the Q_{ao} of God such that neither annuls the other. Since

God is predicated perfection, the Q_{ao} of those substances which are destined to serve the belief system must also reflect this.

Yet it is plain that some ordinary substance, say a rock lying on the road, is not perfect. But through our abstraction processes we can arrive at an ideation of a 'perfect rock', which is what we have to do if we are willing to relate material objects to an ultimate source, in this case God. Due to the particular arrangement of our thought structures (assuming we go along with Aquinas' viewpoint) there is the requirement to imbue the hitherto innocent rock (or at least its mental representation) with something approaching the idea of perfection. 'Perfection' in turn is a PAT^3, derived from some PAT^2 where special attention was given to some specific attribute. It should be obvious that the selection of such an attribute is a function of the mind's knowledge base and its inherent capacity to use it, so that the meaning of 'perfect' is very much a manifestation of the mind's potential. In addition it means the emphasis of one attribute against the diminution of others, since the idea of perfection implies exclusivity. The closer an object's proximity to the ultimate becomes, in this case God, the more pronounced the concept of perfection will be at the same time. Thought structures are evoked which carry the idea of an ever more perfect something, with the aim of eventually arriving at the apex of one's understanding. It matters not *what* such perfection actually means, nor does it matter to the thought processes that, technically speaking, the existence of only perfect things would be impossible (imagine your perfect stomach containing perfect acid). The analogous correctness of the various thought structures, manifesting in PAT^1s, PAT^2s, and PAT^3s, ensures that any mind holding them will not be troubled by real-world impossibilities. It will however prompt its owner to search for ever more suitable examples, inside itself and out, to feed those thought patterns which demand the recognition and eventually the realisation of some kind of perfection. What exactly such perfection entails depends on the particular belief system. In Reith's words,

> Ultimately, the unity and perfection of being come from form and the act of existence that follows form... Matter, on the contrary, is the principle of non-unity. The more a form is immersed in matter the less perfect it is. The more material a being is the more it is divisible and lacking in unity and perfection. Thus the inferior things in nature, being more material, lose less by being divided than do more perfect beings. [18]

The astute reader may reflect on the ultimate benefit of a belief system under which those things that are most readily accessible via observation and analysis (ie PAT^1s and PAT^2s) are relegated to inferior positions, whereas the more removed from reality ideas are (PAT^{3+}s), the more significant they are held to be.

Without PAT^3s such hierarchies would not exist; neither however would considerations about the deeper nature of substance, if by this one means an abstracted concept which itself has its origins in a previous abstraction. When talking about Aristotle in chapter 2 we illustrated the ancient philosopher's system of classifying things according to their degree of perfection, where the highest degree was reserved for those things that could claim to be most natural. Leaving aside the familiar argument about identifying what is 'most natural', we nevertheless remain within the domain of characteristics pertaining to whatever substance it is we try to categorise. Even concepts such as Beauty or Honour are PAT^2 derivatives from events which have been witnessed and judged accordingly. These kind of thought structures do not lend themselves to the proportionality argument such as Aquinas has made use of. A PAT^2 must be relatable to its PAT^1 sources, otherwise it ceases to make sense. A PAT^3 can be transferred as is, because it itself is a principled version of some source. In chapter 4 we touched upon this situation in terms of the ability of more sophisticated religions to be transferable across different cultures, because their doctrines allow for a more abstracted interpretation; same here.

What Aquinas chooses to imbue a particular substance with, generally or specifically, is rather besides the point. The fact that he can do so means he is able to appreciate the potential of a higher abstraction, an advantage he cannot find among the ancients:

> The ignorance of the meaning of substantial form led the ancient philosophers into error in the following way: they had not yet arrived at the point where their intellects were sufficiently raised above sensible reality. Consequently, they investigated only those forms that we call *proper* or *common* sensibles. These are evidently accidents, for example, white and black, great and small, and so on. Substantial form, however, is perceptible by the senses only indirectly. They [the ancient philosophers] had not sufficient knowledge to realise that it is distinct from matter. The total subject that we say is composed of matter and form they said was prime matter, either air or water or something of this kind. But they said that those things are forms that we call accidents such as quantities and qualities whose proper subject is not prime matter but composite substance which is substance in act. [19]

From our perspective, taken at the beginning of the 21st century, the above seems somewhat quaint when placed alongside other, more recent philosophies. The epithet "prime matter" seems outdated, and under contemporary paradigms the concept is practically useless. However, if one considers the absence of further abstractions on one hand, yet the intellectual need on the other to deal with some type of basic material which in principle would lend itself to conceptualise the variety of everyday life, some PAT^2 representation would have to be found. Likewise, for Aquinas who, to his credit, appreciated the usefulness of third-level abstractions in order to construct *his* cosmic model, a PAT^3 could be constructed. In either case it is questionable whether the respective thinkers were fully aware what they were dealing with. But this is human nature: first we discover/invent/build something—only then are we in a position to learn about the consequences of such an act.

Aquinas therefore does not seem to have fully understood the implications of the potential of PAT^3s, otherwise he would not have committed the error of defining a priori the target for a PAT^3 for which then the chosen PAT^2s have to be made to fit. Consider his comments on the abstraction of matter (keeping in mind his 13th century language):

> There are still other objects of speculation [having before talked about flesh and bones, lines and numbers] which do not depend on matter with respect to their existence because they can exist without matter. That is true, whether they never exist in matter, e.g., God and the angels, or whether they exist in matter in some things and in others do not, e.g., substance, quality, being, potency, act, one and many, and the like. Theology or divine science (so called because God is the principle thing known in it) deals with all these. It is called by another name "metaphysics", that is to say, "transphysics", because in the order of learning it comes after physics for us who must rise from sensible things to what is beyond the sensible. It is also called "first philosophy" insofar *as all other sciences take their principles from it and so come after it.* [20] [My italics]

As can be seen, while he applies the abstraction process to ideas, he nevertheless feels comfortable with proceeding from a realm inherently unrelated to reality and using its posited principles with the intent to apply them to those things which had been identified through real-world abstractions. Thus, with our hindsight we can say that as the ancient philosophers lacked the ability to properly deal with certain abstractions but dealt with them anyway, 13th century thinkers had mastered that step but were just as adventurous with the next. Furthermore, having made use of such abstractions without putting them in their proper place, the actual identity of the focus of attention is held to be the content of the thought, rather than the thought itself. Take for instance the following paragraph by Reith, where he explains Aquinas' position on the same theme, the abstraction of matter:

> The "separation" that distinguishes metaphysics from the "abstraction" proper to the other speculative sciences is said to be the act of a "negative judgment". Why is it so named? It is because the intellect in knowing metaphysical truth forms a negative proposition something like the following: "There is being not essentially connected with matter". Thus the judgment is cast in a negative form. However, it really expresses something positive—it asserts the *de jure* independence of matter, but it does say that it is not necessary for being to exist in matter. [21]

And further,

> The subject of metaphysics is real being. It is not grasped in an abstraction of the first operation of the mind but in a judgment which affirms that the objects of metaphysical inquiry exist. In this act of judgment the intellect affirms that there are beings some of which actually do exist apart from matter and motion and that there are others which, though normally found in the world of matter and motion, are such that existence outside of matter does not imply a contradiction. Thus metaphysics has as its subject *being* that can be defined and can exist apart from the material conditions of time, space, and movement. [22]

In any case, as far as thought structures are concerned, the mind does not distinguish between one which has been construed through some recourse to a, say, metaphysical concept, and one which contains the ideations from a real-world scenario. Both are just as 'real' when it comes to their internal representation, and only other thought structures, employing appropriate rule sets, are able to counteract their influence. In the case of absence of such a balance the mind will accept the representation as real and act accordingly.

Once the content of such a thought structure or indeed a thought pattern has become acceptable, it can be used in further contemplations as if it were any item of real substance (which for the mind of course it is). One's logic can be applied, one's known analogies and parables, in fact the integration with otherwise normal, everyday life can be performed seamlessly. It is true that Aquinas advises his readers to study metaphysics *after* the other sciences but, as mentioned earlier, his overall world view is such that certain conclusions are pre-ordained and therefore pose no danger to possibly challenging results from those other sciences. He sheds light upon this process of conceptualisation in the following passage, when he describes the manner in which things are articulated, always with a view towards the highest:

> We give names to things according to the manner in which we receive our knowledge from things. Hence, since those things which come after others in the order of nature are usually the ones that we know first, it frequently happens that in applying names to things, we first use a name of one of two things when the reality it signifies primarily exists in the second. We have a clear example of this in the names that are used of both God and creatures. *Being*, *good*, and words of this sort are first applied to creatures, and then transferred from creatures to God, even though the act of existence and good are found primarily in God. [23]

Not only is the foregoing a demonstration of Aquinas' diligence in articulating how the philosopher arrives at a higher insight through contemplation of baser things, he also hints at the relative bias that is applied to either type of outcome. Because an insight gained through a further abstraction is deemed to be superior than the one more or less immediately derived from external observation, the modifying influence of any rule sets pertaining to some higher realm would have priority over anything the lower level suggests. Or, in Reith's words, preceding the above quote:

> Although the mode of signification of the name is primarily realised in man, the reality it signifies belongs primarily to God and by participation to human beings. By transferring the name *person* from human beings to God, we are led from what is better known to us to what objectively has a greater depth of meaning. Thus by means of analogy we can rise from the knowledge of what is inferior to some kind of understanding of what is above us. Consequently, there is a twofold kind of priority in analogy: first, the priority with respect to the imposition of the analogous term; and second, the priority with respect to the nature of the thing signified by the common name.

Ultimately, the reason for such an allocation of priorities is not found in Aquinas' philosophy, nor in the Catholic church, although both appreciate its existence and use it to underwrite the veracity of their beliefs. On the contrary, the thought structures emanating from an abstraction are inherently more highly valued by the mind than their predecessors. Consider for example someone who has difficulty understanding the aerodynamics of flying. Using the analogy of a vacuum tube the Bernouilli principle is seen in action and consequently its significance on the wing span of an aircraft can be appreciated. In other words, the analogous representation of a particular act, ie a PAT^2, was able to override any inappropriate thought structures that may have existed previously. In addition, the thought patterns representing the entire scenario will be more authoritative in the person's mind than any PAT^1s which may come its way. To the person holding these thoughts the principles of flight are quite significant. Indeed they are, but in the case of that mind their importance is not accepted primarily because of the role they play in physics but because the mind is dealing with a further, created abstraction. First and foremost therefore the relative significance of an idea is not so much dependent upon its contents, but on whether it has been borne out of a higher-level abstraction then some other idea. To the consciousness of the person the idea seems important because of what it means—but its meaning has been given to the consciousness and so that is *its* decision, with the person responding accordingly because he or she is driven by their own consciousness.

One could call it the Bambi-factor. Remember Bambi's excursions into the fields—having learned one word, say, "flower", it calls everything "flower" because it is full of the exuberance of having learned something new. The rabbit has to hop alongside and remind him, "No no, this is not a flower, it's a bee!". "Bee, bee!" Bambi continues in the same fashion, presumably until his vocabulary has grown large enough so that one single, new item has become less outstanding amongst all the rest. Whether in the end such an analogy holds true in terms of the underlying functionalities of both concepts is not the deciding factor—remember the comparison made between a box and one's memory in chapter 3. The comparison *seemed* to fit as far as the mind was concerned, and the conclusion reached, ie that sooner or later the mind would 'burst' if overfilled with information, assumes a governing role in the relevant thought processes.

A more 'adult' version of the same phenomenon can be found in the use of symbolism, where the symbol itself assumes greater importance than the thing it originally came from. For instance the flag of a nation is priced more highly than the people it is meant to represent, and while the deaths of actual humans may be generally commented upon, any damage to the piece of cloth on the other hand inflames the hearts. Not only do the average man and woman not recognise how they are being driven by their neurons, to question them in terms of the ultimate reason behind any resultant behaviour can easily provoke open hostility. But that's another story, and it belongs in Part II.

When Aquinas and Reith call upon an insight which sprang from a "greater depth of meaning", and by doing so exhort their audience to accept its nature as superior to more mundane considerations, they do not necessarily act as 'spin doctors' for Catholicism, to use a modern expression. They are true to their minds, their cognitive dynamics, in which higher-level abstractions are accorded a weightier influence than their lower-level neighbours. Consequently they are driven to declare such ideations as "objective", they are explicitly and implicitly taken as truth, and they are being relied upon to provide knowledge about the intangible.

In the case of names, as alluded to in the quotes above, the relationship between their lower- and the higher-level abstractions is relatively straightforward (for the ordinary user that is, unencumbered as he would be by cognitive dynamics). The situation becomes somewhat more difficult if one deals with logical analogy, because here the principles ascertained, mapped across, imbued etc etc are the very building blocks of the entity being handled. For example, a preacher may hold up a seed and lecture to his congregation about an essence growing into a fully-fledged representation of its principle. Whether or not such a comparison is technically valid, the mind, if suitably prepared, will accept the message because although the higher-level abstraction is not the same as the original object (the seed in this case) nor the final result (the grown person, acquired knowledge, etc), it nevertheless is connected directly to a real-world object. In the case of logic however the internal structure of that system, itself composed of principles, is the one being abstracted. Just as a relationship between a logical element and its higher PAT-counterpart has to be established, so needs there to be a similar workable connection between the counterpart and whatever it is the principle is being compared with.

Reith describes the analogy as practiced by the logician as a "mode of signification of concepts", that is how the mind "attains its concepts". For the metaphysician an analogy is "a question of both the *mode of signification* and the *modes of existence* outside the mind and consequently refers more precisely to the thing signified" [24]. He goes on to say that "logical entities have a basis in reality, but *having a basis in reality* and *being really what the mind conceives* are two very different things." From our perspective we could say that while logic enables us to map out a particular system, once we are moving within the realms of metaphysics any previously ascertained notions should not be taken verbatim but nevertheless should be allowed to extend their principled meaning—provided of course it all accords with the rules contained in those realms.

Which leads to the next question: what are those rules? Here it gets interesting, not only in terms of Aquinas' philosophy but also from our perspective, that is how analogies (the higher-level abstracts) are being treated in relation to their respective sources. Aquinas introduces three types of analogies, which are worthwhile studying because they illustrate the dynamics mentioned [25].

The first is "analogy according to signification only and not according to existence". By this he understands the meaning of a word which can be applied to several subjects provided they can be predicated to the original, albeit to varying degree. He gives the example of "health" as the original word, whose signification can then be linked to an animal as the subject of health, to urine as a sign of health, or to a diet or medicine, being the cause of health. All these things differ in their existence, but can be combined in terms of their predication. From our perspective we can say that "health" is a PAT^2 which can be related to PAT^1s (eg animal, urine, etc) because our powers of observation justify it. There is no problem since the PAT^2 "health" has been derived from PAT^1s in the first place.

The second is an "analogy according to existence but not according to signification". Here he means the existence of several "bodies" which contain the same ascribed characteristics but through this differ in terms of their mode of existence. To avoid confusion with the word "body" Reith uses the example of 'soul'. While the basic meaning of that word remains the same no matter to whom or to what it is applied, to the philosopher, so Reith contends, it differs according to whether we speak of the soul in humans, in animals, or plants. Or, in Aquinas' words, "Therefore, as far as the metaphysician and the philosopher of nature are concerned—since they consider things according to their existence—the name *body* and any other thing like it is not predicated univocally of corruptible and of incorruptible things, as is clear in *Metaphysics X*, according to the Philosopher and the Commentator". Note the phrase "not predicated univocally": this means that the underlying concept of this or that subject will not be unambiguous but can adopt different interpretations. Of course, it is the mind which does the adopting, and so the question of existence is answered through the mind's understanding and classification that is being applied to the subject. Therefore, to remain with Reith's choice of 'soul', when we speak of the soul in humans vs the soul in animals we have not accessed a previously identified thing which in due course we discover to exist in various forms (such as sensory organs in different species for instance). Rather, 'soul' is a derivative of a PAT^2 already (certainly at this stage in contemplative sophistication) and as such can have a variety of meanings subsumed to it because the nature of a PAT^3 allows that to happen, as we have seen. The plausibility of the second type of analogy then is not a function of the subject's inherent nature, but rather a function of the accompanying thought processes, containing as they do the thought patterns and associated rule sets so as to construct the analogy.

Lastly, there is an "analogy according to signification and according to existence". In this case there are things which differ in terms of their meanings as well as their existence, according to "greater or lesser perfection" as Aquinas puts it. He gives the example of truth and goodness and "all things of this kind" which are "predicated analogously of God and of creatures". Compared to the previous type the ambiguity has now been extended so that the interpreted meaning of both, signification *and* existence, is subject to variation.

Using our schema, we can therefore classify the three types of analogies in the following manner:

Type 1: characteristics derived from PAT^1s become PAT^2s which are related back to other PAT^1s.

Type 2: PAT^3s derived from PAT^2 objects are related to other PAT^2s.

Type 3: PAT^3s derived from PAT^2 objects are related to other PAT^2s, and their respective characteristics are treated in the same manner (ie turned into PAT^3s and related back to other PAT^2s).

It is significant that Reith comments, referring to Type 3, "The metaphysician is especially interested in this kind of analogy because of its relevance to the transcendent order of being." And so he would—given the interpretative leeway that kind of thinking affords.

Although the Type 2 and 3 analogies exist in their own realm and are construed according to their conceptual history and so interpretation, it does not mean that *within* that realm the principles of logic cannot be applied. An entire system of existence can be thus created, and as long as none of its member items is being tested in terms of their congruity with a PAT^1 the system itself will not be challenged. On the other hand, the attempt to link any of its PAT^3-PAT^2 connections downwards to a PAT^1 equivalent will meet with resistance because the system relies for its sheer existence on the disconnectedness with PAT^1 thought structures. When Galileo suggested that the earth moves around the sun he did exactly that—and was forced to retract on the pain of death. (An aside: Microsoft Word for Windows 95 does not have the word 'Galileo' in its spell checker; it does however feature 'Galilee'. Even in a late 20th century product religion can take precedence over science. Make of it what you will.)

When we came to the conclusion that the principles of logic can well exist unchallenged within the realm of thought patterns spanning the Type 2 and 3 analogies, then it also has to be said that possible errors in logic can happen as well. A common mistake is to assume a reciprocity of a certain relationship, as in "all cats are animals therefore all animals are cats" for example. Since with cats we are dealing with PAT^1s such a mistake is easily seen. It becomes more difficult if the subjects under discussion are not accessible through direct observation, as would be the case with PAT^2s and PAT^3s being treated exclusively amongst themselves. One such example is demonstrated by Reith when he writes about the metaphysical question, "Is the good toward which a potency naturally tends prior to the potency?" [26].

Reith answers the question by saying that the "goal of any natural tendency must be superior to the potency, because it is the state of perfection as compared to the state of imperfection". The words 'natural', 'potency', 'perfection' and 'imperfection' may have originally been delineated from some identified PAT^1, but at this stage of contemplation any link with the real has been severed. Since the final goal of these intellectual searches has been posited as God, and 'God' being imbued with qualities pertaining to perfection, it is only to be expected that the conceptualisation of things on that path must tend toward such a state. Note that the concept of 'perfection' does not imply anything of substance; it is an idea which suggests itself while musing about what could be in terms of an assumed bias that is discerned amongst the sum total of characteristics. For instance, I could examine a stone and regard its hardness as the most important attribute. A 'perfect' stone therefore would have infinite hardness. But is the purpose of a stone to be hard? I don't know! Still, prompted by my presumption I could envisage an ideal world where stones are indeed 'perfectly' hard. That's the first problem.

Reith goes on to say, "If there is order and intelligence in nature, the act toward which a potency tends has to be superior to the state of the unfulfilled capacity". He concludes that the good which fulfills the potency must already exist. Not only that, but the good must also have an "intentional existence" in whatever entity it is which enables the potency to arise. It is here that the reciprocity occurs. For if one postulates that there is an ultimate power which incorporates 'perfection' (derived as it is from assumptions about a lesser being), it would be wrong to turn the relationship around and declare that power to be drawing some lesser characteristics towards its own state. In Reith's words,

> The good attracts or pulls the efficient cause. And in order to pull, the good must exist in some real way. However, the good that is the fulfillment of a potency does not exist as this particular good, otherwise there would be no true becoming or production. But since this is a contingent good and not an absolute one, *the goodness of this object of striving depends eventually on the existence of a transcendent self-contained good, for the reality of a contingent good is dependent upon the existence of a transcendent good.* The same kind of reasoning applies to truth. [My italics]

In terms of Aquinas' philosophical system the above makes sense. The conditions of the identification of a characteristic, its relationship with what is held to be perfect, its position on the ladder of proximity to the ultimate, and finally its significance regarding the step-by-step improvement towards the highest, they all have been met. It is virtually impossible to argue against such reasoning while restricting oneself to the transcendental; only through analysis from the outside (in our case via the subject of thought structures) can one discover the nature of forced logic.

Just as 'goodness' can be incorporated into a PAT^{2+} thought pattern, so can 'evil'. The principles evoked are the same, and so their functionality depends on the source of definition leading up—or possibly, down—to what is awaiting at the end. Aquinas describes it thus,

> Evil is simply a privation of something which a subject is entitled by its origin to possess and which it ought to have, as we have said. Such is the meaning of the word "evil" among all men. Now, privation is not an essence; it is, rather a negation in a substance. [27]

In the light of what role logic can and does play among the thought patterns spanning several abstraction levels it is worth noting how the nature of logic itself is perceived. Reith, explaining Aquinas' position on the subject, writes that "logical entities have no direct bearing on matter" [28]. He goes on to say, "For example, the notions of genus and species, of subject and predicate, the law of extension and comprehension, the square of opposition, and so on, have no necessary bearing on the realities that exist in matter". This is a somewhat curious statement, but let us remember that we are largely dealing here with the subject of metaphysics, where its items are treated with a seriousness normally reserved for real-world entities (ie material things in modern times). Let us also recall that in the system outlined by Aquinas the immaterial is accorded greater credence than the material. Additionally we have seen how the thought structures pertaining to PAT^{2+} abstracts have been removed from any links which may have tied them to their PAT^{1} sources. Under those auspices it is true that logical entities may not have a "necessary bearing" on matter-related reality— in fact this is how the system works. Considering the cognitive dynamics of such systems however it is equally apparent that any PAT^{2+} abstractions originated from PAT^{1}s, and *they* are very much related to the reality of matter. In Part II we will examine the emergence of logical entities in greater detail, but for now let us note how a wilful disassociation of these entities from any PAT^{1}-related thought patterns brings potential dangers—although the integrity of the logic as such may not suffer. Going back to the quote above, one could justifiably argue that the notions of genus and species for instance are not unrelated to matter at all, but are the abstracted concepts resulting from an examination of the relationships one finds among plants and animals. As such they are not 'made up', but are directly dependent on the evolutionary circumstances viewed through the eyes of an intelligent mind. No wonder the theory of evolution has not exactly been welcomed by the churches!

Like St Augustine, so does Aquinas provide us with a proof of the existence of God. Unlike St Augustine, he has moved the effective functionality of the proof firmly into the field occupied by the PAT^2-PAT^3 relationships. There are five versions but the principles are the same in each one of them. Although he starts by observing motion in the tangible world, the essence is nevertheless based on abstractions of a higher level. He says,

> It is certain, and evident to our senses, that in the world some things are in motion. Now whatever is moved is moved by another, for nothing can be moved except it is in potentiality to that towards which it is moved; whereas a thing moves inasmuch as it is in act. For motion is nothing else than the reduction of something from potentiality to actuality. But nothing can be reduced from potentiality to actuality, except by something in a state of actuality. Thus that which is actually hot, as fire, makes wood, which is potentially hot, to be actually hot, and thereby moves and changes it. Now it is not possible that the same thing should be at once in actuality and potentiality in the same respect, but only in different respects. For what is actually hot cannot simultaneously be potentially hot; but it is simultaneously potentially cold. It is therefore impossible that in the same respect and in the same way a thing should be both mover and moved, *i.e.*, that it should move itself. [29]

He concludes that anything moving is moved by something else and, since this cannot go on for ever, there must be a first mover. This seems reasonable enough but actually this is *not* how he expresses it. Instead, he writes,

> But this cannot go on to infinity, because then there would be no first mover, and, consequently, no other mover, seeing that subsequent movers move only inasmuch as they are moved by the first mover; as the staff moves only because it is moved by the hand. Therefore it is necessary to arrive at a first mover, moved by no other; and this everyone understands to be God.

Observe the reciprocity in that argument, prompted by the dominance of higher-level thought structures combined with their isolation from lower levels. There has to be a first mover as demanded by the higher-level abstracts; since these are not linked to PAT^1s the logic cannot be tested against real-world objects. At the same time however the starting point chosen does belong to the real world and by induction the argument proceeds from there, but is clinched by applying the reverse to the higher-level construct.

The second version is essentially the same as the first, except that 'cause' has been substituted for 'mover'. Aquinas ends it by saying, "Therefore it is necessary to admit a first efficient cause, to which everyone gives the name of God" [30]. The third version employs the concept of 'possibility' and 'necessity', and states that something can be or can not be, although nothing can exist always. Since there must have been a time when it did not exist, and the same being the case for every other, there must have been a stage when nothing existed. But this cannot be true, because then nothing would have come into being. Therefore there must be a being which possesses its own necessity for existing, and passing on this necessity to everything else; this being is God. There follows further text to broaden the explanation, but the gist of the argument is as stated and employs the already mentioned principles [31].

Version 4 reminds one somewhat of St Augustine by the way it uses gradations of value, whether that be "truest", "best", "noblest", etc. Still, the maximum of any standard is also the cause of that standard, so Aquinas maintains, and so there must be an ultimate cause for everything. Once again, we have a starting point in the real, a higher-level abstract overlay, and a conclusion drawn from referring backwards to the erstwhile assumption [32].

The last one concerns itself with the organisation, the "governance", of the world [33]. Natural bodies act for an end, and they do this "so as to obtain the best result". Since they lack knowledge one must conclude that they act by design from above, "as the arrow is directed by the archer". The 'ultimate director' then is God. While this proof is not as explicit as the rest in terms of their dynamics, without the a priori idea about what is the highest and the reference to the tangible in the form of natural bodies and back via the higher-level bridge to the archer it would not be possible.

To summarise: in the philosophy of Thomas Aquinas we find for the first time an explicit reference to several tiers of abstractions together with detailed instructions on how to apply them within the context of Christian church doctrine. They may not be described in exactly those terms we use in this book, but their existence and nature are made clear enough. So we can say that the initial sensory input produces a first-level abstraction through the process of transference—a PAT^1. That PAT^1 is processed further to produce a PAT^2. Various inputs are dealt with in this way, with the result that a similarity in terms of the PAT^2s can be ascertained by the mind and the outcome used to form a thought pattern which can be considered an overall concept, that is a distillation of what the original observations tend to suggest. Further PAT^2s concerned with the issue of finality have evolved over time to form a PAT^3 thought pattern as a response to the cognitive need of the mind to resolve unanswered questions; or, to put it technically, to establish a path back from higher-level abstractions to lower-level ones. Logic, that is a PAT^1/PAT^2 rule set derived initially from outside observations, is used to find the appropriate direction and therefore path to such lower levels. The actual, objective truth value in this operation is not significant; it is the relevance found between the rule set on one hand and the $PAT^2 \rightarrow PAT^3 \rightarrow PAT^2$ paths on the other. As such a link between a PAT^1 and a PAT^2 can be broken (or never made in the first place if the latter happens to be imported from somewhere else), but the assumed correctness of the rule set per se can seem to suggest a similar validity if there had been a link, although the mind would see no inherent need to make one. In the absence of a link to a PAT^1, the PAT^2 representation is held sufficient as the end point of a lower → higher → lower abstract chain. Since a chain of this kind is not any more directly related to the real via its PAT^1s, we can say that as far as the PAT^3s of that scenario are concerned, they exist beyond the logic boundary in terms of any possible references to the real, ie the outside. The 'outside' in this case may not only be the 'real' as in earth, water, plants, animals, and so on, but can also be the information communicated by other humans; as far as thought structures are concerned there is no difference between one or the other except in the possible presence of accompanying rule sets. This is not to deny however the existence of a logic which operates within the PAT^2/PAT^{3+} complexes, since there is nothing to prevent the introduction of rules into such patterns from the outside. It is only when the members of the complex have been severed from their erstwhile PAT^1 connections that the significance of the logic boundary comes into effect. Once a set of rules is allowed to operate unhindered outside the boundary (which is the same as *inside* the higher level abstraction complex), errors of convenience can be introduced into these rules, that is to say hitherto unwarranted modifications and/or extensions may be created simply because they allow a more satisfying conclusion to be reached (ie a link to a

PAT^2). One such error is reciprocity, in other words identifying a certain cause-and-effect relationship and inferring from that a similarly plausible reversal of the sequence. Fig. 5-1 illustrates the aforesaid.

Here is an every-day example where the application of higher-level thought structures in the manner we have just described is fundamental: stereotyping. It is so common because without it we would be in constant trouble crossing the road, driving a car, even tying our shoelaces. Consider the mental dynamics when crossing a road: there is a time when we learn that there are cars travelling along what is called a road. The representation of road, cars, together with the associated rules are PAT^1's. From that we derive PAT^2s as a general symbolisation of like scenarios. Do we always check whether any car travelling along *that particular* road we are just about to cross is indeed similar to those cars we have come across in our initial learning experience? Of course not. We don't even bother re-confirming the status of the road we just happen to be on. Instead we have a 3rd-level construct of a certain type of danger which we map across to any PAT^2 which looks similar, without however establishing a link between that PAT^2 and the PAT^1 of a real-world object. The process saves time, protects us from danger, and serves as a safety mechanism in every situation which can be associated with its core experience.

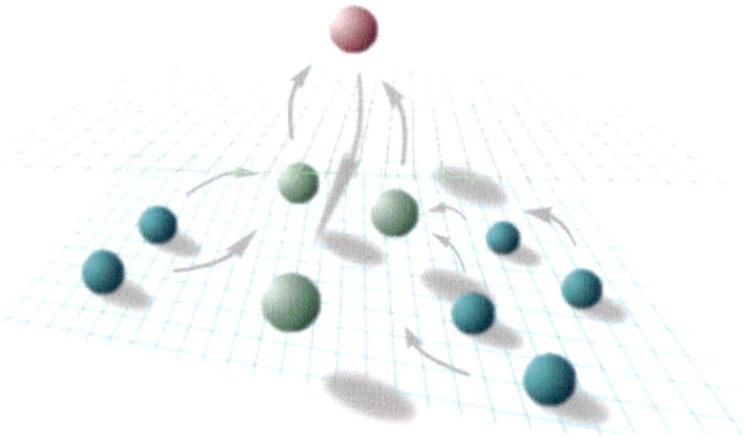

Fig. 5-1: Diagram showing the development of higher-level thought structures. PAT^1's (blue) form a thought pattern if found relevant and are abstracted into PAT^2s (green). These PAT^2s are further abstracted and are used to construct a PAT^3. The rule set may have been imported from PAT^1-PAT^2 relationships but now it exists beyond the logic boundary (here a green plane), and can be applied to another PAT^2 (the green sphere closest to the viewer) even though that abstract does not have a link with the real through its own PAT^1's (their environment being represented by the blue plane).

It can also go wrong. Suppose I wander across Europe and observe dozens of flocks of swans, each one of its members having white feathers. The act of stereotyping prompts me to declare, "There are swans in this world, and they are all white!". Then I find myself in Perth, Western Australia, and lo and behold, *their* swans are all black. Therefore, while stereotyping has a useful, indeed a vital purpose, it is not foolproof. It is important to realise that (1) no transposition to some other PAT^2 is made without there having been a similar PAT^2 at some previous stage (ie stereotyping can be wrong but there is always a reason for it), and (2) the more potentially adverse the effect of overruling a stereotype is the less justifiable it would be to switch it off (ie you can show me as many paper cars as you like, I would *never* cross the road blind-folded).

One formal way of stereotyping is the induction method. Because it is formal, that is being applied in a systematic manner, the potential pitfalls of link-less PAT^2s are overcome by expressly declaring them as such. For example, if a patient presents with headaches and fever the doctor may posit 'flu' as a possible disease, but is nevertheless aware of the need to confirm that diagnosis further; in other words, check whether a PAT^2 can in fact be linked to an existing PAT^1.

So far we have mostly dealt with 'higher-order' cognitive systems, that is belief structures which have been compiled by thinkers with a certain amount of diligence and insight. Because these individuals took care in assembling information, considered their own thoughts in relation to other ideas, and more or less produced a whole which was productive, their impact was to be felt down the ages. Due to such qualities they were able to push the boundaries of the contemporary, existing systems in terms of applicability and usefulness. The result is history—in a very literal sense. This does not mean that the rest of society operated on the same level, indeed the differences were sometimes stark indeed. For the rest of this chapter we will investigate two examples where daily life was played to a different tune, removed from the latest sophistication at the time. The first concerns the Cathars in what is today south-eastern France; for the second we will look at the grassroots level of medieval existence as it was played out on the roads and towns, the inns, and households.

To understand the impact and role the Cathars played we need some idea of the social and political environment of the time. For this purpose I am drawing on Birks' and Gilbert's *The Treasure of Montségur* [34], Madaule's *The Albigensian Crusade* [35], and general history.

In Europe the 12th and 13th centuries were a dynamic era. It could well be said that although most social, political, and cultural implements we are familiar with today had been gradually put together after the Dark Ages had taken their course, it was then that the necessary ground was made fertile for the fermentation to occur. And a fermentation it had been—from the emerging hegemony of the Roman Church to the assertion of secular feudal lords, from the agitations of sects and cults to the political streams engendered by trade and social movements, and all of that being washed over by the ravages of war and disease at regular intervals.

In France Philip Augustus was in the process of re-establishing the greatness of Charlemagne; in Rome Pope Innocent II had similar aspirations on behalf of western Christianity, carried onward by his followers; and across the lands many a count rivalled their kings in power and influence. Then there were the small town officials, drawing their status from the ignorance, superstition and constant fear of the inhabitants; while not far away from the outskirts of fields and ramparts roamed incestuous groups of outlaws who competed with bears and wolves for the fortuitous spoils from travellers, merchants, and mercenaries.

Amongst that cauldron of humanity there emerged the sect of the Cathars, also called Albigensians after the town of Albi. St Thomas Aquinas may have discussed the finer points of intellectual endeavour under the auspices of the Catholic Church, but at the same time the earlier influences of Eastern beliefs were still active within the Christian arena. The movements of people across Europe could not fail to collect those remnants, transport them into other areas of the continent where they found interested listeners and subsequently followers. The process was enhanced by the steadily increasing power of Catholic officialdom and the authority and luxury this made possible. On a psychological level we can observe the gradual advance of official church doctrine to the contemporary heights of an Aquinas, but only after it had fulfilled its more simplistic duties of providing a steadying and civilising influence over the European tribes after the breakdown of the Roman Empire.

Eight hundred years earlier St Augustine extolled the virtues of clean living, underwriting them with direct references to everyday examples. He himself had once been a follower of Manichaeism (see chapter 4) and saw the struggle towards God from a perspective of someone who was at home with the daily battles of coming to terms with one's senses. Similarly, the Cathars, who had adopted Middle Eastern beliefs (amongst them Manichaeism as well), defined human existence as a direct representation of a battle between good and evil, of a one-to-one relationship between one's self and a higher power. Unlike St Augustine, who laid the foundations for transposing such a struggle into the metaphysical realm to be fully articulated still later by Aquinas, the Cathars chose to remain on a level removed from such esoteric musings. Maybe they did not 'choose' at all, but were confined to direct symbolisations by the same mentality which compelled the ancients to represent their gods in terms of bodily, albeit invented, figures. As they moved from town to town, preaching their doctrine, it can be well imagined how different their impact must have been compared to the more distant atmosphere projected by the Roman clergy, socially as well as intellectually.

By the late 12th century practically the whole region of Languedoc had been on the side of the Cathars, who found support not only in the larger towns of Toulouse, Albi, and Béziers, but also in smaller centres such as Lavaur, Avignonnet, Castelnaudary, or Saissac. Naturally they attracted the attention of Rome, and in 1179 the third Lateran Council issued a directive to counter these heretics. The local feudal lords were on the whole sympathetic to the sect, if for no other reason than its hostility towards the material wealth of the Church and ministers, thereby providing less competition against their own appetite for possessions. Step by step the official counter-measures became harsher, the positions more entrenched on both sides. Then, in 1198, Pope Innocent III was elected to the throne; in terms of vision and perceived importance the spiritual equivalent of King Philip in France.

The lands around Toulouse were more accessible from Italy and Spain than from the rest of France. It proved to be yet another reason why the people of Languedoc saw themselves as different from the north. As Madaule comments, they possessed a certain elan, a cosmopolitan worldliness which set them apart, particularly compared to the northern barbarians who lacked the finesse which comes through an exposure to other cultures and through the wilful allocation of time dedicated to intellectual pursuits based on one's own. For the lower classes of that society then, the Cathars provided the contrasting religious flavour; the upper echelons nurtured their own eclectic civilisation which became the focus of outside attention. Together they made for an unhappy alliance that proved no protection against a possibly uncouth, yet organised secular power on one hand and an equally organised as well as intellectually superior church on the other. It is an interesting question to ask what would have happened had those characteristics been matched in reverse order. In other words, had the king of France been pitted against a similarly resolute and rough group of local rulers, and had the Catholic Church been forced to engage its enemy on an intellectual level similar to that of an Aquinas.

Alas, this will have to be conjecture. A possible explanation for the only gradual increase in antagonism towards the sect could have been its very simplicity. As said before, the Cathars derived much of their cosmic perspective from the likes of Manichaeism, but especially from the earliest days of Christianity. To them the ideal of a devout man and woman consisted in a life of poverty, the shunning of material wealth, and utter dedication to one's fellow humans. Their spiritual interpretation of life centred around a stark struggle between good and evil, and to them the Christian Christ had not been significant in terms of his death but acquired his importance through the deeds performed during his life, thereby redeeming humanity. As such their belief system was an altogether more simplistic one, and the subtle points of logic and semantic interpretation presented by Thomas Aquinas would have been quite meaningless to their own minds.

There was never any doubt as to their sincerity, a point not lost on their antagonists. That and their afore-mentioned spiritual earthiness made them attractive to the largely uneducated and emotion-driven masses, a feature alluded to several times in both of our sources. Therefore the threat to Catholicism would not have been seen as an immediate one. A quote from a Cathar bishop, Guilabert of Castres, should illustrate their general attitude towards life [36]:

> I have left father, mother and children. I have given up everything the Gospel commands me to renounce, gold and silver I no longer carry in my purse. I am satisfied with each day's food and am not anxious whether tomorrow I shall have the wherewithal to be clothed and fed. You behold in me the beatitudes which Jesus Christ preached and in which his Gospel consists. You see me poor, meek, peaceable, pure in heart, you see me in tears suffering hunger and thirst, persecution and the world's hatred for the sake of Righteousness.

The above may be apocryphal, so we are told, but the general tone seems authentic enough. Religions, especially in their earlier times, are not noted for their inherent tolerance; nevertheless, sentiments such as these do not readily instill fear—these are not the words of an outwardly belligerent force, spiritual or otherwise. Consequently the initial reaction by the pope consisted in public debates, sending clerics to the provinces to attract people back to the fold, even starting an entire new order to demonstrate the Church's own readiness for piety (the Dominicans).

And yet, for a mind to conjure a severity of such degree, even if towards oneself only, requires a certain strength of conviction which must contain its thought structures in particular isolation from each other. Not for them to ameliorate the concept of 'poverty' for example by taking recourse to a more abstract interpretation according to perceived context, circumstance, or historical relevance. 'Poverty' here means the abrogation of materialism *in toto*, to "hate this world and its works and the things that pertain to it" [37], as we read in the text accompanying the *Consolamentum*, their rite of initiation to priesthood. Further on, reference is made to the Gospels,

> For St. John says in his epistle, "Do not bestow your love on the world, and what the world has to offer; the lover of this world has no love of the Father in him. What does the world offer? Only the gratification of corrupt nature, the gratification of the eye, the empty pomp of living; these things take their being from the world, not from the Father. The world and its gratifications pass away; the man who does God's will outlives them, for ever."

Moreover, the novice declares,

> I promise I give myself to God and to the Gospel, never to lie or take an oath, never again to touch a woman, never to kill an animal or to eat meat, eggs, or anything made with milk, but to take only vegetable nourishment and fish, to do nothing without first saying the Lord's Prayer, never to travel, or spend the night anywhere, or even eat, without a companion; and if I fall into enemy hands and am separated from my brother, to abstain from all nourishment for three days at

> least; never to sleep unclothed and, finally, never to betray my faith no matter what form of death I am faced with.

We already examined a dualistic world view within the context of Manichaeism and pointed out the black-and-white attitudes such a cognitive system creates. The similarity with the Cathars is evident. That style is further enhanced in its severity by the uncompromising perceptions about correct behaviour, particularly as it concerns one's body. (A note on the eating of fish: it was believed that fish do not reproduce sexually and therefore constitute 'clean' food. Another example of anti-sexual thought structures meshing with the demands for one's daily life.) Although these were rather crass even by medieval standards, the underlying stream of thought patterns containing self-loathing and virtue allied with each other has been a recurring feature of Judaeo-Christian-Islamic philosophies throughout the ages and is still influential today, even in the West.

A belief system which is capable of processing third-level abstractions may be dogmatic, severe, inward-looking, etc. Yet at the same time its member minds possess the potential functionality which allows the transposition of concepts from one PAT^1/PAT^2 trunk across to another via its higher-level abstractions. This enables them to think laterally, to go beyond the erstwhile established bounds of perception, analysis and conclusion; all being ameliorated of course by the functional abstracts (ie rules) accompanying those patterns. In the absence of such dexterity any given situation will be judged according to the closest interpretation found and, if deemed impossible to judge, declared unacceptable. A negative outcome therefore is ensured by either an antagonistic rule or no available rule, or both; a positive outcome can only happen if some rule would be an accepting one. It should be obvious that in that kind of scenario the chances for productive resolutions are minimal, while the possibilities for altercation and strife are numerous.

Consider for example the issue of prostitution. A devout Cathar is compelled to declare *all* sexual behaviour as being deviant from the path of righteousness and so any debate about sex workers in particular would be irrelevant. Consequently, neither would be a discussion about hygiene in bath houses or inns. Only when someone is able to admit sexual intercourse as a more or less valid part of human existence would the above questions considered to be acceptable. It is a phenomenon which can be observed over and over again; the more sexually repressive a society is, the more prolific are the destructive side effects associated with sex (disease, unwanted pregnancies, psychological trauma, and so on). *Not*, it needs to be emphasised, because those incidents do not occur, but because nobody dares to address them. In Part II we will see how, in a form of reverse psychology, thought patterns develop which indirectly favour repression so that all those things can occur underground which are officially spurned.

It has been argued that for the Cathars, doctrine came secondary [38]. Such an interpretation is understandable when one differentiates between a set of recorded rules and life as it is lived. From our perspective however the distinction leads to possible errors. A doctrine is the manifestation of certain thought structures and patterns, and the more comprehensive the doctrine the more synchronous the equivalent mentality of a follower will be with that of the originator. When examining a certain behaviour therefore it is the cognitive structure which is significant, not so much its representation (ie through written text or demonstrated actions). It would be wrong to attach a different degree of efficacy to the results of a doctrine as compared to a custom. On the other hand, when it comes to the *efficiency* of a belief system, then one that is underpinned by written records will be in a stronger position than another which is not. This applies especially to those systems which feature higher-level abstractions, because oral transmission and deeds alone (that is PAT^1s) are not capable of capturing the complexity of a $PAT^1 \rightarrow PAT^2 \rightarrow PAT^3 \rightarrow$ etc structure for later usage—it is functionally impossible. (Scholarship is lost on oral history, and societies that exist through oral history do not employ higher-level abstractions.) It should also be remembered that by the 12th and 13th century the intellectual framework of Catholicism had far surpassed the perspective of the early Christians, as the writings of Augustine and Aquinas clearly show; the doctrine espoused by the Cathars was not the same as that held by Rome.

Another example of the confusion and ambiguity which can be introduced by the interactions between lower- and higher-level abstraction systems is the notion of the Grail. It was always understood that the Grail represented a symbol, but a symbol of what? On the Cathar's side one of their priestesses, Esclarmonde, whose castle Montségur had been the sect's last stronghold, was turned into a legendary guardian of the Grail [39]. The general ambience of austere severity projected by the Cathars would also colour the underlying idea of the Grail, no matter how undefined the object itself may be. Yet from the same time, at the beginning of the 13th century, we find already a somewhat different interpretation of the Grail in *Parzival*, where it is a sphere whose keepers were allowed to marry and have children, travel across the countryside, and generally involved themselves with the rest of the world [40]. In still another version, this time from the 20th century, the Grail is a symbolisation of Christ's blood [41] (maybe a sign of a more materialistic age trying its hand). The abstract 'Grail' is quite clearly the representational embodiment of whatever virtues were held appropriate by the respective belief structures, but because the development of that abstract had always been a one-way street (ie moving from PAT^1s upward but never from a PAT^2 or higher downwards), its actual substance would have been modified according to the whim of the times. So it seems that the Holy Grail is a third-level abstract looking for a home!

The phrase quoted above, "never to betray my faith no matter what form of death I am faced with", points to the uncompromising attitude displayed by people such as the Cathars (it is not the only example of course). Given the propensity of humans to inflict pain and punishment, coupled with the contemporary ideas about what punishment ought to consist of, the eventual fate of heretics would have been assured. Since none of the more genteel measures tried by the Church had much effect, in the end the mechanism of the Inquisition was brought to bear on the situation. And a mechanism it was, as it had been much more than merely a process to burn undesirables at the stake. The bureaucracy surrounding the Inquisition had been quite extensive, from the judicial inquiry to the commercial arrangements dealing with confiscated property, to the actual execution of the final verdict. It is true that the Church itself never burned anyone; this task used to be handed over to a secular court. However, the preliminaries were such that once the initial charge had been confirmed the courts had no choice in the matter. The more far-reaching a system, the greater its momentum; in other words, the lesser the chance that any detail along the way would change the course of a procedure which was designed to work its way towards the goal—in this case the elimination of a target group.

One ramification of the above is relatively trivial, but I think worth mentioning. As stated before, the machinery of the Inquisition had been massive, and it permeated every layer of society, from the pope downwards to the bishops, the local priests, to the secular parts of society such as town mayors, inn keepers, even prostitutes and beggars. During periods of heightened intensity, such as at the time of the Cathars in south-eastern France, all of those demographic groups played their part, supported by the added measure of delegating a certain portion of the victim's assets to the accusers if the charge turned out to be proven. For a modern-day cynic that scenario needs no further explanation: individuals who had an axe to grind were more than willing to denounce their enemies, especially if it meant turning an ever-present suspicion away from themselves. There was a momentum therefore sustained by certain aspects of human psychology. In terms of being an efficient eliminator of true heretics the institution fell rather short of such an intent. As a tool of control through fear and innuendo, exemplifying the pervasive power of the Church, it did serve its purpose.

A more deeper-reaching phenomenon is the mutual dependency of action and response. A system may be static in terms of its stated references such as a religious doctrine, however interpreted. The thought processes inside the mind of an individual however are much more fluid, since they not only encompass the official guidelines but also the dynamics belonging to one's self. The object abstracts pertaining to one's view of the self, together with the functional abstracts relating to the type of behaviour which therefore should be performed, will form associative patterns that define a thought structure of a relevant context. If a Cathar then, having ceremoniously proclaimed his or her affinity to the faith, encountered a challenge to the thus formatted status, some kind of response must be expected. In this case it involved the notions of material inferiority against the spiritual, the ultimate goal of merging one's soul with the divine, and the public demonstration of one's steadfastness in the face of adversity. The evident result had been the often observed willingness of the Cathars to go to their death which involved being burned alive. From their point of view it covered all three elements—the body didn't matter anyway, their righteousness would bring them to God, and they certainly demonstrated their unbending character. It is interesting to note the explicit absence of such a virtual death wish among the principles of Manichaeism, although the early Christians were certainly familiar with martyrdom. By the same token, the quote from the *Consolamentum* above came from writings taken down during the inquisitorial examination of the sect's followers, in other words from a time when the burnings had been well under way. To what extent they were originally required to contract their manner of death at the moment of ordination is therefore hard to say; nevertheless, the cognitive dynamics still hold.

On the other hand, it also needs to be remembered that in this confrontation the Roman Church had more to loose overall than the Cathars could ever hope to gain. For the sect its main support base could only be those sectors in any society which were satisfied with a direct, ie PAT^2, imagery; not a promising future wherever an emerging educated middle class transformed the population. Yet for Catholicism to revert back to its more primitive form would have been a degenerative process, resulting in its demise. For the mainstream religion there was no way but forward, and armed with the intellectual strength of its theologians and the accompanying secular influence it could assert over political rulers, such powers were made use of. Although no-one in its ranks had been consciously aware of the effectiveness of a PAT^3 system (as we have seen, even an Aquinas did not recognise its full inherent potential), without it the Jesuits with their schools, the successful conversions of foreign cultures, even a complex mass by Beethoven, would have been questionable at best, but more likely impossible.

Going to one's death willingly creates the martyr. While the immediate observation of that act may evoke feelings of empathy, respect, even reverence, a far greater effect is achieved at a distance; that is to say, within the minds of people who did not have the opportunity to experience the death directly but are the recipients of the story. The phenomenon is explainable in terms of the $PAT^1 \rightarrow PAT^2 \rightarrow PAT^3$ development, where the link downwards is now directed to thought structures which are perceived to have an affinity with whatever the higher level abstracts seem to contain. The degree of affinity varies, indeed it may relate to notions which are quite removed from the original context through time, custom, and understanding. In the case of the Cathars any subsequent observers (ie those individuals who learned about their fate through the reading of texts later on) would ascertain the idea of 'martyr', together with all the surrounding elements of 'suffering', 'sacrifice', 'pain', etc, and delineate from that point downwards to PAT^2 or PAT^1 structures which offer a more personal, contemporary representation of the higher-level elements. By doing so the notion of 'martyr' is made more relatable, more emotionally accessible, but it also veils the actual reality of the initial situation. So, a modern reader may 'empathise' with a heretic who has been burned to death by recollecting his or her personal experience of pain and terror but, I would aver, does really *not* grasp the full experience of waiting two or three weeks in a cell, walking to the square through a throng of staring, weeping, shouting, spitting, cursing spectators, inhaling the sweat of the executioner as one's limbs are tied to the stake, and then having your flesh slowly consumed by the flames.

A more recent example is that of a student who has achieved high grades in a course. His acquaintances may applaud his success and he may even enter the family hall of fame—but unless those outsiders had been in the same course, working under the same circumstances, they would not really appreciate the result. However, because the others' feeling of goodwill is illuminated by their own subjective experiences and therefore the symbol 'high grade' in *their* minds is not retracing the same path than the student's, it has an emotionally higher significance than had the same symbol been linked to more tangible, objective representations. If we regard the up- and then downward progression of abstracts as a loop, we can say that in the above we are faced with an *incongruent PAT loop*, an *i-PAT-l* (ai-pat-ell). It is an extension of an incongruent functionality, which we first discussed in chapter 3. With the latter the mismatch occurs between two neighbouring abstractions, usually a PAT^1 and a PAT^2. With the loop a mismatch may or may not occur between any of the steps leading to the highest abstract, and so the cognitive significance of the highest step may not match that of the ascendant, but certainly never matches that of the lower descendant levels. In other words, while a relationship as such exists, its elements cannot be compared by their own natures; Aquinas would have termed that relationship a "proportionality".

I-PAT-ls are very common. Not only do they occur in the context described in the previous paragraph, every ceremony and ritual is essentially an i-PAT-l. When the British queen touches the recipient of a knighthood with a sword, then that gesture links back to times when swords were used as effective instruments. Today the act is symbolic only, that is it has a PAT^2 which merely holds the idea of a knight's status and does not even feature a PAT^1 at the other end because nothing is actually done with the sword. Or consider the custom of circumcision, male or female. The act itself is rooted in abstracts related to such things as 'pain', 'trial', and 'tribal property' (PAT^1s). The surrounding ceremony represents a PAT^2, embodying the above in symbolistic form. The resultant human being is then considered to be a fully functioning member of the group, where 'fully functioning' is taken to encompass all the capabilities which are expected of an adult (a set of PAT^1's again). To a rational mind, that is one which employs *congruent* PAT-ls (*c-PAT-ls*), the mismatch between something like general maturity and disfigured genitals is pretty obvious. An i-PAT-l mind on the other hand will accept the relationship as good and proper. The earlier an incongruent loop has been established in somebody's mind, the more pronounced the emotional contingent accompanying the links will be (in Part II we will see why), and so the more difficult if not traumatic the exercise of articulating the fact of mismatch will be as well. Therefore i-PAT-ls are hardly ever destroyed, but countless are created throughout one's life. In that sense alone we can say that humans in general do not interact with reality at all; it is more a case of wandering amongst its shadows. For that reason it could well be that the concept of 'illusion' in Buddhism for instance, describing it as a metaphysical phenomenon, has actually been purposefully or unwittingly misunderstood. It is not that reality does not exist—we just can't see it, and we are loath to loose face. Note however that a word which changes its meaning over time does *not* represent an i-PAT-l. While there is a mismatch in the vertical sense—ie over the years—laterally speaking, that is at any given moment, its meaning is congruent. Which is the very reason why so many words lend themselves to historical research: their meanings can get lost literally over time. (A good source of such examples within the Australian context is *Ozwords* from Oxford University Press [42].)

The change in the meaning of words points to one of two possible problems an i-PAT-l can pose for the observer. The other lies in the *intentionality* of the resultant thought structure. Consider for instance the provision of the *Consolamentum* as an assurance for a "good death" [43]. Having partaken of the ritual the devout Cathar can now see him- or herself as a good Christian and ready to approach the divine, particularly at the point of death. For that reason most of these ceremonies were conducted on one's deathbed, often after a life of earthly indulgences. This may seem to be an act of hypocrisy, but it may not necessarily be so. Remember that an i-PAT-l of this kind is a fully fledged thought structure, although its inherent flaws are only apparent to an outside analyst. For the mind which possesses such a notion however it is as valid as any other, since there is no discrepancy within its semantic domain which could alert its bearer to the mismatch. Since the ceremony is a PAT^2, delineated from some PAT^1s which represent positive qualities and themselves being a symbolisation that stands for the idea of accomplished grace, as soon as its existence is evoked through the ritual the symbol's perceived counterpart is accepted also (the incongruent PAT^1). Only someone capable of higher abstractions would be able to theorise about the overall validity of such an act in terms of past and future behaviour, because those incidents and their meanings need to be processed through higher abstractions still. Therefore, while a more intelligent mind may choose to criticise this kind of behaviour on the basis of perfidy, the argument will be lost on people who will or cannot avail themselves of the higher-level abstractions necessary for its understanding.

There are a number of other differences between the Cathars and the Catholic Church which have their origin in the former's inability to form PAT^{2+} abstractions [44]. They rejected the idea of sainthood; for them a 'saint' is merely someone who has been liberated from earthly bonds, whereas in Catholicism a saint represents an intermediate symbolisation between the abstracts of an individual's 'good life' and the divine proper. They also rejected the Mass, since for them the body of Christ was not considered physical while at the same time bread was regarded as pure matter only without any metaphysical aspects; the Church on the other hand *can* accept the body of Christ as physical because it manages its step-wise abstraction towards the higher while doing the same in reverse in the case of bread. Similarly they had no use for Holy Orders (these being a symbolisation) but accepted one's goodness of life as a direct reference of one's standing. On that basis the authors quite rightly reject the idea that the Cathars had been forerunners of Protestantism, although they do so from a different perspective than ours. The Cathars were not part of the emerging complexity of the mainstream church; Protestantism was a divergent development further down the track. One last example is the difference in meaning of 'Holy Spirit'; to a Cathar it is a direct representation of the divine in humans, whereas for a Catholic it is a higher symbol forming the linkage between God-supreme and the various levels leading to the material [45].

It has been pointed out that in the south of France the initial converts to Catharism had been the clerics rather than the lay masses, ie the educated rather than the uneducated [46]. This seems to contradict what we said before about the early Christian church's simplicity as a focus of attraction. Two things need to be pointed out. (1) although Thomas Aquinas wrote extensively about higher-level abstractions throughout his epistles, at that time he represented the avant-garde in Church thinking. It does not follow by any means that provincial church officials would have consummated the full flavour of his intellectualism. (2) furthermore, it had been those in the cities (clerics included) who had been relatively higher educated and so had the opportunity to observe the general life style of those better off—something the peasant on the land could not do, unless and until receiving such information second-hand via those self-same clerics doing their rounds.

The last substantial group of Cathars was defeated when the fortress Montségur fell in 1244 after a siege that lasted over a year. Two hundred and five members of the sect were burned to death, after having been given the choice of recanting their faith. Since they didn't, one *could* say they died willingly. Given their intense belief system, together with the strict routines of constant test and rigour, such an outcome is virtually to be expected. From the viewpoint of thought structures we can identify their positions respective to each other and therefore the levels of priority within the overall pattern. And yet, despite the significance of an idea as the only representation of reality anyone really has, despite the bias the concept of 'death' for instance occupies in relation to 'life', an objective observer can only wonder at the final significance of those structures. In this context that is to say, to what extent does the glory of the symbol override the terror of the fire; or, at what point would the vision of heaven be obscured by the agony of burning flesh. There must surely come a moment—one of the last in one's life—when the immediacy of personal destruction is the only representation left amongst the brain's neurons; or is there something else?

Consider the life style of someone who has received the *Consolamentum* and is henceforth required to follow its tenets to the last detail. The existence is summarised as follows [47],

> They were .. remarkable for their pallor and emaciation, the consequence of the austerity of their diet. Their lives seem to have been divided between periods of itinerant mission and periods of retreat in their convents. Added to the perpetual abstinence already mentioned, they also gave themselves over to long fasts on bread and water, and they had to say the "Our Father" almost incessantly. An original feature of their organisation was that as soon as a minister was ordained he was assigned a companion (male or female) from whom he would never be parted until one or other died.

As indicated before, we will examine the development of thought structures from the bottom up in Part II and visit some of the enhancements and/or impediments of that process. One needs to distinguish however between the influence an already existing environment has on one side and the effects of later deprivation on the other. We can only surmise about childhood conditions in the Middle Ages, but even so they would be cruel parents in any age who treated their infant in the manner quoted above. It is more likely that an individual had focused on certain aspects of one's condition and progressed from there to the eventual total rejection of its perceived source and in consequence turned towards what was held to be the sublime. After all, the physical conditions are unimaginable for an average, modern, first-world citizen. In fact, without being cynical one could say that many such conditions couldn't possible get any worse, since the next step would have meant the demise of the person afflicted. Consider then the constant inconvenience, irritation, and pain of a disease, of parasites infesting your body; of frustration, anguish, and often fury at one's self and the acts of others; the resignation but also fearful sadness at what must seem to be the uselessness of it all. Contrast this with the however simplified abstracts of peace, of comfort undisturbed by other humans and vermin, and believing such an ambience to be the realm of God—to be reached only by having surpassed matter, the flesh, this world. This is not to say that people could not have enjoyed themselves (many certainly did), but to someone with a sufficient enough inclination towards misery and pathos there would have been a sizable contingent of i-PAT-ls, all of which would persuade their holder to abrogate any attempt at amelioration and fully concentrate upon the afterlife.

As we have seen, i-PAT-ls have a significance in their own right, but especially so when there is no conflicting rule set. The ability to cross-refer from one thought structure or pattern to another is not only a function of such another's presence, but is also a function of a brain which is physically capable of these connections. The sparse diet and extreme life style of the Cathars rendered them dysfunctional on a physical as well as psychological level. Given their general environs the former would probably not be readily identified as significant, and as for the latter their brains would have been too addled to be adopting wider perspectives. Sadly, it is one of the profound shortcomings of the human brain to be at once able to have lofty ideals but not necessarily recognise their origins. Or, to put it into our context, to have the ability to form higher abstractions and communicate them but at the same time to import such abstractions into one's own mind and link them indiscriminately to a lower level according to one's whim. Generally speaking we can say that an ideology (spiritual or secular) is the by-product of a mind which is capable of forming higher-level abstractions but is not sufficiently schooled in their understanding and therefore usage.

Before we leave the Middle Ages we should recreate a simile of the mindset of the 'common people' as it had been at that time. Since we do not have the comprehensiveness of an entire philosophy to guide us, we can only assemble some sort of picture through circumscriptive evidence gleaned from contemporary sources, directly or indirectly. While such a mosaic will not be complete it will nevertheless contain a sufficient number of items to provide a general atmosphere, described through the semantics of thought structures. I have made use of fourteen references, each offering a window in its own right. In no particular order they are,

Carnival, an account of a political uprising in France [48];

Uilenspiegel und Lamme Goedzak, a historical novel set in 16th century Flanders, using a persona from the 14th century [49];

The Devils of Loudun, an investigation into the burning of a heretic priest [50];

Saint Lydwine of Schiedam, a reconstruction of the life of a woman later declared a saint [51];

The Autobiography of Benvenuto Cellini, an edited version of the life of one of the greatest sculptors and goldsmiths [52];

The Spanish Inquisition, a re-examination of its origins, purposes, and procedures [53];

The Book of the Sacred Magic of Abramelin the Mage, a translation of a medieval text dealing with magic [54];

M, a reconstruction of the painter Caravaggio's life [55];

Plague, Pox & Pestilence, a description of some of the major diseases throughout the ages [56];

Encyclopedia of Plague and Pestilence, a chronological list of diseases from antiquity to the present around the globe [57];

Psychedelics Encyclopedia, a listing of psychedelic plants from around the world [58];

Torture Through The Ages, a description of the types of torture used throughout history [59];

Complete Poems of François Villon, the collected works of a wandering poet living on the edge of society [60];

last but not least: the insights gained from observing the people on the continents of South America, Europe, Asia and Australia, who in terms of their level of education, belief systems, cultural constraints, and exterior shortcomings, most likely resemble some of the demographics existent during the 13th to the 17th centuries.

I have desisted from making constant references to sections of these books, since their contents supply, however indirectly, the examples of mental phenomena which we illustrate in terms of their principles and not as links in some chain of events. By doing so I avoid peppering the text with seemingly random footnotes coming from all directions. The overall viewpoint is from the present, but the focus is on representative samples within their cognitive domains. I will however reference the underlying mental functionalities in terms of our perspective. Here they are, enumerated accordingly:

{1} the PAT process, moving from any given level to the next higher one;
{2} an object abstract;
{3} a functional abstract;
{4} a default abstract;
{5} an incongruent functionality;
{6} an incongruent PAT loop;
{7} a congruent PAT loop;
{8} the logic boundary;
{9} the qualities of a perceived object - Q_{po};
{10} the qualities of an abstract object - Q_{ao};
{11} a projection.

It may be a useful exercise for the reader to try to find the reasons *why* such and such reference has been attached to a sample.

Let us enter our time warp. We emerge sometime around the 15th/16th century, just outside a medium-sized town of roughly 10,000 inhabitants somewhere in western Europe...

The early sun has begun to touch the roofs of the city, encircled by its wall. A mottled pattern of strong browns and greens emerges amidst the lucent columns of early morning smoke and gradually grows downwards. The hotchpotch of timbered masonry comes into relief, and then disappears behind the stone ramparts. Faint noises of people wakening to the new day spill out. A dirt road winds towards its gate, on it several donkey carts carry farm produce to the market. The whining *creak creak* of coarse wheels hangs in the still air, accompanied by the occasional *thud* of a lazy whip for the animals. A tinny church bell calls across the roofs.

One of the rigs passes what appears to be an oversized bird cage, fastened to the stout branch of a tree. Made of broad, cast-iron bands, it is the size of a child. Every now and then its contents make it jerk a little, otherwise it is quiet. Now a boy and a girl jump from the cart and run towards it. They strain their necks and squint upwards but the half-naked body of the man inside the cage is still alive. His head is jammed between the irons and his knee, the outer eye open and staring at the children, the other a running sore smeared against the thigh. The boy lifts a rock, lets it fly at the stretched buttocks to make the cage move again. They both giggle—for them it is a game {10}. Had the man been dead and rotting away they would have searched for a long stick to prise off a hand to take with them. As it is they have to make do with the lumps of excrement on the ground, and the girl puts some into the folds of her rag skirt.

The man walking with the cart does not even bother to look {3}. He has heard the stories of a beggar been caught thieving; the thing in the gibbet must be him now. While he watches his stall at the market his children will run to the apothecary and get a coin or two for their find. It will be turned into powder for medicine, or a concoction for spells {6}. He glances ahead to see whether the gate has opened yet and notices some people going through. Good; no need to wait then. The guards will be fussy today: rumours about disbanded mercenaries have reached the town, and the citizens are worried about infiltrators. He doesn't care though, they know him well enough.

Now one soldier talks to him and pats the donkey and the cart, playfully cuffs the boy. The man has entered the city. Immediately there is the smell—fouling leftovers, urine, bluish-green bones, mangy dogs, a pig, roasting meat. Nothing good comes from a city, and the man rubs his talisman against bad luck and disease which he knows are waiting everywhere amongst the lanes {3}. He walks past the inn, its door just opened, the barber shop, the still shut boards of the cloth seller, and around a corner he can hear the scraping sounds of a furnace being brought to life.

A few streets away, on the third floor of a squat town house, the mayor sits behind the table in a dark paneled room. His fingers, bent from flea scratching, move across a sheaf of papers containing the tax records of the citizenry. His town sports two doctors of law, both paying taxes, and the physician, also a tax payer—at least he used to be. Then there are a handful of tradesmen, a score of shopkeepers and merchants; in the next lower category, cloth makers, drapers, corders, leather workers, food preparers, butchers, bakers, tailors, metal craftsmen, a couple of knife grinders, candle makers, and the town's soldiery. After that, the rest—labourers, itinerants, handymen; their moneys not worth the paper they are written on. And finally the beggars, the barbers, bloodletters, jugglers, prostitutes, pickpockets, who had swarmed into town like bats to a belfry, attracted by the recent fairs—and the executions. No taxes from them, in fact it seems half the population has been queuing for broadbean-soup handed out by his council.

Ah yes, the executions. Two were burnt at the stake last week, three before that, spectacles for the witch-hungry mob {11}. Rumour is rife about who is to be next, a man sitting in his cell at that very moment. What to do with him—the physician, doctor of medicine, said to have worked a few wonders and correspondingly respected or feared, depending on whom you talked to. Used to be a Jew, but some people claim to have seen light through the shutters every night except the Sabbath—was it a case of relapse? His estate was considerable, impounded at present by the Inquisition, and if convicted would release a not inconsiderable amount for the town's coffers. Particularly important since one of the merchants, the richest, had only recently bought himself the right of freedom from taxes; the shortfall has to be made up somehow. He has two sons, who revel in their new rank of nobility, trampling through the peasants' corn fields, sporting with their young women, starting fights in the inns and late-night streets. But never mind that; what matters are the purity of the Holy Church and taxes {3, 6, 8}. In a week the bishop is due, an honourable man, impeccable when it comes to the high principles of the Church and equally unforgiving towards the heretics who bring hell to earth {8}; he must be appeased.

Yet one has to be careful {1}. Despite his questionable religious background the physician has some followers, even amongst the clerics. It is said he practices magic, which would account for the miracles {7}. Burn him, and certain merchants would get nervous. Let him go, and many people have one more reason to rail against the rich and the clerics—especially the clerics. Only the other day someone came to him with reports about a new song they blared at the inn, ridiculing the servants of God. Those verses were obscene, blasphemous of the highest order {6, 8}. They talked of boiling the men of the frock in Jews' piss, venomous drugs, in a brew of toothless cats, snakes' blood, boils and chancres, filthy nappies! The sores around his neck begin to itch {4}.

On the street below the gaoler crosses over and enters the walls of the watch tower. A narrow space just behind the entrance allows access to the steps which wind downwards a turn into the stony earth. The man carries a taper, its blackish flame the only light, and a jug of water. Below, putrid air fills the passage, barely two shoulders wide, with several iron-rimmed doors set in low openings, each the entrance to a cell. He stops at the first one, frees the lock with his key. Quickly he pushes the jug through the opening and rams the door shut, eager to get away.

He need not have bothered. At the end of a chain the prisoner sits on the slimy floor, his back against the cold rock. The eyes are closed, his body has sunk into the coat he has been allowed to wear, his own. Nothing is permitted to cross the threshold to his mind, not the vermin crawling through the garment, not the rats pushing past his limbs, not even the brief intrusion from the world beyond. The physician's consciousness, an entity which has completely secluded itself from the world at large, the senses and the thoughts of the here and now, this small light flickers in a realm constructed solely by the man himself {1, 10}. What seems like a lifetime of meditation, of cloistered secrecy, shows him the path into that land. Practice has taught him how to still the conscious and so remove the necessity to harness words onto thoughts. Intense concentration has provided the subconscious with the particles of daily life and its emotions, of gestures linked to ideas. Their repeatedly intoned patterns {1} gradually produced a mental machinery {6, 7} that tends to form its pictures mostly in secret, does not relate to distracting senses {9} until the last moment when the vision comes to the fore—a product of pure will {10, 11}. Only, in his cell he does not have the indulgence of billowing incense to offer its shapes to his hungry eye, nor do the essences of herbs and mushrooms bathe his brain. And so the carefully nurtured structures of spiritual symbolism {6, 7} cannot align with the intrusions from the outside and instead produce bewilderment, confusion, and fear {5}. His mental equilibrium will be further tested later that morning when they proceed to crush the bones in his legs {9, 10}.

As the physician in his prisoner's cave struggles to reconcile reality with the abstracts in his arsenal {6}, so does the mayor play the pieces of information against each other, trying to construe a satisfactory outcome {6, 7}. Both men are locked into their representative worlds, and have to make do with those subconscious confluences that they are able to articulate.

In the inn, on the main road leading away from the gate, the first of the day's customers have now settled among the tables. They are a mix; peasants coming for the market in one corner and those few residents who need a drink or two before starting with their work in another. They look at each other warily {11}. Between draughts of beer and wine—only the lowest must settle for water—the quiet talk is about the day ahead, but the unspoken thoughts are with the witches so recently burned. Everyone is aware of the crime of heresy, some may have even known the victims, but it is dangerous to say too much. It is equally dangerous to say too little. The news about the bishop's visit has travelled quickly, and so has the rumour of informers working for the Inquisition. None of the customers understands about the high schemes of officials, of the nobility of Church and State, but they have learned early in life about the power of a wrong word spoken in haste or pride. As every one of them thinks about his own affairs, he gives vent to the kind of frustration common to them all. A child sick, poultry taken by the fox, the high rents charged by the town, the even higher taxes {2, 7}. Without the shared focus of a witch under investigation, their personal problems cannot be offloaded to a greater cause such as sorcery, or God's grace, or a high official, for fear it might give rise to gossip about one's own relationship with such matters {6}. A week ago bad fortune could have been safely linked to the accused heretic, but now with her out of the way what good excuse was left? To speak about the sanctity of God and the angels, to denounce the evils of magic, and to say these things at the right moment when talking about one's chores while at the same time watching the faces of others, sets up a network of values peculiar to an atmosphere stacked with fears and assumptions {3}. After all, if a thing has to be mentioned then to defend oneself is somehow making the thing one's own; but to denounce somebody else guides it away, while at the same time showing one's righteousness to all. Moreover, when subtlety is lost for want of higher abstracts, cause and effect make close neighbours {1 (limited), 5, 6}. A child with fever and a curse spoken the day before become quickly one; so would women's gossip and its victim; so would a painful chancre and a girl who didn't return one's lust.

With such a mentality at work, even trivial events become explosive. If the wine is too sour, it is bad. If the wine is bad, so is the inn keeper. If the inn keeper is bad, so is his inn. Therefore anger does not bother with lengthy explanations or far causes. A short fuse causes the tankards to fly from one moment to the next; an angry word brings out the knifes in a flash; a problem, real or imagined, needs a culprit straight away. Naturally, since none of the responses solves anything except satisfying the projected emotion, the stage is set for endless repetitions, engraving into a limited and confused mind a bizarre tangle of visions. With time, the visions turn into experience, the experience into what is called wisdom; it all makes for the foundations of subcultures.

A couple of prostitutes drop in from the adjoining bath house. It is too early for serious business and they expect none. They are in their thirties, and so find it easier to work during the slow hours when there is less competition. At night it is a different matter. The area around the inn and the small lane to the bath house are full with the crowd. The sweaty men with thirst in their veins and money in their pockets go through the rituals of satisfaction, lured by the smells of food and drink and the bodies of women and girls, some still children if coming from the fields with their elders and wanting to earn some money too. The demonisation of the sex-age complex had not yet begun. Within families, whether high or low, people perform their daily activities openly from eating to defecating, from intercourse to giving birth. If a girl came from the lower ranks she would marry or become a maid or end up amongst the inns; if from a higher one the pressure to marry would be greater if for no other reason than the expectation of a dowry from the suitor and a worthwhile connection from the girl's parents. In either case, the odd unhappy experience with early sex would prompt them to seek shelter from further harassment behind the walls of a cloister {4, 1}. In most cases a choice for life but a convenient one for all concerned, not least of all for the Church which gained a dutiful member who now was unbound by worldly vagaries and imbued with an almost instinctual disdain for fleshly pleasures. This, too, starts a subculture.

The two children of the peasant we watched earlier have a different business in mind. They enter the apothecary so they can sell the half-dried bowel movement of a dying thief. Inside the shop there is mystery. Hidden things sealed in jars and pots; strange shapes hanging from dusty beams; secret instruments standing in shadows. And the scent is the air itself; an odour which comes from wild liquids, herbs, salves, powders and the excretions of humanity, mixed together until they become one. Some customers are as furtive as the wares. They only dare enter when the sun has gone, wearing a cowl for added protection. A rich lady, after an elixir of love maybe or ground diamonds for a murder; a girl needing a tincture of belladonna or a potion of fly agaric; a would-be magician experimenting with blood and henbane with which to attract demons {6, 3}. All this: the vapours, oils, fumes, pieces of hair, skin and mucus; all this seeps into the timbers of the shop over decades and centuries; making them dark, rich, full of character.

The children receive their coin and are chased away. They would like to dash into all the alleys and explore, but they are also afraid of being caught by bandits. Their father has warned them often enough, but until the boy suddenly felt the hand out of an unexpected hollow grab his arm they didn't believe him. That recent experience is still fresh in their minds {3, 7}. Now they are nervous whenever a fork in the street offers a choice {3, 6}. With relief they jump into the open and make for the familiar stall.

The noise out here is different. Jugglers, fortune tellers, masked men on stilts; housewives, servants, onlookers; sellers of cloth, leather and candles; of knifes, wooden things and metal; of fruits, greens, spices, chicken, song birds; of paper saints, holy virgin pictures, talismans; of sweet cakes, marzipans, and honey. On one corner the crowd shouts at an urchin who got put on a tiny seat above a barrel, filled to the brim with sewage. You have to hit the board above with a rock and then the seat drops and in goes the waif. A splash is followed by loud guffaws, but if the rocks don't hit the board they land on the boy and everyone laughs at the thrower for loosing his money {9, 10}. The boy's skin is crusty, and it never dries.

Not far away the cobblestones show a circular mark of soot. It is where last week's stake used to stand, where they burned the witch. A few blackened stones with traces of something lie around, and next to them lean a number of poles with buntings tied to their tips. They are from the Church too, and they remind the people that Sunday's excommunication of the town rats still stands. The priest banned them all, down to the last whelp, and told them never to come back on the pain of God's wrath {6}. Somehow, that spot remains aloof {11}.

Throughout the day the peasant with the two children will stand next to his bags and shout himself hoarse. One hand on his moneybag and the other to talk, one eye on his wares and the other on the crowd. He will watch carefully for the face which shows a sign of interest, he will also mark those who do not glance at the produce but at him, or his children, and just to be sure he will push their heads under the table, away from view. Every now and then he exchanges a word or two with the neighbours. At intervals he reaches under the matting for a large bottle of wine, takes a gulp himself, lets a smaller amount into the children's mouths {4}, and passes the bottle to his left and right, making sure everyone sees it. It's an insurance of goodwill, a guard against malicious gossip, a deposit for the next visit {1}.

When night falls the streets become darker and darker as one bright window after another has its shutters drawn. The noises die out too, except for the occasional rustling of rats or the quick yelp of a dog. Visitors have left the town and the residents returned to their homes, securing themselves behind bolts and beams. Only soldiers free from the restraints of the curfew linger about—they and the outcasts to whom the night means freedom to move through their mournful lives {3}. One of these figures stirs in a cavernous niche. His useless eyes fixed uncaringly, he moves one hand through the wet muck feeling for scraps, sniffing like an animal. He touches something hairy and small. Instinctively his fingers close around the pelt like a trap, but there is no need—the rat is dead already. Later he will dig into its flesh and eat it, any other thought besides hunger long gone {9, 10}. Later still his body will burst into fever and grow black boils, buboes. The plague has arrived—again.

6. Into the light

René Descartes - Benedict de Spinoza

The Middle Ages were like a ponderous monster of an organism. Its mentality grew out of the rawness of European tribes who marched across the ruins of the Roman Empire. Then, before it could settle into a cohesive body politic it was lanced by the protagonists of a near eastern cult. The paucity of its understanding, the flatness of its thought, was no match for the kind of burning conviction that flourishes on the fringes of sanity. Although primitive as well, early Christianity possessed two psychological weapons which have proven themselves throughout all times—the appeal of demonstrated self-denial, and the chains of personal shame and guilt. Once the foreign mindset had arrived, spiritually and politically, it began a never-ceasing battle with the forces that already existed. The symbolism on both sides was straightforward, even simplistic, and higher evaluations did not exist nor were they needed at the beginning. Life, and everything in it, showed as a series of event-pairs; one part outside, the other inside the head, and whatever light they provided was close to the ground.

That organism, a hybrid of ancient pagan beliefs and a newer, antagonistic doctrine, had been driven into an enforced symbiosis where the evolving, abstracting fervour of denial needed the life-force of a living, breathing populace to prove the reason for its existence. The doctrine became its head, the rest a mass of trunks and limbs forever threatening to escape its control. At a glance it may seem like the normal pressures of an emerging power in the process of asserting itself. But the difference goes much deeper.

Religion interprets life; life is birth and growth—and birth again. The genealogy of teutonic myths tells about Wotan, the supreme god, and two of his children, Siegmund and Sieglinde, born to him from a she-wolf. And Siegfried, a main player in many heroic deeds, is the son of these two siblings [1]. Contrast this divine tale of lust with the threatening epistles of a Judeo-Christian mind filled with hate towards the ecstasy of human instinct. It is a chasm which descends to the roots of our innermost being. Over the centuries to come the disgust would seep into the developing symbolism of its doctrinal structure, creating ever more subtle abstracts to define and preserve itself. The imagery from those old gods was left behind, fossilised into empty shells, prevented from evolving into anything higher by being suffocated in its youth.

The general perspective of a religion gives form to the many subtexts used in daily life. Constantly pushing, guiding, and constraining, it exhorts, advises, and prohibits. The result is a world view that *makes* the world, a sum-total of minds that supplies the stage on which its people act. In time the limitations became unbearable to those whose thoughts strained at the confines. Because the system was so ponderous, progress could only be slow. One of the first individuals who dared was René Descartes (1596 - 1650).

In order to observe the shift in thinking we will look at his *Discourse On The Method Of Rightly Conducting The Reason, And Seeking Truth In The Sciences* [2]. One can view that text as a form of introduction to his other works, since he goes to some lengths to explain why and how he chooses to adopt the methods he then applies to the rest. As we shall see, there are some fundamental differences compared to previous modes, and it is quite possible that at least some of the explanations are written in order to forestall potential criticism of the most serious kind. The 17th century may have featured certain enlightened minds, but it was also home to intense convictions and the brutality those tend to produce.

For example, in enumerating the various topics and their sources he decided to deal with, he mentions theology as one of them. He "reveres" it [3], assuring the reader that he "aspires as much as anyone to reach heaven". What follows next is an implicit admission to the existence of the logic boundary, although there is no hint of him being aware of its presence let alone its influence. He continues,

> ..but being given assuredly to understand that the way is not less open to the most ignorant than to the most learned, and that the revealed truths which lead to heaven are above our comprehension, I did not presume to subject them to the impotency of my reason; and I thought that in order competently to undertake their examination, there was need of some special help from heaven, and of being more than man.

For someone from the present to state the above would be a sign of irrationality, but in the context of Descartes' times it can be seen two ways: (1) as a clever way to mask his intent which is to undermine the doctrinaire teachings of the Church, or (2) as a demonstration of his inability to distinguish between reason and belief in his own mind. I do not have enough information to judge either way, and so I will merely present the evidence for both interpretations.

As to the first, to declare one's reverence for the Church and confirming that by pointing to one's desire to reach heaven by 'the approved method' as it were cannot fail to impress a possibly critical clergy. Further, by being deprecating towards one's reasoning faculty the edge is immediately taken off the odd transgression, and by tying such a faux pas to those mental actions which can be attributed to silliness (easy to do since to understand God is beyond human comprehension anyway), would be creating a safe excuse just in case it is needed. Also, calling on heaven to provide help was not unusual in an age when virtually everything that was perceived positive was attributed to an ever watchful god and his angels and everything seen as negative to the devil and his minions; again, it is safe to follow the trend. He also writes somewhere else [4],

> But I resolved by no means to consent to their publication [ie his thoughts on thinking] during my lifetime, lest either the oppositions or the controversies to which they might give rise, or even the reputation, such as it might be, which they would acquire for me, should be any occasion of my losing the time that I had set apart for my own improvement.

Another hint comes from the closing sentence [5],

> ..and I shall always hold myself more obliged to those through whose favor I am permitted to enjoy my retirement without interruption than to any who might offer me the highest earthly preferments.

In other words, what's the point of getting rich and famous at first when as a consequence you spend the rest of your life in prison, or worse.

As to the second, it can be assumed that Descartes was as much a child of his time than anyone else, and so would have a mentality which allows the processing of 1st-level abstracts gained through the senses, but would equally possess a set of thought structures which prevented the internal construction of significant doubts regarding the role of God. The logic boundary, keeping higher-level abstracts away from more productive links to lower levels, would see to that.

It is possible therefore for a mind to employ thought associations amongst rational structures (ie those which allow themselves to be confirmed via recourse to reality), yet at the same time harbour other connections that derive their importance from a different set of abstracts. Just as there are incongruities between adjoining abstraction levels (an incongruent functionality), and there are mismatches within abstraction loops (the i-PAT-ls), we also find inconsistencies between entire thought patterns. We can call them, conveniently enough, *incongruent thought patterns* (ITPs). There are not uncommon; consider the aggression of a dictator and his devotion to his family, or the computer programmer who reads the astrology columns, or anti-globalisation demonstrators who flock to McDonalds after the protests—but also the international businessman who at the end of the day retires to his own culture. We will follow the emergence of ITPs in Part II, but for now let us just say that a thought structure depends on associative abstract complexes for its creation, and there is nothing to prevent the assembly of two or more complexes, each with its own associations and therefore each with its own functional abstracts. What *would* impede the progress of an ITP is the existence of a denominator having associative properties with regards to it, where such a common denominator is able to significantly modify some of the rule sets, ie the functional abstracts, of the erstwhile ITP. Let us call the responsible element within a thought pattern a *denominative thought structure* (DTS), turning the pattern into a *denominative thought pattern* (DTP). If a thought structure therefore incorporates a functional abstract such that its thought pattern becomes a DTP with regards to some other TP, the representational qualities of that other TP will change to a certain degree. Note however that the sheer existence of a DTP does not depend on the degree of veracity of its functional abstracts, whether in relation to the potential target or some other TP. Rather, what determines such a development is the *associative intensity* between one TP and another; in other words, how mutually relevant they are. Examples of DTPs would be: a technology student who develops a liking for a religious cult, the DTP being the feeling of belonging nurtured by the cult, an emotion which underlies the student's view of the world as well as his field of study; or the emergence of a more responsible behaviour in a youth through the growing awareness of cause and effect within the wider society, that awareness (the DTP) again underpinning many other attitudes.

One does not have to be a student of cognitive dynamics to understand that exposure to the wider reality of life would make a change in opinion possible. 'Travel broadens the mind' is not a recent proverb, and Descartes certainly indicates his appreciation of the idea when he says [6],

> Having thus provided myself with these maxims [ie his rules of thinking], and having placed them in reserve along with the truths of faith, which have ever occupied the first place in my belief, I came to the conclusion that I might with freedom set about ridding myself of what remained of my opinions. And, inasmuch as I hoped to be better able successfully to accomplish this work by holding intercourse with mankind, than by remaining longer shut up in the retirement where these thoughts had occurred to me, I betook me again to traveling before the winter was well ended.

On the one hand he tells us that the dictums of faith come first and foremost in his mind, and any other considerations second. But he also announces that he is going to abolish any notions he might have in order to make room for newly developed ones, and take up traveling to expedite his thought processes. In terms of our schema we can say that in Descartes' mind there are TPs dealing with faith, some of which at least exist beyond the logic boundary; there are also further TPs which are congruent (featuring c-PAT-ls), but are disassociative with regards to the former; and lastly, that he is purposefully setting out to assemble DTPs so that some of the ITPs may be improved upon. This desire to better himself is of course another TP, incorporating c-PAT-ls of a quality which ensured his reputation down the centuries. It is those c-PAT-ls which we are now going to investigate.

To begin with, he prefers the opinions of those who interact with daily life to others who contemplate similar problems with detachment [7]. As he sees it, in the dealings of the practical person one encounters less contradictions than in the theorising by "the man of letters" who is preoccupied with "speculative matters that are of no practical moment". One can regard this as an anti-intellectual attitude, but let us consider the resultant TSs in both cases.

Every theory has a starting point somewhere. Over the course of development the initial PAT^1s with their immediate links to outside experiences engender higher-level PATs which due to their nature are becoming more and more removed from actual objects. We have investigated many examples in previous chapters, showing how basic observations in real life gave rise to sometimes quite extraordinary theories about life and the universe. In terms of their associative properties with other abstracts and TSs inside the mind they were certainly functional, but would not stand up to objective scrutiny if 'objective' means having recourse to commonly identifiable characteristics. The entire process usually culminates in a PAT^{3+} representing the divine which becomes imbued with whatever is held valuable at the time. Whether one goes to such extremes or not, the potential danger of creating ITPs always accompanies the formulation of theories and theories alone.

Compare this with the dynamics of practical experience. Once again one starts with PAT^1s which, should the person be of sufficient intelligence, give rise to PAT^2s etc. Since the TPs of the practical person consist of a vast assembly of such PATs however, the outcome is a network of higher-level PATs firmly supported by similarly numerous PAT^1s with their outside links. To draw an analogy from architecture, we can say that the previous example looks like a huge edifice precariously balanced upon a few single columns, whereas the latter resembles a solid vault carried by a multitude of stout pillars. The analogy is not a random one: compare the lofty style of a gothic dome with the solidity of the Roman Coliseum and consider then the differing personalities of their builders—fancy metaphysics vs earthy practicality. For Descartes the choice is clear. In pursuit of knowledge about life, many multifaceted, reality-grounded experiences are preferable to the isolated musings of theorists.

He then addresses the issue of homogeneity [8]. For example, cities which have been designed by a single architect, so he writes, are more impressive to the eye than a hotchpotch of buildings that has grown together over the years. Since this is not a critique of Descartes we are not going to enter the debate of organic vs designed spaces, but let us look at the TSs behind his statement. The creation of TPs relies on the input of information and their PATs, and whatever TSs prove associative will form the relevant TPs. A stream of input consisting of homogenous characteristics will therefore form a single TP, whereas differing, even inconsistent aspects will turn into various clusters of their respective TSs. Which mental configuration is more powerful? Surely a large perspective containing only notions of a similar character, with no other diversions to dilute one's attention, will have a greater impact on the mind than a diffused panorama with ever-changing features from one end to the other. The effectiveness of uniformity or, in a negative sense of monotony, is well understood.

What emerges is the picture of a man who is curious, appreciates diversity as a sign of personal freedom, but is also capable of attentive focus and knows the value of consistency. Add to that 'caution', for he warns against "those restless and busy meddlers who..are always projecting reforms" in the area of public affairs [9]. He would much rather see the "highways which wind among mountains..become gradually so smooth and commodious" through frequent use, trusting time and habit to soften the edges of a problematic situation.

And yet, he recognises the limitations of that approach. It is one thing to avail oneself of a vast array of sometimes conflicting opinions, it is another to sift through such a mixture and find those things that are of value and can be assembled into a productive whole. It takes a mind which can step above the detail and adopt the wider view; in other words, which is capable of abstractions as high as need be. Descartes:

> And although my own satisfaction with my work has led me to present here a draft of it, I do not by any means therefore recommend to every one else to make a similar attempt. Those whom God has endowed with a larger measure of genius will entertain, perhaps, designs still more exalted; but for the many I am much afraid lest even the present undertaking be more than they can safely venture to imitate. [10]

Notice the words "a larger measure of genius". Quite clearly, he talks about inherent qualities of the mind, a notion which does not suggest itself when reading the opening paragraphs of the text. "Good sense is, of all things among men, the most equally distributed" he tells us there [11], underlining it soon after with "..good sense or reason..is by nature equal in all men". Any difference in thinking derives from the quality of data, he maintains, not from the process itself. In another section he talks about certain opinions held by people, some even on the extreme side, which may be considered a hindrance to an intelligent person but which nevertheless provide protection for those whose lack of mental dexterity causes them to make mistakes if unsupervised. He writes [12],

> And I placed in the class of extremes especially all promises by which somewhat of our freedom is abridged; not that I disapproved of the laws which, to provide against the instability of men of feeble resolution, when what is sought to be accomplished is some good, permit engagements by vows and contracts binding the parties to persevere in it, or even, for the security of commerce, sanction similar engagements where the purpose sought to be realized is indifferent.

The foregoing is not an exercise in hair-splitting. For if one believes that everyone's mental capacity is the same and it is just a matter of content, truths and/or falsehoods will be sought amongst ideas of similar complexity, ie abstraction levels. By contrast, if one recognises inequalities in mental states themselves, then it is not only the content which is of interest but what has been done with it, ie the abstraction processes. To be able to adopt that viewpoint entails the awareness of complexity in the first place and the readiness to make use of it, impossible for a mentality that is limited to first- and second-level abstractions.

He presents us with several rules of intellectual conduct [13]. They are more easily understood if seen in the context of what has been said so far across our chapters. The first is only to accept those ideas which he was able to confirm for himself and always limit himself to their actual content without thinking into them what is not there. Given the integrity of his particular TPs, this rule prevents the acceptance of conclusions that have been arrived at through more precarious contemplations.

The second rule is "to divide each of the difficulties under examination into as many parts as possible, and as might be necessary for its adequate solution". Suppose I have an entire pattern which I am in the process of examining. It is not enough to follow the reasoning—

such as it is—along the exact paths found within the pattern; it would result in nothing more than repetition. If however I subject its parts to separate evaluations I set up a number of other TSs, which in turn give rise to associations within my own framework of TPs such that any incongruities will become apparent due to their disassociative nature with the TP in question. We came across a similar insight when discussing Jainism in chapter 3, where its followers were advised to ensure clarity in their own minds before accepting any of the teachings. The method promotes c-PAT-ls and therefore *congruent thought patterns* (CTPs).

Third, it is wise to start with the simple and proceed to the more complex. We can translate that into using PAT^1s as starting points and building higher levels upon them, maintaining congruity along the way. If a PAT^{2+} serves as the beginning and that abstraction happens to be part of an i-PAT-l (which at that point won't be recognised as such) it is obvious that the entire train of thought will suffer.

The fourth should be seen in the light of its predecessors and deals with the necessity to be comprehensive and sufficiently general so as to preclude any details which may harbour inconsistencies. Especially observing rule 2 can only be successful if the nets of the other TSs are cast wide enough, otherwise its very purpose would be defeated.

So much for the rules of logic; now to his morals. He compares his stance to a builder who, while engaged in constructing a new house, occupies another during this time; the morals being his temporary residence while conducting his research [14]. He starts with the desire to live according to the laws of his society. Just as he recognises that people in other cultures may have their different values, it is nevertheless judicious to observe those of one's own—a useful observation for someone living at a time like his. In tandem he has learned to attach greater significance to someone's action, rather than words. By now the idea is nothing original, but it is commonly taken to mean that actions, involving real-life objects, are of greater impact than mere sound waves. From the perspective of cognitive dynamics there is an even better reason to take it to heart.

Words are the *articulated* elements of mental representations, that is to say those elements which the mind was capable of turning into linguistic symbols (the same holds for musical, visual, etc symbols). They do not comprise the *entire* contents of a TS, let alone a pattern. As such it is quite possible to entertain a collective of notions, none of which can be put into words; or where some can be articulated only pictorially, or via gestures, or sounds, or any combination of the above. On the other hand a series of actions, involving as they do a set of acts, responses by others, further acts, etc etc, at least implicitly include virtually all of the abstracts contained within a certain structure or pattern, because they all contribute to the outward expression of such a complex. This is not to deny the possibility of guile, merely to state that hundreds of millions of neurons provide a greater diversity through their dynamic clusters than 500,000 or so words (this applies to English—many languages contain far less). That discrepancy can be problematic, but more of this later. Therefore an examination of somebody's entire behaviour will yield many more clues than a few words are able to convey.

Next he appreciates consistency and the firmness of purpose. Comparing it to a lost traveler in the forest who ought to adhere to a straight direction until picking up further reference points, so will it benefit the mind if it is prevented from meandering around the stated aim. If we consider the cognitive progression through a series of TSs and at the same time want to be on the alert for any inconsistencies along the way, a reference to the theme needs to be maintained no matter how far we are prepared to stray temporarily; otherwise inconsistencies in relation to the theme won't be recognised.

Still on the subject of straying, he advises not to want the impossible. Given a train of thought, ie a TP in which the higher-level abstracts have been supplied their links to lower levels, those links will enable the mind to assess the realism of the TS. In case the links are broken, the respective notion assumes an air of make-believe that may become a source of delusion. Spiritual flights of fancy are of course a grand example, but also consider a business which embarks upon some strategy without adequate input from its accountants. Then decisions are made without recourse to the financial reality of the situation, although the options themselves may appear attractive enough. To attempt the impossible then is a product of TSs with links to the outside world missing but being acted out anyway.

If there is one thing that can be surmised from the character of Descartes' thinking, it is his wilful assertion to maintain control over his thoughts, to remain aware of their contents and their relatedness with each other. To stand aside and observe oneself is a virtue, and the higher perspective and so deeper understanding it produces make for more productive insights. Without this attitude the famous statement, "I think, therefore I am" could not have been made [15]. Let us look at the schematic nature of his words, and why he could arrive at this conclusion.

The filigree of the mind's thought patterns contain a multitude of items, only a few of which find their way into consciousness. We will examine the exact mechanism of this phenomenon in Part II; for now it is sufficient to state that a consciously held idea is only a limited representation of the complex which was responsible for it in the first place (we touched on this before). This means, while such an idea can be accessed as is, its associative parts, held in the subconscious, cannot. These can only be indirectly dealt with by attempting to form associations with the idea's conscious parts, which then in turn *may* make a *productive* recall possible. Trying to remember something through its connections is one example; a person who has worked out a puzzle but cannot describe all the mental steps he has taken is another. Since it is the conscious, ie the visible section of a TS which prompts awareness, it is that which we are bound to react to. We do not react to our subconscious because we are not cognisant of its contents (and if it had not been for certain well-designed experiments we would not have become aware of its existence either). So, who actually is 'we'?

Let us call the visible part of a thought structure the con-TS (conscious thought structure), and the subconscious part the sub-TS. In the course of a conversation, let's say, where thoughts have to be continuously articulated through words and where several sub-topics are dealt with, there exists a constant referential process which goes from the current con-TS to its associated TSs in the subconscious, which in turn spawn con-TSs for further articulation. This takes the form of a recursion, because the current con-TS has to deal with its associatives which seek out similar connections of their own. The recursive process stops when those linkages become exhausted and the highest level at that point moves on to the next TS repeating the procedure. We are not aware of the 'hidden' TSs and their processes, rather our articulated awareness relates to the con-TSs in our mind.

Can we be assured of being able to move on, in other words cut off the recursive process and start anew somewhere else? Not necessarily. I remember an acquaintance of mine who used to be quite a successful industrial chemist, endowed with intelligence and possessing a vast store house of general and particular knowledge. Caused by mental trauma his mind slipped into a continuous recursion from which he had more and more difficulty extracting himself as time went on. It manifested in a possessive tendency to find explanations for usually trivial words, which had to be resolved to his satisfaction before he could continue with the erstwhile topic under discussion. There were occasions when a single expression such as 'moving up' brought the talk to a halt, triggering one recall after another and for the rest of the evening it was impossible to speak productively about anything. In the end he committed suicide—a captive of a mental prison which consisted of isolated shafts reaching deeply into a maze of locked doors.

What we call consciousness then is the physical sensation of being in possession of con-TSs. Without them the representational format of our TSs would be relegated to what we call 'feeling', without being able to communicate their contents in a formal, explicit manner. The recursive process continues more or less throughout our lives; it is a default. We can therefore define consciousness as a *default recursion*, keeping in mind that it is the principle which is carried onwards, not its temporary extent. The genius of Descartes consists in intuitively recognising the fact that the con-TS dynamics are one of the fundamental signs of human life, and can therefore be used as its confirmation. We also realise that without a detailed examination such as the foregoing (that is to say, without a set of separate thought structures of very specific content) any con-TSs are perceived as mental entities that are contrasted with our physicality—the starting point of the mind/body dichotomy. Once we relegate the con-TSs to a phenomenon existing in its own right we are well on our way to combine its representative TP with abstractions containing the concept of ultimate perfection and arrive at a further abstraction: God. A description of that process can already be found in chapter 2.

Descartes alludes somewhat to con-TSs when he writes [16],

> For it is highly deserving of remark, that there are no men so dull and stupid, not even idiots, as to be incapable of joining together different words, and thereby constructing a declaration by which to make their thoughts understood; and that on the other hand, there is no other animal, however perfect or happily circumstanced, which can do the like. Nor does this inability arise from want of organs..

There is a corollary to con-TSs which needs to be made clear. Not only are the sub-TSs effectively unknown to us, but whatever cannot be articulated remains unknown as well. In addition, if a certain notion cannot be accessed because the recursive process does not touch its associations among other TSs, that notion is neither here nor there either. As far as the mind is concerned, it simply doesn't exist. Take for instance the following statement by Descartes [17],

> ..for as to the reason or sense, inasmuch as it is that alone which constitutes us men, and distinguishes us from the brutes, I am disposed to believe that it is to be found complete in each individual; and on this point to adopt the common opinion of philosophers, who say that the difference of greater and less holds only among the accidents, and not among the forms or natures of individuals of the same species.

The topic is the reasoning faculty of humans, but note the opinion regarding a thing's characteristics compared to the thing itself. According to this opinion we can entertain ideas about the varying characteristics (called "accidents" in the quote), but the essence of something is always the same. This is an example of construed abstracts (and there is no reason why the mind could not accomplish such a task), where no thought is given what this actually means in terms of cognitive dynamics (since there are no TSs dealing with cognition). Let's say we have a TS about an Angora cat, and another representing a Siamese cat. Both TSs contain associations within their respective structures defining the characteristics of each species (the thing itself), but they also hold the elements regarding the specifics about size, colour, etc in each case (ie the "accidents"). In other words, for the mind to refer to one or the other entails exactly the same processes in principle (moving through abstraction levels, identifying links) regardless of whether it deals with a species or the appearance of a single cat. A philosopher's opinion which puts the essence into one basket and its characteristics into another merely sets up additional PATs obeying the same functional rules as any others. *It does not change the nature of the thing in the slightest.* What *could* possibly make a difference is the inclusion of other TSs which through their associations enable the thinking process to be more comprehensive, but the level of understanding remains with the mind in question and has nothing to do with the actual object at all.

What does it matter? To set up a mental image of an object and regard this as inviolable (ie 'there is an essential cat and any instance is merely a derivation of the theme') means being in possession of a TS which holds whatever it is that is perceived to be the inherent nature. Such a TS must by definition be a higher-level abstract, as we have seen. Therefore it can be removed from PAT^1 links and is a candidate for being part of an i-PAT-l. Going back to what we said before about con-TSs, there is no way the mind is able to discover possible discrepancies between the real and its abstracts, because no appropriate PAT-loops have been provided for. If, on the other hand, one characteristic is as significant as any other, any cat can be defined in terms of its species, size, colour etc without creating artificial barriers between one set of details and another. Having no barriers allows the mind to draw comparisons where before it could not. The mindset which Descartes makes use of (see also chapters 4 and 5) had not been seriously challenged until Darwin et al, who disregarded such barriers and did away with categorical preconceptions. The theory of evolution is anathema to a mind following Descartes' perspective, and even today antagonism towards it rests with people who abide by medieval world views.

Unwittingly, Descartes gives us a demonstration of separate TPs coexisting in his mind, where each derives its properties from two different sets of functional abstracts. Consider what he says in the context of striving for truth [18],

> In the next place, from reflecting on the circumstance that I doubted, and that consequently my being was not wholly perfect (for I clearly saw that it was a greater perfection to know than to doubt), I was led to inquire whence I had learned to think of something more perfect than myself; and I clearly recognized that I must hold this notion from some nature which in reality was more perfect.

Leaving aside again the question of what constitutes 'perfection', the arrangement of essential categories (ie TSs) with regards to some object leads to the creation of further abstracted objects that have a higher degree of perfection, and so on until one reaches 'God'. So one TP is defined by a series of essence-TSs which is supported by its associations with other TPs containing metaphysical material.

By contrast, compare the present statement about the higher perfection of knowing vs the lesser regarding doubt with what he said about the subject within the context of his logic rules. There we have a TP that includes the observed instances of opinion gathering, has a set of functional abstracts that are based largely on PAT^1s and which thus tell him that it is better to doubt than to have preconceived knowledge. The former TP enshrines the idea of perfection; the latter leaves anything open to question; the rest is history.

There is another aspect to this theme, mentioned here briefly but examined in detail in Part II. If 'perfection' means 'completeness', and given that a con-TS can never be complete compared to the entire structure, then how can the notion of 'perfection' be articulated in the first place? It cannot be done in an explicit way, but it is very likely that the recipient of an articulated TS will have connected it with his own associations. Since the resultant total TS is now dissimilar to the sender's total TS and so on, one articulation after another, it seems that perfection consists of the overlaps shared by those TSs or, to put it more formally, of the intersections of a finite number of structures. The logical implications in reality do not fit the imagery.

To have a TP of a metaphysical nature inhabit one's mind creates constant pitfalls through its set of idiosyncratic functional abstracts. For example, Descartes asserts that the laws relating to the geometric qualities of a triangle (eg sum of its angles, etc) can be ascertained, but in themselves do not prove the existence of a triangle [19]. This is not the place to discuss existentialism, so let us merely observe that there is a TS such that it allows an object's characteristics to be articulated, but not the definition of its existence. However, then he continues [20],

> ..while, on the contrary, recurring to the examination of the idea of a Perfect Being, I found that the existence of the Being was comprised in the idea in the same way that the equality of its three angles to two right angles is comprised in the idea of a triangle, or as in the idea of a sphere, the equidistance of all points on its surface from the center, or even still more clearly; and that consequently it is at least as certain that God, who is this Perfect Being, is, or exists, as any demonstration of geometry can be.

One strong rule for God, a lesser one for triangles? Since in certain quarters both lay claim to reality in equal measure, this is not a facetious question. In any case, it shows once again that higher-level abstractions are deemed to be more significant than lower-level ones, otherwise Descartes would have apportioned the same principles to the contemplation about God as he did when explaining triangles. It is to his credit that he does make a distinction, although his 17th century mentality prevented him yet from asserting the rules with triangles and via the application of logical pragmatism deduce something similar about a god (if at that point its existence is indeed found necessary). His contemporaries were still at a stage where plants, animals and humans were judged according to their idea of the divine.

Descartes' version of the human circulatory system with its vapours and heating heart looks somewhat quaint with hindsight, but is a good example of an intelligence coming up with conclusions derived from a certain quality of knowledge [21]. It reminds us that the mind is essentially a *processor* of information, not a *producer*, and insights are due to its abstractive capacity. It is also worth considering how fortunate we are *not* to possess too powerful a mind, lest its sometimes bizarre interpretations interfere with the running of our bodies. The recognition of the mind's fallibility leads usually to the idea of a grand design, as if all that any thing ever did on this planet had somehow been designed into it from the start (would that be its much-touted essence?). Such an idea is indeed overwhelming, and only a god could ever hope to achieve it. It would become less overwhelming but even more fascinating if one reverses the view and considered that the state of a system is the product of its elements' currently realisable characteristics. We will re-visit this issue in Part II.

The subjective nature of the mind is particularly significant in relation to con-TS dynamics. Since the articulated information is only one part of some con-TS (assuming that no-one is normally granted endless hours to talk about anything), yet blows up to much larger proportions once it has entered the recipient's mind and gets connected to all kinds of TSs there, the total representations in both minds never fully harmonise. The articulated information then is the only part which is shared; it is the only, minor, intersected part by two huge sets of abstraction complexes the depths of which cannot ever be mutually explored. Descartes sure knows the feeling [22],

> ..though I have often explained some of my opinions to persons of much acuteness, who, whilst I was speaking, appeared to understand them very distinctly, yet, when they repeated them, I have observed that they almost always changed them to such an extent that I could no longer acknowledge them as mine.

A sobering thought when it comes to notions about common understanding on anything. A by-product of the hope of being the same in mind with as many people as possible is the significance given to a thought which has occurred to two individuals at the same time. The interpretations range from love to telepathy and beyond, but there is also a more prosaic one. Considering the small number of words which are mutually shared, and the sheer size of what hangs from them which is not, it is quite possible that in circumstances of high profile *both* mindsets will have it within them to come up with a similar response. So, rather than exclaiming, "Isn't it wonderful how we both thought of that!" one should be saying, "It would be amazing if both of us, with our life-long experience and accumulation of incidents on the very same planet and access to the same vocabulary, do *not* have something similar to say to the present occasion." But then again, it's not so mysterious.

In a recent paragraph we have classified the mind as a processor, not producer. When the articulated information passes from person to person and consequently gets connected to the current recipient's thought patterns, its overall property will certainly change in terms of its content. Will it change its quality? If the sub-TSs of its temporary residence include higher-level abstractions similarly to its previous host, chances are the quality will be maintained. Should the sub-TSs be coarser however, the resultant content that is going to be passed on will have suffered from it. The loss of quality cannot be regained easily—even a brilliant mind cannot guarantee finesse out of dross. It seems Descartes experienced these problems himself, for in one passage he mentions the frustration one encounters with people who understand only dimly [23],

> In this they seem to me to be like a blind man, who, in order to fight on equal terms with a person that sees, should have made him descend to the bottom of an intensely dark cave..

The implications for a society in which information sharing is well established are substantial; without continuous, re-energising input from suitable sources the general state of knowledge held in the collective mind is bound to deteriorate.

The schematic diagram of Descartes' thinking appears in Fig. 6-1. TPs which are grounded in reality (ie derive their PAT^{2+}s from many PAT^{1}s) have a greater significance and affinitive TPs create links with each other. Should a particular TP relate to affinitive TSs in a not grounded TP (there are only a few PAT^{1}s but they are used to form many PAT^{2+}s and therefore i-PAT-ls), then the former TP becomes a DTP in relation to the not grounded TP. Thus incursions are made into metaphysical complexes, modifying their extravagant conclusions and bringing them more in line with observed reality. In the *Discourse* references are made to geometry and in particular triangles. Although we are too late to query Descartes in detail about his opinions regarding the relationship between geometric forms and the broader issues of life, we do know that he valued observation and reasoned analysis. His other writings in mathematics confirm this and so we can make the justified assumption that in his mind there would have been numerous thought structures dealing with geometric forms and their inherent relationships.

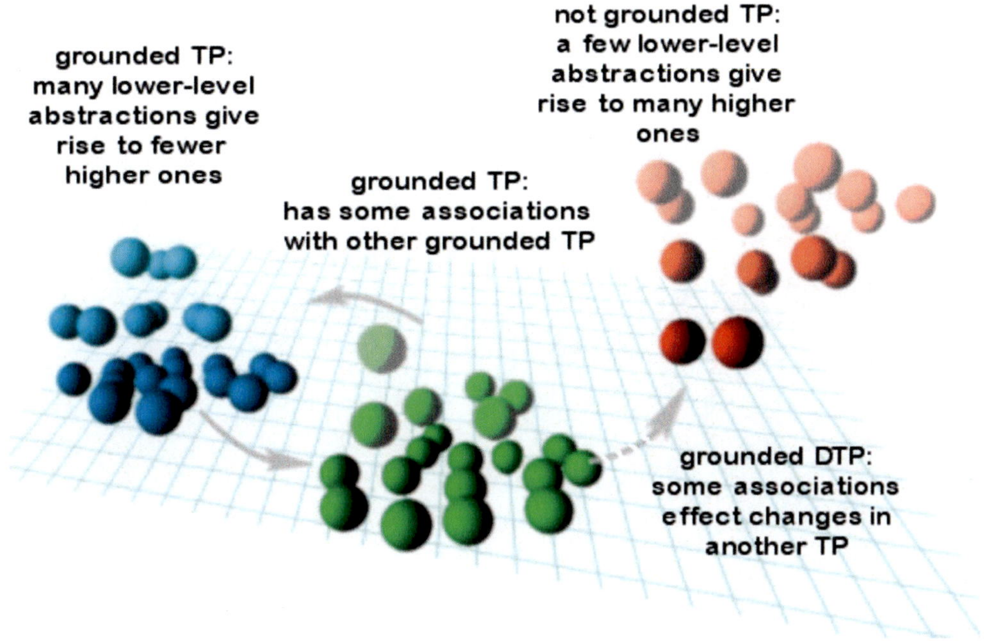

Fig. 6-1: General schematic view of Descartes' thinking. A grounded TP (blue, left) contains TSs relating to geometry where many first-hand observations (PAT[1]s) are turned into some conclusions of a higher abstraction level (lighter colour). Another grounded TP exists (green, centre) containing TSs from his interactions with people. There are also not grounded TPs of the metaphysical kind (red, right) where only a few observations have been used to construct many assumptions used in religion. TSs from grounded TPs which have an affinity with certain religious TSs can become DTSs and so turn their TP into a denominative one.

In Fig. 6-2 we illustrate a simple example of TSs, one about triangles, the other about rectangles. In each case the initial observations lead to general conclusions (ie higher-level abstracts), and they in turn enable the same to happen at a higher level still. His mind also has TSs put there by his religious leanings, and they would have been characterised by a few direct observations, various construed ideas about the observed objects, and many assumptions based on those ideas.

Note however that his mind—or anyone else's for that matter—does not need the entire set of levels for the upper metaphysical conclusions to be operative. It is quite possible, indeed usual, for a top-level assumption about God (eg 'God is a most perfect being') to be imparted to someone as a statement to be learned. That input becomes a 1st level abstract, unless it is accompanied by further explanations which act as the levels underneath. As a 1st-level abstract the idea is both rigid and simple. Rigid, because its anchoring within TPs is accomplished through strength of emotion rather than conceptual items; simple, because there are no lower levels to provide the material for more detailed musings—it is held with 'blind faith'. For someone like Descartes a simple rote learning may not have sufficed, and it may well be that *his* mind contained sufficient material to explain why such and such an article of faith would have come about. Even so, his searching amongst the laws of geometry must have brought him up against the theoretical windows that overlook the doctrinaire confines of religion, to see farther and therefore sense a kind of truth more embedded in nature than in fantasy. At this stage the inroads were only small, more concerned with the general principles than any major rebuttal, but it was a most important start.

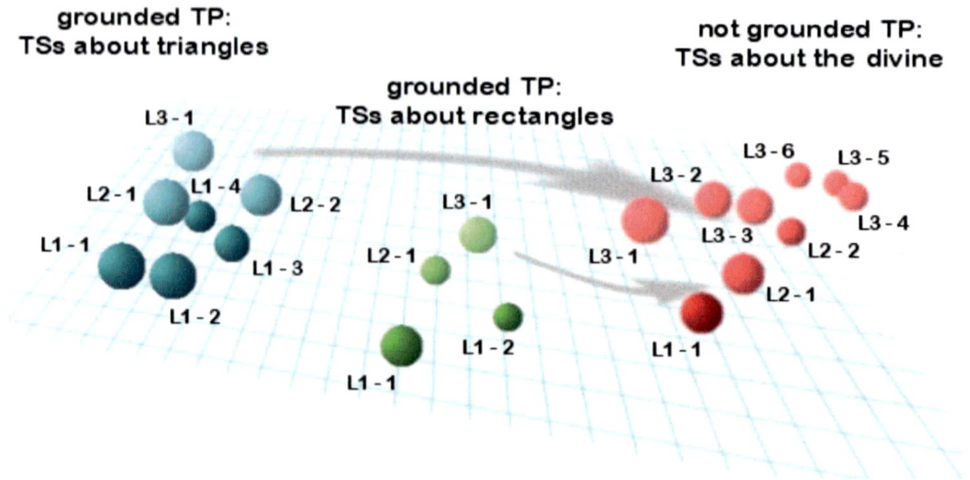

Fig. 6-2: Simplified example of TSs at work. Each sphere represents an idea at various levels of abstraction (levels 1 to 3). Triangles: L1-1 - triangles, L1-2 - angles, L1-3 - sum of angles, L1-4 - squares of sides; L2-1 - understood rule (sum of angles = 180°), L2-2 - understood rule (square of the hypothenuse = sum of squares of sides); L3-1 - there are consistent laws regarding triangles. Rectangles: L1-1 - rectangles, L1-2 - height/length; L2-1 - understood rule (area = height x length); L3-1- there are consistent laws regarding rectangles. Religion: L1-1 - some perceived object; L2-1 - perceived perfection of object, L2-2 - imagined possible perfection; L3-1 - supreme godhead, L3-2 - all of reality, L3-3 - all of life, L3-4 - all plants, L3-5 - all animals, L3-6 - all humans. Higher-level abstracts from grounded TPs have the potential to become denominative in relation to lower-level abstracts belonging to a not grounded TP: here the idea of divine perfection would be changed to a more prosaic image of natural laws.

The essence of formality had undergone a subtle change. For an Augustine or Aquinas the correct framework of thinking consisted in abiding by the doctrine of the Church—in other words, the appropriate positing of higher-level abstracts such that they all culminated in the proclamation of God as the highest and hence the associated principles being made to flow down to the everyday level of people. The effective source had been the metaphysical abstract. In Descartes we have seen the first formal beginnings of a different focus—away from the divine and pointed at the here and now. Note that the *status* of formality retains its severity, only the *content* differs. The idea that rules can rigorously define the quality of geometric figures and therefore the same approach can be taken when it comes to other, more intangible issues, has been pursued ever since. It is a manifestation of an abstract rule set which says it is feasible to identify certain relationships at a relatively high abstraction level and apply similar relationships within contextual relevancies further down. Not surprisingly the actual realisation of such a rule spurned many to explore their applicability across an ever widening field. One of these was Benedict de Spinoza, a contemporary of Descartes although he survived him by 27 years (Spinoza died 1677). We will use two of his works, *On the Improvement of the Understanding* and *The Ethics* [24]. The first, because it demonstrates how in his opinion the mind should organise the processing of data; the second, how such an organisation influences our perceptions of moral conduct. This categorisation refers to the content of the text, ie his thought structures. When it comes to their functionalities such boundaries are neither here nor there however and so we will largely disregard them when explaining the underlying schema.

Spinoza lays down some ground rules which he wants the reader to understand [25]. They are interesting because in themselves they do not explain anything about what is to follow, but define the daily conduct of someone looking for achievement in general. He advises to remain attuned to the outside world, to communicate intelligibly so that others may understand him while he can be open to their response. He also appreciates the nature of pleasure as something enhancing one's life, provided there is moderation. In the same vein he recognises the advantage in having sufficient means to pursue one's efforts, so that health, standard of living, and personal safety do not pose a hindrance. A far cry from the bleak desires expressed by the Cathars (see chapter 5)!

It must be assumed that for someone like Spinoza the outcome of a proposition is not alien territory at all, and the constant cross-checking between cause and effect would have made it easier for him to critically revisit an initial statement. This would be in line with the previously mentioned abstract rule set which allows its owner to apply the principle of logical delineation in areas other than the original training ground. How far a learned rule is permitted to undo already established habits depends on the strength of its denominative thought structures. In the case of Spinoza, the fact that he was brought up under Judaism but excommunicated by the Jews as well as kept at arm's length by the Christians points to a certain flexibility in his thoughts. So while in his trade as a lens grinder the logic of geometry and optics supplied many occasions on which to practise his rationality, we must give him credit beyond that. Otherwise we ought to have serious doubts about the collective sanity of Notre Dame's builders who, while calculating a balanced work in stone were also participating in the medieval mind set (or were they—the beginnings of a masonic underground?).

From one's surroundings Spinoza turns to the individual and starts with the input, the act of perception [26]. He distinguishes four modes, and their systematic description shows to what extent his powers of observation and analysis have been made use of. Mode 1 is the input from the outside, through words, gestures, etc. Mode 2 is the next step, that is information which resides now in memory without having been subjected to any modification by other input; such as practical experience. Mode 3 represents the outcome derived from thinking such as a conclusion or a deduction, but he attaches the rider that while we may know about the result we remain ignorant as to the nature of its cause. The fourth mode deals with what could be described as 'understanding'. For instance, we know the sum of two numbers because we understand the natures of quantity and addition.

With hindsight one might find those definitions somewhat vague, but given his contemporary knowledge base Spinoza's statements are impressive. Since for him they are a starting point it is useful to translate them into our schema to see how they could prepare the ground for someone who has adopted them. Mode 1 is straightforward; they are the familiar PAT^1s, directly abstracted from the outside. The second type is somewhat superfluous if we want to be pedantic and limit the concept to first-levels only. But as we have seen it doesn't take much for a PAT^1 to spawn a PAT^2 and still exhibit no direct signs of what we usually call 'thinking'. The recognition that there is such a thing as a prototypical 'car' say, through observing many cars is virtually automatic and not given much significance.

Mode 3 is more complex. From the description we are given it is not clear whether a conclusion has been reached by going through a series of cognitive steps or whether an act is accepted 'as is', that is without having gone into any technical detail. Spinoza himself gives the perception of the sun as an example, but as it happens it can be interpreted both ways. 'Knowing' about vision, so he says, lets us understand the fact that an object appears smaller when further away and, when applied to our sun, enables us to imagine it to be much larger because it is obviously at a great distance. Still, we might perform astronomical calculations which tell us what we need to know about the sun, and from that infer the phenomenon of perspective—these are the cognitive steps. On the other hand, from early infancy onwards we are accustomed to seeing things in perspective, and therefore we judge sizes through their distance and distance through the size of an object. Some people describe this kind of knowledge as 'intuitive understanding'. The dictionary defines *intuition* as a "direct perception of truths, facts, etc, independently of any reasoning process". In terms of our schema we would prefer to call it a functional abstract, because it is information which has been extracted from a pattern without having to know anything about the pattern's constituents. The word 'understanding' is a misnomer therefore and distracts from the fact that nothing more may be known about some phenomenon except that it exists. For that matter, the entire definition in the dictionary is misleading since without the application of at least some reasoning no "truth" or "fact" can be ascertained at all. The quote rather reflects our cavalier attitude towards the mind, for it accepts an immediate assumption as preferable to a more laborious effort. Such tendency is to be expected when we consider that for millions of years the *only* comprehension of the outside had been based on functional abstracts alone; deductive thinking, which requires higher-level abstracts, came later. Even today the evolutionary baggage makes itself felt. What sounds more convincing to the average citizen: someone declaring, "I do believe that...", or, "According to my calculations there should be..."? So, in the absence of a more precise explanation we can assume then that Spinoza must have meant both versions, however unwittingly.

In any case, for our purposes it becomes necessary to establish two sub-categories of a functional abstract in order to place their respective dynamics within the correct context. To simply accept a situation as a fact because we have experienced it several times in principle is due to a thought structure which we can call an *intuitive functional abstract* (*IFA*). It is an abstract and therefore can exist at any level, modified accordingly. Note however that as a thought structure it merely holds information about the existence of some circumstance—nothing more nor less, and no truth value can be implied. For example, I *know* if I get hit by a car it will hurt, because (a) I actually did get hit by one once and (b) I have come across the outcome of similar events in others and so have acquired a certain image in my mind. I *don't know* it will hurt because I have calculated the kinetic energy of my body mass and done the same with motor vehicles. Or, jealous partners just *know* the other has cheated on them because he or she has come home late; they *don't know* what has happened because they performed a DNA test on the lips.

If the mind has learned not just through perceptions but by employing cognitive processes (eg calculating the kinetic energy), the outcome is also a rule inherent in a thought structure, but this time it is linked to several other TSs which do contain further information: a *cognitive functional abstract* (*CFA*).

The effects of either are significant. An IFA is impervious to modification unless it is done through a denominative TS; a CFA is easily modified because it relies on its associative network of other data for its presence. Both are rule sets, that is they are able to reflect the abstractive complexity at any given level. The former will not only evolve from incongruent thought structures or patterns but can actually strengthen them, whereas the latter enhances the truth value of thought processes due to its links with other information.

Mode 4 goes a step further still. It is the perception "solely through its essence" [27]. More telling is Spinoza's next sentence: "when, from the fact of knowing something, I know what it is to know that thing, or when, from knowing the essence of the mind, I know that it is united to the body." Translated into our nomenclature he is saying that there is a thought structure which enables the mind to conclude that it *knows*. This particular sensation prompts the declaration "I know". Observe that nowhere does the content itself intrude into these considerations (when he talks about the mind it is by way of example only)—the focus is on personal satisfaction in terms of the intellectual outcome. What constitutes personal satisfaction?

Suppose I am a car mechanic and a friend tells me about steering problems he has with his vehicle. Being fully versed in matters mechanical I will ask those questions which enable me to diagnose the source of the problem. In other words, my TSs contain enough information to judge the completeness of the data coming from my friend, because whatever the problem my knowledge of cars surpasses it. Let's assume I have finished asking the questions because I am satisfied with the description of the symptoms and I now take a look under the bonnet. Lo and behold, the car is driven by a newly developed hydrogen engine which I have never seen before.

During that episode my personal satisfaction levels of the kind talked about here have varied considerably. At first I had been comfortable with the conventional knowledge of cars, and why shouldn't I—I had been trained and I had worked on them before. My TSs induced a feeling of satisfaction which enabled me to say "I know". After coming across the new engine I obviously revise my attitude and concede that I really know less than I had supposed. I say "obviously" because my previous learning experience (creating CFAs along the way) has taught me the necessity of being open to new information in order to be effective. Therefore we can say that my current satisfaction is a representation of TSs such that whatever data I hold in terms of that configuration they are linked, firstly, to the CFA which tells me that learning never stops, and secondly, to all those details which make up my image of cars—all interdependent and mutually supporting. However, had I been approached as a bishop for instance and called upon to settle a theological dispute, my reaction would be different. My experience would have been composed of intuitive functional abstracts, my educational process would have been characterised by many assumptions and any subsequent conclusions were directly bound to the accepted veracity of those assumptions. In other words, the feeling of satisfaction did not arise from a repeated patterns of CFAs but a repeated pattern of IFAs, yet it would exist within me nevertheless. *Knowing that one knows* therefore does not reveal anything except just that—knowing that one knows.

While Spinoza's definition of mode 4 certainly shows his astute insight by differentiating those four separate types of thought, we can see that without a further examination of the underlying dynamics any definition of human thinking is precarious at best. It also demonstrates the inherent, fundamental subjectivity we cannot escape from, and which forces us to accept a phenomenon as it manifests itself to *us*, and not recognising it in terms of its actual nature. As we have seen in the previous example, the sensation of knowledge was no more than that—a sensation. Yet it is also evident from one philosophical text after another that the human mind not merely strives to hold some information, but aims for the ultimate understanding of a thing's nature, its essence.

The cognitive process outlined in the above car scenario is essentially an abstractive one, where PAT^1s evolve into PAT^{2+}s and the next higher abstract always relates to its predecessor (rightly in the case of CFAs, wrongly for IFAs). As such it forms a PAT chain and results in a congruent or incongruent PAT loop. Hence postulates such as Spinoza's modes of perception are essentially PAT-ls, where the initial section—based as it is on assumptions—may be an i-PAT-l followed by a c-PAT-l, provided the ongoing process observes valid relations with observed reality *and* remains subject to continuous confirmation. The driving force behind the development of the loop is the mind's need to abstract, and just as the ancients agonised over the meaning of an ideal, thinkers now worry about getting to know a thing's essence. In both cases the act is due to the ability of the mind to ascend to higher abstraction levels; for Socrates it consisted of following the path from 'beautiful' to 'Beauty', for Spinoza it is a matter of formalising thought itself in terms of mode 1 to 4. Since the latter requires adopting the position of an observer, at least the starting point of the resultant TSs will be of an abstraction level that is necessarily higher than what is being observed. Recall our examination of the concept of the divine in previous chapters, culminating in what we termed a *container* which in due course became imbued with all the characteristics that were believed to be appropriate. The cognitive process leading to that result is no different from the process we are faced with here—only its content has changed.

If we compare the number of abstracts necessary to fulfill those four modes of perception, we realise that the first mode requires the least, followed by the second and the third, with mode 4 being the most abstract intensive. That is to say, from the creation of the first PAT[1] until a conclusion is reached that concurs with the nature of the respective mode, the number of abstracts at each level plus the number of levels themselves increases as we go from 1 to 4. Further in the text we learn that mode 1 is the least productive, mode 2 is better, followed by mode 3 and the fourth being the best. This is in line with the already observed propensity of the mind to ascribe a greater degree of importance to things which are represented by higher abstracts. The preference is only relative; whatever level can be achieved will be assigned greater significance than its lower neighbours within the same group of thought structures. Given the variety of abstracts we have encountered so far, from the ideals of the ancients to the cosmos of the Jains to the religious prisms and geometric laws since then, it can be seen that the quality of the outcome is secondary. A conclusion arrived at via an incongruent PAT chain is more important if on a higher abstraction level than an idea at the end of a c-PAT-l which happens to reside on a lower. Furthermore, since only a part of an entire TS becomes conscious and therefore articulated and so is subject to the ongoing process of recursive enumeration across many TSs and TPs, one can expect the effect of such a tendency during the act of articulation itself. The less influence there is from other TSs which might exert some degree of restraint, the more unhindered the tendency will become.

Suppose a group of youths are telling each other stories at a party. For an objective observer who is familiar with the stories' background it will become obvious as time goes on that the content moves more and more away from reality. Each contribution elicits a further emphasis, or exaggeration. Note that in order to exaggerate, one's mind must construct other, possibly higher abstracts. There is also the need for other TSs which evaluate the degree of change against the demands of plausibility (the story can't be too outlandish), of history (there must be a certain continuity between the previous one and the next), and the ascertaining of one's social status (the more respect a person has the more he or she will get away with)—all this is more attraction still. It can also be seen that during the course of the evening fatigue, alcohol, and so on will diminish the influence of any restraining TSs, and so the exaggerations will become wilder. At the same time, further fatigue etc will depress the tendency itself and eventually the flights of fancy are grounded altogether. Of course the phenomenon is not restricted to young people, but maturity is one element contributing to restraint and to witness the playing out of the tendency at various age groups brings no surprises (therefore such an exercise is very productive when there is a need to identify the stature of a person—regardless of age).

One other, in the long term more significant factor, is the overall integrity of the thought pattern which is active at the time. Recall the possibility of having any combination of c-PAT-ls and i-PAT-ls involved in the chain of abstractions. Obviously, the more incongruities there are the more challenged the truth values of any further creations will be. This means that in any type of communication interchange between people, in the absence of any ameliorating influences, the validity of information will suffer in proportion to the weight of incongruities the minds may contain; in other words, it is not only the data themselves which matter but the way their respective abstracts have been organised.

"The fourth mode alone apprehends the adequate essence of a thing without danger of error," Spinoza writes, but concedes earlier that "The things which I have been able to know by this kind of knowledge are as yet very few." He contributes that lack of error to the opportunity for understanding which is inherent in the idea of a thing compared to the thing as such. Since the idea is "something different from its correlate, it is capable of being understood through itself" [28]. Why this should be so is explained next,

> ..in other words, the idea, in so far as its actual essence is concerned, may be the subject of another subjective essence. And again, this second subjective essence will, regarded in itself, be something real, and capable of being understood; and so on, indefinitely.

The extension of the core of an idea into a string of abstracts is by now familiar to us, but it is worthwhile considering Spinoza's attitude towards these creations. The idea of a circle for instance, he says, is not the same as the circle itself; a circle having a circumference and a centre but the idea does not. It is clear that the word 'idea' in this context must be understood as a thought of something excluding anything it actually contains, otherwise his statements would be absurd. As he also recognises the fact that we continually produce ideas, the resultant certainty coming from his preferred perception is "nothing else than the subjective essence of a thing: in other words, the mode in which we perceive an actual reality is certainty" [29].

So, according to Spinoza there is a "true idea" of something which is the starting point for further considerations, which in turn serve as sources to other ideas and so on. The more knowledge we gather along the way the better our understanding and the more certain can we be when we say, "we know that we know." Yet those ideas are also the "subjective essence" of a thing, by which he means the characteristics pertaining to the counterpart of our current idea. That current idea can be subjected to an ongoing development—in this case learning—which enriches our knowledge base. He also points to the fact that we can have an idea of an idea, ie a representation of a representation (and thus "we know that we know that we know etc"). Such a continuation does not contain anything new, it merely mirrors what we have ascertained already. From the point of view of thought structures it represents nothing but the wilful repetition of the previous TS. It seems therefore that the ultimate success of this operation hinges on the "true idea", and we will see shortly how it fits into our schema.

Spinoza uses the example of a burning candle which stands there before us [30]. Knowing about candles, we can surmise of a candle not burning, we can think of a candle somewhere else, and so on, because we are cognisant of its properties and therefore of the possibilities of its existence. Going one step further, he explains that just as we know about speaking and know about a tree, we may well imagine a talking tree. However, this is only possible if we have a rather limited understanding of trees (and what constitutes speech for that matter), otherwise we would well know about such a combination's impossibility. Therefore, the more we know about something, the less we are prone to fantasies. Translating this into our language, we can say that if there is a TS which contains object abstracts and functional abstracts such that they form a c-PAT-l there will not be an image which does not exist in nature. Since, Spinoza states, in the course of the development from one idea to the next, the mutual relationships reflect similar relationships in the real world (provided we have followed reason), the result can be accepted. He also adds via a footnote that the above combinations never produce anything inherently new—each element must have existed in the mind already. Naturally, one has to extend that observation to functional abstracts as well.

Given that any combination is possible to begin with (he calls such a combination "fiction"), and that a detailed investigation into the constituting elements will enable us to think correctly, there must be a point when we have arrived at the smallest detail, the one which is indisputably correct. Such a 'small' idea, that is one which cannot be decomposed any further, must be known correctly by default—the "true idea". Consider however the corollary he offers: such a true idea need not necessarily have its counterpart in nature but can be correct in itself, ie correct as an idea, not its real world object. As an example he suggests a sphere, and to prove its correctness he envisages a semicircle revolving around its centre. Although no sphere in nature has ever been produced in this way, we can nevertheless show through the geometric correspondences that our way of thinking is correct. In other words, once we choose as a starting point of some PAT chain a thought structure which contains a satisfactory set of functional abstracts, then the conclusion of that chain must be accepted. Those functional abstracts require no further affinity with the actual PAT chain in question. Because such a chain can be developed further (eg one could proceed to explain something involving spheres), and because the set of functional abstracts which allowed that train of thought in the first place can be said to be determining the eventual outcome, we can classify such a set as a denominative thought structure.

Next he deals with doubt. For us to have doubt, Spinoza declares, we must have some other idea which interferes with the contents of the first; a single idea cannot induce any doubt. "An idea is in itself nothing else than a certain sensation", he says [31], and therefore does not have any truth value on its own. Recall our discussion about conscious thought structures at the beginning of this chapter. A con-TS will be traversed (possibly ranging across an entire con-TP) if its abstracts can be articulated in a mutually cogent manner. These abstracts cause the sensation of consciousness (and prompting Spinoza's definition of 'idea'). Moreover, they are the ones which recursively make the development of further ideas possible—*provided* we remain within the realm of consciousness. Transposing the aforesaid into our schema, we can say that what makes us certain of some idea's truth is our ability to entertain a series of con-TSs such that their starting point can be articulated either in terms of a single idea or of some configuration which can also be satisfactorily articulated.

If they cannot, we have a problem. For instance, suppose I try to solve the task of calculating the area of a trapezoid. Having never done that before, I draw one and discover the existence of imaginary rectangles and triangles contained in its shape. Since no trapezoid in nature has ever materialised before me through the floating together of rectangles and triangles, the significance of that imagery is a functional abstract (similar to Spinoza's revolving semicircle). Note however that it is my mind's ability to conjure images of rectangles and triangles (object abstracts) plus their respective usefulness in constructing various other shapes (functional abstracts) which enable me to work out a solution. Suppose now I never have dealt with any geometric figures in such a functional manner, and so do not readily conceive their usefulness in finding the area of a composite. Since I cannot articulate their usage my con-TSs will not arrive at a productive result. On the other hand, it could well be that my *sub-TSs* can achieve such intellectual dexterity because they are unhindered by the constraints of articulation—in other words, I may dream about the solution. Can I articulate that? Yes, because I can consciously express the elements inherent in the result. If this is not possible either *and* the eventuating TS happens to be denominative to some other TS, my mind will have changed to a certain degree but I may not know it. More on that dangerous topic in Part II.

For anyone reflecting upon human thought the most obvious manifestation would be the act of expressing our ideas in whatever form. Words is one of them, and for Spinoza they are part of the imagination, by which he means all those thought processes which are open to errors and confusion and to be contrasted with understanding. It is tempting to compare Spinoza's "imagination" to our subconscious, and his "understanding" to our con-TSs, but it would be wrong. The offerings ascribed to imagination may be confused or downright false, but for him they are very conscious indeed, as conscious as that what we hold to be understood. As far as he is concerned the difference lies in the respective laws which either one obeys, where understanding presumably follows the rule of established logic, and imagination some other format that causes errors to occur. With regards to words they do not only express an idea from imagination, but through their meaning wrongly applied contribute to some falsehood.

We know by now that true *and* false ideas are the product of a thought structure comprising various abstracts, and words are the articulated part thereof. Therefore they do not exist arbitrarily, not even confusedly if by 'confused' we mean a random order not subject to any formality. As such Spinoza's alignment of words with his concept of imagination needs to be taken onboard carefully. Consider how he justifies his opinion:

> Moreover, words are formed according to popular fancy and intelligence, and are, therefore, signs of things as existing in the imagination, not as existing in the understanding. This is evident from the fact that to all such things as exist only in the understanding, not in the imagination, negative names are often given, such as incorporeal, infinite, &c. [32]

To be able to fully interpret the quote a comprehensive understanding of Spinoza's mind is necessary, a goal which is impossible to achieve. Apart from an affinity between language and imagination—but not understanding—we read about a propensity for descriptors to be in the affirmative if their source should be fanciful thoughts, but which tend to express negatively if emerging from a formal, rational cognitive framework. If we assume that the above is a statement of fact, then from our perspective there are two fundamental possibilities, although neither can be confirmed due to the lack of more information about Spinoza. They are presented as a hypothetical exercise below. On the other hand, if we take the quote to represent Spinoza's thinking—which in a sense it does of course—then we need to integrate his stated intent with the mental functionalities which exist in every human. So, either that status of words is true and we need to find out to what extent Spinoza is driven by it, or it represents Spinoza's opinion only and we need to check how Spinoza uses it to drive his mind.

To begin with, the hypotheses. We could assume that the initial construction site of every idea is the imagination where they grow, combine, and form structures in a mostly "confused" fashion (for this paragraph I will use the words 'imagination' and 'understanding' according to Spinoza's interpretation). Once we apply rational processes to our thinking the appropriate ideas migrate to the area reserved for understanding, where they undergo further refinement. Because the ideas have been formed in imagination and therefore their corresponding words as well, and because they have to be corrected once they get moved across to understanding, chances are the words have to be negated in order to be useful. This would be Hypothesis 1. Hypothesis 2 is as follows: we again perceive the imagination as the breeding ground of ideas, the true as well as the false ones. But this time the understanding is not a process but an exclusive area only and the actions which stock that area remain extraneous to it, ensuring that only 'correct' ideas and their words may enter; they just happen to be negations.

We can dismiss the hypotheses as just that—assumptions. To think that the mind comes up with words here, there and everywhere and only when being logical suddenly remembers the possibility to attach syllables such as 'in-' to its products is quite absurd. By the same token, truth would be a bizarre creature indeed if it was only made up of negatives. Yet for someone to believe in either version is possible—we certainly *can* marshal our thoughts in very specific ways and still make use of the same old operating system common to us all. Which version would Spinoza have had in mind or, to be more precise, which version *could* he have had in mind given the context of his writings?

The first hint comes immediately after the words cited above, where he says:

> So, also, many conceptions really affirmative are expressed negatively, and *vice versâ*, such as uncreate, independent, infinite, immortal, &c., inasmuch as their contraries are much more easily imagined, and, therefore, occurred first to men, and usurped positive names.

For example, it would be easier for me to conceive of mortality because it is part of my daily experience, and in the process of making up a word for it I would naturally tend towards a positive expression rather than a negative as the first one of its kind ever. But note the kind of words he uses to illustrate what he means, especially against the background of his earlier statements regarding the varying qualities of perceptions. There he preferred one based on the 'true idea', ie incorporating a relatively high number of abstractions. Compared to what can be expected in real life the words chosen above are all products of further abstractions, therefore are all candidates for his favourite type of perception and are associated with "understanding". It appears that for Spinoza a concept has not proven its validity unless it has been subjected to a minimum number of abstraction processes, contains a "true idea" as its starting point, and features at least some negation to demonstrate its independence from that pool of lesser beings, the "imagination". On a personal level, this suggests a person who, in the same order, values intellectual achievement, appreciates a confidence in truth, and feels a certain amount of harassment from people he considers to be uninformed and negligent in their thinking.

Thus we can justifiably assume that Spinoza has identified a certain mental procedure which he considers to be of value and applies it to his own deliberations. Since he is doing so consciously, aware of its purpose and advantage, it does not really matter whether the underlying cognitive system (ie the mind) would act in like manner anyway; considering what we have said so far about the mind's dynamics the process outlined by him is not a given, but it certainly happens now. If he had not been conscious regarding the steps being taken he would not be in a position to recognise any diversion forced upon him by the system following its own preferences.

It becomes apparent that the investigation of a thing requires the definition of its parts. Spinoza appreciates that importance because he is at pains to outline how one may be assured of a good definition. He differentiates between *created* and *uncreated* things, and for the former he advises that firstly, the definition "must comprehend the proximate cause" [33]. That is to say, it has to spell out the circumstance through which the thing establishes its nature which is to be defined. For example, so he writes, if we define a circle as a figure in which all straight lines drawn from the centre to its circumference are equal, we have not defined a circle but only one of its properties. But to say that a circle is a figure "described by any line whereof one end is fixed and the other free" posits the proximate cause on which a circle depends. Secondly, the properties which are to define a thing must be delineated from the definition and must not rely on outside elements as some form of supplement. Using the same example, once we have established the characteristics of the line in terms of a fixed and free endpoint the property of any line drawn thus as being equal necessarily follows, without having to take recourse to anything else.

As far as uncreated things are concerned he lists four rules which I will quote in their entirety, followed by their schematic description:

> I. The exclusion of all ideas of cause—that is, the thing must not need explanation by anything outside itself.
> II. When the definition of the thing has been given, there must be no room for doubt as to whether the thing exists or not.
> III. It must contain, as far as the mind is concerned, no substantives which could be put into an adjectival form; in other words, the object defined must not be explained through abstractions.
> IV. Lastly, though this is not absolutely necessary, it should be possible to deduce from the definition all the properties of the thing defined. [34]

Obeying rule I we have a con-TS which contains only those data (ie abstracts) which we declare to our satisfaction to be concerned with the item in question, and no other. When are we satisfied? We are satisfied with the extent of the TS when any other abstract (object or functional) does not relate to the item as far as we understand the item. When do we understand the item? Rule II tells us that we have achieved understanding when there is no doubt in our mind about the thing's existence, which of course our mind can only realise if and when it has the appropriate abstracts to do so (and it will only ever have those abstracts if they have been presented and something in their nature has prompted the mind to include them in the original TS). As to rule III, we need to remind ourselves that Spinoza's work was written in Latin, and in this language the word 'substantive' can mean the noun *and* its associated adjective (as in "puella bona est"—"the girl is good"). So, to use the statement "the girl is good" as a definition of 'girl' is against the rule because "good" is not a definition, it is a property (similar to the equal lines of a circle mentioned above). In any case, he would not have approved of such a sentence since in his opinion adjectives are largely "modes of thinking" which are formed "from the comparison of things with one another" [35]; hence they are subjective and would go against rule I by drawing on outside parameters. Rule IV may seem to be rather less important, but is significant nevertheless from our perspective. Definition is articulation, hence if the rule is fulfilled we have in our possession a thought structure of which all desired abstracts can be made conscious. But in this case 'desired' means having access to other abstracts which we can articulate to the extent of forming a judgment about something which itself is perceived to be not yet fully articulated. Since articulation is only possible through conscious thought structures, such a judgment is therefore focused on the entire TS, employing the con-TS as an arbiter for deciding which parts will and can be made conscious. And finally, because any abstract can only act upon another (conscious or otherwise) if there is an affinity between the two, the outcome of this selection process is relatively predetermined.

As can be seen, the difference between "created" and "uncreated" things is that in the former the initial representation derives from the outside and so is given to the mind as a PAT[1] and as such as a fait accompli, whereas in the latter the mind has to find the initial abstract amongst its thought structures.

Given what the mind is capable of and what Spinoza requires it to do, we can summarise his approach as follows:

There is a thought pattern such that an incoming abstract (from the outside or from inside the mind itself) finds an affinity with one or more of its TSs. Object and functional abstracts of the resident TSs are applied to discover (again, find affinity with) some abstracts within the incoming complex and if articulated, are stated in terms of the complex. A resultant TS is called a "definition" for as long as the affinitive resident abstracts can be articulated and are perceived to have no further relations with any other TSs which are perceived to be non-affinitive to the incoming TS. Since the definition only uses what can be articulated, the act of perception relies on those functional abstracts which can be made conscious. Note that any sub-TS, however significant in terms of the entire thought pattern, does not constitute the definition and thus cannot be responded to. Put less formally: we define something through the smallest, most obvious detail we can consciously think of and with which we feel ready to justify—anything else we disregard.

The methodology can be called rational because it accepts a formally justifiable relationship between things and rejects all others; as such it represents a vast improvement on more wilful procedures. That it is not without its own pitfalls we will discover in such diverse contexts as pure research, psychology, or domestic violence (amongst others).

One of the truisms which are bandied about and which roll so easily off the tongue is the saying "we know what we know". It seems so obvious, it is certainly true as far as it goes, and it is difficult to argue with precisely because there does not seem to be much more anyone can say about it. It could be seen as a minimal way to express these previous paragraphs—but is it fully understood? The issue rests with our comprehension of the word "know". Spinoza raises that very point indirectly when, after providing us with the elements of a good definition, he goes on to tell us how to find a definition in the first place. It is significant what he has to say about such a need in the preceding paragraph [36].

He begins by saying that as "the chief part of our method is to understand as well as possible the powers of the intellect, and its nature", we need to reach such conclusions by going to the definition of thought and understanding. However, since we do not have any rules for finding the definitions "without a previous knowledge of nature", the definitions must be clear in themselves or else we "can understand nothing". That "previous knowledge of nature" is gained through the understanding of its properties, he writes. What is the problem?

We are after an understanding of the intellect, ie the mind. We are aware that the mind is capable of a number of mental activities, and so the task becomes one of finding the right one for our purposes. Let us illustrate the situation by using an analogy which highlights the current difficulty. Suppose I have a car engine which stalls when on idle. At first I know nothing of its internal functionings and so I begin by examining each one of its parts. Having gone through this process I understand what the fuel pump does, what the carburettor does, the distributor, etc etc. At that point then I have the definitions for the fuel pump, the carburettor, and so on. I posit those definitions against the current definition of the problem—"engine stalls when idling"—and try to see where the nature of the problem fits into the required functionalities of the entire engine. But now I come up against another problem: although I know how a 'good car' works, I still don't know what exactly my 'bad car' is doing except that the engine stops when running at minimum speed. Because there can be a number of reasons why low speed could cause the fuel supply to dry up, I am now forced to define the various properties of *my* car in order to find a discrepancy between it and its normal version. In other words, my mind (a system), examines my car (another system), by taking recourse to what I know of the ideal car (yet another system). This means that I endeavor to define each part of my car and then compare the definitions with my knowledge of 'good cars' and hopefully discover the solution.

Consider now the situation Spinoza refers to. In this case I *am* that faulty car (remember, we need to find a *better* way of using the mind), and the only way I can ever evaluate the performance of possibly better cars is by applying what *I* know of cars (that is to say, what I know about myself) to what I am able to ascertain in others. The only way to do this is by firstly, performing myself, secondly, observing the performance in another, and thirdly, identifying a difference in any one of the corresponding submodules. To put it another way, my system has to define its own parts through an evaluation of their properties and, using these characteristics, compare them with those in a separate system *whose properties I have already defined in terms of my own* (this is all I can do—I cannot 'step outside'). I referred to just that problem in the first chapter (recall the "teacher car"), and now we have come up against one of its manifestations.

There is a solution and to explain it let's remain with our cars. Remember I have only access to my own system, and so I observe the outcome of the various performances I—being the car—am capable of. I can't tell *why* something works differently in myself from others since I don't understand the properties of the responsible submodule, but I can tell from my performance whether a module is up to the expectations. If it is not, I can examine the module in detail in the hope of identifying the reasons for the difference. Note however that such a submodule is a system in itself, and the same considerations as before apply again. I can only continue for as long as my knowledge base provides me with the descriptive tools I need to construct a definition which in turn I can apply to the item under focus. The process stops successfully when I have identified a sufficiently small detail which, when modified, will result in alleviating the problem in terms of the expected performance envelope. It comes to an unsuccessful end when whatever I am examining does not provide me with the definitions I need in order to establish the item's functionality, ie its nature. Such definitions, I repeat, must be able to be articulated to fulfill their purpose.

Here is a further example taken from real life. Suppose I wanted to build a brick house with an unusual type of roof. I consult an architect and he tells me that a brick house will not support my choice of roof. At this stage his mind contains the principles of the house which include the inherent principles of brick, the principles of the roof, and due to his understanding of them all he realises the mismatch. Since the concept of a house contains the building materials and the roof, its nature is senior to the natures of the rest. Therefore the first step is to check whether the rest, ie the submodules, can be modified. I insist upon the roof, so the focus is on the building materials. Considering these in isolation, that is in terms of the functionality they ought to provide with respect to the overall house, it becomes clear that brick is merely one choice out of many. As the architect's mind references itself to the purpose of the basic structure it scans the properties of the range of building materials it has access to and discovers that stressed concrete is the answer to the problem. Ergo, using stressed concrete a house can be built which is capable of supporting my exotic roof fantasy.

Observe the cognitive dynamics: a set of TSs (the house) contains some TSs pertaining to the roof and brick. There are functional abstracts which can be articulated and a problem can be defined. Going through a process of abstracting the con-TSs of the house-structure a wider frame of reference is taken which leads to TSs dealing with affinities to one of the TS (the building materials). The affinitive TSs provide articulation of their definitions and a match is established between their functional abstracts and those of the original TS-set. Due to this match the concept 'brick' is by-passed and 'stressed concrete' is brought in. We have just followed a recursive process and it stops here. Furthermore, the success of the operation depends on the wilful intent to access other thought structures via a functional abstract belonging to a certain TS that is perceived to be the problem. The procedure outlined here represents one of the aspects of creativity, that is the discovery of a solution not directly inherent in the original description. Although Spinoza has not explicitly referred to the issue, it can be seen implicitly through his approach to finding the right definition.

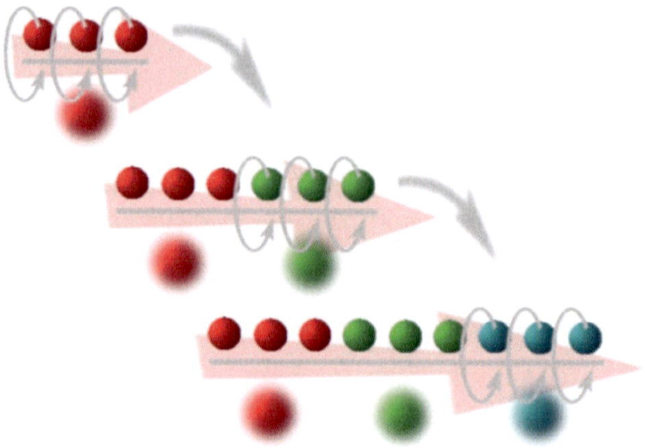

Fig. 6-3: The recursive nature of creating con-TSs, while traversing affinitive thought structures. After one sub-TS has been exhausted, another can be accessed which is mined in turn until no more relevance is found amongst its abstracts. The process repeats itself until it comes to a stop internally (no more relevance), or externally (an abstract is imported which is different altogether). Meanwhile other con-TSs relevant to the waking state continue in the background.

Fig. 6-3 demonstrates the recursive nature of the dynamic. A thought structure is represented by a diffused sphere to symbolise its placement within the subconscious (since its extent is not clearly defined), and its spawned con-TSs are rendered in hard outline (since their extent *can* be clearly defined).

Having established a general methodology of thinking, Spinoza steps into his first application, the concept of God [37]. I am going to present the statements, followed immediately by a translation in our terms. By doing so we have the opportunity to observe the mutual dependencies of abstracts in action. First we have his Definitions:

> D I. By that which is *self-caused*, I mean that of which the essence involves existence, or that of which the nature is only conceivable as existent.

Transl: For something to be recognised as existent it needs to be perceived as existing, therefore it is a PAT^1. Since he focuses on its essence he is talking about a PAT^2 which has been delineated from some particular PAT^1. Therefore we are dealing with a c-PAT-1.

> D II. A thing is called *finite after its kind*, when it can be limited by another thing of the same nature; for instance, a body is called finite because we always conceive another greater body. So, also, a thought is limited by another thought, but a body is not limited by thought, nor a thought by body.

Transl: For something to be recognised in principle, there has to be a higher-level abstract. In his example a particular body (a PAT^1) is recognised as a body in general (a PAT^2), and to say that particular body is limited in its extent means having created a functional abstract (FA) incorporating that characteristic by being in possession of another similar PAT^1 which is being posited against the PAT^2. Since this FA relates to a PAT^1, it is a FA^1. Generally speaking, thoughts about something (to be exact, thoughts about thoughts about something) are higher-level abstracts and thus their FAs reside on the respective abstraction levels of their source. Therefore, a FA^1 about some material item cannot relate to a PAT^2 dealing with a certain thought, and a similar FA^2 cannot relate to some other PAT^1.

> D III. By *substance*, I mean that which is in itself, and is conceived through itself: in other words, that of which a conception can be formed independently of any other conception.

Transl: At this stage it is not clear whether he is talking about an object perceived directly from the outside (this would be a PAT^1), or some other abstract (a PAT^{2+}) which has been discretely identified. Either way, it forms the starting point of a PAT chain.

> D IV. By *attribute*, I mean that which the intellect perceives as constituting the essence of substance.

Transl: The essence of a substance is the PAT^{n+1} of some PAT^n, ie the second abstraction in a c-PAT-l (the first one being the PAT^n).

> D V. By *mode*, I mean the modifications of substance, or that which exists in, and is conceived through, something other than itself.

Transl: If a substance is represented through some thought structure, its modification (also a TS) has a distinguishable relationship with its source. Since the modification has been productively accessed via the source, that source has to be a denominative thought structure in relation to the modification.

> D VI. By *God*, I mean a being absolutely infinite—that is, a substance consisting in infinite attributes, of which each expresses eternal and infinite essentiality.

Explanation.—I say absolutely infinite, not infinite after its kind: for, of a thing infinite only after its kind, infinite attributes may be denied; but that which is absolutely infinite, contains in its essence whatever expresses reality, and involves no negation.

Transl: The concept "being" initially needs to be derived from some observation, hence it exists as a higher-level abstraction containing similar FAs which were equally delineated from lower-level FAs that denote finiteness. Therefore "God" is a TS which contains higher-level abstracts arrived at via an i-PAT-l (none of the infinity-related abstracts could have been part of a c-PAT-l). In his *Explanation* he emphatically denies the infinity-FA the same history as given to it under D II, therefore stressing the fact that whatever FAs the God-concept may contain, they have not been created via the $PAT^n \rightarrow PAT^{n+1} \rightarrow PAT^n$ process (described under D II); rather, they are the result of incongruent abstractions taking us beyond the logic boundary.

> D VII. That thing is called free, which exists solely by the necessity of its own nature, and of which the action is determined by itself alone. On the other hand, that thing is necessary, or rather constrained, which is determined by something external to itself to a fixed and definite method of existence or action.

Transl: For the mind to call something "free", it needs a TS from which any articulated statement (any con-TS) is derived and from nowhere else. For something to be called "constrained" there must be an affinitive TS from which the articulated statements can be derived. Therefore a "constrained" thing takes its con-TSs from a denominative thought structure because in the process the receiving TS will be modified and/or extended.

> D VIII. By *eternity*, I mean existence itself, in so far as it is conceived necessarily to follow solely from the definition of that which is eternal.

Explanation.—Existence of this kind is conceived as an eternal truth, like the essence of a thing, and, therefore, cannot be explained by means of continuance or time, though continuance may be conceived without a beginning or end.

Transl: To conceive 'existence' there needs to be a PAT^2 or higher since its abstraction has been derived from something existent (usually a PAT^1). Therefore "eternity" is a further abstraction of 'existence', making it at least a PAT^3. As such it can be a candidate for any TSs which exist beyond the logic boundary by virtue of its abstraction level alone, particularly since D II insists on pairing only abstractions occupying the same level. The *Explanation* says, because the concept of "eternity" is a PAT^{3+} and furthermore has been derived via an i-PAT-l (nobody is able to confirm the eternal state of anything), any FAs relating to time will be contrived (therefore they can lead to the mistaken conclusion that just because some abstract can be imagined over and over again, generation after generation, its counterpart in the real must be equally permanent. It is a classic case of "I think *of* X, therefore there must *be* an X"). Hence "eternity" cannot be explained by taking recourse to any c-PAT-l that contains abstracts about 'time', although 'continuance' itself may be a higher abstract and so can be situated beyond the logic boundary.

Spinoza continues with his Axioms [38] which I present in the same fashion.

> A I. Everything which exists, exists either in itself or in something else.

Transl: A reiteration of what has been said before—either there is a TS which serves to articulate a thing, or there is an additional TS to which the previous structure serves as a DTS.

> A II. That which cannot be conceived through anything else must be conceived through itself.

Transl: A corollary to Axiom I. If the recursive articulation process cannot find a TS to which the current TS can be applied, the current TS must provide its own con-TS. Or, to express it conversely, "If you can't articulate something clearly, don't use it!"—which leads to,

> A III. From a given definite cause an effect necessarily follows; and, on the other hand, if no definite cause be granted, it is impossible that an effect can follow.

Transl: In order to describe (ie articulate) an effect there must be a con-TS capable of stating its cause. If there is no such con-TS, no relevant effect can be articulated either.

> A IV. The knowledge of an effect depends on and involves the knowledge of a cause.

Transl: See above. Although Spinoza may have felt it necessary to state this axiom, from our perspective there is no need, as can be seen.

> A V. Things which have nothing in common cannot be understood, the one by means of the other; the conception of one does not involve the conception of the other.

Transl: Again, our schema made this obvious under A II—no DTS, no connection.

> A VI. A true idea must correspond with its ideate or object.

Transl: We have seen that by "true idea" Spinoza means the starting point of a concept, that is a PAT chain. It cannot contain anything which gets picked up later along the way, and so it must be the minimal representation of a concept. Therefore it has to match its ideate, otherwise it would not be representative.

> A VII. If a thing can be conceived as non-existing, its essence does not involve existence.

Transl: If there is a TS capable of forming a con-TS through which the quality "non-existing" can be articulated, then a further abstraction of the con-TS contents will not contain the notion "existing" either. This is in contrast to the contents of a sub-TS which may be abstracted, this time without the regulatory influence of articulated expressions and so the resultant abstracts may contain ideas which subsequently turn out to be idiosyncratic or inexpressible altogether. Since they are problematic, they cannot be properly formulated and are subject to misunderstandings and confusion. Although not mentioned here, this is true in general. Hence A VII is not as trivial as it looks.

Next come the Propositions. Unlike his Axioms and Definitions they are accompanied by Proofs and Notes. In the translations to follow I will mostly concentrate on the propositions alone; to explain the others via thought structures etc does not shed any more light upon matters than the translation does already, with a possible exception here or there.

> P I. Substance is by nature prior to its modifications.

Transl: Since the abstract of some item (which of course is the source of a perception leading to the articulation of "substance") forms the beginning of a PAT chain (see D III), any other abstracts pertaining to the represented item must follow later. This goes especially for those relating to a modification, given that the "substance"-TS is denominative with regards to the "modification"-TS, therefore exists already.

> P II. Two substances, whose attributes are different, have nothing in common.

Transl: Because each "substance"-abstract is the starting point of a PAT chain, any subsequent abstracts (of whatever type) of one chain are kept separate from the other chain.

> P III. Things which have nothing in common cannot be one the cause of the other.

Transl: If the recursive articulation of a con-TS cannot present something which a similar process related to another TS has presented as well, then there is no justification for saying that one is in any way the cause of the other. Note however that this proposition depends upon the ability of the mind to recurse to a sufficient extent; in other words, it depends on how 'deeply' we understand something. For example, refer to the scenario about building a house with a strange roof we have described earlier; if the architect did not know about the properties of building materials, then to all intents and purposes concrete would have nothing to do with my house.

> P IV. Two or more distinct things are distinguished one from the other, either by the difference of the attributes of the substances, or by the difference of their modifications.

Transl: This follows from D III and D V. If one were to examine the abstracts of two distinct PAT chains, not only would one find differences amongst the abstracts of the "substance"-chain, they would also be apparent amongst the "modification"-chain.

> P V. There cannot exist in the universe two or more substances having the same nature or attribute.

Transl: Again, a follow-on from his position on "substance". We need to remind ourselves that statements such as "exist in the universe" rest on our interpretation of "exist" as well as "universe". Since both are articulations from some PATs which are made use of at a certain abstraction level, their ultimate meaning for the person saying such words must be interpreted with caution (note that the interpretation is done outside the articulating person's mind and as such is subject to the same criteria, only in another mind).

> P VI. One substance cannot be produced by another substance.

Transl: This goes back to earlier statements concerning the exclusiveness of distinct PAT chains (particularly P III). If I cannot find something similar between two items because their representations in my mind are held to be different, then I cannot justifiably claim any cause-and-effect relationship between them either.

> P VII. Existence belongs to the nature of substance.

Transl: Can be delineated from the previous proposition. If one maintains a thought structure which allows the articulation of "substance", then to be rational one must also articulate the fact that such a thing exist. It is yet another example of Spinoza's intent to demonstrate the functional interdependency of statements—if X then a certain Y must be accepted if a relationship between the two is identifiable, otherwise Y cannot be accepted. One may argue about the finer detail in any particular case, but there is no doubt that his method forces one to think clearly and carefully.

> P VIII. Every substance is necessarily infinite.

Transl: Similar to D VIII, where we pointed to the abstractive nature of the concept 'eternity'. Although "substance" could be any item, in the philosophical sense it means more the inherent nature of something, and since the latter is necessarily a higher-level abstract the conclusion can be reached (employing concepts residing at those higher levels), that such an idea is not limited to its actual manifestation but exists for all time. As is obvious, the idea of "infinite" is the result of an i-PAT-l, therefore its TS has to be denominative to any other structure where that concept is applied. In other words, the concept of "substance" in this case really implies the existence of an incongruent thought structure. In simple terms this might be repudiated; after all, if I say for instance that the "substance" of 'cup' is a small container which can hold liquids, then this seems perfectly reasonable. True; but what is the "substance" of a stone, or a flower, or of a human? We need to confess that we don't understand enough about those things in order to fully articulate the content of such an idea, and sooner or later we are forced to surmise—which creates i-PAT-ls.

> P IX. The more reality or being a thing has the greater the number of its attributes.

Transl: Spinoza appends "(Def. iv)" to the above, so we can assume that for him a thing has the more "reality" the more definitions can be articulated. This may seem to be stating the same thing the other way around, but if an attribute is the essence of some substance, that is a PAT^{n+1} of some PAT^{n}, then a greater number of such abstracts will make the thing "more real" to the holder of such TSs. There cannot be an a priori TS telling us that a certain thing has more reality, unless that same TS has provided us with the appropriate abstracts to make such a sensation possible. Therefore, while Spinoza can express the phenomenon in terms of the above proposition, the underlying cognitive dynamic is actually of the opposite sequence.

> P X. Each particular attribute of the one substance must be conceived through itself.

Transl: Since an attribute must be understood as the essence of some "substance", and the responsible abstract for the latter is the starting point of a PAT chain, then the abstract of its "essence" can also conceivably be the starting point for a chain (keeping in mind that we are dealing with a purposefully designed scheme for thinking). In a *Note* to the above Spinoza warns us about taking "attribute" and "substance" to stand for one and the same thing, in which case a second attribute would necessarily mean another substance; on the contrary, there may be a substance with several attributes, which would make it even more real according to P IX. All these considerations happen on a relatively high abstraction level, and so it doesn't take many recursions along these articulated con-TSs to arrive at the next proposition, namely:

> P XI. God, or substance, consisting of infinite attributes, of which each expresses eternal and infinite essentiality, necessarily exists.

Transl: After all of the above he would say that, wouldn't he? In any case, here we have, for the first time, a quite comprehensively articulated reasoning process which arrives at the supposition of God's existence. So far we could only conclude from circumscriptive evidence, provided by hints expressed through more or less vague descriptions of what the writer thought would be the reality. Notice the subtle cognitive circumstance which made Spinoza's statement possible and indeed logical in terms of his format: were it not for the posited need to declare an abstraction of an essence as unattachable to any other quality (ie it must maintain itself as a starting point of a PAT chain), yet be still linkable to its "substance"-abstract (and that, too, must be a starting point), then the idea of a "*greater* reality" could not be attached to the meaning of any substance. That is to say, we can only talk about the inherent nature of something as being quite different from the inherent nature of something else, if and when our mind does not allow us to create commonalities between the two items. If we could, if we were allowed to view two items as part and parcel of a greater whole, then there would be attributes which have now become interchangeable and therefore the notion of "greater than" would become largely insupportable. That principle is applied quite commonly; consider the building-up (ie enhanced differentiation) of a persona in order to make him or her more demanding of respect, or the opposite expressed in the saying, "familiarity breeds contempt".

> P XII. No attribute of substance can be conceived from which it would follow that substance can be divided.

Transl: The above is in line with previous assertions regarding the starting point of any abstract pertaining to a "substance". If "substance" and/or its "attribute" is taken to be the starting point then any further abstractions should only go forward, and not backwards to some more detailed subdivision.

> P XIII. Substance absolutely infinite is indivisible.

Transl: The combination of "substance" and "infinite" resides at a relatively high abstraction level where such notions can be entertained. It therefore represents an i-PAT-l, and so, from the viewpoint of cognitive dynamics, anything is possible, provided those con-TSs are brought together which are firstly, considered relevant and secondly, fulfill the demands of consistency to the holder's satisfaction. In our case the holder is Spinoza and therefore his demands are more difficult to satisfy. Nevertheless, regarding the above we are beyond the logic boundary and although the mutual relationships of the present abstracts may be rational in themselves there is no guarantee the same quality can be maintained if links to lower-level abstracts were to be explored. Naturally enough, Spinoza remains at these levels in his *Proof*, where he points to the basic indivisibility of something which is infinite, especially if that 'something' represents an inherent nature.

> P XIV. Besides God no substance can be granted or conceived.

Transl: The validity of this statement rests solely upon the content and organisation of the TS which Spinoza constructed in order to justify his idea of God. From an outsider's point of view it may be argued that the particular status thus conferred upon God is not really necessary if one felt the need to imbue the notion of God with omnipotence. On the other hand, there is an implied grandeur which the proposition evokes in the mind of a religious recipient, and therefore would not be rejected by such a mind. As far as the TS itself is concerned, its holder draws certain con-TSs from it which enable him to articulate the existence of a substance with an infinite number of attributes so that the substance has now become most real in comparison with any other substance. The idea of 'infinite number of attributes' is also an abstract, as is its—constructed—corollary 'most real'. According to other articulated con-TSs it is not possible to posit another like substance because to do so would be a contradiction (see for instance propositions V, VIII, XII, XIII). There is yet another message which the above provides. Just as the articulated abstracts from some TS present a certain meaning which has been derived from its TS source, so the configuration of the elements belonging to the medium of articulation have their own meaning. In the present case the medium is words, and there are combinations of words which humans have learned to associate with particular situations, quite apart from the representational value of the words themselves—let's refer to it as the *format* vs the *content*. The two usually support each other, but they may not. Compare for example two sentences, "There is a guy in the classroom standing on the desk" and "There is *some guy* in the classroom standing on the desk". Which sentence sounds more disconcerting? I think most people would agree that the second invokes a feeling of unease through the phrase "some guy", because we are used to that phrase occurring in similar circumstances—it is an example of an intuitive functional abstract (ie no real facts available). Consider now P XIV. The following parts are associated with IFAs: "besides .. no", "can", and "be granted". Since we have several in the same sentence, there are also combinations of the three which evoke IFAs too: "besides .. no .. can", "can be granted", and of course the entire set together. Looking at the singles, IFA 1 says, "here is an authority so powerful that nothing else comes or may come within reach". IFA 2 says, "here is something which is effective". And lastly, "here is something which has the authority to confer, and in a formal manner at that." As can be seen, through their subconscious influence IFAs can be very powerful no matter what the medium, and there is no effective speech writer, marketeer, politician, etc, who does not make use of them.

> P XV. Whatsoever is, is in God, and without God nothing can be, or can be conceived.

Transl: Given there is a TS (beyond the logic boundary) which holds abstracts to the tune of an infinite extent of something, the conclusion can be reached through an articulation of the appropriate con-TSs that therefore any subset of attributes must be part of the greater whole, a kind of universal set. Conversely, if the abstract relating to such a universal set is declared invalid, then all the subsets must be invalid too. Whether Spinoza availed himself of the principles of set theory is debatable, given they were formally expressed by Georg Cantor in the late 19th century, although his *Proof* makes use of them in a roundabout way.

> P XVI. From the necessity of the divine nature must follow an infinite number of things in infinite ways—that is, all things which can fall within the sphere of infinite intellect.

Transl: By now it is becoming obvious that the following propositions are delineations which have been made possible by the initial statements, including the *Definitions* and the *Axioms*. They also demonstrate the ease with which conclusions can be drawn once the logic boundary has been breached. Since none of the assertions hence have to conform to any c-PAT-l link to lower-level abstracts, the need for grounded TSs has been removed. Note that abstracts such as 'infinite' and 'intellect' can quite comfortably sit within high abstraction levels and can therefore be imbued with notions which otherwise would not be possible. Thus it is possible to establish a network of interdependent ideas which through their presence alone create IFAs and even seemingly cognitive functional abstracts, all of which in turn reinforce the intended, overall meaning. Suppose I wanted to prove the existence of the divine through the previous assertions and finally bring in Cantor's set theory to show the correctness of proposition XV. Because I apply Cantor's theory within the domain (and *within* it only) of the higher-level abstracts beyond the logic boundary, it is possible to point to the principles of the theory (which can be proven correct by more down-to-earth examples) and declare the existence of the divine as correct. I have used CFAs, but only in an apparent fashion—there are no connections with the identifiable real. Some people would even go as far as to say that Cantor has now proved the existence of God! We can say therefore that once a CFA has been moved across the logic boundary it has become an IFA.

> P XVII. God acts solely by the laws of his own nature, and is not constrained by anyone.

Transl: Any law which is being spelled out is the articulated version of a functional abstract. Since in this case all the relevant abstracts exist on higher levels beyond the logic boundary, whatever can be articulated through a recursive process which mines the appropriate TSs, is going to reflect the functional abstracts resident there. Those functional abstracts are IFAs, and Spinoza's system makes it clear that they are not to be interfered with by any more realistic considerations, ie CFAs. He puts it thus, ".. solely from the necessity of the divine nature .. an infinite number of things absolutely follow in an infinite number of ways" [39].

> P XVIII. God is the indwelling and not the transient cause of all things.

Transl: If we have a TS which postulates the sum total of attributes pertaining to some representative abstract—in this case God—one can conclude that any cause such attributes may eventuate must also reside in the representative abstract.

> P XIX. God, and all the attributes of God, are eternal.

Transl: From our point of view this is merely a repetition of what has been said before; namely an abstract, being linked with the notions of "substance", "infinite", and "all attributes", can be understood as having an eternal nature. Such an understanding represents an abstract to which the others ("substance", etc) are denominative.

> P XX. The existence of God and his essence are one and the same.

Transl: It has been posited that an infinite number of attributes makes something most real. By employing certain IFAs which reinforce the notion of general grandeur (nurtured by previous abstracts), one could therefore conclude that when it comes to the existence of such a being, even its further abstract (ie essence) is not much different either. If one was being technical in terms of our schema, one could say that at this point the abstract "existence" and the abstract "essence" have been fused together, in other words, they are at the same level. On a more pragmatic note it could be claimed that Spinoza's mind had reached its limit of creating abstractions.

> P XXI. All things which follow from the absolute nature of any attribute of God must always exist and be infinite, or, in other words, are eternal and infinite through the said attribute.

Transl: Nothing new there—any PAT chain at that level of abstraction contains similar notions of "infinity", etc etc.

> P XXII. Whatever follows from any attribute of God, in so far as it is modified by a modification, which exists necessarily and as infinite, through the said attribute, must also exist necessarily and as infinite.

Transl: Same as before, except such a PAT chain includes now any denominative abstracts and/or DTSs.

> P XXIII. Every mode, which exists both necessarily and as infinite, must necessarily follow either from the absolute nature of some attribute of God, or from an attribute modified by a modification which exists necessarily, and as infinite.

Transl: Similar to the above, except that the perspective has been turned around and one looks now from some TS which, by virtue of God's infinite attributes, must be denominative to any TS or sub-TS relevant to the divine, towards such a relevant TS. It is an example of a functional abstract being made affinitive with some other thought structure (even if that TS exists this side of the logic boundary).

> P XXIV. The essence of things produced by God does not involve existence.

Transl: In terms of what has been said under the previous proposition, the fusion of the existence-essence abstract is taken to represent essence alone. Some people may baulk at the rationality of that conclusion, but it shows what is possible beyond the logic boundary. If there is a functional abstract which stipulates that "essence" is taken to have a more elevated status than "existence", than in the case of such a fusion where the functional abstract has become an IFA the element "existence" may be dropped altogether and still retain the presence of the basic abstract.

> P XXV. God is the efficient cause not only of the existence of things, but also of their essence.

Transl: If the "existence" abstract is synonymous with the "essence" abstract, whatever has been related to the former can now also be related to the latter.

> P XXVI. A thing which is conditioned to act in a particular manner, has necessarily been thus conditioned by God; and that which has not been conditioned by God cannot condition itself to act.

Transl: Whatever a con-TS can articulate with relation to a particular thing, must contain qualities which can also be found in the infinity-set pertaining to the God-TS. It is a necessary conclusion derived from the functional abstracts postulated so far.

> P XXVII. A thing, which has been conditioned by God to act in a particular way, cannot render itself unconditioned.

Transl: As Spinoza himself declares, "This proposition is evident from the third axiom", [40]. Once a con-TS linked to a structure regarding God articulates some condition, one cannot 'switch off' the item under focus.

> P XXVIII. Every individual thing, or everything which is finite and has a conditioned existence, cannot exist or be conditioned to act, unless it be conditioned for existence and action by a cause other than itself, which also is finite, and has a conditioned existence; and likewise this cause cannot in its turn exist, or be conditioned to act, unless it be conditioned for existence and action by another cause, which also is finite, and has a conditioned existence, and so on to infinity.

Transl: No matter how comfortably one is able to argue along the lines of high-level abstracts beyond the logic boundary, sooner or later a connection has to be made with reality, that is those PATs which undoubtedly have been derived from real-world observations (the c-PAT-ls). *How* such a connection is carried out depends on the individual mind—sometimes they are more bizarre, sometimes less so, and there are individuals who do not even bother with any connection at all. In Spinoza's case, proposition XXVIII is it. Without the above one would have to postulate that since everything is somehow caused by God with his infinite qualities, the results of those causations must be infinite in nature too. Well, in order to accommodate the everyday experience of finality, an i-PAT-l is set up for just that purpose.

> P XXIX. Nothing in the universe is contingent, but all things are conditioned to exist and operate in a particular manner by the necessity of the divine nature.

Transl: By now, a sentence containing such elements as "nothing", "in the universe", and "all things", should alert the student of thought structures to the fact that we are dealing with IFAs at the very least. What remains to be examined are the contexts of "contingent" and "divine nature". Since we know in this case that the "divine nature" is held to be the cause of all things, the above should not be surprising.

> P XXX. Intellect, in function finite, or in function infinite, must comprehend the attributes of God and the modifications of God, and nothing else.

Transl: The statement is a functional abstract related to what has been said previously about the all-including nature of God. Words such as "must", "comprehend", and "nothing else" invoke secondary IFAs of the type described under P XIV.

> P XXXI. The intellect in function, whether finite or infinite, as will, desire, love, &c., should be referred to passive nature and not to active nature.

Transl: Spinoza emphasises that by intellect he does *not* mean "absolute thought", and therefore we can take it to mean a thinking process modified by the absolute kind; thus it is passive. We can also say that it is the target of some denominative TSs, where the DTSs would most likely originate from thought patterns dealing with the divine. The associated functional abstract is significant; the holder of such a rule would constantly remind himself that ultimately his thoughts are not the product of his own self, but are subject to control by other thoughts, which are part of the TP containing the contextual details of the God-concept. One common name given to such TSs is 'one's higher self'. It is also a further instance of the segregating influence of the logic boundary, where anything emanating from there is automatically held to be not really one's own and therefore represents some kind of separate being that resides within. Given the functional abstracts that are specific to the TSs beyond the boundary (eg, representations there are held to be a reflection of something bigger than humans), it is easy to see why. One of the ramifications is the next proposition.

> P XXXII. Will cannot be called a free cause, but only a necessary cause.

Transl: Spinoza calls "free" an event which is not linked in any way or form to the outside of its domain, in contrast to "necessary" where at least some of its part have an exterior connection (see D VII). The expression "will" represents an articulated form of a relatively high abstract. It comes from an array of TSs which in one form or another give rise to the sensation that one's actions, decisions, preferences, etc are the product of some active agent. The notion of 'agent' must be an abstract at a higher level than the observation which was the original trigger to the notion. The level's status is relative in the sense that the observation itself can be a simple or a more sophisticated thought process (we will encounter samples of it in Part II). Thus the concept of 'will' is a candidate for an association that is not necessarily based on context, but on the functionality a particular abstraction level implies. So far we have encountered the former, that is *contextual associations* between thought structures; now we are dealing with a *functional association.* (An everyday example of a contextual association is someone who thinks of their upcoming holiday when watching a tourism commercial on TV. A functional association would be the idea that just as there are classes of bees within one hive, there ought to be similar social differentiations amongst the members in a society.) With regards to the present proposition, the existence of the 'will'-abstract at a certain level will make it functionally associative with similarly positioned abstracts beyond the logic boundary (or anywhere else for that matter). As it happens, the targets are the ideas a person has about God as an agent working as an authority in the background, just as we can view our 'will' as some controlling force underlying our affairs. Hence it is possible for Spinoza to arrive at the above conclusion. From this perspective his comments under *Corollary II* are more interesting than the actual *Proof* he furnishes, for he writes, "It follows, secondly, that will and intellect stand in the same relation to the nature of God as do motion, and rest, and absolutely all natural phenomena, which must be conditioned by God (Prop. XXIX) to exist and act in a particular manner" [41]. A corollary of our own would be the recognition in this case that for Spinoza the concept of 'will' would have the same intellectual status than 'motion', 'rest' and like phenomena.

> P XXXIII. Things could not have been brought into being by God in any manner or in any order different from that which has in fact obtained.

Transl: A follow-on from his previous thought pattern which contained the idea, expressed in so many ways, that God is the ultimate definer. Therefore there cannot be any functional abstract associated with anything which represents a circumstance that the FAs associated with the God-TP do not represent themselves.

> P XXXIV. God's power is identical with his essence.

Transl: A phenomenon similar to P XX, where the abstracts "existence" and "essence" had been fused together. Here it is "power" and "essence".

> P XXXV. Whatsoever we conceive in the power of God, necessarily exists.

Transl: After the previous proposition, he can hardly say otherwise. Still, it is an example of a reciprocity.

> P XXXVI. There is no cause from whose nature some effect does not follow.

Transl: A further ramification from the nature of Spinoza's God-TP, as far as a recipient of its articulation would be concerned—everything relates to God, therefore there is no cause etc etc. From the purely cognitive perspective however it is a case of evoked IFAs, similar to the case we already discussed under P XIV. The mere word 'cause' implies the existence of an effect, no matter how well either of them have been observed, analysed, and/or understood. Since many of the TSs associated with God refer to some kind of action, the idea 'cause' suggests itself sooner or later and becomes transferred to TSs this side of the logic boundary by virtue of its functional associativity with the latter. There is also an element of reciprocity—IFAs about God present the notion of 'cause' and so 'cause' is now justifiably applicable to the rest of the world. Thus end his propositions regarding the divine.

Given a statement such as P XXXVI above could lead one to assume that there are functional abstracts which influence Spinoza's thinking to the extent that God is seen as some kind of deliberate architect and planner of the universe and especially our affairs. This is not necessarily so. Consider proposition XXVIII which serves as a conduit between high-level abstracts beyond the logic boundary and those derived from more realistic PAT chains this side of the border. There, other rule sets are invoked which explicitly deal with the concept of finiteness and so are in direct contrast to those abstracts which contain the notion of infinity. Therefore any comparable TSs (comparable in terms of such things as 'influence' or 'cause-and-effect' phenomena) are *not* denominative as far as infinity and/or finiteness are concerned—which is of course their whole point. One by-product of their difference is the ability by their holder to conduct thought processes in either domain without one automatically influencing the other. Such a possibility must be kept in mind when evaluating the state of someone's mentality if prompted to do so by a quaint pronouncement of theirs. It may well be that there are IFAs which stretch right across several TSs or even TPs, but it may also be the case that the person is in possession of a particular IFA which does *not* carry the 'quaintness-feature', in fact deliberately rejects it, and so helps to protect associative TSs in other areas from being tainted. There is an analogy with genes: suppose I have a gene block which is concerned with the hair growth on one's head. One set of genes deals with hair colour, another with baldness. One could transfer the entire block to another person and colour as well as baldness would be reproduced. One could also switch off the baldness set and then transfer the block—same hair colour but no more baldness, yet there is still integration with the rest of the recipient's gene chain. To put it formally, we can say therefore that if there is a TP which contains some rule set in its TSs including a particular functional abstract X, then there can be another TP containing a similar rule set but *without* X if there is an associative TS between the two TPs such that it contains a functional abstract which is incompatible with X. The more functional abstracts of the incompatibility-producing kind a person possesses, the more complex the mind will be. The higher the abstraction levels at which those abstracts exist the more room there will be for any TSs below, thereby increasing the informational value of the structures themselves. Higher-level abstracts mean greater intelligence—a familiar corollary to complex personalities.

Such a principle occurs through P XXVIII and Spinoza, being ever the thoughtful writer, did not forget to address a possible misunderstanding caused by a too literal a reading of statements such as the last proposition. He says,

> All such opinions [ie certain misconceptions] spring from the notion commonly entertained, that all things in nature act as men themselves act, namely, with an end in view. It is accepted as certain that God himself directs all things to a definite goal (for it is said that God made all things for man, and man that he might worship him). [42]

He goes on to say that because humans tend to view the world through their own, personal perspective (which includes planning and design), such an attitude is transferred to the greater scheme of things which involves God being posited as the origin of all things. As such, so the opinion goes, God is taken to have planned the particular nature of existence, even in detail—for example, a rose is beautiful because God wanted to create beautiful flowers. However, he warns, a rose is not inherently beautiful—it is only our opinion that makes it so.

We have seen that the presence of certain thought structures, particularly some functional abstracts, lead to a transfer of assumptions to other, similarly perceived occurrences because there exist functional and/or contextual associations between the structures. We have also seen—especially through examining P XXVIII—that there are means by which such a transfer can be avoided. In Part II, where we will look at the evolution of thought structures from the ground up, the ramifications of these TSs can be seen in greater detail; they are by no means insignificant.

There is much more to be found in Spinoza's writings and, if so inclined, the reader will find it an interesting exercise to apply the cognitive schema to what the philosopher has to say about such things as the mind, our emotions, and freedom.

We continue our investigation with a philosophy that directly influenced the political landscape of Europe, and indeed the rest of the world. The English Revolution of 1688, also called the "Glorious Revolution", drew its intellectual principles from one man: John Locke. It is to him that we turn in the next chapter.

7. The mind as government

John Locke

The Middle Ages are behind us. True, one could have descended on a kind of time-parachute into a rural hamlet somewhere in Europe and find the conditions not much different from two or three hundred years ago. It would have been equally possible to sneak into a soiree of aristocrats and listen to conversations about King and God and Honourable Lineages and be none the wiser as to the precise date. And yet, there was an increasing number of people who possessed the intellect and opportunity to discuss topics which would have been literally unthinkable in days gone by. Such concepts as gravity, calculus, the telescope and the microscope, the paths of the planets, the calculating machine, and many more, extended the range of minds and gave them the vocabulary to express in new and unfamiliar ways. These people, who later hindsight categorised as the 'middle class', found a niche between wrestling existence from hostile dirt and exercising control in precarious splendour to pursue a life which, generally speaking but true nonetheless, enjoyed a self-created buffer against mind-numbing labour on one side and the energy-sapping wiles to maintain one's authority on the other. More and more that class put *its* stamp on society, and because of the different function it invited denigration from below and suspicion from above.

Such changes did not escape the attention of philosophers. Ever positing one phenomenon against another, weighing its significance in relation to this, then that aspect, they naturally imported and made use of the new concepts of their day. In earlier chapters we drew on the possibility to make analogies between thought structures and manifest actions (for example, see chapter 6 regarding Roman and Gothic architecture). The possibility exists because ideas do not remain within the confines of one's brain; they find their way into the daily conduct of life, whether that be defining a business, a state, or an urban landscape.

Just as a mind becomes powerful when its thought structures not only perform higher and higher abstractions but at the same time maintain vibrant links with direct representations from the outside, so does human organisation at large compete better if its institutions mirror the mental alacrity. Such achievements are relative of course, and when everything is said and done we, with the benefit of hindsight, can see the limitations every age, despite its optimism, had to endure.

In England the revolution of 1688 established government as a trustee of the people, manifest in the parliament; it laid the foundations for an intellectual re-organisation that, on a broad scale, made the British Empire possible. In terms of the embodiment of a political idea it was an abstract, placed between the lower levels—the Populace—and the higher ones—King and God. The architect of its philosophical material was John Locke (1632 - 1704), and we are going to examine his major work, *An Essay concerning Human Understanding* [1].

A witness to an after-dinner conversation in a middle-class salon in 17/18th century London could have listened to a debate about the solar system, the flow of blood in the human body, or the mathematics underlying the definition of population changes in a society—and would have walked away thinking, "Yes, but where does all that leave God?". It is not that most people had turned away from Christian tenets, it's just that amongst the multitude of fascinating topics the question of religion would not necessarily have been broached—and for good reason. The mental dynamics which produced the flow of conversation had no direct need to incorporate metaphysics because the themes at hand did not need them. To directly articulate the position of God in relation to the contemporary understanding of life, at this stage of history, required a degree of daring not yet achieved—

this would come later. Still, those questions were there, hovering in the background, and a philosopher had to address them sooner or later. Therefore, in order to map out the intellectual space of the man John Locke let us deal with his perspectives on God first, so that we may define the entire field in terms of the irrational on one side and what has been set aside on the other for the material which contributed so effectively to the Age of Reason, because this is where we are now.

Any concept of the divine has its origins in the extrapolation of human qualities; it is the nature of these qualities, expressed within a society's culture, which defines the intellectual path that leads from Man to God. Our soul, mind, etc, feature under those terms. Such terms influence how we see human will, desire, and so on, act themselves out. For Locke the mind is an entity which contains faculties, namely "understanding" and "will". Moreover, he wants us to know that by the word "faculties" he does not mean some outside agent but an intrinsic functionality, ready to operate according to the mind's preference. Here is how he puts it,

> ..And the ordinary way of Speaking is, That the *Understanding* and *Will* are two *Faculties* of the mind; a word proper enough, if it be used as all Words should be, so as not to breed any confusion in Men's Thoughts, by being supposed (as I suspect it has been) to stand for some real Beings in the Soul, that performed those Actions of Understanding and Volition. [2]

And a few sentences later,

> ..Yet I suspect, I say, that this way of Speaking of *Faculties*, has mislead many into a confused Notion of so many distinct Agents in us, which had their several Provinces and Authorities, and did command, obey, and perform several Actions, as so many distinct Beings; which has been no small occasion of wrangling, obscurity, and uncertainty in Questions relating to them. [All the italics in these and subsequent quotes are those of the quoted author]

While he does not mention any metaphysical beings as those agents, he clearly organises his thought structures such that under the auspices of 'mind' any identifiable faculties are well positioned under *its* sphere, and no-one else's. Once this is done, it leads to the ascribing of potential to the faculties; not one that derives its inherent characteristics from on high (whatever they may be), but one which is produced and implemented by, and ultimately responsible to the entity to which it belongs—our mind.

Once we allow our minds to relegate responsibility to our own thinking, we are in a position to define what those responsibilities can be. In Locke's case they are the ideas of liberty and necessity, a step which opens up a whole set of perspectives focused on the individual person rather than a supreme being.

Another collection of functional abstracts gets evoked by how Locke interprets the perception of substances. For him there is such a thing as a "complex idea", which is the result of a combination of "simple ideas", and so anything that possesses several attributes becomes represented by a complex version of the basic process. The "incomprehensible supreme Being" is such a complex idea, and the concept is derived in exactly the same way. There is a set of FAs then which encompass the notion that whatever God means to us has been derived from observations and perceptions that we, humans, were able to coalesce into a greater whole: God as being defined by humans. For example, our perception of Existence, of Knowledge, Power, Pleasure and Happiness; they all get enlarged "with our *Idea* of Infinity" and thereby form our picture we have of God. Note the fundamental shift of the logic boundary—from a self-contained system of the world placed wholly beyond its line to a conscious developmental path of abstracts which only at their point of extremism are placed on the other side. All this of course does not invalidate the mechanics of the abstraction processes described in earlier chapters; the difference now is the conscious recognition of what the mind has actually been doing.

The foregoing has not been written by Locke for the express purpose of proving the existence of God. I have used the statements to illustrate certain basic considerations that have wider implications, for it is the inherent potential of functional abstracts to influence other, even not directly related PAT chains. Later in his book there is however a chapter dedicated to "our Knowledge and Existence of God", where he sees the need to apply his general concept to our understanding of the incomprehensible. By the way, it would be an interesting exercise to investigate the reasons *why* philosophers feel compelled to delve into areas about which no however posited human faculty can make any productive sense. A Mani or an Aquinas, given the context of their times, would have had a justifiable urge to do so, but even Locke is not the last in the line and the idea suggests itself that just possibly, the dedication comes from a nagging doubt about the veracity of their output per se, and so calming their uncertainty through an exercise in *quantitative* completeness, if not a *qualitative* one. A task for a sensitive psychiatrist to be sure, but hardly realisable for want of willing, real-life subjects.

In any case, Locke's proof of the existence of God comes initially from the realisation of our own existence [3]. Just as we know in our innermost selves that we exist, he says, so must we also accept that there is some power which has brought us into existence. Therefore there must be an all-powerful being which has brought the entire universe into existence. He goes one step further by declaring that while we do not know everything there is to know about the world, we can conceptualise its totality and therefore apply a certainty to the knowledge of God we could not gather in relation to the world's innumerous detail. Hardly a revolutionary idea, but one which someone of Locke's caliber (and assumed Christian beliefs) would have to entertain. After all, everything else he says about mental dynamics is placed well and truly this side of the logic boundary.

On that note it is significant what he writes about the proof's ultimate efficacy:

> *How far the* Idea *of a most perfect Being*, which a Man may frame in his Mind, does, or does not prove the *Existence of a God*, I will not here examine. For in the different Make of Men's Tempers, and Application of their Thoughts, some Arguments prevail more on one, and some on another, for the Confirmation of the same Truth. [4]

A far cry from the "Truth is one and common to all" of St Augustine (see chapter 4)!

A major factor in deciding on an all-powerful being which brought the universe into existence is the assumption that all the perceivable complexity and hence interdependence ought to be due to a plan, a design from the very beginning. We have come across the idea already and find another hint in Locke's contention on the same page that it would be "senslesly arrogant to suppose Man alone knowing and wise, but yet the product of mere ignorance and chance". He continues to develop this thought by suggesting that while there might have been Matter and Motion at the beginning of time it could not have produced Thought, unless there had been a "cogitative being" present which imbued matter with it, and such a being must have existed for all eternity [5]. Such an assumption is based on the fundamental differentiation between Thought and Matter, a problem we have followed from antiquity onwards and whose final resolution will have to wait until Part II.

To conclude the section on Locke's perspective on God, here is a truncated passage full of subtle hints about what awaits the active, curious mind employing reason, however religious it may be at the outset:

> No Man is so wholly taken up with the Attendance on the Means of Living, as to have no spare Time at all to think of his Soul, and inform himself in Matters of Religion.
>
> ..Besides those, whose Improvements and informations are straitned by the narrowness of their Fortunes, there are others, whose largeness of Fortune would plentifully enough supply Books, and other Requisites for clearing of Doubts, and discovering of Truth: But they are *cooped in* close, *by the Laws* of their Countries, and the strict guards of those, whose Interest it is to keep them ignorant, lest, knowing more, they should believe the less in them.
>
> ..And, however they may seem high and great, are confined to narrowness of Thought, and enslaved in that which should be the freest part of Man, their Understandings. This is generally the Case of all those, who live in Places where Care is taken to propagate Truth, without Knowledge; where Men are forced, at a venture, to be of the Religion of the Country, and must therefore swallow down Opinions, as silly People do Empiricks Pills, without knowing what they are made of, or how they will work, and have nothing to do, but believe that they will do the Cure.. [6]

One of the nicest pieces of philosophical subterfuge in Western literature, if I may say so.

Towards the end of his *Essay* Locke presents a set of guidelines, one dealing with the potential failing of reason and the other with the intellectual ingredients of discoveries. As he uses descriptives which a reader needs to know in order to understand the current discourse, their particular placement in the text makes sense. For us, translating his explanations directly into our specific vocabulary, not only is the sequence irrelevant but it is in fact advantageous to first understand his view of how the mind produces and possibly errs in the process before going any further. Whatever he had to say on previous pages is of course the outcome of such an assumed activity.

We begin with the list of pitfalls related to reasoning [7]. Reason fails when our ideas fail, he writes. In other words, there is no reasoning without proper content. When it comes to words, if the words are not backed by ideas, then "'tis only about those Sounds, and nothing else". We know that a thinking process, that is a PAT-chain or an affinity-seeking activity, would not come into being unless there are abstracts to begin with. In Locke's case however we are not only concerned with the content of general TSs, but those parts which become articulated in one way or another. Therefore there have to be conscious sensations that can be fed back into the system to be further processed—it is not a matter of passing through mists of feelings but of effectively handling concrete chunks of cognitive detail.

The second cause of failure is the confusion associated with certain ideas. Keeping in mind that it is only those abstracts which can eventually be articulated that keep the articulation process going, it is easy to see that in the absence of these the recursion, relying as it does on con-TSs, breaks down if it comes across structures which cannot be mined productively. Confusion arises when a TS is sensed but cannot furnish any concrete detail. At this stage other factors come into play: the need to continue one's speech for instance (FAs related to the ego), the personal goal of the conversation (some concept residing in a TS which needs to be brought forth), and any affinitive TSs that are perceived to be supportive to the goal (ie abstracts which may not be part of c-PAT-ls but are used nonetheless). Here it is significant that Locke refers to thoughts about God as being much more prone to confusion—naturally, since it is these which represent most likely the content of incongruent abstraction chains leading beyond the logic boundary.

Thirdly, the reasoning process stops because the articulation process is unable to link to those abstracts which *could* provide a satisfactory outcome—ie the inability to traverse TSs either laterally (along the same level of abstraction) or vertically, upwards or downwards an abstraction level (Locke himself does not address such detail however). Such an inability is more pronounced in some people than others he says, which relates to our definition of intelligence.

A similar problem (Locke's fourth cause) arises when i-PAT-ls are followed without the mind being able to extract itself from such a sequence and jump into a c-PAT-l. Since any sequence relies on the current abstract's affinity to what should follow, an incongruent chain is unlikely to be affinitive to TSs that are congruent in terms of the entire flow of ideas. A phenomenon which can be observed any time a problem becomes intractable and one is forced to retrace one's steps to some previous concept and start anew. In certain cases however there is no realisation of that need and so the process continues unproductively or, worse still, produces results which are not recognised as being flawed. Such phantom logic occurs mostly where the processes are protracted, include many details, and are made use of by minds that are predisposed to particular outcomes at all costs.

On a similar note Locke mentions the further possibility of having chosen words which by themselves do not provide a clear idea and so lead to obscure trains of thought. This is akin to problem 3, in that the word's entire TS cannot either come up with links to productive abstracts or contains affinities to those TSs which do not have usable con-TSs (or none at all).

Once these obstacles are overcome, the mind is led to discoveries. According to Locke there are two types [8]. The first consists in the participation of other ideas which, although not addressed directly at first, are nevertheless included via the evaluation of the initial concept. We can translate that by saying that two con-TSs are able to access some other affinitive TSs in a congruent manner, and thereby providing us with the confirmation we are seeking—it is the positive outcome of the scenario underlying the second problem mentioned before. Several conditions have to be met: the inherent congruency of the original TS, the additional abstracts belonging to some c-PAT-l chain, and the mind's ability to consciously retain the details along the way. The last point Locke addresses by mentioning the functioning memory, in the absence of which it becomes "too hard for those who have not strength of Parts distinctly to perceive, and exactly carry so many particulars orderly in their Heads".

The other type concerns those discoveries which rely on similar affinities, but where their contents are not the exact detail wished for but suggest themselves in the form of probabilities. For us the difference is not as clear as it appears to Locke: there are degrees of affinity in any case, ie associative intensities, which guide the process one way or another (for an explanation of associative intensity see the section on Descartes in chapter 6). For Locke this kind does not constitute knowledge as such, rather an outcome which presents itself on the basis of its relative strength of argument.

Comparing the general perspective of a John Locke as we have encountered it so far to the assumptions made in previous ages, we find an insight which could only have been sharpened once certain traditional FAs had been closely examined and, if found wanting, dispensed with. The fundamental ability to question a train of thought is not an absolute power; it grows incrementally, by drawing on one example of possibility after another, and even the experience of newness itself becomes a light towards which a curious intellect is drawn. To uncover a hidden principle, to find it again somewhere else, to then abstract the act of finding per se and search anew, now armed with even greater powers of cognition, and so on and on until a previously unassailable intellectual edifice remains a shadow of its former self; we are looking at a period of human history where these very processes achieved a unique intensity.

Consider Locke's description of learning. He talks about basic ideas we carry in our minds, some so basic that we are unaware of them ever having been taught. This leads many people to assume that the ideas had been put there by God. He comes to the conclusion that such an opinion can only happen if the mind never consciously queried the idea in the first place, and so accepted it as something given and made present by an outside entity. Not so, he says. Having applied his logic to considerations of simple vs complex ideas, to what constitutes a discovery, and how one's thinking can make mistakes, he is able to be critical [9]:

> ..That when *Men*, so instructed, are grown up, and reflect on their own Minds, they cannot find any thing more ancient there, than those Opinions, which were taught them, before their Memory began to keep a Register of their Actions, or date the time, when any new thing appeared to them; and therefore make no scruple to *conclude, That those Propositions, of whose knowledge they can find in themselves no original, were certainly the impress of God and Nature* upon their Minds; and not taught them by anyone else.

While some awareness about children's learning had certainly existed for a long time—otherwise there would have been no schools of any kind—the conscious spelling out of the absorption process does not come naturally, unless one is open to some useful concept of ideas and how they could have been formed in the first place. Once again there are ramifications; if I come to accept that an idea, *any* idea, has been the result of an act of importation by the mind, and needs further ideas to enable meaning and understanding, then my mental doors are equally open to the task of finding instances where ideas exist *without* the accompanying meaning and understanding, and how this could have come about. The pertinent word in the situation is "if", and the pertinent FA in the mind is "what if"—the most powerful of them all.

So, where *do* all the ideas come from? In Locke's words, "From *Experience*" [10]. By this he means the act of observing on the outside as well as living through the thoughts on the inside, these two being "the Fountains of Knowledge, from whence all the *Ideas* we have, or can naturally have, do spring". Outside experiences provide the "sensations", he says, and build up a picture of the world in terms of their perceived qualities. Internally we also observe, this time our mental activities such as thinking, doubting, believing, etc etc, culminating in "reflections". Note however that Locke's differentiation aside, all these thoughts are the result of abstracts which coalesced into TSs, parts of which provide the conscious articulation for anyone to pronounce the word "thinking" or "believing" in a meaningful way.

"Observe" then the obviously conscious act of articulating something like 'thinking'. There needs to be an abstract (keeping in mind any other TSs that contributed to the formulation of the entire concept) which finds its way into the con-TS to be later spoken out as so many words. Since such an abstract does not come from the sub-TS the words will only refer to something which the mind is conscious of. Therefore the process, entertained by Locke's mind and which gave rise to the declaration, "Man thinks", also contains the message that whatever is not made conscious does not constitute thinking. It is understandable when one considers the preliminary steps needed in order to come that far; given the context of the times, not a negligible task. Further FAs leading to musings about activities beyond the conscious had not yet been created, and so Locke is compelled to ask,

> Can the Soul think, and not the Man? Or a Man think, and not be conscious of it? [11]

But, having contemplated the origin of ideas and what we do with them, he is at a stage where he can write soon after, quite logically,

> I see no Reason therefore to believe, that the *Soul thinks before the Senses have furnish'd it with Ideas* to think on; and as those are increased, and retained; so it comes, by Exercise, to improve its Faculty of thinking in the several parts of it, as well as afterwards, by compounding those *Ideas*, and reflecting on its own Operations, it increases its Stock as well as Facility, in remembering, imagining, reasoning, and other modes of thinking. [12]

Of course, it all depends on what one means by the word 'thinking'. It is worthwhile pointing out that Locke, for all his astuteness and insight, follows the trail of cognitive elements to the point of simple, *conscious* ideas and their inner workings—but no further. His TSs prompt him to advise,

> A sufficient impulse there may on the Organ [eg the eye, the ear]; but it not reaching the observation of the Mind, there follows no perception: And though the motion, that uses to produce the *Idea* of Sound, be made in the Ear, yet no sound is heard. [13]

As we will see later, the above conclusion is true according to the context of Locke's thought structures; it is false according to the actual dynamics of the mind.

To summarise the current process: a TS has been developed which includes the notion that while there is a God, the *manner* in which the entity is perceived depends on the individual mind's history, that is exposure to experiences and subsequent thought processes. FAs are set up containing the rule set about thoughts associated with specific input, specific perception, specific interpretation, specific conclusion. Associative TSs emerge which hold notions of 'idea' and its variances—the representations of the content of input, perception, etc. Other FAs exist, again associative to the former, holding the principles of such concepts as 'the simple towards the complex', 'the juxtaposition of items which are similar in one way yet dissimilar in another', 'the validity in the derivation of several items, all of which are valid themselves', 'the significance of words as symbolic representations of ideas', 'the connection between consciousness, thought, and words, and their mutual interdependency'. Consequently there are now TSs where the experiential development of humans is held to be important, the existence of thought processes is acknowledged but mutually tied to the existence of consciousness, and only that is accepted as known in any way or form which can be consciously accessed and articulated—though its ambivalent truth value is understood.

Although the articulated results of these TSs do not necessarily reflect reality (for example, the fact that the brain *does* receive information without always making it available to one's consciousness has not been demonstrated until the 20th century—see Part II), their principle format leaves the door open for future updates should the appropriate information come to hand. They furnish Locke with sufficiently effective FAs to state,

> The best way to come to the Truth, being to examine Things as they really are, and not to conclude they are, as we fancy of our selves, or have been taught by others to imagine. [14]

Once the concept has formed to see ideas as separate elements of the thinking process, it will find its way into various considerations if the degree of affinity allows it to do so. One topic which becomes accessible is our appreciation of time, and how it relates to the flow of ideas. Here is Locke's view, together with an example of how related TSs enable him to apply additional thoughts which lead to further conclusions:

> 'Tis evident to any one who will but observe what passes in his own Mind, that there is a train of *Ideas*, which constantly succeed one another in his Understanding, as long as he is awake. *Reflection* on these appearances of several *Ideas* one after another in our Minds, is that which furnishes us with the *Idea* of *Succession*: And the distance between any parts of that Succession, or between the appearance of any two *Ideas* in our Minds, is that we call *Duration*. [15]

Several things in that passage are interesting. First of all we have the opinion that ideas can only be seen as passing one another if the person is awake. There is a grammatical ambiguity—the first sentence can be interpreted as saying that the act of observation is contingent upon the observer being awake, which is true enough; it could also mean that the act of succession is due to the person being awake. Knowing what we do about Locke's stance on perception (see quote [13] above) we can safely assume that he meant the second version.

Secondly, we know there is a TP in Locke's mind dealing with the concept of "reflection", containing very definite thoughts on what that description entails. Applying the set of FAs which go with the particular TP he arrives at the conclusion that there is a sequential process where one thought after another is presented to our consciousness.

Thirdly, having established the incremental time line, and taken the ideas themselves as its reference points, the hitherto undefined quality of a time-sensation is now able to accentuate itself through the continual appearance of separate thoughts.

We have dealt with the underlying dynamics of the first two, so let us concentrate on the third. The creation, aligning, identification, and modification of abstracts do not have to occur in linear sequence, that is one abstract and its associated function do not have to wait until another has been dealt with. Distributed processing, the existence of several distinct processes at one and the same time, is by now an accepted way to handle large-scale computations and all indications are that the mind represents a distributed system (more on this in Part II). What does occur in linear fashion is the accompanying articulation of elements of con-TSs, whether that happens through speaking, making music, or conscious thinking (well, more or less—see Part II again!). The occurrence of abstracts during articulation is a set of sensations fed back into the system—after all, it is a source of perception just as an observation or a conscious but unexpressed thought is. This set of sensations provides another set of FAs, one of which concerns the fact that now this, now that idea has been handled, with another following and so on; the resultant TS we verbalise (ie bring into consciousness and articulate via language) as 'time'. If the markers on the so far undefined time line are the sensations our recursive, conscious process of thinking generates, their frequency would determine how time is perceived—slow, fast, but also frustrating, boring, or exhilarating in the case of more complex associations.

Consider now the nature of low-level abstracts compared to higher-level ones: the former being easily assembled and accessed, the latter a result of more work. Since the process of articulation involves the continuous mining of abstracts, then in the case of PAT^1s the frequency of access would be higher than if they had to be extracted from PAT^2s or above. This means that a process of articulation relying mostly—if not exclusively—on PAT^1 chains will produce a sensation in which time is subjectively perceived to be passing slower, that is to say any given objective period of time (say, 24 hrs) is perceived to be more substantial than if the articulation process concerned itself with PAT^{2+} chains. In the latter case the markers are further apart and so provide the subjective experience of a lesser quantitative significance for the same 24 hours. The ultimate, *personal* reference is of course the subconscious, internal clock, that is the subconscious experience of all the continuous, autonomous processes inside our bodies—and these are normally (I emphasise, *normally*) not affected by one's thoughts.

Is there any other, corroborating evidence of such a view? Yes, there is. Take the perception of time manifest by children and compare it with the perception held by older persons. To a ten-year old a single day seems more significant than does a week to someone who is fifty years old. It is also true that a child's mind contains less higher-level abstracts then someone of more advanced age—a feature which plays a significant role when it comes to the conscious experience of the same. Consider also a state of fear, in which any span of time feels "longer" than during a period of calm consideration. Fear of course is one emotion which depresses the mind's ability to access higher abstracts, for which reason a training involving the disciplining of one's stress levels ensures a more successful outcome in otherwise dangerous situations—higher abstracts being necessary for finding solutions to a problem. More of this later.

Locke seems to have sensed the underlying relationship between subjective and objective time experience, for he writes,

> The diurnal, and annual *Revolution of the Sun*, as having been from the beginning of Nature, constant, regular, and universally observable by all Mankind, and supposed equal to one another, have been with Reason *made us of for the measure of Duration*. But the distinction of Days and Years, having depended on the motion of the Sun, it has brought this mistake with it, that it has been thought, that Motion and Duration were the measure one of another. [16]

The representations of such things as a sun rise or a night fall are PAT^1s. Concepts such as 'hour', 'day' etc, are PAT^{2+}s (depending on the concept's inherent complexity). Members of cultures where the predominant representation of time is based on PAT^1s will have a different appreciation of the passing of time than other people to whom time has been introduced on a more formal level from early childhood onwards, even if the former are superficially familiar with hours, minutes and seconds. A characteristic one disregards at one's risk when travelling to certain regions.

All this links back to the subjectivity of the idea-sensation. In his chapter on true and false ideas, Locke says,

> To conclude, a Man having no notion of any Thing without him, but by the *Idea* he has of it in his Mind (which *Idea*, he has a power to call by what Name he pleases) he may, indeed, make an *Idea* neither answering the reality of Things, nor agreeing to the *Ideas* commonly signified by other People's words;

> but cannot make a wrong, or *false Idea* of a Thing, which is no otherwise known to him, but by the *Idea* he has of it. [17]

An abstract, any kind of abstract, represents a sensation inside the mind which to the holder is real, regardless of whether such an abstract actually exists on the outside. It is not the appearance of the abstract itself which tells us about its ultimate veracity, rather further abstracts that confirm or deny its objective reality. Without these additional abstracts the mind has no way of ascertaining the truth value of its ideas; indeed, there may be other abstracts which seem to confirm the abstract anyway—leading to a phantasmagoria from the quaint right up to the bizarre. Locke continues,

> When I frame an *Idea* of the Legs, Arms, and Body of a Man, and join to this a Horse's Head and Neck, I do not make a *false Idea* of any thing; because it represents nothing without me. But when I call it a *Man*, or *Tartar*, and imagine it either to represent some real Being without me, or to be the same *Idea*, that others call by the same name; in either of these cases, I may err.

In other words, an idea can never be wrong as such; its attains its qualification for being true by being associable with other ideas that have an affirmative relationship with reality. Once again, without the thought structures about "sensation", "reflection", and the development of ideas in general the foregoing would not be possible to be thought out.

As an aside, in the context of the previous paragraphs and similar ones the reader may ask, "All this is well and good; but what actually *is* an idea, and how does its content present to our minds?" The explanation has to wait until we come to Part II. While stoking one's curiosity it is not essential for the present purposes. But back to John Locke.

He returns to the theme of association with other ideas later on when he talks about the learning curve of children. *What* and *how* information is introduced to a child's mind is very much significant, and by now we have enough knowledge of mental dynamics to immediately understand why. Having mentioned the effective power of associations which can induce positive or negative thoughts in a person, he writes,

> I mention this not out of any great necessity there is in this present Argument, to distinguish nicely between Natural and Acquired Antipathies, but I take notice of it for another purpose, (*viz.*) that those who have Children, or the charge of their Education, would think it worth their while diligently to watch, and carefully to prevent the undue Connexion of *Ideas* in the Minds of young People. [18]

And further,

> This wrong Connexion in our Minds of *Ideas* in themselves, loose and independent one of another, has such an influence, and is of so great a force to set us awry in our Actions, as well as Moral as Natural, Passions, Reasonings, and Notions themselves, that, perhaps, there is not any one thing that deserves more to be looked after.

And especially,

> The *Ideas* of *Goblines* and *Sprights* have really no more to do with Darkness than Light; yet let but a foolish Maid inculcate these often on the Mind of a Child, and raise them there together, possibly he shall never be able to separate them again so long as he lives, but Darkness shall ever afterwards bring with it those frightful *Ideas*, and they shall be so joined that he can no more bear the one than the other.

Substitute "Goblines" with anything you like; make the appropriate adjustments to "Darkness", and instead of "Maid" read some other profession; and you have a scenario which is still prevalent today.

I referred to the association with further abstracts before, and how their objective veracity cannot be assured through the act of associating per se. We find a reference to this situation where Locke speaks about "mixed Modes", by which he means the thoughts surrounding or accompanying our ideas of things. They are not the direct representations standing for 'sun' or 'house' or whatever but are further abstractions according to our understanding of the original item. 'Understanding' means being able to distill the intrinsic quality, the essence, of some thing, and so constitutes the result of an abstraction process which could not have eventuated had there not been associative TSs to help the process along. This is how Locke expresses the phenomenon, keeping in mind the level of articulation achieved so far and how adequately therefore his words are able to describe what he has sensed:

> In the next place, these *Essences of the Species of mixed Modes, are* not only *made* by the Mind, but made *very arbitrarily*, made without Patterns, or reference to any real Existence. [19]

Since he focuses on the directly identifiable thoughts, that is those which can be ascertained and followed along the trail of conscious articulations, the contents of many associations on a theme are likely to be confusing. The other, sub-conscious parts are hidden and so do not become apparent as associative elements when contexts are created. Thus it may seem that they are made "very arbitrarily", although he is quite correct in his assertions that there is no reference to any real existence. It is PAT^1s which have that link, but further abstractions derive their material from other PAT^1s and higher and so are now removed from the outside. The resultant PAT chains and their qualities have been discussed already. The outward manifestations of these chains, ie those parts which can become articulated and therefore noticeable to an observer, are basically independent from any counterpart in nature—if indeed there should be one at all. Locke, acting as such an observer, would therefore be driven to ascribe an element of independence to those associations, as he makes clear in subsequent paragraphs. The serious consequence of having thought structures which are held to be significant yet built upon abstracts has been recognised, for he says,

> And we cannot doubt, but Law-makers have often made Laws about Species of Actions, which were only the Creatures of their own Understanding; Beings that had no other existence, but in their own Minds. And, I think, no body can deny, but that the *Resurrection* was a Species of mixed Modes in the Mind, before it really existed. [20]

Quite so. As the truth value of any chain does not primarily depend upon its links to c-PAT-ls, rather on links made use of in terms of their associative intensity, we can understand how difficult it can be to undo such associations, especially if the intensity happens to be a function of the emotional values imbedded in them. The mind proceeds "as may best serve its own Purposes", Locke warns, "without tying it self to a precise imitation of any thing that really exists."

However a PAT chain may turn out to be, at its beginning there must have been some input represented in the mind through first-level abstracts. Such abstracts are stored in memory and provide the basis for further abstractions, as we have seen. Therefore, the extent of one's memory determines the efficacy of subsequent structures. Part of the initial TSs are the ways in which the data have been imported—these too are now abstracts. Locke refers to the role of memory when he distinguishes between the actual experience and one's recollection of it. While remembering (ie articulating the data internally) may reproduce the content, it occurs without the immediacy of the original input. For example, one may remember having been burnt by fire, although one does not live through the physical trauma all over again (note however, that because of its associative intensity a great deal of the pain could be recollected as well). A dream, so Locke says, is identifiable as a dream because its representation occurs minus the sensory experience which links us to the outside world [21].

To what degree the contents of a TS are accepted as true is a function of their associative intensity, as said before. Associative in relation to what, one may ask. Locke gives us an example [22]:

> ..if all *English*-men, who have occasion to mention it, should affirm, that it froze in *England* the last Winter, or that there were Swallows seen there in the Summer, I think a Man could almost as little doubt of it, as that Seven and Four are Eleven.

For him there are "degrees of assent", a somewhat more superficial description of the dynamics categorised by our phrase 'associative intensity'. Having identified an occasion on which assent has been given, he uses the occasion itself as a means to categorise the result. This leads him to arrive at several degrees, which he circumscribes by the word "probability". Common instances serve his purpose rather well, as long as they are—well, *common*. In other words, if the result of a particular association can be seen in similar fashion amongst many people, and the circumstances leading to such associations are also repetitive, a definition via outward examples is justified. Consider the quote above—were the assertion to be transferred to an equatorial region it would not work. Underlying this seemingly trivial fact is the difference between lower and higher-level abstracts. Locke's demonstration of the phenomenon of association uses lower-level abstracts because the necessary thought structures are a new development in philosophical discourse; the higher-level versions come later. Because the elements are PAT^1s (Englishmen, frost, swallows, etc), they need to find resonance with other PAT^1s amongst associated TSs for them to be meaningful. If I had never been to England nor had any idea about the weather there, Locke's example would pass me by, although this does not mean that I am not capable of being instructed in thought associations. On the other hand, presented with a sentence such as, "A statement will be believed if its content finds resonance in the listener through the content's associative nature with information that the listener has already accepted", it does not matter what the actual details are, it will be understood provided the recipient is able to make use of the generalised format of the message. Which of course is another way of saying what that sentence expresses anyway. Had I placed those same words within the first chapter of this book, not too many readers would have understood what I was saying; now, this being chapter 7, the ratio is more in their—and so hopefully my—favour. We can term the result of an associative intensity, *associative resonance*. Note that while associative intensity refers to the nature of a

TS, associative resonance relates to its product, that is it determines the quality of the resultant conclusion, another TS. We can call such a TS a *resonant thought structure* (*RTS*), not only because it represents the effect of a resonance (from a semantic-technical point of view, the phrase would not be entirely correct), but because it also reinforces the acceptability of its contents (and in principle, this *is* exactly what 'resonant' means).

Armed with the new set of abstracts, we can cut through Locke's descriptions of the types of assent. While he explains the second kind via the nature of historical texts (one generally accepts what is written in history books because of their inherent authority), and the third through the example of eyewitness accounts, all these being PAT^1s, we are able to state that the perceived veracity of some information is due to the quality of the RTS to which it belongs, whatever the specific circumstances might be (a PAT^{2+} statement).

What does the associative resonance depend on? Clearly, it must be the nature of the TSs leading to it; how significant they are already, what emotional values they have, and how many abstracts they posses which do the associating. Consider once more Locke's example of the English talking about their weather. To him, being an Englishman himself, a compatriot mentioning frost at a time which everyone there would naturally associate with low temperatures, obviously represents an assent of the first magnitude. To an intellectual a scholarly work would come a close second. And the tale of witnesses, possibly carried by people in whom he had often occasion to identify confusion and ignorance, would have to come third. As far as the contents as such are concerned, none of the above has anything definite to say about their inherent truth value; they acquire a truth value due to the quality of the RTS of which they are a part. Ideally then, in terms of our system the veracity of a statement cannot, should not, be ascertained unless and until the same can be done for its progenitors; and the same goes for them, and so on until the entire thought pattern has been evaluated. In real life the complete procedure is impossible, not the least reason being the subconscious nature of many of the elements. There is however one technique which gets close, at least as far as one's inner imagery is concerned: meditation.

A thought structure consisting of only PAT^1s is therefore rich in detail but lacks the facility to generalise; a thought structure consisting of mostly PAT^2s and higher is poor in detail but generalises. In terms of its associative quality a low-level TS will give rise to RTSs at the same low level, whereas a high-level TS is able to associate with equivalent abstracts and so will evoke high-level RTSs. Because higher-level abstracts are less detail-specific the scope for RTSs is greater, which means more flexibility but also more likelihood for errors since associative resonance is not restricted to c-PAT-1s.

Remember the simile about a box in chapter 3, comparing it to the memory and then believing that just as a box can be filled to bursting point so can one's memory? The box and its contents represents a PAT^1 scenario; generalising it to a container holding things ramps it up to PAT^2. The message is still there, but the detail has gone. Doing the same to an image of memory arrives at a similar PAT^2 TS, and now it becomes possible to create an RTS of a PAT^2 character which claims, "Learn too much and you'll burst your memory!" Going through the qualities of the various abstracts in both versions (the Q_{po}s and Q_{ao}s) will of course highlight the mistake, but unless this is done the RTS itself will not point out any errors.

So, the "simple ideas"—using Locke's terminology—need to be provided for any following train of thought to be productive. It is the "complex ideas" which can be generated in turn anytime, anywhere, by anyone with sufficient mental energy. In his chapter "Faith and Reason" he expresses it thus:

> *First*, Then, I say, That *no Man inspired by God, can by any Revelation communicate to others any new simple Ideas* which they had not before from Sensation or Reflexion. For whatever Impressions he himself may have from the immediate hand of GOD, this Revelation, if it be of new simple *Ideas*, cannot be conveyed to another, either by Words, or any other signs. Because Words, by their immediate Operation on us, cause no other *Ideas*, but of their natural Sounds: and 'tis by the Custom of using them for Signs, that they excite, and revive in our Minds latent *Ideas*; but yet only such *Ideas* as were there before. For Words seen or heard, re-call to our Thoughts those *Ideas* only, which to us they have been wont to be Signs of: But cannot introduce any perfectly new, and formerly unknown simple *Ideas*. [23]

In our previous example about box and memory the RTS has been accompanied by an FA which makes the conclusion ("I might burst my memory") possible. Note that a proper evaluation must encompass both PAT chains, the one relating to the box and the other dealing with one's memory. Since in the ordinary course of life the existence of a box-chain is more probable than that of a memory-chain, at this point the person will have no recourse to be critical of the resultant RTS—its FAs are associative with boxes, but there is nothing significant about memory. Such a case will occur most likely with children. Consider what Locke has to say on that subject:

> There is nothing more ordinary, than that *Children* should receive into their Minds Propositions (especially about Matters of Religion) from their Parents, Nurses, or those about them: which being insinuated into their unwary, as well as unbiass'd Understandings, and fastened by degrees, are at least (equally, whether true or false) riveted there by long Custom and Education beyond all possibility of being pull'd out again. For Men, when they are grown up, reflecting upon their Opinions, and finding those of this sort to be as ancient in their Minds as their very Memories, not having observed their early Insinuation, nor by what means they got them, they are apt to reverence them as sacred Things, and not to suffer them to be prophaned, touched, or questioned: They look on them as the *Urim* and *Thummim* set up in their Minds immediately by GOD Himself, to be great and unerring Deciders of Truth and Falshood, and the Judges to which they are to appeal in all manner of Controversies. [24]

It could be assumed that a subsequent addition of a TP about memory will make it possible to revise one's ideas about learning. This is possible, *provided* the relevant memory TSs are associative with whatever else there is relating to the false conclusion. What diminishes their chances of productive association are content, the nature of FAs, and any emotional components they may have (which is what Locke alluded to in the quote above). As we have seen, associative intensity is independent of the truth value, and particularly emotional elements can easily override abstracts which have nothing else going for them but truth. Locke again:

> Would it not be an insufferable thing for a learned Professor, and that which his Scarlet would blush at, to have his Authority of forty years standing wrought out of hard Rock Greek and Latin, with no small expence of Time and Candle, and confirmed by general Tradition, and a reverend Beard, in an instant overturned by an upstart Novelist? Can anyone expect that he should be made to

> confess, That what he taught his Scholars thirty years ago, was Errour and Mistake; and that he sold them hard Words and Ignorance at a very dear rate? [25]

Not only children, but also adults are subject to the relative influences of the associative resonance exhibited by their thoughts. Although we have largely concentrated on identifiable data in the contents of TSs, throughout this chapter another, more indeterminable factor has been raised: emotions. When Locke speaks about the influence of tradition in the previous quote, or of peer pressure when it comes to classify types of erroneous assent [26], or the arbitrariness of "complex ideas" earlier on, one of the participating elements in all these situations has been human emotion.

In his chapter "Modes of Pleasure and Pain" he details several types of emotions, such as hatred, love, desire, joy, sorrow, and so on. Like in his treatment of assent he uses direct examples—PAT^1s—in order to illustrate what he means. It is of course worthwhile to bring the presence of a person's emotional status to the attention of the reader while discussing human thought, but as in the previous case the nature of his TSs compel him to stop short of a more general identification. The following statement is as far as he goes:

> But it suffices to note, that our *Ideas* of *Love* and *Hatred*, are but the Dispositions of the Mind, in respect of Pleasure and Pain in general, however caused in us. [27]

The time has come to be more descriptive. First, let us remember that we are engaged in an examination of the *mind*—not the neuro-biological machinery which constitutes the brain but the system that does our thinking. Second, our methodology consists of reverse-engineering any functional modules we are able to identify, followed by further analysis of each module to see whether there are any sub-modules, and so on. Therefore, we know—have identified—abstracts which coalesce into thought structures that in turn form collectives we have called thought patterns. There are mutual influences that organise the forming and disbanding of the various combinations which we have labeled according to their functionalities. Throughout the ages their contents have varied widely, but the principle relationships remained the same—hence the possibility for us to name them in a formal manner, regardless of the specific circumstances. This suggests that there is a system which incorporates the potential to manifest in certain ways (TSs, DTSs, FAs, etc etc) such that these outcomes are producible by its inherent mechanism.

Likewise the emotions have remained the same, and once again we can deduce the distinct possibility that their basic functionality is not a construct of some particular experience (ie input), but that instead the presence of the modules is given and they are being activated by the input. For the present purpose it is not necessary to go into the physical detail of these modules (this would be the domain of neuro-biologists), only to ascertain that each emotion (eg love, hate, joy, etc) represents a functional entity, sometimes active, sometimes not. Since their presence is not an attribute of a specific experience, they must exist a priori, and therefore accompany, indeed shape a thought structure—from the most simple to the most complex version. We also know that a living organism has responses which trigger at the most basic level, in terms of represented cause as well as manifest effect. Such things as survival, procreation, the seeking of company and others have been existent long before more abstract concepts could have been entertained, and so, from an evolutionary point of view, the emotions would relate to such demands in tandem with the organism's need. In terms of a human individual's development, from a new-born to the adult, the demands on the overall system would be commensurate with the growing complexity of the system itself, especially its mind. In that sense emotions can be viewed as the first, primitive thought structures of a cognitive system.

This leads to the following conclusion: the human mind incorporates a set of distinct modules, commonly called emotions, which are present at all times but can be activated separately by stimuli in order to manifest a certain behaviour; the type of behaviour being dependent on the demands of the moment, and relevant to the intrinsic needs of the individual. As such, with regards to any input, an emotive element participates in the dynamic mental configuration to perceive, associate, abstract, and recursively access said input. The type and degree of participation relates to the capacity of the mind to meaningfully evaluate its environment to a greater or lesser extent.

For example, an infant is hungry but is prevented from eating because its parents are driving along the freeway. The emotion can be anger because to eat is a basic instinct, and the feeling is not ameliorated by the fact of driving on the freeway because the infant mind's cannot grasp the implications of stopping right there and then, and so it cries anyway. A mother is concerned for her child to be included in the school's activities, but the feeling is not ameliorated by the fact that an inclusion may cause undue hardship for everybody else, and so she demands it anyway.

Let us call emotions *emotive modules* (*EMs*), a name which better reflects their functional modularity in conjunction with the other dynamics. The capacity of the mind to evaluate its position in terms of the wider social context we will call its *social horizon*. The social horizon is dependent on the presence of appropriate FAs which suggest the possibility of a wider perspective, and the mind's ability to abstract to higher levels to implement that view.

EMs become significant when we examine the autonomous activities dealing with the creation of TSs; they are 'autonomous' in the sense that they are not a direct result of some input, whether from the outside or as a consequence of some train of thought. Instead they are a function of the constant dynamic underpinning the state of being alive and with their character rely on the combined attendance of PATs and FAs in conjunction with those EMs which have been triggered by the perceived meaning of the PAT/FA mechanism. The EMs are drawn from the particular mind's reservoir, the general state of which reflects the individual's personality.

When Locke discusses the will he alludes to the above in his circumscriptive, PAT^1-type manner we have become familiar with by now. He recognises that will is "an act of the Mind directing its thought to the production of any Action" [28]. He also realises that the question of "what determines the will" can be answered by saying, "the mind". If this is not satisfactory, he extends his answer by including "the motive", by which he means the urge to change a current state to something different. Given the low-level abstraction of Locke's TSs in this matter he is forced to differentiate between 'will' and 'desire', the will being the actual force which results in an action, whereas desire being the personal preference which may or may not be actualised. From our perspective it is clear that there is no difference as far as the thought process is concerned—the final outcome being a function of the relevant abstracts in tandem with triggered EMs, and determined by the associative resonances being evoked.

His use of the two concepts, "will" and "desire", leads him to discard the notion of 'free will'. Since his "will" is the cause of the mind's decision (and whether there is an action or not is still the result of a decision), there has to be a "will" in any case. Therefore freedom consists "in our being able to act, or not to act, according as we shall chuse, or *will*". The question of 'free will' has occupied thinkers for a long time, and considering the ramifications flowing from one or the other conclusion it is no wonder. From the attitudes towards our daily affairs to the weighty judgments of a metaphysical nature, whether one holds Man to possess an independent willing faculty or not matters a great deal.

The problem is the question's intended absolutist nature: either there is a free will that makes us completely independent from any outside source, or there isn't and so we are merely reactive agents to some grand plan. If we translate the meaning of the word 'will' or any symbol standing in lieu thereof into a mental picture of a singular entity, then our thoughts are led towards the black-and-white view—there either is such an entity or there is not. Now consider the phenomenon in terms of thought structures. There is an input (an experience from the outside or a new thought), and the process of TS association begins, utilising the mind's contingent of functional abstracts, affinitive PAT chains, all accompanied and influenced by a set of applicable EMs. It becomes obvious there is no singularly identifiable block of mental substance which causes the response. Yes, there are the abstracts held in memory, and so one could call their effect predetermined; there are also the given emotional modules, equally existent. Yet there are also the myriad combinations of affinities, pushed this way and that by the eventual effect of specific set of EMs, transforming the qualities of emerging abstracts as the action proceeds. Such is their total combinatorial power that even the person doing the thinking cannot predict the result beforehand. It is akin to meteorological phenomena—we know the physics to a great extent but the sheer number of possible incremental steps make it difficult to foresee what is actually going to happen. We can predict tendencies, given the identification of the main factors. Therefore to speak of 'free will' is misleading because, to summarise the above, the 'freedom' in this case is the acted-out potential at varying levels of directness in terms of mutual interdependency between dynamic elements. The question of free will then is a prime example of the effects of semantic obfuscation; in reality the problem does not exist.

Still on the theme of "will", Locke argues that "those *superior Beings* above us" are "more steadily *determined in their choice of Good* than we" because "they enjoy perfect happiness" [29]. Leaving aside what he could mean by "superior Beings", his statement is characteristic of a thought pattern that posits 'will' as an entity. If 'will' is singular, its quality must be reflective of itself and as such discretely identifiable. Hence, if there are superior beings their higher status will mean a better implementation of their will. From the point of view of TS dynamics however this is not necessarily so.

The mental complexes which are being created in the process of evaluating a situation are, amongst other things, dependent upon the inherent ability of the mind. If higher abstractions can be formed (and we have seen the conditions that need to be met) then they will, and therefore the interpretation by the mind will be more complex also. Generally speaking it can be seen that while a sophisticated mind is more likely to come up with a good response, it will also ascertain the problem on a deeper level, in other words enhance the problem's perceived difficulty. (Whether this means that God must be infinitely worried I leave to the theologians.)

Apart from our metaphysical superiors Locke also speaks of "finite intellectual Beings" who deliberate about their course of action and possibly suspend their desires so that the result will be of a high standard. This is our duty, he writes [30]. In other words, there can be intentionally created TSs to attract certain EMs to counter other emotions that are triggered by less attractive abstracts ("attractive" in every sense of the word!). For example, I may feel like ramming someone who cut in front of my car—these are TSs attracted by EMs representing anger and egotistic wish fulfillment. I can also remind myself that the traffic *is* unnerving and that at the end of the day an undamaged car and body is preferable after all—somewhat more complex TSs which invite EMs of satisfaction and of being in control.

So, emotions can be controlled and directed if the driving force of the thought process is able to overcome their own influence; many times the situation is the reverse. The emotive power directs the flow of PAT chains, especially when there is an existing affinity along certain TSs. I-PAT-ls have been laid down not through the force of logic (which of course they lack), but through the affinity with ideas that derive their significance from the quality of accompanying EMs. Religious sentiments are a prime candidate for such resonance, where Locke comments:

> And whatever odd Action they find in themselves a strong Inclination to do, that impulse is concluded to be a call or direction from Heaven, and must be obeyed; 'tis a Commission from above, and they cannot err in executing it. [31]

We have talked about abstractions and how they carry their contents along certain PAT chains. The direction of those chains is determined by the affinity the abstracts have with other TSs. It is these affinities which characterise the outcome, whether it represents a useful solution or not. Let us look at an example.

In his chapter "Simple Modes of Space" Locke writes about the "modification" of simple ideas and how their nature becomes more complex by being aligned with other ideas. One such modification involves the concept of space. 'Space' is used by "Men, for their common use, that by it they might be able to design the particular Position of Things" [32]. Here we have an idea—space—which in itself is not useful unless it can be placed within a particular domain. He mentions a chess board, in which space the pieces have their positions referenced through the 8×8 grid. However, he says, if the pieces were in a bag, their locale would be better described by the room they are in, and not via the chess board. Similarly, if one wanted to know where to find the story of Nisus and Euralius it would be better in the first instance to give their place as Virgil's works, and not as Bodley's Library. To remain with Virgil's verses, if the information about them has been introduced in a formal manner, the geographical position would, in the main, be considered irrelevant. Why?

Suppose a child learns that there are such things as books. By the time it gets to know about Virgil—if ever!—it will have come across many books in many places. It knows therefore that the existence of books is not restricted to one single location, and so the question of which room, house, or library a particular volume is in has been recognised as being of a more general nature than the name of an author. If from that point onwards it should be asked, "Where can you find the story of Nisus and Euralius?" it will focus on the most particular information it has, rather than the more general. Formal rules have been designed in order to describe the process in algorithmic form, and one version would be the identification of domains which are categorised in terms of physical extent; another way would be to classify a domain according to its functionality, where a more discrete functionality takes precedence over a more encompassing one. In such an algorithm a book is a more specific object than a library and so would be chosen over the latter.

We are back at functional abstracts. The formal algorithm above would be represented by *congruent* functional abstracts, although this is not how a child would learn about its environment, at least not in the initial stages. Rather, it is the experience of looking for and finding an item together with the associated information that was considered useful at the time which imbed in its mind the nature of the most productive choice. The learning experience creates *intuitive* functional abstracts, in other words samples of instances that become tied to sensations of success or otherwise. Suppose a child knows next to nothing about engines and is sent to various hardware stores to look for spare parts. If questioned, "Where do you find the carburettor?" it won't talk about the engine's fuel inlet but would give the name of the store where the item was bought. Thus the answer is not necessarily logical, but it does relate to the child's own experience. IFAs therefore do not represent a logical format but a set of circumstances, as pointed out in chapter 6.

Furthermore, in the context of a person's general intellectual development, the TSs created through the application of IFAs are the structures which exist first and foremost, and only later can their contents be abstracted to form CFAs which in turn give rise to our algorithms. Just as in our single example an IFA existed before its TP could be harvested for articulating general rules, so does the cumulative total of life's experiences provide the individual with a set of IFAs which, if so desired, can be translated into formal rules. Although such rules can be taught at any time, they would not exist unless there had been their intuitive precursors held by someone, somewhere. Since the experience of success or otherwise accompanies the IFAs, the more exposed the mind has been to whatever activities, the more profound the experience would be and therefore the higher the mind's potential for the creation of formal, productive rules.

If we extend our considerations even further we can say that the more active, diverse, and intelligent a society as a whole, the greater its capacity for successful solutions to problem situations. This capacity should be understood in a short as well as a large time frame; that is to say, while a vigorous society will be more productive at any given instant, its true value will accumulate gradually over time as it amasses a multitude of IFAs and stores them in its collective memory. Because IFAs do not derive their perceived significance from their logic value but are defined by the importance the associated outcome presents to their users, the appreciation of their value needs to be nurtured, that is the underlying experiences need to be repeated over and over again. If the relevant activities are not being undertaken, their IFAs loose their meaning and are replaced by others, possibly less effective ones.

For example, if a person is forced to consume his money sparingly, skills will be developed that enable the person to survive in poor circumstances. Should he win a million dollars, those skills may become forgotten and sooner or later the person is not only back where he started from but will be actually worse off. Although he knows how to add and subtract (the CFAs), the daily experience of living in luxury (the IFAs) will be the dominating factor.

There is then an added relevance to Locke's work. As history has shown, many of the philosopher's ideas found their appreciation and acceptance at the beginning of the 18th century in England. Together with the flowering of science the implementation of logical principles in government and society in general led to the country's hegemony for years to come. Her past provided the training ground for the demonstration of the effects of activities, approaches, and policies. The resultant IFAs in their sum-total fed her capacity for competitiveness.

As stated, the development of formal rules needs an uninterrupted path of c-PAT-ls, from the 1st-level abstracts which define the items being dealt with to a level high enough for the generalisation to occur. Part of the chain is the array of object abstracts contributing to the definition. Note that a higher level of abstraction does not necessarily mean an absence of straightforward objects. For instance, the higher-level abstraction of a tea cup is a container, and both are solid, tangible things. A problem occurs when, in the course of ascertaining a rule, the existence of such objects looses its perceived importance while at the same time the abstraction process continues. Locke refers to that problem when he writes,

> If anyone should be asked, what is the subject wherein Colour or Weight inheres, he would have nothing to say, but the solid extended parts: And if he were demanded, what is it, that that Solidity and Extension inhere in, he would

> not be in a much better case, than the *Indian* before mentioned; who, saying that the World was supported by a great Elephant, was asked, what the Elephant rested on; to which his answer was, a great Tortoise: But being again pressed to know what gave support to the broad-back'd Tortoise, replied, something, he knew not what. [33]

While the idea of an elephant supporting the world etc seems quaint to us, the TS itself may not be incongruent at all; it could have been simply the result of wrong data. The i-PAT-l is shown in the last sentence: once the Indian bases his cosmic view on a train of thoughts which ends in 'thin air' as it were, he changes the TS to an incongruent one.

Locke gives us an unintentional demonstration of that principle when he outlines his three versions of Law [34]. For him the first type is the "*Divine* Law", derived from religious tenets; the second, "*Civil* Law", defined under the Commonwealth; and the third, the "*Law of Opinion or Reputation*", created by society at large. If we examine their underlying dynamics, we see that the last exists on the basis of daily actions and experiences, a largely PAT[1] scenario of direct instances. What governments decide should be right and wrong is not only based on societal perceptions but is also the product of administrative and political expediency—already a set of higher-level abstractions which may or may not concur with the expectations of the ordinary citizen. Finally, religions ultimately draw their conclusions not from the here and now at all but from high-level abstracts beyond the logic boundary, and therefore *their* laws include the products of i-PAT-ls. It is also worthwhile observing that the speed with which each type can and does change follows the same sequence, from type 1 (the fastest), to type 2 to type 3 (the slowest). This is in line with the respective inertia of different abstraction levels; the higher the abstraction, the less easily is it modified and/or replaced, a phenomenon in accord with the underlying dynamics of each kind.

We can observe the differing speeds in action in the chapter on "General Terms", where Locke shows how the names of ideas can change in contrast to their "essences", their underlying character. For Locke the topic is the general abstraction of things, and how we perceive them through the names we apply [35]. For example, 'grass' changes into 'sheep', and when the sheep is consumed it becomes part of 'Man'; nevertheless, our ideas of 'grass', etc remain the same. One can extend this further and say that, from our perspective, the abstraction 'grass' can be modified, but less so the notion of 'plant matter', less still the meaning of 'organic substance', and probably even less the concept of 'food chain'. In concrete terms, we could quibble about the name 'grass' in relation to what sheep eat, but there would be more resistance doing the same with 'plant matter', and to question the existence of a food chain as such would be just about foolhardy.

Locke gives us another example when he talks about the types of truth [36]. Similar to his categorisation of laws and listed under the same sequencing, there is the "Metaphysical Truth", followed by the "Moral Truth", and lastly the truth in "the marking down in Words". We can recognise the same principle structure, and the dynamics can equally be compared to the TPs relating to the laws.

When 'grass' get generalised to 'plant matter' and onwards to something like 'organic substance', we observe a PAT chain in action. It is a *vertical PAT* since it occurs across differing abstraction levels. Because of its generalising nature the principle underlying the content of the previous abstract is brought into relief and now serves to enable associations with other, relevant abstracts. These can reside on relatively lower levels, where it is the principle contained within some detail which is the attractor, or on the same level where there may be an affinity with another, also exclusively represented principle. PAT chains can also occur along the same abstraction level, in which case we can call them *horizontal PATs*. How useful are they?

Locke presents various instances of horizontal PATs [37]. Take the following two, taken from his book: (1) "What is, is", and (2), "It is impossible for the same to be and not to be". "What" stands for any item and in its representative sense is just another word for 'house', 'dog', 'tree', etc. The first "is" constitutes a FA and together with "what" forms a TS. The second "is" is also a FA, coming from a different TS in memory and now joined with the first. Both exist on the same, low abstraction level because the degree of detail and/or generalisation of either content is the same. Therefore the information values of both TSs are equal, and together they tell us nothing new, nor do they even provide us with another perspective on the theme. In the second sentence the concept "to be" is on the same level as "not to be", although one is the opposite from the other. We will deal with opposites later but once again, the degree of detail is the same in either version. We call these horizontal PATs maxims, and Locke quite rightly observes, "'Tis plain that they are not, nor have been the Foundations whereon any Science hath been built". He compares them with the discoveries of Isaac Newton and tells us that it was not the maxims which made that man's insights possible.

In order to understand their functional differences let us look first at vertical PATs to see what makes them so useful. The following puzzle is given: you have a wooden door, a candle, and a box of tacks. Using these three items and no more, attach the candle to the door such that it can be lit in an upright position. For any reader who wants to try their hand it is a useful idea to observe the thinking process along the way. The solution comes in the next paragraph.

The following is the solution in terms of the abstracts' functionalities. The three objects are PAT^1s. Going one level higher we can identify the door as a panel which is vertical and where things can be attached to its surface (note that the definition of the abstracts here is a simplified version—in reality the modules are far more fine-grained). Doing the same with the candle we see it is an object which needs to stand on a flat surface to be upright. Abstracting from the box of tacks we see first that it is not one single object at all but a composite, consisting of the box itself and the tacks inside. Having the concept of 'box' we realise that it can be emptied, and what comes out are the tacks—these have become many separate items. The box is identifiable as an object consisting of its bottom, a flat surface, and its walls, through which a tack can be pushed. So now we have, in generalised form: a solid, vertical surface to which things can be attached; a box which has a flat, horizontal surface able to support a small object and having four thin walls that can be easily penetrated; a candle which is the thing to be placed upright on some surface; and the tacks, some of which can be used to push through something thin and by doing so fix it to something else. All that remains to be done is to arrange the characteristics in the right order and there is our solution: empty the box, use some of the tacks to attach the box to the door by one of its side panels, and stick the candle onto the box. Fig. 7-1 illustrates the procedure.

Without the ability to elevate any one of those concepts to a higher level, the solution is impossible. Note that the higher abstracts, while derived from their lower-level source, contain different data in terms of object content *as well as* information value. For instance, a 'box' is a box, but a 'platform with four thin walls capable of supporting some light object' is a different thing again, certainly as far as usefulness is concerned. Therefore, vertical PATs lead to an association with other representations of a more and more generalised format, where the principle functionality of each lower-level abstract is a subset of its higher-level neighbour. There will be a solution if the lower-level subset is indeed part of a higher-level domain; there will be no solution if the subset belongs to a different domain, one which has not been accessed.

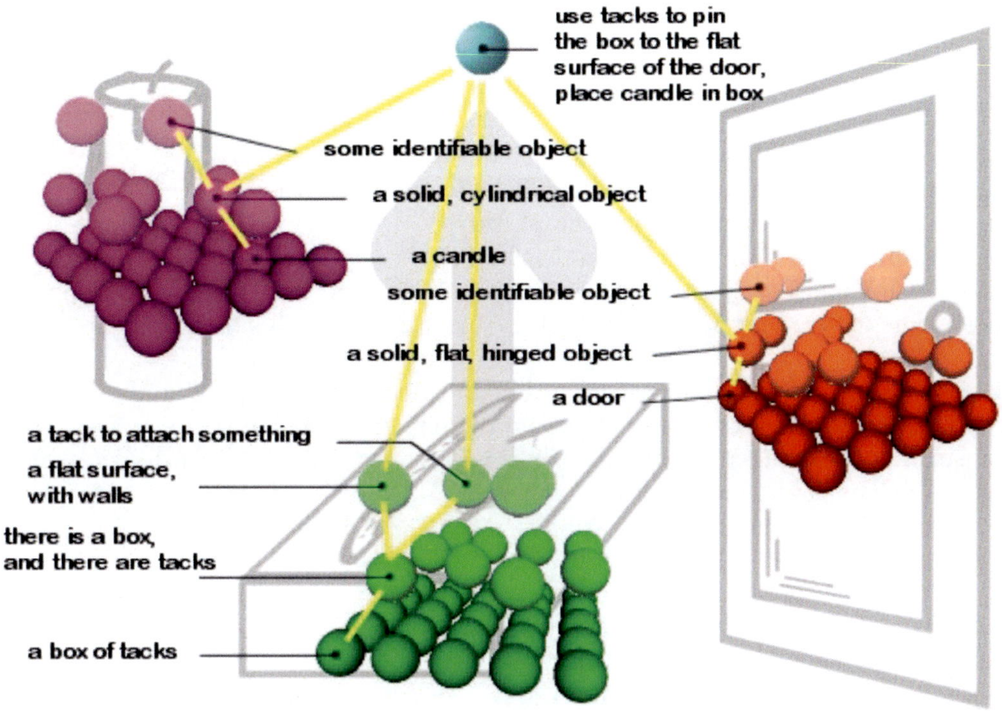

Fig. 7-1: A graphic depiction of the mental procedures necessary to solve a puzzle. For details see the text.

On the other hand, a horizontal PAT does not feature varying abstraction levels and so every link in its chain does not hold qualitatively different information, although the quantity may change. Compare our puzzle with a chess problem. There we have many possibilities to be contemplated, but in each and every case the rules are the same and none of the chess pieces will change their nature in any way or form, no matter how difficult the question. Chess players utilise the brain's capacity for horizontal PATs, whereas a scientist or engineer needs to employ the mind's ability to evoke vertical PAT chains. Both types demonstrate their extremes in familiar scenarios: the above average chess player who falls short in other areas because there are no profound solutions to be applied; and the creative genius who is disorganised in every-day affairs because their details are not strung together properly.

There is also a more sinister ramification, particularly with people who have difficulties with vertical abstractions. Here is Locke, still on the subject of trivial maxims:

> ..a Child having framed the *Idea* of a *Man*, it is probable, that his *Idea* is just like that Picture, which the Painter makes of the visible Appearances joyned together; and such a Complication of *Ideas* together in his Understanding, makes up the single complex *Idea* which he calls *Man*, whereof White or Flesh-colour in *England* being one, the Child can demonstrate to you, that *a Negro is not a Man*, because White-colour was one of the constant simple *Ideas* of the complex *Idea* he calls *Man*; [38]

In other words, there are PAT^{1}s which form certain TSs, but when juxtaposed to another TS the abstraction level remains the same. 'Man' then is not more general than 'White' or 'Negro', therefore the differences between the first two and the last cannot be overcome because it is not realised that *in principle* there is no difference anyway. Hence, "*a Negro is not a Man*".

This leads Locke eventually to consider human reason. He identifies several degrees, and keeping in mind how vertical PATs work it is easy to see the parallels:

> So that we may in *Reason* consider these *four Degrees*; the first and highest, is the discovering, and finding out of Proofs; the second, the regular and methodical Disposition of them, and laying them in a clear and fit Order, to make their Connexion and Force be plainly and easily perceived; the third is the perceiving their Connexion; and the fourth, the making a right conclusion. [39]

In terms of the complete sequence in a thought process which goes up and down the abstraction ladder his degrees are somewhat out of step. They are interesting when it comes to the abstraction levels themselves however. What he sees as the highest represents the culmination of having generalised some principles and thus assembled the proofs. Their details need to be arraigned in the correct order so that they can be related to the representative items—the next lower abstraction. Then comes the cross reference, that is realising their mutual significance which means making the connections amongst same-level abstracts. And the fourth consists of applying the generalisations in terms of the actual objects in question, which are usually PAT^{1}s.

Another interesting observation of his concerns the usefulness of syllogisms. Just as he questions the information value of a maxim, so does he dismiss the syllogism as not telling us anything we hadn't known before. Here is a true and tried example of one: "All men are mortal; Socrates is a man; therefore Socrates is mortal." In formal terms, we have the major premise ("All men.."), the minor premise ("Socrates is.."), and the conclusion ("therefore.."). Yet Locke makes the point that

> ..there are many Men that Reason exceeding clear and rightly, who know not how to make a Syllogism. He that will look into many parts of *Asia* and *America*, will find Men reason there, perhaps, as acutely as himself, who yet never heard of a Syllogism, nor can reduce any one Argument to those Forms.. [40]

We can leave the issue of the syllogism's ultimate importance to the logicians, who after all haven't been sitting still since. He is certainly correct when he contends that knowing how to construct a major and/or minor premise is not essential to the mastery of life. But we need to be careful: there is a difference between the understanding and so the ability to write a sentence which accords with the definition of a premise, and the understanding and ability to express the notion that all men are mortal and that Socrates happens to be male. The former is an outcome of an articulated rule set which sets down the format belonging to premises. CFAs have been learned, they are part of TSs of a certain kind, and therefore they can be evoked by the need to come up with a syllogism. However, to do just that the underlying information needs to be configured such that it can be packaged into the major and minor premise while at the same time incorporating the material for the conclusion. It is not a 'natural' process in the sense that one's knowledge comes pre-arranged in favour of syllogisms but is more like a jumble of data imported throughout any 24hr period.

The second kind, that is merely expressing the notion, may not have the clear-cut formality of the first, but nevertheless is arranged and processed by the mind according to dynamics that apply in any case, whether we feel like being logical or not. To say that all men are mortal is an observation and constitutes a TS holding PAT^1s; to say that Socrates is a man falls into the same category—not too many people have problems with that. The interesting part comes next, with the question, "So, what's the conclusion?" Those individuals who have come across logical formats before do not hesitate and give the answer, "Socrates is mortal", which is to be expected. It can also be observed that people who are exposed to melodious or rhythmic presentations will say the same thing, not out of logical considerations but because the difference between the first and second statement is one of incomplete sameness: men - mortal vs Socrates - man leaves 'mortal' hanging, a situation that can be resolved by saying a sentence which includes it and preferably in a similar format to the one where the word occurs. So 'Socrates' being the lead-in of sentence 2 is used for the same purpose in sentence 3, and 'mortal' being the lead-out in sentence 1 is given the same role in sentence 3. This has nothing to do with the formal logic taught by logicians, but is the system used by affinity-seeking thought structures. They mostly work—not because of their logical functionality (which they don't have), but because the preparation includes the wanted information already. The conclusion, syllogistic or otherwise, would never suggest itself by saying, "All men are mortal" and "Socrates is a philosopher", because the ingredients do not match. Which prompts Locke to declare that it doesn't need a syllogism for us to be able to reason:

> If Syllogisms must be taken for the only proper instrument of reason and means of Knowledge, it will follow, that before *Aristotle* there was not one Man that did or could know any thing by Reason; and that since the invention of Syllogisms, there is not one of Ten Thousand that doth. [41]

To show how strong some attractions in the forming of associations are, consider the content of children's rhymes. Here is one from the German language, chosen because the English-speaking reader can concentrate on the sounds of the words and will not be distracted by the meaning (I have used a phonetic hybrid to get the sounds of the nonsense words right).

> "Eenie meenie moo, draus bist du,
> draus bist du noch lange nicht,
> musst erst sag'n wie alt du bist."

Given that sentences and verses evolve due to some need, it becomes obvious here that the need was rhythmic and phonetic conformity, together with the creation of a social context. Therefore some words are strung together in order to fit "moo" - "du" and "nicht" - "bist" together (in some regions, the last two have very similar pronunciations). The overriding factor of sound affinity becomes even more evident if one does consider what the words actually mean. Here is a translation:

> "Eenie meenie moo, out are you,
> you are not out by a long shot,
> first you have to say how old you are."

By the way, the rhyme is used by one child standing in a circle with others and pointing from one to the next with each word, with the finger stopping at "bist" and that child having to say its age, whereupon the count starts in a similar fashion using the numbers from 1 to the age; only then is a child properly identified and dismissed from the circle. And so on.

Apart from the sound there are other factors. The purpose of the rhyme is to get rid of the kids in the circle until two are left, the caller and the target. This could be a problem in terms of power and social in- and/or exclusion, but it is ameliorated by firstly, using rather long-winded verses, thus giving ample opportunity for everybody to be included and then excluded (when you have 23 words to go through it becomes rather difficult to pre-arrange the person on whom the final word will fall). Secondly there is the element of 'second chance', whereby the first identification ("draus bist du") is immediately followed by another, far more elaborate one which, as it happens, hardly ever returns to the same person either (you only have to watch the face to realise to what extent this is appreciated). Mathematically this is easily confirmed by the fact that the first line consists of 6 words, the other two of 17 words. 17 not being a multiple of 6 ensures that whoever 'is' the 6th word will certainly not end up being the 17th either, unless there happen to be 17 children (or 34, 51, 68...). However, children are not mathematicians; instead there are sub-TSs which, in terms of their eventual implementation, demonstrate their effects one way or another. The underlying reason, the abstract rule, comes through the act of generalisation—the CFA; the sensory experience through acting out the ideas provides the material for the IFA. There is also the element of age, and its context as it applies to children. One of the most often asked questions for a child would be its age, right next after the name. So, not only is the age as such an identifier, even the act of being asked contributes to the state of being a child. Asking for each others' ages then becomes an exercise in status assertion, which simultaneously contributes to the fairness of the elimination process; effecting exclusion through inclusion—a good balance.

The purpose of that little diversion was to show the importance of aesthetics when it comes to TS associations. Despite other influences, if it looks or sounds good or, conversely, if something seems to be missing, the mind will come up with an appropriate response. Whether it is about children's rhymes, a piece of music, or an assembly of words which induce longing, the dynamics are the same. It is important to understand that, by themselves, these dynamics are *not* concerned with any inherent truth value or any logic or congruency of the participating TSs.

A more general angle to the above is the question of affinity in relation to abstraction levels. The IFAs which produce the conclusion, "Socrates is mortal" are on a lower level than the CFAs which enable the statement of major and minor premise and their conclusion. For that reason an IFA-induced resolution may be more attractive than a CFA-induced one, because there are more minds relating to the former. Since the significance of IFA-promoted associations contain a sizable component of EMs (the aesthetic value would see to that), to disentangle oneself from views derived from such sources is difficult unless the emotional quality of the viewpoint is being modified.

The degree of abstraction occupied by FAs is therefore significant, especially when it comes to the intuitive variety. Whether they are of the PAT[1] type or at a higher level, determines the plausibility of their overall TS. Locke writes of the incidence of revelation and testimony in the context of the mind's readiness for new information, or "enthusiasm" as he calls it. Observe his attitude towards revelation against our perspective of abstraction levels:

> Thus we see the holy Men of old, who had *Revelations* from GOD, had something else besides that internal Light of assurance in their own Minds, to testify to them, that it was from GOD. They were not left to their own

> Perswasions alone, that those Perswasions were from GOD; But had outward Signs to convince them of the Author of those Revelations...
>
> *Moses* saw the Bush burn without being consumed, and heard a Voice out of it...
>
> ..and yet he thought not this enough to authorise him to go with that Message, till GOD by another Miracle, of his Rod turned into a Serpent...
>
> *Gideon* was sent by an Angel to deliver *Israel* from the *Mideanites*... [42]

In those days of old then, it was not the abstraction process which yielded the hoped for insights and testimonies, but presentations of the PAT^1 kind, eg burning bushes, rods that turn into serpents, and the manifestation of angels. If we posit the resultant TS in general as the desired outcome of some train of thought and follow its emergence from the start until its finish, then in the case of rising abstraction levels the attraction represented by relevant EMs and a certain inspiring imagery would define for us the ingredients and the steps leading to the result. There the steps would consist of insights and intermediary conclusions, all at varying levels of abstraction. We have seen many examples in previous chapters. What if the mind had some difficulty in reaching higher-level abstracts—would the process itself be abolished, and if not, what would take the place of the insights? Well, a possible answer is provided in Locke's quote above. Instead of esoteric musings the mind furnishes intermediary steps of the PAT^1 type, that is actual scenarios, seemingly played out before one's very eyes. It may not be without coincidence that the visions described before occur in humanity's past, when the general abstractive tone was lower than in the centuries to come. Also, take your mind back to chapter 5 where Thomas Aquinas laments about the ancients' inability to take concepts to a higher level ("they had not yet arrived at the point where their intellects were sufficiently raised above sensible reality"). It may equally not be without significance that people who have visions of what are perceived to be the profound kind are not sophisticated thinkers (for example, the appearances of the mother of Jesus to a peasant girl in 1858 in Lourdes, France). All this, I emphasise, is a matter of conjecture, although the elements leading up to it are sound enough.

Having examined thought structures from the ground up, and followed the creation and organisation of PAT^1s to PAT^2s and beyond, we can better appreciate Locke's approach when he sat down to write his *Essay*. The following are quotes from the *Introduction*, interspersed with our translations.

> *First*, I shall enquire into the *Originals* of those *Ideas*, Notions, or whatever else you please to call them, which a Man observes, and is conscious to himself he has in his Mind; and the ways whereby the Understanding comes to be furnished with them.

Transl: He starts with PAT^1s. Derived from observation, they are assembled in an affinitive fashion and become the building blocs of more advanced TSs.

> *Secondly*, I shall endeavour to shew, what *Knowledge* the Understanding hath by those *Ideas*; and the Certainty, Evidence, and Extent of it.

Transl: From the objects he follows the processes through which the mind makes sense out the data elements; the understanding comes from the abstraction and generalisation of said items.

> *Thirdly*, I shall make some Enquiry into the Nature and Grounds of *Faith*, or *Opinion*: whereby I mean that Assent, which we give to any Proposition as true, of whose Truth yet we have no certain Knowledge: And here we shall have Occasion to examine the Reasons and Degrees of *Assent*. [43]

Transl: Faith and Opinion are two versions of a conclusion, and interestingly Locke not only lists them last, but separately. Both are end points of a thinking process, where faith draws on the mind's desire to idealise and extrapolate in its own image, and opinion being a function of the mind's desire to resolve open-ended PAT chains.

Cast your mind back to Socrates, and how he argued for certain ideas to be pre-known, certainly in terms of one's current life (eg Beauty; see chapter 2). To him that meant there had to be some existence before birth where such knowledge could have been acquired. Significantly, Socrates' examples are higher-level abstracts, the very entities which develop sub-consciously and hence out of conscious sight; whereas simple objects (eg a beautiful rose) find their way into the mind during our present life—in other words, conscious acts of perception which are being ascertained in the here and now.

Locke's view is different. For him there are no "*innate Principles*", resident in one's soul from the very beginning. Whatever we know needed to be learned at some stage, and if there *were* innate ideas they surely must be known.

> If therefore Children and *Ideots* have Souls, have Minds, with those Impressions upon them, they must unavoidably perceive them, and necessarily know and assent to these Truths, which since they do not, it is evident that there are no such Impressions. For if they are not Notions naturally imprinted, How can they be innate? And if they are Notions imprinted, How can they be unknown? [44]

To possess a TS which postulates that one has innate ideas about something, prompts two further TSs that incorporate FAs gleaned from daily life. One FA is due to the realisation that whatever exists does so as a result of a cause, the complexity of the situation touching both sides, cause and effect (a stone may have been washed up by a river; an epistle had been written by an intelligent being). The other stems from the notion that a cause not only influences but pre-organises its effect, again through mutual complexity (a mindless river deposits a mindless stone, and that's all there is to it; to write an epistle one needs articulated information which is noteworthy in itself). Apart from demonstrating the affinity with similar abstraction levels binding cause and effect, the existence of these particular FAs in just about anyone's mind predisposes their application within the context of innate ideas. If we accept innate ideas then, given the FAs, we are driven to attach the abstract 'cause' to something, and since the level of complexity has to keep pace with the character of an example's content, it becomes natural to link the cause of human knowledge to other high-level abstracts denoting a supreme being. *We* can see the incongruency in such a PAT chain, but only because we have the necessary intellectual tools (ie TSs) to help us. In a sense, so did Locke, although his train of thought was more laborious due to its newness. Obviously, once we know that abstracts are built by the mind as it goes through its life, we are looking for their causes amongst contemporary experiences and view them as something to be dealt with by *us*. As Locke says,

> The senses at first let in any particular *Ideas*, and furnish the yet empty Cabinet: And the Mind by degrees growing familiar with some of them, they are lodged in the Memory, and Names got to them. Afterwards the Mind proceeding farther, abstracts them, and by Degrees learns the use of general Names. In this manner the Mind comes to be furnish'd with *Ideas* and Language, the Materials about which to exercise its discursive Faculty: And the use of Reason becomes daily more visible, as these Materials, that give it Employment, increase. [45]

Now that his TSs allow the continuous assembly and organisation of abstracts, Locke can turn his mind to the question of how that process goes through its paces. We mentioned the necessity of PAT^1s before any generalisation of them can take place. To the ancients this was by no means clear, as we have seen repeatedly. Not so for Locke, who recognises the processing of basic data (the PAT^1s) as a prerequisite for PAT^{2+}s. For example, he says, a child will know that an apple is not a fire as soon as it has learned to hold the two notions separately in his mind. Yet the proposition, "It is impossible for the same thing to be and not to be" may not be understood until much later, when higher-level processes have been mastered. They can only be mastered however if the mind's TSs contain the appropriate FAs, and for their presence the direct, concrete experience is needed. Therefore, the content of higher abstractions can only be fully understood if and when there has been the opportunity for the relevant FAs to be developed, otherwise that content will be largely meaningless. As a less trivial example take the need for a child to be exposed to music for it to be appreciated later on; or, on a more complex level, the need for language students to rote-learn the new vocabulary and its grammar, only after which the ability to write creatively can be nurtured. The absence of PAT^1s invites an additional problem which we discussed already: since higher abstractions are rich in generalisation but poor in detail, to find your way down to realistic representations from an initial high-level abstraction is fraught with danger if the upward path had never been taken. A point that seems to be lost on certain zealous educational theorists of our time. Locke explains it thus,

> For having once established this Tenet, That there are innate Principles, it put their Followers upon a necessity of receiving some Doctrines as such; which was to take them off from the use of their own Reason and Judgment, and put them upon believing and taking them upon trust, without farther examination: In which posture of blind Credulity, they might be more easily governed by, and made useful to some sort of Men, who had the skill and office to principle and guide them. [46]

He continues in the following chapters to describe the creation of ideas, and we have touched upon the details from the perspective of our own intent. Because the only way to transmit our ideas about ideas is through the written word (in this case), the role of words as symbolic representations cannot be left untreated. Before we have a word for something we need to have its concept, and so Locke observes,

> ..the *Ideas of primary Qualities* of Bodies, *are Resemblances* of them, and their Patterns do really exist in the Bodies themselves; but the *Ideas*, *produced* in

us *by* these *Secondary Qualities*, *have no resemblance* of them at all. There is nothing like our *Ideas*, existing in the Bodies themselves. [47]

Without pre-emptying this topic in Part II we can say that some sensory experience (and everything our mind receives is, in one way or another, a sensory experience) generates the abstracts, say, A, B, C in the receiving mind. They are what they are because of the current state of TSs and the accompanying EMs. Since human brains are not synchronised machines, each mind will be in a somewhat different state from another, so the same incoming data might produce say, B, D, E somewhere else. Nevertheless, an agreed upon symbol for these data serves to communicate the idea, in whatever way it may have been represented. However, since the word is a result of the articulated con-TS only, it will never truly reflect the combined significance of the entire source TS. *The way* in which the structure has been formed is as important as the contents themselves if not more so, given the latter are dependent upon the former.

Locke furnishes an example, relatively simple but still intriguing [48]. Suppose a blind man was taught to differentiate between a sphere and a cube by making him feel the surfaces of both with his hands. Suppose later the man could be made to see, and a sphere and a cube was placed before him: could he tell the difference by sight alone? To which Locke answers, "No". There is conflicting evidence from modern research which we will go into later, but considering TS dynamics we can agree with him, provided the knowledge derived from FAs is insufficient to assist. The mode of building one TP (via the sense of touch) being different from the mode of the other TP (via the sense of sight) would create separate TPs to the extent that any of their object abstracts will not, by themselves, have associative properties with each other. Functional abstracts are another matter: if a set of FAs consists of the information representing the arrangement of surfaces and their edges, cognitively evaluated as 'cube', then the blind man should by inference make sense of the visual representation. The reason people with both senses do not have any inherent difficulties cross-referencing the contents of either TP is the fact that they collected their experiences in concert with any of the senses available at the time—they know what they see is the same what they touch is the same what they smell is the same what they taste... the typical exploratory approach of an infant (and now we can see why).

Once higher-level abstracts have been created, their presence is usually deemed more important than their source. "Jumping to conclusions" is one result, the expectations of certain geometric figures if parts of them are hidden, another. This is especially so if there is no distinctive alternative to one particular PAT chain and so the next most suitable is chosen, although there is no guarantee as to the validity of the outcome. See Fig. 7-2 as a simple example of expected and unexpected abstraction processes.

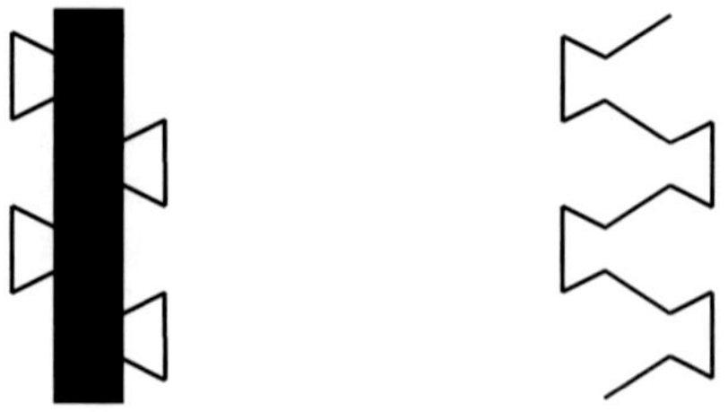

Fig. 7-2: Example of an expected abstraction which has been generated by four apparent triangles, partly hidden. Actually, they are not triangles at all but a continuing pattern of a single line.

In most cases we are not even aware of the path taken, and so the presentation of a high-level abstract can cause misunderstanding in the recipient because the steps leading up to it are not being compared with the overall context of the abstract's meaning. "*Habits*", Locke explains, “especially such as are begun very early, come, at last, to *produce actions in us, which often escape our observation*” [49].

Although the erstwhile PAT[1] idea becomes modified as the process continues, the initial integration within a TP is not random either. We have touched upon the role of EMs already, and to Locke it is clear that pleasure and pain make the deepest and most lasting impressions. We will explore the relative significance of EM types later on, but for now let us note that their accompanying influence has been deemed important enough to be listed alongside a description of thought formations. As usual, once a particular TS has been born, the possibility of associations within the mind is enhanced. From a purely mathematical point of view it is obvious that the increase is an exponential one; from an epistemological perspective the scenario is even more fascinating, suggesting as it does a fanning out of knowledge acquisition which is forever (?) on the increase. For example, the idea that emotions accompany and so influence one’s data input gradually determines the policies towards education, once the comprehensive perception of learning, understanding, repetition and implementation has run its course through the public milieu—there will also be something to be said about past practices. Similarly, Locke’s recognition that not only abstracts within one’s mind but access to them is consequential to one’s performance [50] leads to the methodical ordering of information. Furthermore, to enable outsiders (ie outside one’s own mind) to use the data, measures of conformity have to be introduced in line with efficient procedures to digest them. Without some relatively simple insights about how ideas are formed the need for these measures will not be recognised. It comes as no surprise that the Internet was not developed in the Vatican.

Locke’s interpretation of how combinations of several “simple ideas” are formed may sound superficial if not trivial. However, as an exercise in openly articulating an activity, even if it should be performed unthinkingly by millions of people at any given day, and so as an example of a semantic standard therefore it is of great value. So when he says,

> Under this of Composition, may be reckon’d also that of ENLARGING; wherein though the Composition does not so much appear as in more complex ones, yet it is nevertheless a putting several *Ideas* together, though of the same kind. [51]

it is arguable whether an idea as such gets bigger if numerical values therein are being increased. At the same time, the spelling out of a particular thought which comprises one, then several units of the same item and thus demonstrating the category, is useful to set down a common colloquial representation everyone can relate to. The process of articulation (and let us not forget, there are intellectual advantages in establishing con-TSs in the first place), the broadcasting of such information, and the reception and subsequent integration into other minds sets up IFAs and possibly even CFAs which contribute to the social fabric. A conglomeration of people alone does not make a society; it is the conscious awareness of shared and mutually usable information in each and every member of the group which enables, no allows them to identify themselves as a whole. The abstraction process which facilitates the creation of manageable symbols so that a common domain may be established, is crucial to that process—hence Locke’s considerable efforts devoted to it.

A shared pool of information, offering a multitude of associations and ‘what ifs’, combined with the capacity to articulate and so pass on the seeds for further explorations, these characteristics make for a potent mix, and so a potent society. Locke presents the process of thinking one’s way to infinity, by taking any length and adding to it until the act of adding itself is turned into an abstract to be multiplied [52]. While he realises that “our *Ideas* are not always Proofs of the Existence of Things", the very act of thinking into whatever territory evokes at least the conjecture of something. How far can the utility of an idea—however outlandish—be stretched?

Consider what happens when the concept of infinity is subjected to negation. The development of a TS along ever increasing magnitudes, represented by the appropriate abstracts (CFAs), presents no difficulty: the possibility of a thinker suffering internal trauma would be rather remote. But there also would be other abstracts, this time IFAs, collected whenever some state and its 'not' had been experienced (an expected item was looked for but it was *not* there; somebody was supposed to arrive but did *not*; someone got selected but it was a mistake, so he was *not*). Such a universal IFA would easily be applied to the CFAs comprising infinity, but the results would very much depend upon one's disposition. Locke writes about his frustration at this point:

> [Having arrived at infinity-] And such an *Idea* of infinite Duration, I am sure I have. But this *Negation of a Beginning*, being but the Negation of a positive thing, *scarce gives* me *a positive* Idea *of Infinity*; which whenever I endeavour to extend my Thoughts on, I confess myself at a loss, and find I cannot attain any clear comprehension of it. [53]

We need to remind ourselves that 'infinity', 'negation', etc are first and foremost abstracts in our mind, and so have no concrete counterpart in reality at that point. Any links with PAT^1s are only possible if the PAT chain featured them in the first place, and they and their derivatives (in the form of PAT^{2+}s) are associative with whatever the 'infinity' etc TS offers. In this context note the approach taken by the Jainas when dealing with Reality (see chapter 3). The ideas of space, matter, rest, motion and so on were firmly and consciously relegated to the abstract, and so they became Space, Matter, Rest, and Motion. They were also recognised as empty ideas in themselves, yet necessary intellectual elements to formulate complex concepts of our existence. To a Jaina the sudden emergence of a perplexing juxtaposition amongst two abstracts, such as infinity and its negation, would not constitute a problem. If one empty idea says "A" and another empty idea says "not A", well, so what? If in the case of Space and Matter etc there eventuates the possibility to imbue these abstracts with concrete connections to the real (as can be done of course), the problem is solved. If there are no such connections we are simply left with a bunch of empty abstracts, the non-appearance of a link in itself being a negation played out wholly inside our mind.

The difference then between the sensation of amusement and frustration in the face of such phenomena lies in the existence of TSs which capture run-away FAs and their objects and relegate them to their own type of domain and keep them there. It is a matter of creating sufficiently powerful RTSs to balance another RTS. With the Jainas the productive RTS was the recognition of the abstractive quality residing within the notions of Space and onwards, acting positively for their holders. For Locke the unproductive RTS was his idea of God connected to infinity and now suddenly annulled by the 'not'-IFA; there was no other to show a way out. The rise and fall of RTSs is more or less a technicality of cognitive dynamics, were it not for the acute discomfort they provide in some situations. What follows are two examples including their remedies, but readers who find it difficult controlling their thoughts are advised to skip the entire next paragraph.

Let's assume it is night and you are walking in a lonely part of town; there is no-one else besides you and even the lights are dim and far in between. Do the following: as you walk, put it into your mind that there is a chance somebody is following you, and if only you turned your head you will see the figure. Reiterate the thought a few times and then turn your head.

Most likely the street will be empty anyway, but the act of looking has set up a separate TS which says that the only way to know if someone is there is to look back. So, TS-1 holds the belief you are alone, and TS-2 tells you that the only way to confirm that belief is to look back. There is also a more general TS-3 which holds the fear of dark places and of more or less defined details what the fear is actually about. TS-1 and TS-2 are RTSs in relation to TS-3, with the addition of TS-2 being resonant towards TS-1. As you continue to walk down that street the need to look over your shoulder increases because instead of being calmed by *not* seeing anyone the interaction between the background fear and being alone on one side and the fear and the compulsion to ascertain you really *are* alone sets up an IFA which combines fear of being alone with the fear of not being alone into one vicious circle. The remedy: either disengage the apprehension about being alone so seeing someone will not be a problem, or disengage the thought of not being alone so you can stop turning your head all the time. In other words, the trick is to purposefully create another TS which is definitely *not* resonant with whatever is uncomfortable—it just takes some willpower. The second example is a variation on the theme but can be more disturbing for some individuals because there are no open spaces. Suppose you lie in bed at night and the room is dark and you are alone, preferably alone in the entire apartment. Stretch out your leg until the toes just reach over the edge of the mattress, outside of any cover. Now imagine: what would it be like if, at this moment, you felt *something* gripping one of your toes? At that thought you could quickly pull in your leg; the next moment you probably chide yourself for having given in to a silly fear—but would you stretch out the leg again? Whether you actually do it or not, the question itself together with a repeat memory of what you just did can take their hold on you. If you stay quiet, the imagery will haunt you; if you act and stick out your leg the images will haunt you even more because now the toes actually *are* over the edge of the bed...waiting. There is the background fear of things under the bed, similar to the TS-3 before. The TS-1 is the imaginary scene of the entire episode, and TS-2 is the act of reaching out with your toes. TS-1 and TS-2 are fed by TS-3, but the first two also reinforce each other. Again, the solution is to introduce a circuit breaker in the form of another TS. One effective way is to turn around and face the other side. Not that ghosts can't tell one side of the bed from another, but *your* ghost is a mental abstract which sits inside a framework of certain objects, one of which happens to be the particular side of the bed. Remove parts of the TS, and its associative powers to resonate are diminished. Another way is to push out your toes but now with the covers wrapped around. Once more the abstractive details of the structure have been interfered with and its resonant power scaled down. These two examples may not evoke the desired chills in everyone, but they contain the principles which are effective in cases of mounting hysteria: self-feeding resonance turning a basic non-affinity into an affinity.

There are no guarantees that an evolving thought pattern will not contain elements which ought not to go together. If an abstraction chain has due recourse to directly perceived objects and can maintain productive links with them then, no matter how high the flights of fancy, the mind can be at rest—reality will see to it. If however ideas are invited without bothering what they can lead to, one's mental state can become precarious—especially when their inherent nature is not understood. As far as the mind is concerned, it can represent anything, or, as Locke says,

> For there are no Things so remote, nor so contrary, which the Mind cannot, by this art of Composition, bring into one *Idea*, as is visible in that signified by the name *Universe*. [54]

Such versatility comes at a cost. There is an etching by the painter Francisco Goya, showing a man asleep at a desk while chimeras from a dark corner flutter around his figure. The picture carries a warning: when reason sleeps the monsters awake. Locke died forty-two years before Goya was born, and his discourses on ideas and their processes made it possible to actually talk about these kinds of monsters. Not the fantasies of religious mania, justified as a bulwark against the antagonists of a god but as real entities against reason, living right there behind the face before you.

Not all people in this world know about Locke—then as now. Only minds which had access and mobility made use of his thoughts and allowed their inner panorama to be enriched. Through gradual dissemination and the disposition of others the articulation of reason manifested in daily acts of living and gradually changed a society. And those effects did eventually touch all through the outward projections of their nature. But not everybody was enthralled.

8. God by any other name

Jean-Jacques Rousseau - Immanuel Kant

An idea forms from the contents of its environment. It is observed, gets dissected, is responded to. The response updates the environment. A new idea forms and the process continues.

A policy is released by a political entity. It is observed, gets dissected, is responded to. The response updates the environment. A new policy is released and the process continues.

The repetitive wording highlights the similarity of the two scenarios. It should come as no surprise that the realm of ideas and concepts in human society is a reflection of the dynamics of the single mind—which is, after all, the source of material for our cultural contents. A person's articulated desire (or plan, attitude, view point, decision, refusal, question...) represents the culmination of certain data and a particular set of processes, all held within that individual's intellectual domain. The result can be directly relatable to the domain, or it can be derived indirectly through the abstractive modification of the data. An example of the former is the choosing of a certain flight according to given departure times; an example of the latter would be the rejection of the death penalty because instances of wrong convictions are a concern. *Why* and *how* the result comes about is due to the individual's intellectual nature, and so a desire or a plan or an attitude demonstrates the make-up of its owner.

Allow the actions of individuals to become apparent to all; let them be compared, juxtaposed, and judged against each other; and there is a confluence of material which reflects the dexterity, the understanding, the leanings, of the populace. In the previous chapters we have concentrated on the functional modules as far as we have been able to identify them from an outside observer's perspective. Naturally, all these functionalities add their product to the daily life of a people, and as such they, together with what they produce, create the stage on which society conducts its affairs.

We are now in the 18th century. Books had been around for a while and their utility of holders and distributors of information has lost its novelty value. Being able to keep information in terms of content and format allows both to grow more complex and more structured, while at the same time permitting a greater part of the population to partake of its value. We have seen how abstractions loose their tendency to float freely away from their lower-level sources, provided the source data and their affinities can continue to be linked with the higher levels—the concepts become more grounded. To achieve this aim a purposeful reminder was placed every now and then amid the text of the Pañcastikayasara in the case of the Jains, advising the reader not to loose touch with reality. A more general outcome can also be realised by placing information about anything within reach of the citizen. Gradually, a significant pool of what has come to be known as common knowledge widened and deepened and made it possible to refer, revisit, and confirm. An overall infrastructure consisting of the physical space to house books in private and public places, the associated means of their maintenance and handling, the necessary transport for the readers together with the people's opportunity and will, it all contributed to an environment which grew, changed, and spawned sub-systems that in turn continued the evolution.

It is time therefore to place some of the mental phenomena of a single mind next to their wider counterparts, those that operate in society at large, to see how far the similarities would go. Their inclusion at this stage does not mean earlier times did not feature such public cognitive functionalities—we simply lacked the descriptive tools for dealing with the issue in a precise manner.

One of the philosophers who not only enabled these comparisons to be made but actively included societal mechanisms as an important topic to be discussed was Jean-Jacques Rousseau (1712 - 1778). His *Political Economy* and *Social Contract* [1], on which we will draw during the following pages, contain articulated ideals only a relatively comprehensive collection of social instances could assist in formulating. In terms of the listed content and its possible reverberations throughout society then the text could not have been written before its time; nevertheless, it often uses material from the past in order to lend it weight. Since our agenda is about underlying thought structures we will move between both works as the occasion demands.

When Rousseau talks about society he compares it to the body of a man [2]. Although he acknowledges the lack of precision in such a model, the perceived roles the various functions play are significant because they tell us something about how one looks at society at large. In his view,

> The sovereign power corresponds to the head; laws and custom are the brain, which controls the nerves, and is the seat of the understanding, the will, and the senses, while the organs of sense are the judges and public officers; commerce, industry, and agriculture are the mouth and stomach, making nourishment available to all; public finance is the blood which economic wisdom, performing the functions of the heart, guides throughout the body, distributing life and subsistence; the citizens are the limbs and body that make the whole machine move, live, and work, and which cannot be injured in any part without a sensation of pain being transmitted to the brain, provided that the animal is in a healthy state.

To us the comparison seems quite reasonable. Despite using abstracted versions of the physical body functions as well as those of the body politic, both refrain from overstepping the logic boundary since the stated characteristics have been derived from events that can be observed in reality. Of course, this is just the point. Let us remind ourselves that it was the English physician William Harvey who is credited with demonstrating the circulation of the blood through the body; this happened in 1628. During Rousseau's time Luigi Galvani, an Italian physiologist, conducted experiments on animal muscles to ascertain the role of electricity in nerves. The analogies presented by Rousseau may seem imprecise to him, but with hindsight we can see that whatever inaccuracy can be found is not so much due to the crudeness of the abstractions (after all, they *are* grounded) but has more to do with the state of science in its infancy.

It was not the first time the role of a head of state has been identified with overall authority, nor were people unaware of the controlling nature of public bureaucracy; these notions go right back to antiquity. Still, to *know* for a fact that there is such a thing as a brain, and nerves, and blood, and furthermore have a basic understanding of their functions, leads to an image in one's mind which radiates a degree of comprehensiveness and mutual interdependency that cannot but influence one's perception of society, given that *its* image is now affinitive with the one about the body. Thus there is an emerging interpretation of human activities which rely on science for its stated principles, together with the actual fact of a certain dynamic equivalence between the thought structures in a single mind and those conducted beyond in society at large.

In principle two TSs are affinitive if their components can be related to each other—in other words, it is their characteristics which determine the relationship and not any truth value in terms of reality. Therefore any image can be created and, if found acceptable to the mind, be used to draw analogies of one kind or another. We have come across instances where the conceptual model was based on religious interpretation and social necessities, usually with one competing against the other (ie wishful thinking vs reality), and witnessed the compelling nature of the resultant behaviour, often leading to extreme actions. In Rousseau we find a new type of source material in the form of largely grounded assumptions, and so through him and especially the many scientists and explorers which were to follow the educated citizenry of Europe had at its disposal a combination of cognitive functional abstracts and IFAs to aid its thinking processes.

As with any affinity, both TSs enhance their significance in the process. If someone like Rousseau looks at society with the aid of his knowledge about the body, he is prompted to expect similes which otherwise might have remained unstated. "Every political society is composed of other smaller societies", he writes therefore [3], using the idea of modularity to define human gatherings. With him it is not only a matter of positing something that can be observed "through their official outward form" however, but the idea of a large conglomerate of organs (as in the body) also suggests other sub-groups which are "no less real for being less perceptible". This is a sentiment produced by a state of awareness which allows glimpses of what-is in addition to expectations of what is still to follow—something that science is bound to evoke. Note the role of IFAs here: the experience of a steady stream of discoveries being made in so many areas of nature, coupled with the repeated occurrence of something newer still and unknown up to now. The result is the perception that there are readily identifiable groups in society as well as even more groups even though nobody has seen them—yet. This is an example of how fine a line the logic boundary can become: grounded assumptions producing IFAs in the form of c-PAT-ls, leading to inductions (again quite logical) which nevertheless could end up as i-PAT-ls (such as conspiracy theories in this case).

Note further the last part in the sentence quoted above, "provided that the animal is in a healthy state", although initially the analogy concerns the body "of a man". Given the mechanics of affinitive TSs the dissimilar expressions hint at a certain mismatch between what Rousseau tends to think in rational terms and what comprises his thought patterns dealing with personal musings. The TP holding the notions of science provides the material for the realisation that the functionalities of the *human* body can be used to explain what goes on in society—a process clearly built upon c-PAT-ls. However, the following *interpretation* of what society is doing so that its parts *can* be a useful comparison comes from Rousseau's own inner workings, and *they* have affinities which are not necessarily limited to whatever scientific literature he had access to but include numerous other experiences and so TSs which can draw on any combination of emotive elements, IFAs, i-PAT-ls and hence their configured states within his mind—in other words, his personality and character. Thus when it is about the initial logic of phenomena the body is human, but when it comes to the interpretations the body changes to that of an animal. When his mind ranges more freely the affinity is with the wilder aspects of nature. Shortly we will see how well that first hint becomes confirmed.

The previous relationship between individual- and society-related TSs has been expressly stated by Rousseau. As such it was subject to the wilful articulation by his mind as his words prove. An example where the connection is more unwitting comes when he writes about public deliberations, that is group processes. First he states that "during any public deliberation" it can be seen that "the general will is always for the public good" [4]. Then he identifies "secret divisions and tacit alliances" which often turn the "natural disposition of the assembly" towards "particular purposes", with the result that the one common focus has changed into several of different priorities. What is the "public good"? Since it is public it must contain things which have been made known to all and sundry, the openly expressed sentiments of the people. They are not simply a conglomerate of many thoughts, bundled together; rather they are influenced by constant modifications and alignments that are being demanded by society and responded to by its members under the auspices of social belonging and adaptation, and once that has been achieved they are deemed to be 'good'. Therefore they are not the true will of individuals at all but represent what is expected of them, and so the phrase "public good" is only half correct—public, yes, but not necessarily 'good' (on the other hand, 'good' can also mean 'possession' or 'material' in the plural sense, and so constitutes exactly what has been said just now; still, the underlying meaning is not only a matter of literary semantics but also of social mechanics, which was to be shown). Once these public utterances come up against private dealings, the overall cohesion of the group suffers.

Now consider the thought structures in a single mind. There are con-TSs which are the articulated part of TSs, and they are able to be articulated because the mind possesses a repertoire of words and signs which stand for commonly agreed meanings. The contents of con-TSs are therefore not the actual elements coming from a wider TS, but are the result of conforming such content to the public format of one's language. Does that mean words are by nature untrue? Not necessarily, but because they *must* abide by the common mould of language (otherwise they can't be articulated in the first place) they sit atop a more comprehensive complex containing elements which are not expressed, but which nevertheless can and do form affinities with other TSs—the "tacit alliances" of the sub-TS realm. Moreover, the less discrepancy there is between what can or should be said and what is being thought privately the truer a person's statement will be. Taking the wider view again it can be said that the more incongruent public perception in general is, and the more rational individual people's views are in general, the less reliable public utterances will be as well. Similarly, the more rational public perception is made to be, and the more idiosyncratic people are in their private opinions, the less reliable any open statements will be also. An example of the first is an outdated legislation on moral matters controlling a more mature population, prompting stilted debates; an example of the second is children acting under the supervision of a school, where truth is often hard to come by because although a child may know very well what ought to be done it acts out its own wishes in a furtive manner.

Rousseau discusses laws and how they ought to be instituted—their number, their severity, privileges and exceptions, and what to do about transgressions. Since he compares judges and public officers to the body's senses, that is those physical functions which receive and hold information (which includes the brain), then one should be justified applying similar considerations to the law as the basis on which the judiciary and executive perform. In his opinion the fundamental issue is the respect for the law. He continues,

> ..rigorous punishments are a futile expedient that was invented by small minds, in the aim of using terror as a replacement for the respect which they are unable to earn. It has constantly been observed that countries which have the most terrible penalties are the countries which resort to them the most frequently. So that cruelty in punishment merely indicates that the number of infringements is high, and when equally severe penalties are applied to every case, offenders are forced to commit crimes in order to escape being punished for their misdeeds. [5]

His comments can readily be tested even in our times. One only needs to go to nations that are regulated on religious grounds and confirm the often severe penalties reserved for deeds which in secular societies may not be classified as a crime at all (as usual, sex and drugs feature most prominently in such situations and we will see why in Part II).

When it comes to the single mind we find there are two functional entities that most closely resemble the laws in society, namely emotive modules and functional abstracts. We can extend our comparison by saying that the EMs represent the pre-existing moral values among citizens whereas the FAs are the purposefully constructed laws. The paragraphs of the legal code are not changed ad hoc, but they are subject to change; public morals do change as well, but at a slower pace and usually not as overtly expressed items but as a consequence of other contingencies. Similarly, CFAs are purposefully constructed and so are the IFAs, except that the latter are taken on 'as is' and so do not lend themselves to concurrent manipulation. Emotions, in most cases, are not controlled as such but it is the actions and thoughts that are influenced by them which get manipulated.

So, if there is a TS which is in conflict with another TS because that other follows certain rules then the outcome shows certain similarities with society at large where people run foul of existing regulations. Let's say there is a TS which basically says 'yes' and another which says 'no' then the resultant conflict must be resolved one way or another. If the no-TS happens to be part of a broader TP that encompasses a series of ideas all of which are a reflection of strictest attitudes, then for the yes-TS to carry the day it must be able to match its adversary in terms of significant associations to lend it weight. The heavier the no-TS, the more powerful the yes-TS will have to be and the consequence is a more ruthless overall response. We will investigate the detailed mechanics in Part II, for now it suffices to point out another instance of the equivalence between the dynamics of the single mind and what can be termed the group mind.

In chapter 7 we introduced the concept of social horizon, that is the extent to which the mind appreciates its actions in terms of the influence they have on others. Rousseau writes about it as follows:

> It seems as though our feelings of humanity evaporate and weaken as they extend across the earth, as though we cannot be as sensitive to calamities in Tartary or Japan as to those that are suffered by a European people. Concern and compassion have in some way to be limited and compressed, in order that they should be active. [6]

Although in this case the single mind is dealing with the outside already, the equivalence occurs in the way society's institutions are organised such that its social horizon is being defined, just as inside the mind its TSs configure the degree of extension. A few sentences later we find an example of how this is handled in society:

> Certainly, the greatest marvels of virtue have been done out of patriotism: a vigorous and pleasurable feeling which joins the power of self-love to virtue in all its beauty, giving it energy without disfiguring it, and so creating the most heroic of all passions.

The altruistic act is done because there are TSs which supply the necessary affinities and reciprocity to neutralise possible conflicts with other TSs, and thus forming denominative structures that become TPs in their own right (in essence, the scenario is comparable with the conflict dynamic described above—what has been *disassociative* there, is now *associative* here). In society the group mind creates features which enable a TS of its own (ie the citizen) to become affinitive across a certain span of associations with other TSs (ie other citizens).

To do that the citizens are being defined correspondingly under the category of 'nation'. Conversely, the extension is stopped by implementing a negative definition, thereby disrupting the associative process with further TSs (citizens). Therefore a 'nation' will represent some people but not others, a class will constitute certain members but not others, a group will include some individuals but reject everyone else. Furthermore, just as in the single mind one set of affinities relates to a particular content but not another, so are the types of activities performed within the different social horizons distinguishable from each other (for instance, members of a nation will share the same kind of food; members of a group will share the same meal; and members of a family will share the same spoon). How the types of affinities change from the small to the wider radius will be discussed in Part II.

Note the comment regarding the need to "compress" people's concern so that it may be "active". Although the associations which prompt one's expression of concern can be of any number, they do not act in isolation. Emotions play a significant part in making a particular TS attractive to another (or unattractive for that matter), and so we have to include the EMs in the equation. Recall chapter 7 where we first identified these modules and postulated a kind of reservoir from which they are drawn. Well, here we find further evidence that there seems indeed to be a finite number of modules. Given the virtually automatic allocation of EMs to various thought structures, and also given that one's self is a construct ranging in significance from the centre outwards, it stands to reason that the outer peripheries attract a lesser number of any particular EMs than would the inner circles. When Rousseau speaks of "compressing" one's concern we can translate that into our nomenclature by saying that, in order for the social horizon to be inclusive of some TSs, their potentially affinitive characteristics need to be shifted towards the centre of the general thought pattern which constitutes one's self. Only then will the available number of specific EMs be sufficient for the necessary affinities to be achieved.

In another passage Rousseau criticises the use of mercenaries as a substitute for the actual citizens because a mercenary does not share the common bond a society confers [7]. If people can be compared to distinct thought structures, the difference between a citizen and a mercenary would be reflected in the difference between a TS having an affinity with TPs representing one's own preferences and another TS which had been imported and does not share the same reciprocity. Depending on the circumstances a TP will be constructed to accommodate the alien TS with sometimes precarious consequences. An example would be someone falling in with a bad crowd which makes demands on that person's behaviour and so causes inner conflict. Every society has stories that are full of such themes.

The main constituent in our comparisons is the functional abstract. It provides the rules according to which the content is to come into effect, and any affinity between FAs transposes the efficacy of one set of data across to another TS and therefore another content. Regarding society, Rousseau defines the role of ruler vs that of the legislator [8]. The ruler "has only to follow a pattern", whereas the other "must devise it". It is significant that more primitive societies combine both figures in one and the same person, and to separate the two powers is rightly regarded as an achievement and a sign of maturity. Keeping in mind that societies are the results of minds, we must look within the latter for the reasons.

FAs, whether the cognitive or intuitive variety, are first and foremost the result of constructed abstracts in action. As such any affinities they possess relate to other FAs, since this is what they are. Pure (ie object) abstracts however are representations of data, and therefore *their* affinities relate to similar information, not acts. A vertical PAT chain, whether congruent or not, is the processing of abstracts and can be identified by the presence of different abstraction levels. We have seen the significance of lower and higher abstraction levels, where lower levels exhibit more detail and the higher ones a more general, principled picture. Solving a problem involves the utilisation of higher levels in order to tease out the inherent principle of a scenario (ie the content), and therefore the effectiveness of a cognitive process can be measured in terms of its ability to access higher abstraction levels. The functional associations which may occur in any given set of TSs can only be productive if the appropriate levels have been attained, for the pure as well as the functional abstracts. Since higher abstraction levels mean longer PAT chains, the possibility of contextual associations further up the chains between two or more strands of PATs becomes more and more remote the further along a chain one progresses. The result is less chance of interference between rule and content in the case of a more capable mind. Does the similarity between our TSs and their PAT chains and the coexistence of powers in society and its instrumentalities emerge now? The greater the effective separation, the greater the integrity of the output of the involved modules will be—on the inside as well as on the outside.

The positive effect of elongated PAT chains is therefore the separation it affords to strands of thought structures so that they may develop more or less independently of each other. That is to say, the advantage relates to the contextual difference they enjoy. Somewhat ironically that same extension has also the potential for negativity, and Rousseau touches on this point when he explains the destructive influence of a lawless citizenry [9]. Laws, he explains, ought to be respected, not only for fear of punishment but because people are exhorted to obey through their inner moral configuration. General lawlessness is an insidious process, where first remorse becomes stultified, then priorities are skewed to escape the influence of jurisdiction, and eventually the entire society has lost its cohesion to the point where even government becomes corrupt. In such a case the counter measures, that is more and more laws, contribute to the decay because now there are even more opportunities for disrespect. Rousseau:

> In such circumstances edict is piled on edict, regulation on regulation, but in vain; it all serves only to bring in new abuses, without curing the old ones. The more laws you add, the more they fall into disrepute, and all those whom you appoint as supervisors simply become new offenders, who either share with the present ones or take their plunder separately. Soon it is brigandage that gains the price for virtue; the most degraded men are those who win the greatest acclaim; the greater their position the more despicable they are; their infamy is made manifest by their distinctions, and the honours they receive dishonour them.

Considering the role functional abstracts play, it is easy to see the reason for such a development. Every act is represented by an FA, whether it be an observed act on the outside or a representational thought inside one's mind. Since FAs represent the act itself but not its contents (for these we have the pure abstracts), the mere existence of a certain FA lends itself to an affinitive influence with another TS, regardless of the actual contents. Thus to disobey a law—any law—creates an FA, and the possibility to disobey some other law exists because there is an affinity between law-1 and its FA and law-2. That FA becomes the affinitive bridge between the two TSs representing the law-1 and law-2. So, if one law can be disregarded then a precedent has been set to disregard some other law, and so the structure grows. The process is not negative as such; after all, if it were not for FAs we would not be able to learn by example. Nevertheless, here the build-up of PAT chains (ie the growing number of laws) promotes the proliferation of IFAs which happen to be destructive to society at large (if they happen to be CFAs we would be dealing with professional criminals). An accompanying danger of elongated PAT chains is the disappearance of groundedness, that is the link between abstracts and their sources in reality. In other words, not only will a decaying society dig its own grave, its members will be the last ones to understand what they are doing! (It is tempting to immediately draw parallels between such sentiments and the current state of affairs in this or that region of the world. But as we will see later, the world of today is substantially different from anything the past had to offer. While not invalidating the aforesaid, there are other factors as well.)

One issue which features large in any legislature is that of property. One can define it in terms of appearance or value, that is by objectively assessable measures, but even here we run into problems. Compare the stated value of a table with that of its weight: provided we remain on the surface of this planet the actual weight will basically be the same, regardless of the type of measuring units we choose to employ. A price of $1000 however reflects its source, manufacturing, and eventual appearance only to a certain point. The kind of timber is subject to fashion and availability; the manufacturing can be done by machines or by hand, in a developed nation or in a poor country; the appearance will be appraised according to period and demographics. In the end, as always in such matters, it is our perception which defines an object, and so the definition of property and its value is dependent on the characteristics of thought structures. No philosopher or ideologue has ever argued about the chemical composition of material possessions, but some have roused the masses even by their interpretation of what these things *mean*. In Rousseau's words it is important for the law to deal with property because,

> It is certain that the right of property is the most sacred of all citizens' rights, and in some respects more important than freedom itself, whether because it is more closely connected with the preservation of life; or because, a man's property being easier to appropriate and harder to defend than his person, the thing that is more readily taken should be the more respected; or finally because property is the true foundation of civil society and the true pledge of the citizen's fidelity in fulfilling their obligations: for if possessions did not answer for a person's acts, nothing would be simpler than to evade one's duties and flout the law. [10]

There are a number of contexts in this statement, all of which are a delineation of the underlying mental dynamics in the single mind.

Firstly, the right of property and therefore property rights are indeed held to be very important, and they would not have gained that position were the associated thought processes not of such a fundamental nature. A particular content becomes significant if the associative intensity of the relevant TS happens to be high. It is high when the points of association with other TSs are numerous, so that one can speak of a significant context. This means there exists a high degree of probability that a new TS will experience a functional or contextual association with some other TS in that context's TP. Since the mind has the ability to abstract input, that is to represent to itself the inherent principles of what it has observed, the resultant entity becomes a participating element of a structure. The consequence is that the human mind is predisposed towards the utilisation of outside objects as a means to enact the abstracts their principles allowed it to evoke. Note the mutual relationship between the existence of objects and their representative nature as held by the mind: capable minds collect objects, and objects facilitate the actions by capable minds. The more sophisticated the society, the more complex the array of possessions will be also. It is unavoidable therefore that societies have developed social structures through which to administer the taking, holding and giving of objects (and depending upon the degree of sophistication the objects themselves can be abstracts, such as intellectual property for example).

The FAs alone which are associated with the issue of property are thus many. Hence the chance of two TSs with similar contents, where the one deals with the self and the other with someone else, is high too. The opportunity presents itself then of having IFAs derived from the experience with others to become associative with elements of more self-related TSs and thereby furnishing the insights of observed rules in action. In other words, an event having to do with property is witnessed and now supplies the experience to be used in the handling of one's own possessions. The result is a general consensus amongst citizens to enact laws regarding property rights because all know how the law affects everyone. Note again the mutual relationship between the experiences derived through the self and others; should there be individuals who for some reason do not share those experiences, the perception of what the law ought to be looses its homogeneity. So, property issues are high on a society's agenda because the mind needs the usage of objects to occupy itself—wherever, whenever; they are the foundations of civilised societies because the more complex the usage of objects the more complex the users can be, and vice versa; and finally, the issues tend to take precedence over one's life because common experience has taught the citizens that what can befall someone can befall oneself.

So far we have mainly dealt with functional abstracts, that is some kind of action which undoubtedly occurred in one form or another. Rousseau does not just observe however, he delves into his own imagery to come up with abstracts with which to furnish his constructs. Consider the following passage:

> Man was born free, and everywhere he is in chains. There are some who may believe themselves masters of others, and are no less enslaved than they. How has this change come about? I do not know. How can it be made legitimate? That is a question which I believe I can resolve. [11]

The above serves as the opening to the *Social Contract* and can justifiably be taken as one of Rousseau's basic perspectives. As such it should be examined more closely.

As we have seen in our discussion of property rights, the mutual interdependence of TSs is an important feature of their existence. Through it the constant feedback helps to create PAT chains, assures their reciprocity and evokes denominative structures and patterns. Keeping this in mind, read the words, "Man was born free". There are enough IFAs in one's TPs relating to the concept of freedom for a person to have a relevant opinion on just about anything, and so the phrase is surely evocative in the extreme. Note however that there is nothing in the *content* which actually makes it so—it is the *functional* abstracts which, through their associations here and there, provide its cognitive weight. As such they are unlikely to be grounded, but occupy a realm which extends beyond the logic boundary. After all, how can anyone be born free? Is it freedom to require a midwife or a doctor, have somebody cut one's umbilical cord, offer a breast to supply one's nourishment, and take care of the most fundamental functions of one's body—and all this within the very first few minutes of one's life? Not to mention of course the pre-existing fabric of social dependencies occupied by the parents of the child, about which the newborn can do absolutely nothing. We can safely assume therefore that the phrase "Man was born free" is not the result of a c-PAT-l but rather the outcome of a process which drew on comfortable elements within one's inner life and summarised them accordingly.

No wonder then that Rousseau does not know how the change from something that has never even existed to a general state comes about—there simply is no change to begin with. Yet to treat the content of some set of TSs which suggest such a change as a reflection of reality and then to commit himself to continuous attempts at reconciling other abstracts with such a construct leads to a type of dynamic we have encountered before. Recall Socrates' concern with ideals, prompting him to invent the notion of life before death; or St Augustine's vision of his god, leading him to subsume human endeavours under its auspices; and so on. What these types of thinking have in common is the creation of a TS which does not have a link to 1st-level abstractions (ie it is ungrounded), but nevertheless gets conjoined with others by establishing the necessary affinities. Thus there are incongruent functionalities that form a container. For a detailed description of how they come into being refer back to chapters 2 and 4.

Given the existence of such a container (ie "Man was born free") immediately certain IFAs emerge. One is the notion of origin, of what-comes-first, and consequently another is the idea of naturalness. When Rousseau thinks of the family, the affinity between its abstraction and that IFA springs into action. We now have 'family' being connected to 'what-comes-first' being connected to 'nature' being connected to 'free'. Consider what he has to say about the family:

> The most ancient of all societies, and the only one that is natural, is the family. Even in this case, the bond between children and father persists only so long as they have need of him for their conservation. As soon as this need ceases, the natural bond is dissolved. The children are released from the obedience they owe to their father, the father is released from the duty of care to the children, and all become equally independent. If they continue to remain living together, it is not by nature but voluntarily, and the family itself is maintained only through convention.
>
> This shared freedom is a result of man's nature. [12]

Note that despite the word 'freedom' as a semantic anchor every single instance of living arrangements cited above is the result of utter necessity: the initial bond between the children and their father (need for conservation), the release from obedience (the need to look after themselves), and even the eventuality of continuing to live together (by force of convention, arguably the strongest demand of them all). Only a container, created out of incongruent functionalities, is capable of furnishing such a train of thought, all the while intentionally projecting the image of 'freedom'.

History, and indeed the very present provide us with a multitude of social and political formats where the degree of freedom varies, to the extent of the word itself becoming open to quite varying interpretations. Yet Rousseau does not enter the debate on specific implementations, positing one kind against a societal circumstance in question. For him the concept of 'freedom' is not so much a technical term expressed through objective measures, but it constitutes the articulation of a general idea which is being applied through its emotional value rather than some defined scale. The quote above illustrates the effect this has on his use of the word, and by which the phenomenon itself has been identified within his cognitive system. Let us be clear on the underlying mechanism and the sequence of events leading up to such a state, so that its full ramifications can be recognised.

There is a thought structure then which holds the notion of 'freedom', having been created at some time in the past through various experiences which are hidden from us at this point in time. We *can* say however that this TS has demonstrated a sufficient number of affinities to establish connections with other TSs relating to 'family', 'society', 'nature', and other notions which have to do with what is held to be fair and lawful (all this can be gleaned from the quotes provided above—and which, by the way, is not contradicted by the rest of the text *not* shown here).

There are generally two ways in which the formulation of such a TS can be accomplished: firstly, a systematic assembling of pieces that are taken from formal presentations which happen to feature the element 'freedom', and therefore have been grouped together based on such a commonality. The result in this case would be a congruent TS or even TP, where each and every part would contain links to some neighbouring concept. Because it is those concepts which provided the reason (ie the content 'freedom') for getting connected by the encompassing TS, the direction of influence points from the contributing TSs towards their combining centre. Therefore, the mind speaks of 'freedom' in some context because it has encountered the idea there and identified a pattern linking it to somewhere else.

Secondly, there is a TS representing the notion 'freedom' which at first is not strengthened by other connections but instead by emotive modules. Their assorted combinations can in our language be described as love, hate, greed, anguish, envy, altruism... the only limit to the potential list being one's preference for articulated detail. This second type of TS will also seek out affinities, but this time in terms of the EMs it is imbued with. Some other TS will not be conjoined because that other TS features a 'freedom'-related grounding, but because its own EM-set is reciprocal to the 'freedom'-TS and nothing more. The direction of influence points from the centre outwards. Therefore, the mind speaks of 'freedom' in some context because the emotional quality of that context happens to be similar to the emotional quality found in the central TS.

Both 'freedom'-TSs have the functional appearance of a container. The first type we can call a congruent container, the second an incongruent one. An example of the first type would be a TP comprising the concept of 'metal'. Only a material that is identified as having certain characteristics will be classified as 'metal' and from then on makes up the overall thought pattern. A contemporary example of the second type is an urban gang. A person, or a piece of music, or a fashion accessory will only then be accepted by the group if the EMs found in these items concur with the emotional make-up of the gang.

In Rousseau's case the idea of 'freedom' is an incongruent container because the TSs it is connected with are not about freedom but have similar emotional qualities. The justification from saying so comes from an analysis of those other TSs as shown in the previous paragraphs (it could also have come from an investigation into Rousseau's personal attitudes, that is focusing on the EMs instead, but I guess we are too late for that).

Having constructed a TP which holds the general idea of being free together with such diverse items as 'family' and 'nature', and further associations reaching into 'original state' and a vague concept of 'pureness' produces attitudes which do not hone themselves according to what is under focus but do the opposite: they influence the object under attention. Take Rousseau's opinions on war, and in what manner it occurs at various levels of society:

> For the simple reason that men who are living in their original condition of independence are not in a sufficiently continuous relationship with each other for a state either of peace or war to exist, they are not naturally enemies. [13]

Note the assumption regarding men's original condition. Since Rousseau feels the need to present his case in some kind of chronological order, the perspective therefore is from earlier times towards later events. This connects again to his perceptions of the original state, and with it the surrounding abstracted entities of freedom and naturalness. Thus the phrase "original condition of independence" is used almost automatically, providing a starting point that has to be accepted as given. It follows that 'independence' does not infer any relationships, and so there are none. Since war—or peace—relies on some kind of relationship to exist it does not eventuate because that basis is deemed to be missing from the equation. To be an enemy therefore supposes the advent of all those things which manage to sully the pristine, initial condition of humanity, in other words all the notions which are represented by TSs outside the incongruent container-TP. So to be an enemy means being a victim of outside forces—in more ways than one—and the state of being natural has been lost. In Part II we will follow the creation of dissonant situations between minds, but for now it suffices to observe the dynamics in terms of the container in question to illustrate the problem. He continues,

> It is the relationship of things, not of men, that constitutes a state of war, and since the state of war cannot be engendered merely by personal relationships but only by relationships between things, a private war between man and man cannot exist—either in the state of nature, in which there is no permanent possession of property, or in the social state, in which everything is controlled by laws.

For the TSs holding 'men' to become affinitive with something negative like 'war', it has to be extended such that the necessary EMs can be put into place. From the original state (still positive) we move towards a form of deterioration in order to import the negativity, and moving along the timeline away from the 'origin' brings us to TSs connected with 'later', 'growth', 'increase', 'development' and the like—all ideas which suggest themselves associatively when contemplating the flow of time with regards to human existence. People in pre-agricultural societies have very few possessions (recall the mutual relationship between mind and object mentioned earlier) and, given the emotive reciprocities evoked by the container-TP, the further along the evolutionary path of human development we move, the more pronounced the existence of artifacts and the more significant their contribution to life in general becomes. For Rousseau then it is the things and their relationships which drives Man away from a state of purity and leads to war.

Observe further his stated grouping of TSs: 'no war' ↔ 'state of nature', 'state of nature' ↔ 'no permanent possessions', 'no war' ↔ 'social state', 'social state' ↔ 'laws control everything'. All these items making up the groups are elements of the container-TP, and as such are subject to the same sets of EMs which are used to seek out potential candidates for affinity. Now an affinity is an affinity—whether across a TP or within a pattern amongst its constituent structures. The stage is set for a frame of mind in which the social state is considered close to nature, will ensure the absence of war, and where personal possessions are at least frowned upon but in any case are regulated by law to a high degree. Given the incongruent nature of the container the result is a rather absolutist perspective on society, and in principle that development is no different from what we have already encountered during our examination of religious constructs. There the container was about 'god', here it is about 'nature'.

One interesting role containers play in the affairs of society is in the substitution of effect for cause, and vice versa. While writing about the task of the legislator Rousseau warns about the difficulty wise men have in communicating their ideas to the common people. It is not just a matter of idiom, but there are ideas which by their very nature do not lend themselves to over-simplified representations. For the reason and purpose of some project to be understood by everybody, the basic concept of it needs to be communicated, although the actual measures may seem more obscure. In his words,

> In order that a people in the process of formation should be capable of appreciating the principles of sound policy and follow the fundamental rules of reason of state, it would be necessary for the effect to become the cause; the spirit of community, which should be the result of the constitution, would have to have guided the constitution itself; before the existence of laws, men would have to be what the laws have made them. [14]

The laws, the guiding principles according to which citizens conduct their affairs, create through their capacity the overall standard of the societal environment. The expected nature of the environment (the view of which prompted the creation of laws in the first place) is the container, under whose auspices the public affairs are being conducted. But it is also true that the affairs make the container, since it is through their affinities that the overall environment becomes defined. So, which comes first? Clearly, without some stated aim (ie the expected nature of the environment) no project will be undertaken, and therefore the creation of the container takes precedent over its eventual definition. Since at that point there are no congruent links to actual occurrences (they haven't been enacted yet), in its initial phase the container would have to be an incongruent one. It follows, as Rousseau remarks, that the legislator "is unable to employ either force or argument", and instead has to "have recourse to another order of authority":

> That is why the founders of nations have been forced in every period to resort to divine authority and attribute their own wisdom to the gods, in order that their peoples, who are subject both to the laws of the state and those of nature, should recognise the same power in the creator of man and in the creator of society, obeying freely and submitting meekly to the enforcement of public felicity.

We can say therefore that there can be situations in which it is necessary to establish an incongruent container, even if later on it should be converted to a congruent one (note once again the similarity between single mind and group mind). In the case of the single mind it would be a hypothesis which one sets out to confirm, and if successful is turned into a theory, ie a congruent container. On a wider scale it is a political party platform which, so the theory (hypothesis?) goes, gets implemented during the party's term. Regardless to what extent the policies become reality, at the very least the process of discussion, media exposure, and public absorption ensures a certain measure of groundedness during the container's lifetime, and at the end the incongruent container could be seen for what it is—namely incongruent (then the other party gets elected and the process probably repeats itself). By contrast, in a society ruled in an absolutist fashion (whether secular or religious) the expected environment will always remain just that—expected. That is to say, the container will stay incongruent and only those measures will be taken which conform to the container's standards. To return to the single mind, a hypothesis can become a viable theory if supportive elements are found on the outside and brought in to substantiate the container; a hypothesis turns into an obsession if the container is allowed to select its ingredients on the basis of emotional qualities only. Both are common themes.

Once a container has come into existence there is no inherent limit as to what can be subsumed under its structure. In Rousseau's case the most likely targets would be TSs relating to society. If the container happens to be an incongruent one, the direction of significance points from the centre outwards, that is from the container towards any potential affinities beyond. The effect is a construction of further TSs to accommodate the potential targets in terms of the container's nature, and we can observe another example of such a process when Rousseau talks about the definition of political bodies. They can be identified in two ways, he says, by the size of their territories or by their populations [15]. He begins by making two assertions, which are expected given his TP regarding nature, original states, and so on, followed by his interpretation of those assertions:

> It is men that make a state, and it is the land that feeds men: the relationship, then, is that the land should be sufficient to provide sustenance for its inhabitants, and that there should be as many inhabitants as the land can feed. The maximum strength of a given population lies in this ratio; for if there is too much land, it is burdensome to guard, it will not be fully cultivated, and some of its produce will be superfluous; it will be a direct cause of defensive war. If there is not enough land, the state depends, in order to make up the deficiency, on the goodwill of its neighbours, which is a direct cause of offensive war.

The observations about men making a state and the land feeding a population are IFAs which Rousseau would have had ample chance of collecting during his life. The next one, that produce must be sufficient to feed whoever lives where the produce is being made, is also an IFA. There can be some doubt however whether my previous sentence is in fact truly representative of Rousseau's IFA. The first part in the quote above, regarding sufficient land, is only correct when taken as a corollary to the statement about land feeding men; it is not necessarily true as such. There are many examples in the past as well as in the present where the inhabitants of some stretch of land were clearly not comprehensively catered for by their local produce, but had to import a substantial part to make up the shortfall. One of the more successful places of that nature in the past had been Malta, and nowadays it is Singapore. That also goes for the second part of the quote. Rousseau's delineation into the subject of war is bound to be flawed because of his precarious starting point.

Furthermore, he imports a TS from the field of mathematics to underline his argument (in other words, to satisfy the characteristics of his container-TP): the ratio between an area's agricultural productivity and the number of its inhabitants. A ratio, in mathematical terms, is a definite value, where the precision of its quotient is determined by the precision of the antecedent as well as the consequent. That is to say, there needs to be a set of two identifiable real numbers, so that one can be divided into the other. But according to Rousseau,

> It is impossible to give a fixed numerical ratio for the area of territory and the number of inhabitants that adequately support each other, not only because of the variables that are found in the qualities of the land, its degree of fertility, the nature of its crops, and the effect of the climate, but also because of those that are observable in the temperament of those who live on it, some consuming little in a fertile land, others much on difficult soil.

He then continues the list of factors by mentioning the fertility of women, the topographical layout of the country, and legislation-induced increases in population. Note this particular incongruent nature of the TP: on one hand we have a TS of a very definite item (a ratio), but on the other we are pairing it with a range of ambiguities which mostly derive their importance from wilful perception, and so these ratios can never be produced in reality. This is not an exercise in nit-picking; the combination of a rational element with emotional intangibles is the stock-in-trade of ideologues, whose effect on human misery through these kinds of pseudo-rationales surely does not have to be argued any longer.

Such enumeration of detail with which to fill a container, without due concern about such details' groundedness, invites ad hoc definitions to imbue each element with a certain significance. It is the general dynamic of the container-TP on a smaller scale, since it seeks to create links between the container's subordinate ingredients and some other TSs perceived to be appropriate. Take Rousseau's description of the three types of wills as exhibited by an officer of government [16]. First he identifies the individual will, then the will pertaining to the members of his government, and thirdly the will as manifested by the people at large. He then goes on to characterise each version and comment on its respective usefulness. Since our priority is not a dissection of the philosopher's opinions we pass over those remarks, but let us consider how he sees the inappropriate application of these wills for, as far as he is concerned, it is bound to be inappropriate:

> In the natural way of things, however, these different kinds of will become more vigorous in proportion as they are more concentrated. Thus the general will is always weakest, the corporate will has second place, and the particular will comes first of all; so that, in the government, each person is primarily himself, then a member of government, and then a citizen—an order of priorities which is the exact contrary of the one demanded by the social order.

As said before, one may argue about the degree and utility of one's will (we did just that when discussing social horizons), but it is one thing to debate an issue using convenient descriptive symbols (ie individual will, corporate will, etc), yet quite another to seriously posit a political system (the "social order") and determine how this should be exercised by its officials (following the "demands" made by their master), with the constant understanding that these self-same officials are bound to get it wrong anyway. The concept of 'guilty until proven innocent' is waiting in the wings here.

Rousseau's liking for the past, seeing it as some kind of pristine condition from which humanity has fallen since (note the shades of religiosity—a mark of incongruent containers), leads him to view historical events under the influence of certain emotive modules. For example, a substantial reason for the strength of ancient Rome, so he writes, was the decision of her leaders to elevate the social and political status of rural dwellers and to diminish those of the cities. He expresses himself through the following words:

> The rural way of life was followed by the best of the patricians and was held in honour by everyone; the simple, industrious life of a villager was rated higher than the craven idleness of the townsmen of Rome, and a man who in the town might have been no more than a wretched member of the rabble became, through working on the land, a respected citizen. [17]

Recall our discussion of IFAs in a container-pattern and the outward-flowing influence of its resident sentiments: fine if, when the spotlight rests on you, you happen to be of an acceptable ilk, too bad if you are not. There are records of people who lived under such regimes, and their sufferings should never be forgotten.

As mentioned above, there is no inherent limit to what a container might apportion to itself, and religion—another container in itself—is not immune to such affinity seeking either. By now we can recognise the reason why Rousseau distinguishes between two types of religion, that of the man and the other of the citizen [18]. 'Naturally', the former is the "simple and pure religion of the Gospel, true theism, or what may be called natural divine law"; whereas the latter has "its dogmas, its rites, its visible form of worship, ordained by law", and therefore "it regards everything, apart from the one nation which follows it, as faithless, alien, and barbarian". Once again, our concern in this context is not the historical validity of Rousseau's ideas, but what cognitive dynamics it takes to produce them.

On that note it is interesting to read his opinion on the "Christian republic" [19]. Rousseau:

> Christianity preaches nothing but servitude and dependence. Its spirit so favours tyranny that the tyrant benefits from it. True Christians are made to be slaves; they know it and are hardly bothered by it; since this short life, in their eyes, is worth so little.

Such a stance is worth debating, but knowing the nature of his container-TP, holding as it does the notions of 'purity' and 'simplicity', it becomes obvious that his criticism is not based on some c-PAT-ls that mine any instances of ill-advised servility, political and/or psychological, but is derived from the fact that Christianity has become an established, formal religion, together with all the attributes such an organisation will have collected over the centuries. As such it is far removed from the "pure and simple religion of the Gospel".

One significant ramification of incongruent containers is the need for their owners to stock them with fitting contents. We have seen the characterisation of officials and their wills, the descriptions of rural and city dwellers, and the shading of religious practices. What happens when reality refuses to come to the party, as it sooner or later does? Rousseau provides the answer:

> Although it [ie the sovereign] cannot force anyone to believe them, it can banish from the state anyone who does not believe them; he can be banished, not for impiety, but for being unsociable, and for being incapable for cherishing the laws and justice sincerely, or of sacrificing, when necessary, his life for his duty. And if, having publicly accepted these same dogmas, any person conducts himself as if he did not believe them, let him be punished by death; he has committed the greatest of crimes: he has lied before the law. [20]

Once we have a society where the severest punishment is not reserved for a materially identifiable act but for showing disrespect towards an abstract, especially if that abstract is dependent upon wilful and emotive affinities, it is time to be afraid—very afraid.

The existence of a container which can hold sway over one's plans and actions because of its sheer overall significance does not therefore restrict itself to the inner mental dynamics. There are very real acts which are committed in order to fulfill the requirements of the concept thus represented. Rousseau's "social order" for instance demands a certain type of behaviour of its members, and the justification is the concept itself. 'In the interest of society' then becomes the rule of law and as we have seen, the direction of concern is not from the citizen towards the concept, but from the concept towards the citizen; that is to say, the law ceases to exist for the benefit of the citizen, and instead the citizen exists for the benefit of the law. Contemporary examples are cases where 'in the interest of ...' is the basis for any planned approach. Note the direction taken—it is not the concern for what- or whoever which is the problem, but the mutual relativity of the importances of all the issues concerned. In effect then the intention is not really to act 'in the interest of...' at all, but serves the preferences and character of the container-TP first and foremost.

Now that we understand the nature of containers more fully, the following passage may be seen in a different light. It has been taken from the first few pages of the *Social Contract* and sets out the premise from which Rousseau proceeds.

> 'Find a form of association which will defend and protect, with the whole of its joint strength, the person and property of each associate, and under which each of them, uniting himself to all, will obey himself alone, and remain as free as before.' This is the fundamental problem to which the social contract gives the answer. [21]

The system we are being promised is designed to be effective (ie there is strength right across its range); it will be capable of aggression (ie it can defend and protect); it demands the individual member to be at one with his community (ie "uniting himself to all"); the responsibility of each individual rests with him alone (ie "will obey himself alone"); and whatever it permits its members to do will be classified as "freedom" (ie the member will remain "as free as before"). Or, using our own nomenclature, there is an entire thought pattern which is deemed to be more prominent than any other structures; it holds emotive modules enabling it to be belligerent to the point of destruction; it is denominative in relation to any other TSs relating to the self, no matter how personal and therefore self-defining those may be; and in terms of articulated description the word will be 'freedom'.

I would like to repeat: we are not concerned with the *contents* of Rousseau's assertions, but rather the *way* in which thought structures are being built in order to represent the ideas. Just as a tool is not merely a thing by itself but is also defined through the way it is being handled, so does a concept not only exist as a point in a vacuum but acquires its life by being thought and acted upon in a particular manner. Consider his advice on property and what conditions pertain to the right of ownership:

> First, the land must as yet be uninhabited; secondly, no more must be occupied than is needed for subsistence; and in the third place, possession must be taken not by empty ceremonies, but by work and cultivation, the only mark of ownership which ought, in default of juridical title, to be respected by others. [22]

Since incongruent containers contain ungrounded abstracts, there are no links with appropriate TSs which would lead one to the contingencies of reality and so enable some productive feedback. In brief, the inevitable problems with an attempt at implementation of the aforesaid are:

what does 'uninhabited' mean—no animals at all (for instance certain species which are useful for neighbouring inhabitants), no humans at all, or certain types of humans?

what does subsistence mean—different economies require different levels of support, from hunter-gatherers to farmers to industrialised societies, and is water more important than soil, a river more important than a harbour, a valley more important then a mountain?

how does one define "work and cultivation"—is the hunting of game 'work' or is it the plowing of a field, or the damming of a river, the mining of a hill, or the erection of a factory somewhere?

A congruent container on the other hand would be able to supply the definitions, because the existent links will point to them. In addition, those definitions will be objectively ascertainable because their sources are part of the common knowledge. The contents of an incongruent container however are not a reflection of what can be gleaned from the outside, but are the construct of the container's inner make-up. Therefore they are created on the basis of that particular pattern. Take Rousseau's comments on 'general will' [23]. There is the initial assertion, followed by an interpretation, and then a ramification:

> It follows from what precedes that the general will is always in the right [assertion], and always tends to the public welfare [interpretation]; but it does not follow that decisions made by the people have equal rightness [ramification]. One always desires one's own good, but one does not always see what it is; the people can never be corrupted, but it can often be led into error, and it is only in this case that it seems to desire the bad.

The assertion reflects the character of the TP; the interpretation must follow since there is a fundamental quality to society in general (ie 'good') and so one of its properties must *be* and *do* likewise; and the ramification springs from the need to reconcile the significance of the TP (the concept of society) with the less straightforward conditions found in its counterpart in the real world (the people out there). According to Rousseau there may be "small differences" of opinion among the citizens, but the general will "would always emerge" as their common denominator and so decisions "would always be good"—*provided* the people have been "properly informed". Again, it is not the content as such which requires attention, but the manner in which it is put together. To say that people should be listened to and whatever consensus emerges ought to be the accepted wisdom is fine as far as it goes, but then comes the proviso and if it exists amidst a framework of an idealistic scenario the stage is set for compatibility problems with reality.

The need for reconciliation between the overall concept and what the world offers exists all along. Just as the idea of 'ratio' was brought in to justify the perspective on land usage, 'sets' are used to qualify the role of law in relation to the will under discussion. The general will, so he writes, can never relate to a particular object because to do so would strip it of its commonality—it would no longer be the *general* will. The point in question therefore is the definition of the "particular object", since it is its particularity which determines the correct type of will. Consider what he has to say about that object:

> For any particular object of will is either inside the state, or outside. If it is outside the state, a will foreign to it is not general with respect to it; and if it is a thing inside the state, it forms part of it, and in that case, the relationship created between the whole and its part is such that they are two separate things, one being the part and the other the whole minus that part. But a whole lacking a part is not a whole, and as long as this relationship subsists there is no longer a whole, but two unequal parts; whence it follows also that the will of the one is no longer general with respect to the other. [24]

Note how the debate hinges on the quality of the object. It might not be too difficult to decide whether the focus of attention is relevant to one's society or not, in which case the question of foreignness could be resolved. But who decides how "general" a local issue really is? If it is found to be applicable to the entire society, its constituency presumably melts into the greater whole and so does not form a part of it; yet should it be more specific, it "forms part of it" and the two are "two separate things". At the same time, the whole is now "minus that part" it was separate from in the first place. Apart from the faulty logic regarding sets and their members, the ambiguity of the major players in this argument is what makes the issue so disconcerting. If a society follows a regimen which is defined in such a loose way, the life of a citizen can become dangerous very quickly, once the authorities insist that he or she has transgressed their definition of 'national interest'. Indistinct outlines of illogically argued principles that are nevertheless taken very seriously are a hallmark of human rights-abusing nations.

Not only are basically rational concepts misused by importing them into an incongruent TP, their own affinities lead to misleading associations which the original theme as held within the container does not deserve. For example, when Rousseau talks about the role of the people within the social contract, he compares their activity to a centrifugal force "by which they act upon each other constantly, tending to increase their size at the expense of their neighbours" ('centrifugal force' should be understood in the sense of Descartes' circular motions everywhere in the universe) [25]. For someone who has read Descartes (something Rousseau clearly invites) the idea of such a force evokes thought associations which have to do with objects straining to fly away as well as a counter force preventing them from doing just that (ie the centripetal force). If brought to bear upon the subject at hand, there suggests itself the necessity of having such a balance in society, acting continuously to preserve the integrity of the nation. Technically speaking we can say that for someone to possess a TP concerning centrifugal and centripetal forces, there are IFAs (since most likely, the subject of forces would have been learned about somehow) which encompass the notion of constant forces, one acting in opposition to the other, and one being simultaneous to expansion, the other to contraction. Since the content of the forces-TP is made affinitive with the content of the society-TP (the incongruent container), the functional abstracts are being transposed also. The result is a perspective which says that society oversees the expansion done by its people, and therefore has to apply means by which the opposite, the contraction, can be realised. In Rousseau's words,

It will be apparent from this that there are reasons for expansion and reasons for contraction; and it is not the least of the political thinker's talents to find, in considering both, the proportion best suited to the preservation of the state.

The importance of having an abstract grounded in reality, that is providing it with links to PAT[1]s so that something tangible can be used to give it meaning, is highlighted when Rousseau writes about general laws which govern, amongst other things, a nation's assets. His opinions on people's eating habits according to the climate they live in leads him to postulate that a "barren .. land, where the produce is not worth the labour" should "remain uncultivated and uninhabited, or populated only by savages"; barely productive lands ought to be inhabited by "barbarian tribes"; areas where the surplus is adequate are "suitable for free people"; and places of abundance "ought to have a monarchical government" [26]. He does not inform us how one should go about making an a priori decision regarding the use of a certain tract of land by a certain group of people, when neither the eventual yield nor the inherent quality of the producers had a chance to establish themselves (unless a community is given the land and an amount of time during which it has to demonstrate each other's usefulness, after which they either are allowed to stay or told to leave). In the end it all seems to depend on the surplus, so how does he define it?

> Suppose two regions equal in extent, one having a yield of five units and the other of ten. If the inhabitants of the first consume four units, and those of the second nine, the surplus in the first case will be a fifth of the yield, and in the second a tenth. The surpluses being in inverse ratio to the yields, the region producing five units will give a surplus double that of the region producing ten. [27]

First, the mathematics:

case A:		case B:		
yield	5	yield	10	
usage	- 4	usage	- 9	
surplus	1	surplus	1	
equals	20%	equals	10%	of the respective totals.

In each case there is a surplus of 1 unit, but expressed as a percentage of the *respective* total, case A is left with a result which numerically is twice as much as the one of case B. Note that the percentage value is of the particular size with respect to *its* total, and not the total of the other. Therefore, as a comparative item between A and B the figure is useless because it expresses something different in each example. By the same token, to claim that the surpluses are in inverse ratio to the yields, is implying these yields together with their surplus-percentages can be compared with each other, and this is not the case. Because "20" and "10" are percentages of their respective totals, it is incorrect to state that one surplus is double that of the other—in fact they are both exactly the same, namely the size of 1 unit. It is their percentage terms which differ from each other, but the percentages do not cross-relate but are specific to their own situations.

Now for the cognitive dynamics. Reading the above quote induces a TP in the mind which puts the TS of "20" alongside that of "10" and indeed makes the first double the size of the second. Add to that the—false—conclusion that region A is subject to an inverse ratio of surpluses due to its lesser output and the impression is one of a certain amount of unevenness with regard to lower yield-producing regions. How the unevenness becomes interpreted further down the line depends on the context in which the demonstration is being used, but whatever the slant it is based on falsehoods. The meaning of "20" as a percentage of the total yield is different from the meaning of "20" as a comparable representative of a surplus. In other words, there is a mismatch between the Q_{po} and the Q_{ao} of the same object, ie an incongruent functionality.

In the previous example we have used mathematics to point out the fallacy in an argument. This was necessary because the paragraph alone did not contain sufficient hints of a different nature. Of course, having examined the general perspective under which Rousseau approaches society and its questions, we have already been alerted to the incongruity of the container-TP, and so his words should have been taken with a pinch of salt anyway; testing the mathematical integrity only confirmed our suspicion. Generally speaking however, the use of absolutist language should warn the observer of the distinct possibility that an i-PAT-l is in progress. Not because the words themselves can be wrong (although in most cases they would be), but because an absolutist expression is the result of a TS which does not possess links to other TSs which could ameliorate the expressed significance of the former. For instance, to declare "all Australians are larrikins" is an absolutist statement because (a) not every single inhabitant of that continent can possibly be of the same character and (b), to think in this fashion induces associations which bring in the idea of larrikinism every time the word 'Australian' comes up. Clearly, neither are realistic nor desirable.

Here is a final example of absolutist expressions and, keeping in mind what we have seen in terms of Rousseau's cognitive structures, it should not be too difficult to discern their origins (each expression is marked by italics):

> *So long as* a number of men gathered together consider themselves as a single body, *they have* a single will also, which is directed to their common

> conservation and to the general welfare. *All* the mechanisms of the state are strong and simple, and its maxims clear and luminous; *there is no* tangle of contradictory interests; the common good *is obvious everywhere*, and *all that is* required to perceive it is good sense. *Peace, unity, and equality are the enemies of political subtlety*. Simple, upright men *are difficult* to deceive because of their simplicity: elaborate pretexts and allurements *fail to impress* them; *they are not* sophisticated enough to be dupes. When, among *the world's most fortunate* nation, groups of peasants are to be seen under an oak tree, deciding on matters of state and governing with *unfailing* wisdom, how can we *not despise* the refinements found among other peoples, who gain themselves *glory and unhappiness* with such *ingenuity* and such an *air of mystery*? [28]

Although checking for absolutist wording is not a perfect (*perfect*!) means to spot incongruities, it's a good start.

Throughout the text there are many sections dealing with the law—the role it plays in society, the influence it should have at various levels, and how the citizens ought to use it. Knowing about the overall thought pattern which is responsible for Rousseau's attitudes in that area, it is instructive to investigate how he categorises the basic concept. Recall Locke's similar endeavour (see chapter 7) and the three main groups he arrived at: the divine law, the civil version, and rules based on people's opinion; we related these groups to the heights of the abstraction levels employed by them.

For Rousseau the law constitutes a relationship between entities, and so the categories have been derived from the nature of their underlying relationships [29]. The first type represents the "relationship of the whole to itself, or of the sovereign authority to the state". He calls it the "political law". It can influence any other jurisdiction and the people may hold on to it if found acceptable or change it if found wanting. The second type is the "civil law" and refers to the relationship between "the members of the state themselves, or between them and the whole body". Ideally it is designed in such a way that "each citizen should be perfectly independent of all the others, and extremely dependent on the state". The third kind of law is in the relationship between "disobedience and punishment", which he calls the "criminal law". Above everything however stands one category which to him is the most important of them all. Here is how Rousseau defines it:

> ..it is not graven in marble or bronze, but in citizens' hearts; in it lies the true constitution of the state; its strength augments day by day; when other laws decay or become extinct it revives or replaces them, it maintains in the nation the spirit of its constitution, and imperceptibly changes the force of authority into the force of habit. I refer to moral standards, to custom, and above all to public opinion..

These definitions do not involve the actual contents, such as what constitutes an unlawful act and what kind of punishments should be available at any given time. They are significant due to their cognitive format, their placement within the mental scheme of things. When Locke distinguishes between divine, civil, and customary law, he establishes TSs which first and foremost concern themselves with the actual laws themselves; it is their specific flavour that puts them into a particular category. As such their enunciation relies upon the perception of their content. Rousseau on the other hand bases his classification on the relationships the law seeks to address, therefore a proper definition of a given law requires not only ascertaining its content but also incorporates an interpretation of the participating parties. What is the effective difference?

Consider a dispute about the applicability of a certain rule. Under Locke the argument focuses first on the broad jurisdiction (eg does it come under the auspices of the Church, the parliament, or is it some native custom—all relatively easily established), and so the air is cleared to concentrate on the question of guilt. Under Rousseau, not only is there the need to argue about the domain, but then the problem becomes one of identifying the partners in the relationship. The outcome of that process will also influence further proceedings and we can end up with a situation in which it is not only important as to *what* type of law applies, but to *whom* it will apply. Regardless of the actual paragraph itself, the initial conceptual framework of such an entity is therefore instrumental for the outcome once the action starts. Generally speaking, a set of TSs will be more productive in terms of its implementation if their affinities establish bridges to other TSs which tend to lead back towards the original set; it will be less productive if any associated TSs lead away to other patterns whose content is not denominative with respect to the initial set. This particularly applies to TSs that form containers, because it is in a container's nature to predetermine the processing of information addressed via its associations. Since affinities are not only restricted to object abstracts but also to the functional kind, the direction such affinities take will have a great influence on the type of behaviour which results from the acting out of the container-TSs.

The writings of Rousseau show the role of containers from a content-perspective different from what we have been used to so far. For the first thousand years the most powerful cognitive patterns in European minds had been put together by the concepts provided through Christian teachings (the absence of similarly informative texts before the advent of Christianity makes it difficult to judge earlier times). This does not mean it takes an entire religious doctrine to evoke a functionality such as a container—any preliminary assumption that serves to direct the flow of thoughts in a particular manner will exhibit similar dynamics. At the same time it can be expected that the more fundamental perspectives of human thinking will be based on equally comprehensive structures, and therefore the influence the whole assembly can exert through its affinities and denominative derivatives will be considerable.

Does the existence of a thought pattern prove the presence of an intent to create it? It is an interesting question, and in many ways it goes to the heart of how we understand the mind to be. Do we cause a thought to happen so that we may experience a further idea, and thereby exercise some mysterious faculty of foreknowledge, enabling us to place one building block upon another just so until the result happens to be what we had wanted all along? The following investigation will go a certain distance towards an answer, precisely because we are going to deal with a purposefully created collective of abstracts on one hand and more autonomous, detailed abstracts on the other. By comparing the two sets their mutual interdependencies may tell us something about the cause and effect relationships at work.

During the 18th century the effects of scientific discoveries did not stop with the developments themselves. Questions were being asked about science's ultimate efficacy, its ability to shed light not just on matters of medicine, physics, astronomy and so on, but also to assist with all those age-old problems that so far had come under the auspices of metaphysics. In the rush to uncover nature's secrets it was believed that the obvious successes being enjoyed would also lead to a similarly analytical deciphering of her deeper mysteries. Yet just as Rousseau tended to identify general happiness in society with the assumed advantages of earlier customs related to contemporary needs, there came another philosopher who strove to restore metaphysics with the help of analytical thinking and thereby get as close as possible to the explanations of such things as the universe, God, and our ultimate existence. Immanuel Kant (1724 - 1804) set out, in his *Critique of Pure Reason* [30], to assemble a framework of interlocking ideas which would furnish the mind with the means as well as the exactly defined limitations of travelling the often attempted paths towards those answers. To arrive at that framework he needed to create a structure which served his purposes, and because its elements had to interact with each other his personal preferences he could bestow upon them were tempered by the need for their sheer utility: the classic scenario of a container.

For us, as travellers through this or that philosophy, it has become a common experience to be guided by their authors at first to pass by what the writers perceived to be the starting points. It can also be said that it is a sign of the philosophy's overall quality of how well those initial assumptions agree with reality, and therefore how adept they are to supply the productive questions so necessary for what is to come. In the case of the *Critique* the introductory sentences advise us about the intractable nature of questions which are posited by the mind but prove unanswerable when pursuing the strands of their context. The way the picture is painted is significant:

> Our reason has this peculiar fate that, with reference to one class of its knowledge, it is always troubled with questions which cannot be ignored, because they spring from the very nature of reason, and which cannot be answered because they transcend the powers of human reason. [31]

But, according to Kant, reason cannot be blamed for it, since these questions arise from principles which the mind has no choice but to follow because they stem from experience. It is from those beginnings that complications arise and, when no solutions present themselves, "refuge" is taken in "principles which exceed every possible experimental application"—at the same time such principles "seem so unobjectionable that even ordinary common sense agrees with them." He goes on to say that reason eventually finds itself amidst darkness, and while there is a recognition that errors are "lurking somewhere", there nevertheless is no facility to discover what they are and thus find a way out of the situation. This, in a word, is the "battlefield" of metaphysics.

It is an astute observation, although couched in the 'soft' language so characteristic of the initial stages of perception. Using more precise terminology we can translate the above and say that there exist thought structures which are built by using first-level object abstracts and their associated functional counterparts (ie the "principles" coming from "experience"). Further abstractions which, through their affinities with other TSs, create ideational options (ie the further "questions") which lead to the associations with affinitive FAs such that imaginary concepts are evoked. These affinitive FAs originally come from like experiences (otherwise they couldn't be affinitive) and therefore they do not seem out of place. However, since at that point they are no longer grounded the incongruities creep in and we end up with i-PAT-ls. Unless one is satisfied with such a state of affairs (and many people would be) attempts will be made to somehow resolve the "lurking errors", but the point at which the logic boundary has been overstepped is unrecognisable and so the attempts are futile. All in all, a familiar development we have encountered many times before.

Kant makes the point that the above refers to one particular class of knowledge, not directly identified in those initial pages. We, at this stage of our journey, can of course make a pretty good guess as to what sort of problems are being dealt with in that fashion, but Kant's perspective on how to get to any answers is interesting in itself. Somewhat later in his book we read,

> Metaphysics has for the real object of its investigations three ideas only, *God*, *Freedom*, and *Immortality*; the second concept connected with the first leading by necessity to the third as conclusion. Everything else treated by that science is a means only in order to establish those ideas and their reality. [32]

It is too early to go into the sequence implied by the first sentence, but the reader will readily identify the second one as an obvious description of a container. As observers we need to be aware of the metaphysician's propensity to turn concepts which are hitherto outside their domain into suitable members of their overall concept.

We know that a major source of factual errors in the emergence of a thought structure is the tendency to place its abstractions beyond the reach of the links that could provide the connections to PAT^1s, that is the incongruent functionalities resulting from a mismatch between the qualities of perceived vs abstract objects and culminating in i-PAT-ls. The fundamental reason for those chains is the sheer ability of the mind to do such a thing—higher abstracts are just so attractive for those of us who can make them. Historically it had been philosophy which provided the arena, since there had been no cognitive safeguards to curb the flights of fancy. Only science, through its insistence of analysis and confirmation, furnishes the much-needed brakes (keeping in mind of course that even scientists have essentially the same brains as anyone else and the seductiveness of self-serving tangents are by no means unknown to them). Kant's attitude in such matters is demonstrated later in the book by his comments when dealing with pure reason:

> In their application [ie the assertions of reason], and in the progressive extension of the employment of reason, beginning from the field of experience, and gradually soaring up to those sublime ideas, philosophy displays a grandeur which, if it could only establish its pretensions, would leave all other kinds of human knowledge far behind, promising to us a safe foundation for our highest expectations and hopes for the attainment of the highest aims, towards which all the exertions of reason must finally converge. [33]

The deployment of a structure in the form of a container, acknowledging simple principles derived from experience as possible candidates for basic support, but allowing its abstractions to reach for the sky to touch on themes any other conceptualisation would shrink away from, while bolstering such a construct with ingredients for the express purpose of giving it authority—this is what we have to expect from the *Critique*.

Every structure needs some basic items for it to be built. It is not a trivial question to ask how such items come about. Consider a house made out of brick and concrete and featuring doors, windows etc. A brick is a purposefully made object, as is concrete and as are window and door frames. The main constituent of a brick is clay, such as kaolin, which in turn is the product of the decomposition of granite; window frames usually consist of timber. How far does the overall concept of 'house' as an example of intent stretch to its ingredients? Granite does not decompose so that humans have something to make bricks of, and even life forms such as trees do not grow in order to provide building materials for another life form. Rather, it is the organising aptitude of some entity which happened to choose clay and timber as a starting point for its endeavours, and it is at that point only that one is justified in ascribing a context-related intent to the effect of such action (the effect must be context-related otherwise the weathering effects on rock for example get placed into a relationship with the construction of a house, which is not desirable). Therefore, to seriously show gratitude to the sun and the rain because they caused clay to appear would be misguided, and the same goes for trees. To answer our question then becomes a matter of identifying the stage at which the organising entity takes control away from the previous system and proceeds to impose its own standards on the process.

Returning to thought structures, our basic elements are the abstracts (default, object, and functional). If the purpose of a particular TS is a container that satisfies Kant, at what point do those abstracts assume the controlling role under the auspices of that very intent? When one considers the problems metaphysics deals with the question virtually suggests itself, as it certainly does for Kant, albeit indirectly. Talking about our need for explanations he writes:

> What then can be the cause that hitherto no secure method of science has been discovered? Shall we say that it is impossible? Then why should nature have visited our reason with restless aspiration to look for it, as if it were its most important concern? [34]

In our drive to find the reasons for our existence and other esoteric matters we make use of the mental configuration 'given' to us by nature, yet it is that very system that in the end lets us down. It would be illogical to assume an explanatory intent on behalf of nature if its representative inside our heads is demonstrably unsuited to carry it out. On the other hand, if the mind is inherently incapable of a solution, how then can it prompt us to search for one? So much is clear: for us to be able to ask the question—any question for that matter—there must be object abstracts to provide its content as well as functional abstracts to give us the optional reasons for asking it. For instance, in order for me to ask, "Why do we need a house?" I must have a concept of 'house' plus a set of ideas comprising the things houses can be used for in general; it is from those ideas that I derive the need to ask because my personal knowledge does not include a match between such possibilities and whatever I make of the present situation.

That there are pre-existing ideas in our minds is indirectly acknowledged by referring to our reason's "restless aspiration", without whose particular character the aspiration could not be formed. The general possibilities are to be found in the same TSs, without whose functionalities the conceptualisation of 'God', 'us' and the 'universe' are not possible. As Kant writes a few pages later,

> Yet, in a certain sense, this *kind of knowledge* also must be looked upon as given, and though not as a science, yet as a natural disposition (*metaphysica naturalis*) metaphysic is real. For human reason, without being moved merely by the conceit of omniscience, advances irresistibly, and urged on by its own need, to questions such as cannot be answered by any empirical employment of reason, or by principles thence derived, so that we may really say, that all men, as soon as their reason became ripe for speculation, have at all times possessed some kind of metaphysic, and will always continue to possess it. [35]

Note the words, "..as soon as their reason became ripe for speculation..". For Kant to identify a certain point in time he must have the information to do so, which in this case is the perceived act of speculating. Since he can only perceive what has been articulated, the information imported into his mind represents the spelled out con-TSs of his fellow men. Further, since con-TSs are only one part of an entire structure and are created largely during the actual event of articulation, all the other elements necessary for that event must have been in place before such an act could take place. As the combined emergence of all the elements at one single moment is highly unlikely, there must have been some which had been part of a

TSs for a period prior to the speculation longer than others. Indeed, one can say that there would have been TSs in a person's mind at a time when that person had no interest at all in any higher abstracts, let alone ones of that nature. Many would have come about during childhood, and therefore their composition would not only *not* reflect the context of much later musings but would be of a disproportionate simplicity. Therefore, not only does the mind possess abstracts whose design does not relate forwards to their later uses, the great majority of them are part of sub-TSs and so would not even be consciously recognised as existing. It is these abstracts, residing amongst the sub-TSs of the mind, which Kant labels as "a priori" (for the rest of this chapter we will use the term 'a priori abstract' for an abstract which has moved from a sub-TS to a con-TS; this is done more out of deference to Kant's system rather than a technical necessity in terms of our own—hence the restriction to the current section only). The notion of a priori concepts forms an essential part of his container. A priori concepts, that is the abstracts taken from the sub-TSs, are being articulated and are now part of Kant's con-TSs. The realisation of that phenomenon (the emergence of some idea from unknown sources), is a concept in itself and as such represents a further abstraction. Note that neither the concept, nor the concept of the concept, has any links to PAT[1]s, although these may come later. At the point of their evocation they stand alone.

Nevertheless, their position in the subconscious realm is due to mental activity triggered by the mind's surroundings, and so they do not exist ad hoc. They feature in structures, have their associations, create denominative TSs, and participate in PAT chains—all part of sub-TSs and all away from the conscious evaluation of the mind which is now in the process of using this or that element for its own declared purposes. Generally speaking, two possibilities suggest themselves: (1) should an object abstract be drawn into a con-TS its associated FA is imported also, in which case the outcome may be congruent; or (2), any abstract used for a con-TS stands purely on its own, in which case the chances for an incongruence are very much higher.

On that note it is interesting what Kant has to say about dogmatism:

> Our critique is not opposed to the *dogmatical procedure* of reason, as a science of pure knowledge (for this must always be dogmatical, that is, derive its proof from sure principles *a priori*), but to *dogmatism* only, that is, to the presumption that it is possible to make any progress with pure (philosophical) knowledge, consisting of concepts, and guided by principles, such as reason has long been in the habit of employing, without first enquiring in what way, and by what right, it has come possessed of them. Dogmatism is therefore the dogmatical procedure of pure reason, *without a previous criticism of its own powers*;.. [36]

In other words, while he accepts the usage of abstracts themselves about which he has no further knowledge, he also recognises the need for testing their integrity. In the process of such a test he is aware of "deceptive representations" [37], and counters them with the rule that "whatever is connected according to empirical laws with a perception, is real". Recall the provision in Jainism to allow for a perception to be relatable to the outside world, otherwise the pure idea of the Real would be meaningless (see chapter 3). Kant's rule seems reasonable enough, yet we need to remind ourselves that he does not distinguish between object and functional abstracts, an omission which is bound to interfere with the relationship between a priori ideas and resultant conclusions, as we are about to see.

"Experience is no doubt the first product of our understanding," Kant writes [38], and in the light of his quote in the previous paragraph we can assume that the resultant perceptions are taken seriously (these perceptions he calls "sensations"—no doubt alluding to their import via the senses but as an expression also interesting in terms of what we have said about consciousness being a sensation). "Experience tells us what is," he goes on to say, "but not that it must be necessarily as it is, and not otherwise." According to him, our sensations only provide us with the imagery of the object itself, but not its generalised form, and yet it is such "general truths" for which our mind is "particularly anxious"—the PAT^{2+}s that emerge from the field of first-level abstracts and are carrying information which seems to come from nowhere and so prompt Kant to posit the existence of a priori knowledge. His a priori abstracts therefore are the product of a typical PAT loop: a PAT^{1} has been created through an outside event (note again—outside the mind) and resides within the subconscious part of a TS; it becomes articulated during a thought process and is now part of a con-TS; at that point it has been made conscious and abstracted further, creating a PAT^{2} or higher; it can also be ascertained in terms of the underlying PAT^{1} and as such made affinitive with other abstracts amongst the con-TS and thus forms associations of a possibly different context compared to its original placement. From Kant's perspective the nature of the a priori abstract is sufficiently mysterious for him to declare on the same page,

> Now it appears, and this is extremely curious, that even with our experiences different kinds of knowledge are mixed up, which must have their origin *a priori*, and which perhaps serve only to produce a certain connection between our sensuous representations.

Suppose we remove from our inner view all those items which we have learned through sensory experience, he continues, then we are still left with their principles and "the judgments derived from them". This prompts him to assume their origin as being a priori and "independent of all experiences". We can see how, by being unaware of such things as object and functional abstracts, the rule of the latter can be attached to abstracted derivatives of the former, all the while having no inkling about such a source's original nature. If such ignorance is an obstacle on the road to understanding the mind, it is nevertheless true that Kant, in his own search for solutions, senses that he has hit something although not being cognisant about what it could be. Consider the following passage, much later in the book:

> No one can definitely think a negation, unless he founds it on the opposite affirmation. A man born blind cannot frame the smallest conception of darkness, because he has none of light. The savage knows nothing of poverty, because he does not know ease, and the ignorant has no conception of his ignorance, because he has none of knowledge, etc. All negative concepts are therefore derivative, and it is the realities which contain the data and, so to speak, the material, or the transcendental content, by which a complete determination of all things becomes possible. [39]

For us the above is nothing new. A certain state is being perceived, and through the experience of an opposite (the *perceived* experience of course) there is now not only an object abstract of a contrary nature but also a functional abstract that contains the act of switching from one version to the other. Once such an intuitive FA exists its affinity with the actual experience is sufficient to enable further associations with other object abstracts, so that from now on the concept of 'opposite' can be applied in the abstract, that is without having to be confronted with the real thing on the outside—the perception of one version is enough. It is ironic that from Kant's perspective there are groundless a priori representations when in fact they do draw on real events, yet when it comes to negations, which in themselves do not require real counterparts but only an IFA, there is the assumption that they spring from some material, and as such invite the purposefully created nomenclature of "transcendental content". (See chapter 7 for a more detailed treatment of negations.)

Another hint that abstracts are not taken as pure, sensating representations but are a somewhat direct counterpart of the thing on the outside they purport to stand for, comes from Kant's view of mathematics. At one point he says,

> The science of mathematics presents the most brilliant example of how pure reason may successfully enlarge its domain without the aid of experience. [40]

Only someone who believes in a substantial relationship between an actual object and its perception can be led to regard a pure abstract that does not have a 'real' counterpart as being a priori. As far as the mental dynamics are concerned, the abstraction of a wheel producing a cog is in principle no different from looking at a bunch of numbers and coming up with set operations. That assumed connectedness also leads him to write,

> ..our reason can only use the conditions of possible experience as the conditions of the possibility of things; it cannot invent them independently, because such concepts, although not self-contradictory, would always be without an object. [41]

His statement contains a curious truth. While there are PAT chains that hold IFAs which provide the material for later theoretical contemplations, and therefore such contemplations could not be called independent, nevertheless there can be parts of these structures which are so far removed from any PAT^1s that they could well be termed object-less. We know these developments as i-PAT-ls, yet not recognising their possibility to the point of not creating i-PAT-ls in the first place seems like a case of being right for the wrong reason. The tendency to accept one's inner images as a confirmation of some real, outside existence is not new; it is particularly strong if the mind is incapable of and therefore unfamiliar with higher-level abstractions. The inner sensation of PAT^2s are taken to mean the same as the inner sensation of PAT^1s, since it is the further abstraction process which informs the mind about the nature of perception. That is to say, a PAT^2 cannot be understood as a 2nd level abstraction because there is no 3rd level through which to identify the former, just as in wider society a custom will not be identified as a custom unless the person has had the opportunity to step outside the familiar domain and realise that what has been believed to be a given is very much subject to variance. Take the following quote from a Brisbane metropolitan newspaper:

> "The ancient Egyptians believed that every time you say a man's name, he's still alive. Jim is always with me. Every day, somewhere in the world, a Doors song is played. The energy of Jim Morrison is still with us, in the ether"—Doors keyboard player Ray Manzarek, visiting the grave of Jim Morrison on the 30th anniversary of the singer's death on July 3. [42]

Other examples are the donning of masks or costumes by priests, and indeed the acting out of any situation by people in theatre, film and TV which is meant to stand in for the real thing. Although we can rally our reasoning faculty in order to understand the intended subterfuge, without such a conscious effort the general impact by the imagery is quite considerable. In terms of cognitive dynamics we can say that the representation of an object, particularly if its abstraction does not extend beyond a PAT^2, has the same perceived truth value as if the representation had been the result of a direct import via the senses; the truth value can be modified if there exist other structures which provide the affinities with the represented object. The truth value will be enhanced if there exists a denominative TS supporting the image, and it will be downgraded if there exists a DTS or pattern which satisfies the owner as to the image's unreal status. An example of enhancement is the portrayal of an illegal drug user in the media by pushing all the right buttons (ie preconceived notions) as the actor dresses and acts as antisocial as he/she is expected to by conservative individuals; the image will be downgraded in the minds of all those who have a more realistic experience with users and whose TSs therefore counterbalance the presented view.

The degree to which an image is perceived to be a counterpart to the real and the line that has to be overstepped for this to happen, is by no means a negligible issue. We will deal with its social implications in Part II, but right now it forces us to be careful when analysing Kant's structure. Not only does he accept the existence of a priori knowledge, but he is also aware of the influence previous experiences have on current considerations. A person who undermines the foundations of his house, he writes [43], may not have experienced an entire house falling down, but his appreciation of unsupported, heavy bodies is enough to make him act carefully. Such knowledge however, so he says, is not a priori. Only knowledge which is "*absolutely* independent of all experience", can be a priori and, if "mixed up with nothing empirical, can be called *pure*".

How can a state of 'absolute independence' be identified? Suppose I muse about creatures in distant galaxies and come up with an undefined shape which I call a "moki" (note to the reader: if you have never heard of a "moki" either you might find the following an interesting experience by putting yourself in my position). Having come up with a name I still find nothing better to do and proceed to give the creature a body, maybe legs and a pair of wings. That "moki" might now resemble a dragon and despite my protestations to the contrary I will be unable to convince anyone that I have never seen nor heard about dragons—they may be purely fictitious but there are a sufficient number of stories and pictures of them for my mind to draw on those, whether I am willing to admit this or not. According to Kant then I *can* say that I have never seen a "moki", although I sound less convincing when it comes to dragon-shaped beasts—in other words, dragons are not a priori, but "mokis" are. Let's say I change the "moki's" appearance until it resembles wafts of mist. Now the *form* of this alien is a priori, albeit not purely so because I have learned about fogs and clouds and therefore include something empirical. However, the *name* "moki" is still purely an a priori item, isn't it?

No it's not. Having turned prudent I check the dictionary and discover that 'moki' is used in New Zealand: it's a Maori word for several common food fishes, such as the blue moki (Latridopsis ciliaris). So, as far as some observer is concerned I could be, (a) creative, because I can think of extraordinary beings to which I attach a strange-sounding name; (b) a complete fantast, because I make up some impossible (?) organism with a name to boot; or (c) plain silly, for imagining that a fish which anyone can buy in New Zealand lives in a far distant galaxy (I might add that at the beginning of that thought experiment I had no idea that mokis exist—*honest!*). Note that none of these qualifiers gets applied without some notion as to the content of what I've said. In other words, the definition of 'absolute independence' or otherwise cannot be made unless there is *some* TS which possesses the appropriate affinities with the content of the idea in question. Even if there are no affinities one cannot claim such a fact unless I can formulate the thought "there is no X", where X stands for something which I am in the process of thinking (recall Kant's remark "A man born blind cannot.." a few moments ago). Therefore the question of a priori abstracts is not so much a matter of non-existent abstracts, but of the abstracts' visibility from the probing set of con-TSs.

How Kant expresses the notion of independence is significant. There is such a thing as "empirical universality", by which he means a generally applicable rule derived from experience. There is also a "strict universality", by which he understands a generally applicable rule that allows no possible exception and which, according to him, could not have been learned from experience; moreover, it is strict universality and necessity (ie the need for something to be such and such) that are criteria for a priori knowledge. Note the type of abstract domain with which the idea of 'strict universality' can be associated. Not only must there be a PAT^2 having been formed from some initial first-level abstraction, that PAT^2 must have been abstracted so that it can serve as underlying principle to further, similarly abstracted notions. Hence for any idea to be a candidate for an a priori knowledge it must exist in the form of PAT^{3+}s. For Kant any mathematical proposition falls into that category, as does the idea of change needing an effect for it to happen. He goes as far as to ask, "For whence should experience take its certainty, if all the rules which it follows were always again and again empirical, and therefore contingent and hardly fit to serve as first principles?" [44]; a good example of how higher-level abstractions are considered to be more significant than lower-level ones. It is important to remember that the creation of any abstraction is not due to the nature of that what is being thought of—it is an act of the mind because the mind is capable of doing so.

The main subject matter of these PAT^3s and higher are the "inevitable problems of pure reason", namely God, freedom, and immortality [45]. Being familiar with other philosophical systems we can understand the propensity of thinkers to apply their abstractive abilities to these realms, since the ideas in themselves are sufficiently removed from real experiences to allow i-PAT-ls to occur without causing their owners undue scruples.

Next we have the distinction between analytical and synthetical judgments [46]. The former are assessments in which the object and a predicate are part of the same thing (Kant uses the example of "all bodies are extended"). Synthetical judgments are those in which the predicate is outside the object's concept (Kant suggests, "all bodies are heavy"). Analytical judgments teach us nothing really new, he says, but the synthetical variety requires us to have some other idea through which we can express the predicate. The combinatorial concept of object and outside predicate he calls synthesis, something which occurs all the time with empirical judgments.

In terms of TSs we can say that the above distinction is one of perception only. The TP which contains the information to the statement, "all bodies are extended" is a collection of abstracts which cover 'all', 'bodies', 'are', and 'extended' (to put it more laboriously, but technically correct: there is a concept which I can articulate through calling it in the plural 'bodies' by which I mean the complete set of all I could think of which can possess a quality that I can articulate by calling it 'extended'). In order for me to say such things I need to have the abstracts as well as utilise their affinities. In the case of, "all bodies are heavy" the functional situation is exactly the same. It is impossible for anyone to have the notion 'bodies' and have the abstract 'extended' inseparably linked with it. If that were so, one would be unable to think of 'non-extended' bodies since the idea is unimaginable by definition—even Kant would agree with this given his remarks on negation. When we say, "all bodies are heavy", there are again TSs with their affinities forming a particular pattern. The only reason one could view the abstract 'heavy' as more remote than 'extended' is the longer range of the PAT chain responsible for it. This has nothing to do with the quality of content, but much with the manner in which the information has been imparted (ie 'en bloc' vs 'bit-by-bit').

Using our terminology the difference then between 'analytical' and 'synthetical' is one of conceptual distance. As an experience the connection between 'bodies' and 'extended' represent a contextual association, since both are imported through the senses as part of a particular event. In terms of Kant's interpretation however the two ideas form a functional association, because he implies a general universality which in itself forms an abstraction, and thus 'bodies' and 'extended' need to fulfill their explanatory purposes on an abstraction level greater than 1; that is to say, they relate to each other in terms of their abstraction rather than their content. If 'bodies' and 'heavy' on the other hand are deemed to be two separate events in one greater experience and matters are left at that, their abstract association is a contextual one. We can see how a contextual association is more likely to invite a critical evaluation because their common abstraction level makes productive links in both cases more possible. A functional association is less likely to attract critical thinking because one or both items have been moved from another abstraction level and thus any links may have been broken. Therefore, when Kant speaks of "bodies are extended" and regards the statement as an analytical judgment, he tends to accept its truth value, whereas "bodies are heavy", being synthetical judgments, are seen as imparting new information and would therefore be tested for their correctness. Such a differentiation takes place all the time when people express themselves with either more or with less confidence, but to classify an object-predicate relationship as either one or the other from the very beginning organises one's TSs in a particular manner, hence promoting containers. Therefore, what Kant calls analytical judgments are the functional sources of incongruent containers, and a synthesis promotes congruent ones.

Incongruent containers make it difficult for their owners to access the underlying relationship between the participating TSs. This becomes apparent when Kant talks about empirical judgments in general:

> Empirical judgments, as such, are all synthetical; for it would be absurd to found an analytical judgment on experience, because, in order to form such a judgment, I need not at all to step out of my concept, or appeal to the testimony of experience. That a body is extended, is a proposition perfectly certain *a priori*, and not an empirical judgment.

And especially the next sentence,

> For, before I call in experience, I am already in possession of all the conditions of my judgment in the concept of body itself. [47]

The statement clearly shows his predisposition according to what he perceives to be the correct category of relationship. Now such relationships do not exist in one's mind as active inhabitants, telling us what to do; rather, they constitute a particular configuration of abstracts which, in the process of their recursive mining by some con-TS, becomes accepted either as a unit not to be taken apart or as something more tenuous. As the last sentence above indicates, the decision to accept or not accept is taken before the relationship as such is visible to the con-TS, and hence forms part of the sub-TS dynamics. "Analytical" judgments are a form of presumption, and the reason such preconceived notions are so difficult to counter lies in their generation within the subconscious part of a pattern: presumptuous persons are usually the last to recognise their fault.

Let's have a brief summary before rejoining Kant. Incoming information is placed within thought structures where it forms affinities with abstracts on the same abstraction level, but it also gives rise to PAT chains which continue through to 2nd and 3rd level abstraction levels and possibly higher. These complexes establish patterns of many different chains, some longer, some shorter than others, and their lengths is one factor that determines the types of associations the elements within these chains are able to make. They can be functional, that is their affinity is based on the functional potential of the abstraction level at which the association is about to occur; they can be contextual, that is their association is based on the content of the participating items, where 'content' refers to the quality of the represented object (and where 'object' can stand for a tangible item on the outside or another abstract generated by the mind itself). Longer chains are more likely to include varying abstraction levels and therefore are more likely to feature functional associations than would shorter chains. During the process of articulation the mind, through con-TSs, passes through the associated links of the entire TSs, with some links visible, others not visible to the con-TS active at the time. If the links are not visible, the represented set of abstracts is treated as an undissolvable unit, whereas in the case of their visibility the mined abstracts are seen as discrete elements, thereby allowing further associations in terms of each element's characteristics. If the emerging, associative structure is related to a container, the nature of the container, that is to say its qualities made significant through further linkages and/or emotive modules, predisposes the establishment of associations with the emerging structure such that the overall quality of the container will be maintained or even enhanced. It is the governing quality of the container which determines the integration of incoming abstracts with the existing framework, and there is no inherent restriction as to what the bias may entail. In Kant's case there are TSs which had not been subject to further abstractions, forming contextual associations, and as such are deemed to be pre-existent in terms of their composition (the a priori abstracts). There are also other TSs which had been subject to further abstraction processes, forming functional associations, and are now used to form separate links under the auspices of the overall container. Since higher abstractions are generally more attractive than lower level ones, they receive higher priority when it comes to the creation of further structures. Note however that a priori abstracts can also be moved to higher levels as single units and then become subject to associations under the auspices of the container. Therefore at the higher levels, the difference between the results of functional and contextual associations is that the former are more likely to be grounded through still existing links than the latter, and so are less manageable for the purposes of the container. It is the contextual associations which we can expect to play a significant role in the development of Kant's TP.

Back to the *Critique*. Kant is now faced with a dilemma. On one hand he identifies something he calls a priori knowledge, which consists of insights gained seemingly without the aid of experience. Yet on the other he sees empirical knowledge which enables synthetical judgments to be made, that is statements which reliably add to the knowledge one has about some item. Thus, in order to learn more about something synthesis is required, yet it is the a priori variety which for him opens the doors to understanding the deeper mysteries of existence, given that there is nothing in human experience that could shed any light on such questions.

His con-TS stream produces two sets of abstracts: objects which do not possess any materiality but are defined by their functional characteristics already known by his mind ("a priori knowledge"); and objects which are representations of the outside world and have been paired with characteristics, thereby teaching him about the veracity of the conclusions reached ("synthesis"). At this stage it becomes a matter of recombining these elements to achieve the most productive effect for his purposes, which is the investigation into immaterial objects under a reliable authority. The present combination is (A) immaterial object ↔ pre-known function, and (B) material object ↔ evaluated function. (A) is also associated with the concept of necessity—relatively easy to do if we remember that an abstraction, being basically a generalised principle, virtually invites the principle's applicability in a formal, authoritative manner. What Kant wants is the combination immaterial object ↔ evaluated function, in other words an a priori synthesis. Having seen the dynamics of functional and contextual associations we know that to achieve this aim he needs a container where contextual associations are mapped across PAT chains in the way functional associations would be. What's more, the lowest abstraction levels in these operations would be PAT^2s. As he writes,

> I call all knowledge *transcendental* which is occupied not so much with objects, *as with our manner of knowing objects, so far as this is meant to be possible* a priori. A system of such concepts might be called Transcendental Philosophy. [48]

So "transcendental philosophy" deals with the generalised principles of procedures, ie the abstractions of functional abstracts, or FA^{2+}s. They would be IFA^{2+}s in the case of elevated contextual associations, yet would more likely be CFA^{2+}s when it comes to functional associations.

Keeping in mind that in all these explanations we are using linguistic symbols which stand for functionalities of the mind *only*, the arrangements and configurations of the symbols themselves have an impact on what they eventually convey (this goes for Kant's system as well as ours). It is therefore significant how the elements are named and put together. To begin with, Kant sees the mind as an active entity which reaches out to understand its surroundings:

> Whatever the process and the means may be by which knowledge reaches its objects, there is one that reaches them directly, and forms the ultimate material of all thought, viz. intuition. [49]

The above suggests two possible locations for the objects at the moment of access, outside the mind or inside. He means 'inside' because the very next sentence reads,

> This is possible only when the object is given, and the object can be given only (to human beings at least) through a certain affection of the mind.

If the objects were on the outside, the state of the mind would be irrelevant at the time of access. Since it is not (a "certain affection of the mind" being necessary), they already must exist in the form of some representation within the mind. Of course, the ultimate place of them is somewhere out there in reality anyway but there is a distinction, although a rather subtle one. By recognising the fact that one deals with a representation of a thing and not with the thing itself, yet at the same time accepts its perceived nature as sufficiently representative of the real thing to be used in considerations of the manner in which such representations are being handled, indicates a pre-emptive attitude regarding the perceived nature. As we are dealing with containers, this should not surprise us.

If we take "knowledge" to mean the general experience of sensating a train of thoughts, then the naming of "intuition" tells us that Kant must be aware of the inherent subjectivity of "knowledge". In philosophy 'intuition' means pure, untaught, non-inferential knowledge, and so we are told that first and foremost our thoughts consist of representations of objects which are of a certain quality, where the quality depends on the entity which does the representing, namely the mind. (Note: the word 'affection' has been translated from the German 'Gemüt', and a more modern English version might be 'nature' or 'mentality'; similarly, the present 'intuition' comes from the German 'Anschauung', which today would be translated as 'viewpoint'. Our *Critique* has been translated by Max Müller in 1881, in a linguistic sense placing it more closely to its source which appeared in 1781, compared to more modern versions).

The capacity of the mind to hold representations in this way Kant calls "sensibility". These representations turn to "thoughts" through the application of "understanding", giving their owner "conceptions" of things. The effect a certain representation has on the mind he terms "sensation". "Empirical" therefore are all those ideas which exist in our mind due to the processes of sensation. In turn, the "undefined object of such an empirical intuition" he calls "phenomenon". He goes on to say,

> In a phenomenon I call that which corresponds to the sensation its *matter*; but that which causes the manifold matter of the phenomenon to be perceived as arranged in a certain order, I call its *form*.

Clearly he already distinguishes between the *content* of a thought and *how* such content is represented. Because he views the content as being dependent upon the actual object that is being represented, its manner of representation must be due to how the mind works. This leads him to declare all "matter" to be a posteriori, ie a product of "sensation", whereas the "form" is a priori, ie a given inherent in the mind before any sensation has even taken place. "Pure intuition" then is information which exists without the participation of "sensation", and it precedes any knowledge derived (or completed) through the act of representing an object imported from the outside. Equally, one can deduct from the entire representation all that which is a product of the understanding as well as of sensation, and one is left with what he labels "pure intuition"—and it exists without any objects. Below is a diagram illustrating the basic process of content creation through the application of "intuition" and "understanding", resulting in "thoughts".

We need to remind ourselves that the picture does not include the subconscious, and whatever is being ascertained by Kant is an articulation of the conscious processes. Hence we can see why he insists on a priori faculties: suppose one does trace the development backwards as he suggests in order to pare away anything empirical, then sooner or later there would be nothing left of the mind itself unless it contained the "form" all along, that is its inherent ability to process what may come its way.

Note the question mark before the addition of Intuition. If Intuition begets Sensation, and Understanding brought to bear upon it causes Thoughts which become a further complement of the Mind, then what makes the Mind ready for further input? For Kant the solution lies in positing "form", and while—as usual—we are not concerned with the ultimate truth value of his philosophy, the question is relevant because the absence of an explanatory item leads to the creation of substitutes that serve the assumed purpose of what is missing. The situation is typical of i-PAT-ls, and in Kant's initial definition of his system the conditions for these kinds of chains have been laid down already (we might even say, a priori!).

In his search for something which does not arrive via experience but is, indeed must be, ascertained in any case, and which therefore fulfills his condition of necessity, he hits upon Space and Time. These two concepts, "pure forms of sensuous intuition" as he calls them, are of course abstracted versions of experience, but because of their considerable degree of universality in daily life he is lead to declare them as a priori concepts. If we consider that the events depicted in Fig. 8-1 only relate to *conscious* mental processes, it is understandable that a virtually continuous idea such as space or time can be assumed to be part of the system (*something* has to stand in for the question mark).

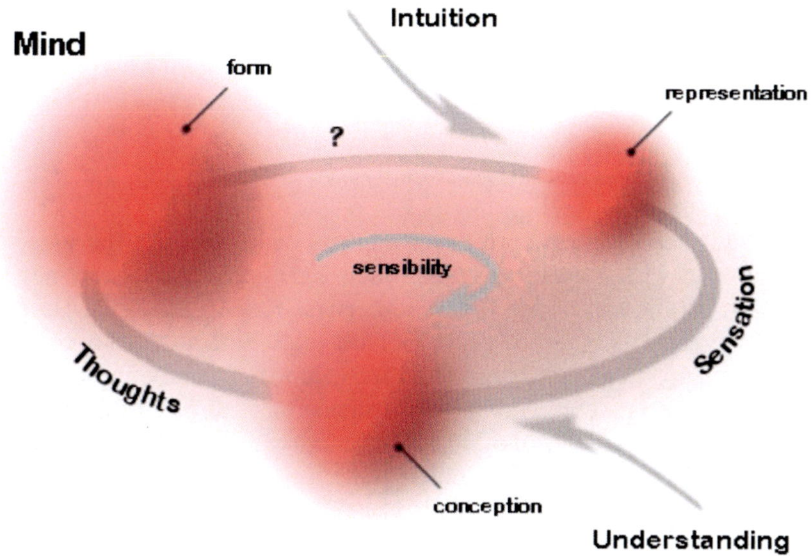

Fig. 8-1: Representation of our basic thinking processes according to Kant. For details see text.

Based on the information available it is a reasonable decision, especially so when we read somewhat later about his recognition that the experience of anything contains the functional elements of space in its framework—a fundamental ingredient to our knowledge in general:

> Although we know therefore a great deal *a priori* in synthetical judgments with reference to space in general, or to the figures which productive imagination traces in it, without requiring for it any experience, this our knowledge would nevertheless be nothing but a plaything with the cobwebs of our brain, if space were not to be considered as the condition of phenomena which supply the material for external experience. [50]

Not only that, but there cannot be thoughts which do not abide by the functional prerequisites delineated from such a priori concepts:

> The highest principle of all synthetical judgments is therefore this, that every object is subject to the necessary conditions of a synthetical unity of the manifold of intuition in a possible experience.

To put it into our own words, there are ultimate functional abstracts related to space and time, which must play a part in every thought structure created through lower-level abstracts for such TSs to have any validity; and further, that once these FAs do form part of a PAT chain its contents enjoy the same potential for realisation as would their counterparts in the real world. As he puts it,

> Thus synthetical judgments *a priori* are possible, if we refer the formal conditions of intuition *a priori*, the synthesis of imagination, and the necessary unity of it in a transcendental apperception, to a possible knowledge in general, given in experience, and if we say that the conditions of the possibility of experience in general are at the same time conditions of the possibility of the objects of experience themselves, and thus possess objective validity in a synthetical judgment *a priori*.

In Part II we will see how close Kant actually got to describing the process of TS development, his main obstacle being the unawareness of subconscious structures and patterns. Right now however let us observe the circularity of his picture: there exists a set of TPs which come equipped with a specific set of functionalities such that any newly formed TSs must contain them in addition to any others, otherwise those TSs do not have any claim to objective validity. The verification of such a claim rests with identifying the particular abstract and its structure by using said functionalities, and ultimately these functionalities only, because it is they and no others which possess an adequate degree of necessity.

Kant is explicit in his statements denying the functionalities any experience as their source, and that would be enough for us to identify them as IFAs (CFAs being rules that have been consciously learned). However, since we also know of Kant's ignorance (or neglect?) of the subconscious, their thus stipulated existence immediately categorises them as intuitive. Being IFAs therefore places them first and foremost within the sub-TSs of any thought processes, and their presence becomes acknowledged only when accessed by the recursive events of a con-TS. Since at that point any possible links to PAT[1]s have long since been severed (Kant's con-TSs have seen to that), it is up to the holder of such TSs to ascribe their significance in terms of the purposes of the con-TS. Kant gives us an example via the field of geometry, where

> all geometrical propositions are apodictic, that is, connected with the consciousness of their necessity, as for instance the proposition, that space has only three dimensions; and such propositions cannot be empirical judgments, nor conclusions from them. [51]

..that is to say, there are no TSs which offer an alternative to the readily discovered FAs defining geometrical relationships, and as such they exist of necessity due to the absence of differing rules imported via experience as well as the requirement that they feature in the current TS. In the absence of any alternatives our space-IFAs are the ones which possess affinities with the elements of other structures, the "subjective condition of our sensibility", and they contribute in the processing of any incoming information without having been introduced by such information itself. Hence they "must necessarily precede all intuition of objects", and the form thus given, the functional/organisational matter of the mind, "may be, in fact, *a priori* in the soul" [52].

For us these sentiments contain echoes of ancient musings. Recall Socrates' view that the soul must already know about ideals because it is impossible to experience an ideal version of something in one's life; to him this was proof that we have lived before our—current—birth (see chapter 2). But compare also with the Jains' opinion that the qualities of an object are derived from a common pool, to which our mind has a more or less tenuous access (see chapter 3). In both of these cases the information is 'out there', and it is our mind which finds and collects it. No thought—quite literally—is given to the *act* of collecting. To use the analogy of fossicking, the ancients saw the process of thinking as wandering around in the wild and picking up interesting stones here and there (in the case of Socrates this might have been done under cover of darkness), and whatever one is going to find is a matter of being in the right place at the right time. For Kant it is accepted that the stones are there waiting to be found, but it is our equipment that determines the success of each foray, although there are some essential pieces which we have from the very beginning. Still, the focus is now on how well prepared the collector is, and it is principally the quality of one's efforts that matter. Ancient fatalism vs modern self-empowerment—the results are there to see.

Fig. 8-1 represents the mental processes as a whirl, suggesting a continuous adding-on of content, whether that be in the form of objects or their associated functions. The fundamental problem is not so much the complimentary nature of its dynamics, but the validity of the entire event when it comes to self-consciousness. It is one thing to observe—however objectively or subjectively—the understanding of an outside object; it is quite another to describe the conceptualisation of oneself using the same ingredients. If my idea of something is composed of "intuition", which is now added to my previous state of knowledge, what then is being added about myself that I didn't know already, unless by 'ego' one means a summary name for all that I know and so the more I know the more conscious I become? Kant solves the problem by taking the consciousness of self as the "simple representation of the ego", requiring "internal perception of the manifold", ie the variance of thought material, and thus calls this faculty our "sensibility" [53]. "Sensibility" then is the act of sensating the actions of one's mind, nothing more, and so constitutes consciousness. Note how Kant employs a circuit-breaker to interrupt the form → representation → conception sequence when it comes to consciousness. Since self-awareness has been moved from the path and placed in the centre to act as the fulcrum around which the gathering of knowledge etc revolves, the characteristics of the latter will not impinge on its own nature. In our language, this allows him to deal with functional abstracts only when it comes to the mind and not having to consider the object abstracts which take part in the unfolding sequence surrounding such a symbolisation. What is the benefit?

Take for example his comments on reason. Reason, Kant says, is the "faculty of deducing the particular from the general" [54]. Now the general can either be given because it is "certain in itself" or it has to be ascertained through the particular when that is certain. If we were to apply this proposition to the concept of consciousness, then in the latter case the concept itself would be indeterminable because the particulars are derived from the sheer variety and unpredictability of experience; but even in the case of the former the generality can only be defined if the principle of operation is the content, otherwise the delineation to the particular is subject to the assumed object-characteristics contained within consciousness, and that would be putting the intuitive cart before the formative horse.

If we translate the thinking processes according to Kant into our own system we arrive at the scenario shown in Fig. 8-2. As usual the abstracts are symbolised by coloured spheres; the elliptical arrow denotes the act of recursive mining for affinities within the structure's domain, which is represented by the fuzzy colour spreading out from the centre.

The "Mind" in Fig. 8-1 has been replaced by a container which serves as the conscious attractor of all those thought structures containing affinities with the abstractions of the space-time concept. The affinities relate to functional abstracts, some of which are in congruence with the rule sets inherent in the space-time TS. As the articulation of ideas proceeds, the affinities find their way via the con-TSs to the accumulated pool of expressible knowledge, with the result that not only is there a centre lending its articulated notions a specific format, but at the same time the centre is acting as a conduit between TSs due to their shared affinities. When Kant presents his examples from geometry or some time-related information he articulates the results of such an activity. He can do so because his examinations of ideas in general produced the common space-time characteristics, which prompted him to formulate his container, which in turn enabled him to uncover further concepts, enhancing the mutuality of the container, supporting the discovery of even more appropriate ideas, thus strengthening the container... etc etc.

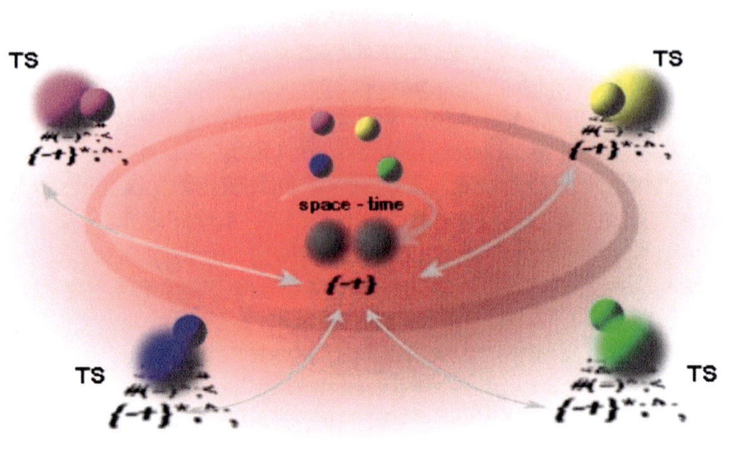

Fig. 8-2: Kant's container, creating associations with other thought structures in terms of the affinity with those functionalities (rule sets) which pertain to the concepts of space and time. The full contingent of rule sets is only accessible to the sub-TSs (the large fuzzy spheres), but in this case it is the con-TSs which are absorbed into the container structure. As such the association process relies on Kant's conscious train of thought.

In the end he feels justified to declare the space-time abstraction as *the* underlying a priori knowledge, something he could not have done had he been stopped in the early stages of the exercise. As we have seen, the space-time IFAs are at least PAT^2s, otherwise Kant would have identified them as being connected to empirical events. As such they do not have any links to first-level abstracts and are therefore 'floating freely' as it were, representing the typical initial building blocks of a container. In his case the driving force of the articulating con-TSs is his desire to demonstrate the validity of an overall concept (and to write a book saving metaphysics from the dangers of empiricism). It could have been emotive modules only, or some other combination of TSs and EMs. In any case, the link-less nature of these IFA^{2+}s ensures that any possible limit to the container's affinity-seeking endeavour does not eventuate. For example, suppose the space abstractions were connected to Kant's knowledge of triangles and nothing else. Any musings about spherical bodies moving through space would not be affinitive with thoughts about the straight lines of a triangle and so would remain meaningless in terms of the fundamental question about one's inherent understanding of reality. That functional relationship between link-less higher FAs and their potential associations constitutes a major component of creativity as we humans know it. In this case the FAs were intuitive but they could just as easily have been derived from the conscious study of something, and they had to be second level or higher in order to relinquish any links they would have had. A focused, productive act of creativity therefore involves a purposefully constructed container comprising the target of the search, one or more functional abstracts at least at the second level of abstraction—and the absence of any negatively denominative thought structure which could bring the process to a halt. In more prosaic language one can say then that creativity requires determination, intelligence, *and* the absence of censorship.

The cognitive elements inside the container may not sustain their linkages with grounded PAT chains, but that does not mean there are no affinities amongst each other. The associations thus produced serve to identify the productive potential of these TSs to the mind; in other words, they allow us to play around with their abstracted nature and so form hypothetical scenarios. Since the TSs are essentially ungrounded, they will only ever be hypothetical.

We find examples of these kinds of ungrounded associations and what can be done with them in the thesis and antithesis spiel Kant presents in the chapter *The Antinomy of Pure Reason*. By placing positive and negative assertions alongside each other he shows how such a descriptive system fails to teach us anything because it cannot reach a new conclusion. As he writes,

> These apparently rational, but really sophistical assertions open a dialectical battlefield, where that side always obtains the victory which is allowed to make the attack, and where those must certainly succumb who are obliged to keep on the defensive. [55]

Let us examine the dynamics of one such thesis-antithesis pair in terms of the thought structures involved. Under the *First Conflict of the Transcendental Ideas* the Thesis states:

> The world has a beginning in time, and is limited also with regards to space [56].

Here and in the antithesis we will deal with 'time' only, since the principles remain the same (as they do for the other theses).

The concept 'world' represents a TS which contains all the imagery one is able to arraign under its auspices. In the present case it is sufficiently large to avoid conflicts with other TSs due to mutually incompatible characteristics (for instance, 'world' does not insist on a specific astronomical location thereby rendering it useless in relation to some other star system), and so it can be used as a single unit.

'Time' is a notion for which a set of TSs exist, albeit mostly in the form of sub-TSs which can however be articulated for the current purpose. Associated with 'time' are the abstracted ideas such as 'duration', 'interval', 'sequence', 'chronological step'. Each one of those will have had other associations dating from the time the mind has become acquainted with them and therefore reflect the individual's life.

'Beginning' is another abstract, and there would be links with representations such as 'start', 'end' (ie the beginning of something marks the end of something else), and 'new'. Because all the three concepts are drawn together in the sentence of the thesis, that is to say they are pulled in by the thesis con-TS for the purpose of grouping them, the tendency exists for any affinities to be invoked. As such 'world', 'time' and 'beginning' are made to share their respective affinities and a thought pattern gets formed. Let's group these surrounding associations under the name of a TS's *associative halo*. Note that this operation is by no means contrived, in the sense that 'time' and 'beginning' for example are forced to coexist against a better judgment. On the contrary, in the initial set-up phase the TP could be considered harmonious in terms of its constituents and in the absence of interfering groundings—a congruent pattern. (Let us also remember that our selected concepts are mainly summaries of filigreed strands and connections, only the sum-total of which is capable of making sense to us as a single idea. After all, we don't have 10^{11} neurons for nothing)

The thesis starts with a congruent TP, characterised by seconded TSs which share their object and functional abstracts with each other. The proof consists in assuming for the moment the negation of the original statement and as a consequence to observe the obliteration of the statement's validity. Thus we posit that the "world had no beginning in time". 'No beginning' is an abstract which in itself has an associative halo, and as soon as we use the concept the halo's contents are brought into play. Given the relationship between 'no beginning' and 'eternity' there is now the need to reconcile 'eternity' and its halo with 'time', which prompts Kant to say,

> For if we assumed that the world had no beginning in time, then an eternity must have elapsed up to every given point of time, and therefore an infinite series of successive states of things must have passed in the world.

Now that the 'no beginning-eternity-time' complex has been brought into conjunction with 'world', certain ramifications appear which need to be incorporated for the model to retain its validity. So Kant observes that

> The infinity of a series, however, consists in this, that it can never be completed by means of a successive synthesis.

And therefore,

> Hence an infinite past series of worlds is impossible, and the beginning of the world a necessary condition of its existence.

Note that the proof does not function by exposing a certain observable situation and pointing to its impossibility in relation to some stated goal, but solely by matching the associative halos of certain abstracts, negating one abstract and then recognising the similarly disjunctive situation with regards to that modified set of associative halos: from 'beginning' to 'no beginning', to 'eternity' and 'infinite series', put against 'world' and its assumed normalcy in terms of its possible states. At that point we have a clash and so we have proved the veracity of the thesis.

The antithesis states:

> The world has no beginning and no limits in space, but is infinite, in respect both to time and space. [57]

The proof proceeds in a similar fashion. Against the background of stated concepts and their associations we assume that the world *has* a beginning. The abstract of 'starting point' within the associative halo of 'beginning' gets attached to 'world' and elicits the sentiment,

> ..as beginning is an existence which is preceded by a time in which the thing is not, it would follow that antecedently there was a time in which the world was not, that is, an empty time.

'Empty' now brings its own affinities to bear upon 'time', and in the mind of Kant the latter acts as a denominative TS such that the resultant complex implies a set of conditions which are clearly outside the bounds of the original assertion:

> In an empty time, however, it is impossible that anything should take its beginning, because of such a time no part possesses any condition as to existence rather than non-existence, which condition could distinguish that part from any other (whether provided by itself or through another cause).

We are left with an incongruent TP, and since incongruent TPs are to be rejected the original statement in the antithesis must be accepted.

In summary, we have a set of TSs, each with its associative halo, exhibiting affinities that can be mutually shared. One TS gets replaced by its negation and *its* connections, and the state of harmony between the affinity sets is gone. As there are no further associations with grounded TSs where an actuality could be ascertained once and for all, the entire system relies purely on the relationships between the sets of abstracts and nothing more. Thus it does not matter whether one labels one statement a thesis and the other the antithesis, or vice versa (Kant's personal purposes notwithstanding), since it is the congruence inherent within the initial TP which sets the tone. If there *were* grounded links (and provided of course they are not dismissed out of hand) the whole exercise would be conducted along the lines of c-PAT-ls, with the result becoming a function of the associative intensities of denominative thought structures. As it is, we are dealing with a system in isolation, a fact recognised by Kant.

It is possible to view the process from a slightly different perspective in order to highlight its truth value, where 'truth' means the perceived satisfaction derived from the particular configuration of some thought structure. We start with a TP enabling an assertion, all the abstracts having been settled within a framework of associations; next comes a new, critical assertion (this can be a simple negation or something more sophisticated), which is met by another, positive assertion derived from the original complex, to show how 'false' such a newcomer is; this can be followed with a further attack, again taken from the main TP; and so on and on until one side becomes exhausted.

The schema is presented in Fig. 8-3. The plus-sign denotes the accepted, established, overall concept, and the critical response is shown by a minus-sign. Initially the 'plus' dominates, and hence constitutes the fundamental, overriding perception. At any given time therefore the positive outnumbers the negative, unless the supply of affirmatives happens to cease.

All in all Kant's creativity, although being ultimately stymied by his neglect of the subconscious, can be considered productive because it enabled him to posit a significant commonality shared by most of our concepts, namely our awareness of spatial and temporal relationships. In the end it faltered because he only considered the conscious part of our thoughts, and so the deeper connectivities which contained the representations of just those relationships remained hidden from his view. His mind only answered the question when it was capable of stating it, by which time the relevant abstracts were already in place.

Since the conceptual relationships we establish, dismantle, and re-establish as we conduct our trains of thought are only accessible to Kant via the brief appearances of their con-TSs, he is forced to erect a complex structure of descriptive elements and their interplay to which the greater part of the *Critique* is dedicated. It is not dissimilar to an explorer who finds himself in a vast forest and who is limited to investigate the wildlife by deducing its presence from the traces it leaves. Much can be gleaned from a spoor here, a dropping there; much more though can be learned if our explorer only dared to leave his outpost and enter the foliage, day or night. Such readiness would come later.

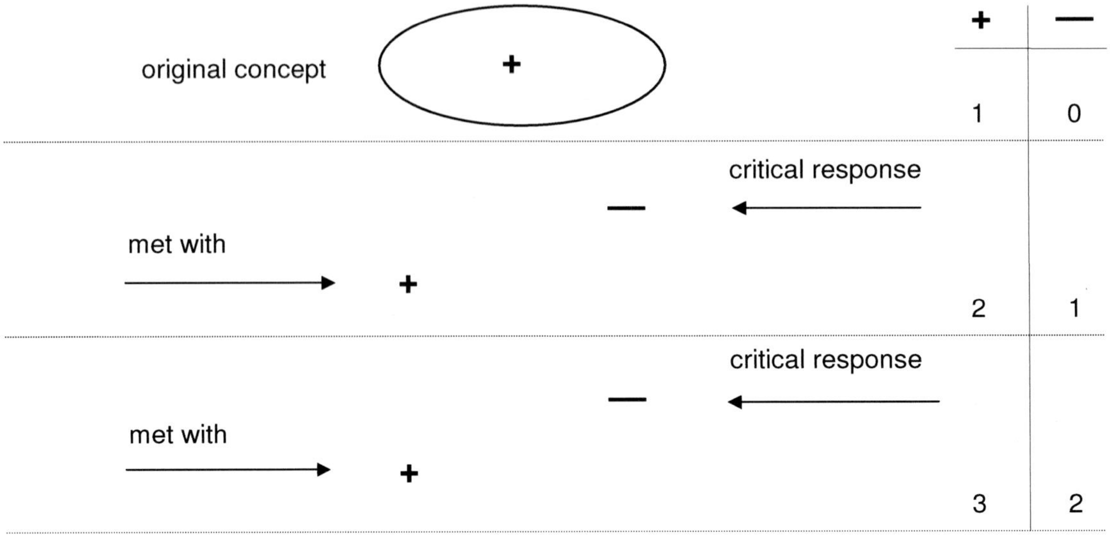

Fig. 8-3: Schematic diagram of assertion and refutation within the overall dynamics of containers. For details see text.

Nevertheless, Kant's expressly stated intent to abide by the principles of reason, coupled with the conscious realisation that their potential is not unlimited, prevented his mind from straying into the illusionary landscape of ungrounded assumptions. While it is true that he did not rely on direct representations of objects—indeed, mostly spurned them—the artificial constructs put in their place were endowed with the functionalities of logic, and *they* derived from the greater reality of his surroundings, acknowledged or not. Kant's mental structure of a priori concepts is a typical manifestation of a container, carried to sometimes extreme heights of abstraction by the intelligence of its creator. It is also a tribute to this philosopher's integrity that despite his incomplete knowledge of cognitive matters the exercise did not deteriorate into phantasms. Against the background of our general lack of omniscience such a demonstration of productive thinking must be appreciated.

9. The will for independence

Georg W.F. Hegel - (Charles Darwin) - Friedrich Nietzsche

It is the second decade of the 19th century. A cursory glance across the political map of Europe shows three general bands of governance. Starting in the north-west there is Britain with her parliamentary monarchy; adjoined on her south-eastern side by several states featuring a constitutional version such as Sweden, Belgium, France; followed again to the east by the absolute monarchies of Prussia, Austria, Russia, among others; and Switzerland nestled in the middle, the sole republican exception. The French Revolution had come and gone. Its cauldron of ideas, escaping from a process of singular concepts poured into the seemingly bottomless vessel of society where they accumulated steadily and began to ferment, presented an abrupt confrontation with their varied nature. The appearance of so many strands and forms now suddenly brought into the open had been too much an experience for the group mind, which hitherto existed with only a select few of societal abstractions allowed into public consciousness at any one time. Snippets of democracy, republicanism, monarchical tendencies, outright rebellion against religion and spectacles like the veneration of the goddess of Reason in the nave of Notre Dame no less, invited shock, fear, and the inevitable backlash. Libertines were executed, fanaticism met with fanaticism, and the monarchy was put back in place. The mixture of novelty and familiar tradition eventuated in the attempt to accommodate the new age, the New Order, through conventional means. The decades to come would show how naive such a perspective had been, with the current period already experiencing further revolutions in Germany (1819), Italy (1821), Spain (1823), Poland (1831)—all in the eastern-most band of our view.

So far Europe had been largely agricultural, but we find ourselves on the threshold of the Industrial Age, the result of a confluence of insights and expectations. The development of railways, telegraph services, postal communications, the advances in medicine and hygiene, and the modernisation of agriculture towards an intensive, organised cultivation of yields brought a dramatic increase in activity, carried on the back of similarly significant increases in population.

When trains of thoughts are projected freely, that is to say unencumbered by their original associations with those concepts that at the time were considered to be appropriate, the effect is not chaotic in the conversational sense. Rather, affinities which before were not acknowledged or had not even entered the consciousness of their bearers, now sought and found connections in different quarters and produced novel conclusions. These *are* legitimate linkages and productive connections, but their constituents appear in unfamiliar settings and therefore render the entire process unpredictable. Just as such overturning redefines the single mind, so does the consciousness of society at large deepen. The observing self, using the abstracts at its disposal, employs their nature and functionalities in order to identify the outer scenario thus presented. To accomplish this, descriptors have to be articulated, concepts established and mapped out, and all the while their inherent dynamics related to already existing ideations.

In those years a German university lecturer surveyed the state of history as it appeared to him. During his own life he had experienced the major historical events of the period and was exposed to the intellectual tendencies around him as well as the material changes some of them effected. His name was Georg W.F. Hegel (1770 - 1831), and the series of lectures dealing with the philosophical aspects of implemented history were issued after his death by his son under the name of *The Philosophy of History* [1]. To understand how his mind arranged itself to pursue the agenda of its owner and what kind of perspectives it generated, we are going to examine that work.

As we have seen in previous chapters, the understanding of some item involves having a representation of its content plus a representation of what that content does—its functionality. Both representations can occur at various abstraction levels, and while a generalised version of something might be perceived, its associated functionality may not have undergone a similar modification. For example, recall the box-memory metaphor from chapter 3, inducing some people to worry about exceeding the capacity of their brains just as one would overfill a box. There we had a generalised form of the concept 'memory' which produced 'container', and it was paired with the functional abstract of 'bursting'. However, 'bursting' stayed on as the same kind of abstraction it appeared under when it was attached to 'box', and so we were left with a mismatch between the Q_{po}-Q_{ao} relationship when it came to 'bursting' and 'memory'. It is possible to elevate 'bursting' to another level and so turn the image from one of flying bits and pieces to another of an abstract picture of some 'destruction due to overload', but that would entail a ramping-up of the 'burst' FA which would not have been the case in our example. One instance of a further abstracted FA is the contemporary expression of 'burning a CD', meaning the writing of data to a compact disc. The word 'burning' is an abstraction because a laser beam heats up a tiny spot on the disc's surface to over 300°C to change its magnetic state and not to perform a process of oxidisation on the material; hence the understanding of the word involves the correct interpretation of the event since no actual 'burning' takes place. Note that while it is possible for someone to copy a file to a CD and all the time wondering where the smoke has gone, this has more to do with the general safety and fool-proofness of electrical appliances rather than the wisdom of the user—those safeguards of course having been put in place by people who *do* know what is going on. Without safeguards the implementation of an abstract concept that happens to incorporate an incongruent Q_{po}-Q_{ao} relationship which in turn places that concept within the context of an incongruent functionality is fraught with danger.

The 1820's were a time when new ideas presented themselves on an almost daily basis. For the general population the information value of a novelty would have gone through a certain process of modification, and the cases would have been rare indeed where the instigator of something new had been the self-same person who introduced—and explained!—the thing to an appropriate audience. In other words, a breakthrough in technology, an advanced piece of surgery, or the discovery of some unknown part of the world, they all ended up as object abstracts at varying degrees of sophistication. They had to be associated with usable metaphors, that is to say with functionalities whose abstractive nature was pressed into the service of turning something alien into something familiar. During the transformation the new object abstract becomes embedded within a network of links, some of which connect to FAs such that the nature of the object acquires meaning. The meaning will be productive if the association between the object and its functionalities is a congruent one; it will be misleading and destructive if the association is incongruent.

For someone like Hegel to look across the centuries and contemplate history, the picture would have been vastly different to what an Augustine or even a John Locke saw. Not only were there the original texts, but the person also had a significantly extended arsenal of ideas, words, meanings, and interpretations to play with. From our perspective that means more comprehensive thought patterns, PAT chains covering several abstraction layers, and a framework of linkages with functional representations allowing for an ever greater variety of re-representations. Since the basic goal of the exercise is to make sense of the world, there is the tendency to create containers, that is thought complexes which provide the denominative influence for those TSs that are in the process of being integrated with the rest of one's mind.

In order to be denominative the two TSs in question must exist within each other's associative halo, and to do that the connections between the right abstracts have to be made. Given the sheer availability of abstraction levels the chance of such connections being congruent is enhanced because the maximal abstraction level in some denominative TS is likely to be higher than the abstraction level of the newly arrived object and its functionality. For example, if I am capable of thinking in terms of 'container' when I am shown a vessel I have never seen before, I am more likely to relate to that new item than if all I knew was 'amphora' and anything other had to be explained laboriously.

Although it is impossible to give a detailed analysis of Hegel's mind, we can nevertheless deduce a certain complexity judging by his overall categorisations of the "methods of treating" history. He identifies three types, arranged in order of abstraction (these are my words, coming from a still higher perspective) [2]. Note his choice of words: there are not different types of history, but one single entity that simply *is* and of which the perception gets processed in one of several manners. Compare the present attitude to that of Aquinas, who also categorised subjects but who saw them as sufficiently distinct to warrant a particular treatment in terms of the subject's nature (see the quote "If there were no other substance beyond those in nature of which physics treats, physics would be first science.." in chapter 5). For Aquinas the subject is the perceived object and thus gets treated in terms of the perceived qualities of the object ("nature" through "physics", "immobile substance" through "philosophy", and so on)—he did not have the advantages of higher abstracts which permitted modifying one's approach and still remain with the same topic. What more abstraction levels demonstrate, apart from hinting yet again at the differences in basic intelligence between people living then and now, are the possibilities opening through a higher cognitive vantage point and making themselves felt as we move towards the present (regarding the higher vantage points recall the puzzle example and the dynamics in Kant's space-time container in chapter 7 and 8 respectively).

Hegel tells us that the first method of treating history is "Original History":

> To this category belong *Herodotus*, *Thucydides*, and other historians of the same order, whose descriptions are for the most part limited to deeds, events, and states of society, which they had before their eyes, and whose spirits they shared. They simply transferred what was passing in the world around them, to the realm of re-representative intellect.

We can almost hear the cognitive scientist talking—describing the external source of perceptions (the deeds and events), moving on to the type of individual importing those sources (the historians), and finally identifying the format the source information ends up with (the respective, re-representative intellect). Of interest therefore are Hegel's descriptive tools used to communicate his theme, his personal understanding of what he is looking at, the formatting of the subject, and what imagery his sentences evoke, particularly in the present view of the technical details involved. "Original History" then stands for a straightforward recounting of events in the form of PAT^1s, where the relationships of such abstracts are simply the perceived version of what occurred before the writer's eyes (or at least before someone's eyes who then passed the knowledge on to the historian). There is no interpretation, no generalised principle that can be sought out, and "the author's spirit, and that of the action he narrates, is one and the same", that is "he has not attained an elevation above it".

The second type of treatment Hegel calls "Reflective History", which he divides into three sub-categories [3]. The first of those, "Universal History", is the investigation of a historical epoch in order to gain some understanding of the times. The writer "approaches his task with *his own* spirit", which is "distinct from that of the element he is to manipulate". The resultant text is a function of the historian's intent, be it to merely record events or to present a certain point of view. The second sub-category is "Pragmatical History", by which Hegel means the re-telling in the form of a moral lesson for the present. A historical development is considered in terms of contemporary values so that its outcome may teach us something in the here and now. There is a problem though:

> But what experience and history teach is this—that peoples and governments never have learned anything from history, or acted on principles deduced from it. Each period is involved in such peculiar circumstances, exhibits a condition of things so strictly idiosyncratic, that its conduct must be regulated by considerations connected with itself, and itself alone. [4]

It is more than a mismatch in specific circumstances however—the benefits of hindsight plus a more mature view some centuries later makes a direct relevance useless, he says. The ambience of the times changes, and to address a distant historical event from the perspective of the present will say nothing productive about the past nor can it pronounce anything useful about the present. Sub-category 3 is called "Critical" if the aim of the exercise is an "investigation of their [the historical narratives] truth and credibility". It is done through extorting "something from the records which was not in the matters recorded", resulting in having the past resurrected in all vividness but also "putting subjective fancies in the place of historical data; fancies whose merit is measured by their boldness". The last sub-category is described as possessing a "fragmentary character" and adopting an "abstract position"; it forms a bridge to the third main category of treatment and is concerned with a particular type of human enterprise such as the Arts, Law, or Religion. Once the observer has attained a generalised point of view in such matters the doors have opened to an understanding of a nation's psyche.

Which brings us to the third type, "Philosophical History". Hegel did not find it necessary to offer further explanations for the previous two types, but here an "exposition or justification" is needed. In very general terms, it represents the "thoughtful consideration" of history [5]. On one hand Hegel recognises that Thought, being "essential to humanity", must be secondary to the actual reality of things; yet on the other, "Philosophy dwells in the region of self-produced ideas". Regarding the former, dealing with the past requires the observance of facts, and therefore would be in opposition to the auspices of the latter. Clearly, some form of reconciliation is needed.

Since philosophical Thought concerns itself with Reason, the "Sovereign of the World", when applied to the subject of history it focuses on the "rational process" being observed. In order to know what exactly one ought to observe under such a heading we need a definition of Reason, and for Hegel it is delineated from his personal, theological view of the world. In his own words,

> It is there proved by speculative cognition, that Reason—and this term may here suffice us, without investigating the relation sustained by the Universe to the Divine Being—is *Substance*, as well as *Infinite Power*; its own *Infinite Material* underlying all the natural and spiritual life which it originates, as also the *Infinite Form*—that which sets this material in motion. [6]

Having all these qualities, Reason therefore determines the course a people enact for themselves, and the historian, availing himself of the present type of methodology, recounts the past against the backdrop of this universal, all-pervading medium.

We can summarise the various categories in the following terms:

Original History. A set of TPs featuring not more than two abstraction levels, where the object, ie the event, is represented as a perceived version of the original without any further processing.

Reflective History. A set of TPs featuring possibly more than two abstraction levels, but where the additional layers are reserved for the TSs representing the observer, not the observed. These take four forms:

a. Universal History. Similarly abstractive TSs are organised as the writer's own viewpoints, and they act as denominative TSs with regards to the event-TSs.

b. Pragmatical History. Additional TSs are organised as the writer's viewpoints, but it is their higher abstractions which act as the denominative TSs with regards to the event-TSs.

c. Critical History. Not only are there more highly abstracted additional TSs, they are in fact substituted for the event-TSs in order to serve the denominative processes of their owner.

d. Abstract History. Somewhat similar to (c), except that now the contents of the event-TSs are being substituted for the historian's TSs, such that the former have been ramped up to serve as the analytical grist for the observer-TSs' mill.

Philosophical History. A set of TPs which have been expressly formed to accommodate any event-TSs, such that the content of what is observed gets related to highly abstracted notions of the considered centre of human history.

The list indicates a certain progression from the simple to the more sophisticated (if not necessarily well intentioned); it also demonstrates Hegel's ability to move between a sufficient number of abstractions for that categorisation to have been contemplated in the first place. Type 3 is of course a fully fledged container, but note the intentional placement of that type within the rest. While there is no mentioning of container dynamics per se, the special influential nature of a set of ideas that are dedicated to the purpose of explaining the world in terms of metaphysical concepts has clearly been appreciated. Even under (2) Hegel recognises the essential differences in outcome when the same item is being handled in certain subjective ways.

To link the notion of 'reason' with the set of ideas comprising the divine has an outcome depending on which element has played the denominative role. In Hegel's case it is his concept of God which determined his view of 'reason', hence this idea's elevation to "infinite power", "infinite material", and "infinite form". As such we are dealing with a God-container, and the familiar abstractions of ungrounded objects (ie 'power', 'material', 'form') can be seen. Nevertheless, the entire complex has firstly been identified as an operation of a particular nature, and secondly been shifted to a band of abstractions positioned sufficiently distant from more mundane TPs so that incongruent connections are at least not a matter of course. As much as the God-substitute like 'reason' is the "*True*, the *Eternal*, the absolutely *powerful* essence", "revealing itself to the World", its abstraction level is so high that one must be a dedicated Hegelian indeed for some street scene to shimmer in unearthly splendour.

Still, there *are* connections. When Hegel writes about historic data, he refers to the sun's rising in the east, personifying the ascent of "Light" [7]. He compares the scenario to a "blind man suddenly possessed of sight", which causes him "boundless forgetfulness of his individuality" at first, but later, beginning to perceive the objects around him in the new light, proceeds to a "contemplation of his own inner being". The result of all this is an awakening of consciousness and therefore, he suggests, a free relation to one's spirit; where 'free' is meant as an expression of confidence and control. The thus construed imagery takes us from a picture of a circling planet presenting its revolving surface to the rays of the local star, to an interpretation of that planet's inhabitants as having evolved in similarly sequenced blocks:

> The History of the World travels from East to West, for Europe is absolutely the end of History, Asia the beginning.

We can recognise the above as the product of a denominative TP, the DTP holding Judeo-Christian religious material, which for all its abstractive span has built associations between mismatching abstraction levels, ie contextual associations leading to incongruent PAT chains. But as we have seen in the previous chapter, the existence of a container, actively modifying its sphere of influence to suit some cognitive configuration, is in itself not a recipe for total nonsense. Provided any grounded links are recognised during the process of association building, it is possible to create a framework of c-PAT-ls sufficiently extensive for its contents to be used in a productive, rational manner. For example, Hegel advises that one "must proceed historically—empirically" [8]. He goes on to denounce the belief in "original, primeval people, taught immediately by God", who are the holders of "all natural laws and spiritual truth" and similar wishful thinking. In Part II we will have the opportunity to examine such contrived attitudes as the product of a particular, IFA-led fear of maturity, but at the moment it is enough to realise that despite Hegel's predisposition towards the Christian religion the beginnings of his logic boundary are relatively far up the abstractive ladder of concepts.

In that context it is also interesting to note what Hegel considers *not* to be a useful material for his purpose. Referring to "original history" earlier on he dismisses "Legends, Ballad-stories, Traditions" as "dim and hazy forms of historical apprehension", because they stem from people with only a "half-awakened" intelligence. Since he doesn't reject legends, etc outright—as becomes apparent throughout the book—we can say that it is the unwritten, ie undocumented, continuously re-told and therefore re-interpreted story which in his opinion shrouds a people's background in mystery, unless there are balancing factors in the form of addressable and lasting records. If we consider the role containers play, we know that without grounded links the perceptive quality of thought patterns can come up with literally anything the mind is able to construe. Together with a lack of c-PAT-ls (ie a rational understanding of one's environment), the constraints of a culture which seeks to perpetuate itself (ie the morals and customs of one's society), and a low number of abstraction levels in the single and group mind (ie a 'one-to-one' perception of things without the interpretative quality of generalisations), oral history on its own can indeed be seen as a sign of intellectual backwardness (we will examine the dynamics of such scenarios more closely in the second part).

The advances in science during Hegel's time and the on-going revelations about the world outside Europe is certainly ground for positing 'reason' as perhaps *the* guiding force in the greater human scheme of things. At the same time there is the still existent intellectual demand to align one's thinking with the accepted dogma of society's spiritual rules, and so the need for reconciliation in terms of the denominative influences on offer makes itself felt. While it is God who designs the fate of the world, it is the entity of 'reason' that has to act in the role of the interlocutor between the general spirit and the human performers in the grand cosmic play. For Hegel this means,

> Our intellectual striving aims at realising the conviction that what was *intended* by external wisdom, is actually *accomplished* in the domain of existent, active Spirit, as well as in that of mere Nature. [9]

A cynic might suggest that it does not really matter *what* one calls a god, as long as the same actions are subscribed to such an idea. Yet it does matter in terms of the possible associations one or the other concept abstract is able to evoke. In the present case the labelling of divine manifestation as 'reason' makes further links with other TSs possible which *do* have a justifiable connection with a more rational meaning of the word. While therefore,

> Our mode of treating the subject is, in this aspect, a Theodicæa—a justification of the ways of God— ..so that the ill that is found in the World may be comprehended, and the thinking Spirit reconciled with the fact of the existence of evil..

the inclusion of empirical data and the sheer necessity to relate one's considerations to the overall ambience of the early nineteenth century make the exercise more grounded than would otherwise be the case. Consider also what Hegel sees as the final achievement of 'reason':

> The Spiritual becomes reconnected with the Secular, and develops this latter as an independently organic existence. The State no longer occupies a position of real inferiority to the Church, and is no longer subordinate to it. [10]

Because the infusion of this kind of reason into the minds of people enables true freedom to be gained, "Freedom has found the means of realising its Ideal", which is its "true existence". Again we have the jostling of position between one set of TSs and another. The religious set, ensconced beyond the logic boundary and acting as the driving force of Hegel's container, selects for suitable connections from the stream of imported TSs on the basis of high-abstract affinities; yet those higher abstractions come from PAT chains evolved from the observation of historical data. In terms of metaphysical suitability such goods carry a price: the diminution of the God-TP's territory and the accompanying retraction of its logic boundary. The "ultimate result which the process of History is intended to accomplish" is for Hegel the fusion between the two. Given the manner in which the process is going through its paces the outcome may be quite ironic in the end: a highly abstracted perspective spanning from mundane event-objects to the universal principles of existence. The future will tell.

So far we haven't arrived there yet, and in the meantime we are restricted to interpret the present through less perfect means. How productive can those means be? Consider the situation as Hegel paints it. There is the distant past, hidden in mystery and unarticulated sensations; followed by a gradual awakening of the senses and through them the intellect; until at the end we can gaze at life itself through the clear and bracing atmosphere of fulfilled reason. The intermediate stages represent the tentative steps taken by the members of a society and moving them gradually towards the goal. Some get there earlier than others (in Hegel's view the Europeans and particularly the Germans are closest), but each progress builds upon the previous one. In other words, although only the final result will provide total insight, we are already on the path towards it—unaccomplished as we are at present. Does this mean that somehow an incremental, imperfect achievement nevertheless possesses enough perfection to make the *right* step, so that the journey can represent an improvement in the long run; that is to say, the generalised representations of our own selves residing in our minds contain enough information to implement one's course, although there exists no internal representation of the final outcome?

Hegel does not say directly; we find a certain allusion to that point in the preamble to his treatise on the ancient Roman world [11]. There he tells of a conversation between Napoleon and Goethe on the nature of tragedy, with the former venturing that its modern version concentrates on policy, whereas the ancients looked to fate. Hence it is policy which ought to define that "irresistible power of circumstances to which individuality must bend". On the basis of our system of thought structures we can explain the perceived difference by pointing to the necessary levels of abstractions required in each case. If fate is accepted as the ruler, whatever happens can be taken as its own representation—the cognitive flatness of Hegel's "original history" in action (quite literally, 'whatever happens happens'). If it is policy which determines our actions, there is an immediate evocation of active and passive participants, the required concepts for evaluation and interpretation, and the eventual feedback causing even more abstracts to proliferate. The difference then consists in the inherent sophistication of the process, so that the availability of higher abstractions enables a more flexible and more profound understanding of the status quo, with a subsequent improvement in the response. In both cases however the material comes from the mind, and must have been put there as a resource emanating from the respective constituency. Therefore, while it is the intrinsic nature of the mind (ie single and group mind) that determines the eventual quality of the historical development, the fact that there is a development at all is due to the situatedness of a people within their realm.

According to Hegel the situation in a culture in terms of TSs is this: once History is possible (the members of the culture having moved on from oral traditions) the characteristics of their minds permits them to interpret their circumstances, creating other circumstances that are reacted to in turn. At some stage the process will be completed, at which point the members of that culture have fulfilled their destiny. That destiny, being a function of the circumstantial processes, has been laid down long before the temporal progression of members' actions are able to make it manifest. Because the circumstances are defined by the culture, their power is indeed irresistible—a certain culture moving towards its very own destiny. But note the lack of consciousness on the individual level: it is only ever enough to advance by one step, not more.

We can say therefore that Hegel's container holds a pattern in which there is an ultimate definition of someone's history, arrived at through the interpretation of abstractions beyond the logic boundary, yet where the representations leading up to it feature grounded links to observations of a lower abstractive nature. As we have seen before, lower abstraction levels have much detail, less generalisations; higher levels have less detail, more generalisations (see chapter 7). The inclusion of divine concepts at the higher levels (the container engine, as it were) ensures incongruent PAT chains; they act denominatively towards the TSs at the lower levels so that on one hand the TSs' detail makes for extensive functional associations along horizontal lines, on the other the enforced associative halo coming from the metaphysical abstracts engenders contextual associations between them and the lower-level details, creating incongruent TSs. To put this in less technical language, if I was a follower of Hegel then I would consider my society and my culture the perfect vehicle to achieve my own vision of final destiny, knowing fully well that whatever my present opinion of the state of the world may be, it is the sanctioned perspective of my people compelling me to act.

And yet I cannot rest easy. Hegel's driving force is the metaphysics of Christianity, derived from Judaic tradition. My desire for knowledge as a means to reach total insight is tempered by that tradition's story of the Fall:

> Man, created in the image of God, lost, it is said, his state of absolute contentment, by eating of the tree of the Knowledge of Good and Evil. Sin consists here only in Knowledge: this is the sinful element, and by it man is stated to have trifled away his Natural happiness. This is the deep truth, that evil lies in consciousness: for the brutes are neither evil nor good; the merely Natural Man quite as little. [12]

He goes on to describe 'consciousness' as the separated element from the "Divine Spirit", the unification with which ought to be the driving force behind human action. One must be mindful of the level of abstraction on which the above is contemplated. If the negative concept of "Sin" is associated with the hitherto positive concept of "Knowledge", then under the control of the denominative influence of Hegel's container "Knowledge" will become tainted. That is to say, the abstraction levels of both concepts are the same, and so we are dealing with functional associations (we know they cannot be contextual associations because their representational contents differ from each other). If recognition per se is perceived as a function of the mind (which of course it is), and a non-compliance with God's design therefore constitutes a form of disapproving awareness, in the absence of such awareness one cannot talk about disobedience either (that conclusion is not automatic but comes from the particular set of associations). For the current argument the actual level of abstraction is a moot point—knowing our steady progress in such matters it may well be that the people of the Old Testament had *their* concepts placed on a lower rung than Hegel two thousand years or more later; the important issue is the relative position of these abstracts.

Now "Sin" and "Knowledge" each have their associative halos. The basin holding "Sin" does not change because it is part of the container centre and hence denominative. Elements linked to "Knowledge" however are subject to the former's influence, resulting in two possible outcomes. Either "Knowledge" is turning denominative as well and modifies whatever is linked to it, or there arises a discrepancy between it and the rest causing i-PAT-ls. If we accept the first case then we have to assume that for Hegel the facts of history, which after all ought to be appreciated, are in themselves sinful—something which is clearly inconsistent with his theory on Philosophical History. Hence the second case suggests itself and we can expect incongruent PAT chains. Note however that "Knowledge" co-exists on the same abstraction level as "Sin", and being the centre of its associative halo will either be in the form of a functional association with some member of its halo (ie still on the same level), or will connect contextually (ie on a different, lower level). If such an abstract on the same level is held to be part of "Knowledge" it will not be grounded amongst the network of representations of historical events, but will feature an incongruent nature similar to "Knowledge"; its accessibility depending on a comparable influence such as another container. If on a lower level, the item will be classified directly as "Knowledge" also (being associated contextually) and will be linked to abstracts pertaining to the subject of history.

From an evolutionary perspective we can say that the idea of the "Fall" represents the result of a container operation conducted across a low number of abstraction levels, where everyday knowledge is represented as PAT^1s and PAT^2s leaving no room for more profound generalisations between them and the logic boundary. From Hegel's perspective, being situated amongst a culture of far more developed insights which utilise a greater number of abstraction levels, the "Fall" appears as the symbolic endpoint of a steady progression of phenomena that are meant to define the metaphysical format of the cosmos.

Since "Knowledge" at its high level is consciousness disengaged from "Spirit", its accompanying associations with more mundane but positive versions of itself lower down pose a danger to the supposedly negative character it is meant to portray. In Hegel's mind the problem is solved by the definition of "Spirit":

> It is the one immutably homogeneous Infinite—pure Identity—which in its second phase separates itself from itself and makes this second aspect its own polar opposite, viz. as existence for and in self as contrasted with the Universal. But this separation is annulled by the fact that atomistic Subjectivity, as simple relation to itself [as occupied with itself alone] is itself the Universal, the Identical with self. [13]

Firstly, all the players in the quote above occupy the highest abstraction levels. As such they are not concerned with any contextual associations but only functional ones, one of which happens to be "Knowledge". Secondly, note the splitting of "Spirit" into two aspects, one the opposite of the other. If the "Knowledge" TS, through its contextual reach down towards useful abstracts, needs an affinity with a non-negative quality (careful: 'non-negative' is *not* the same as 'positive') within the associative halo of "Spirit", the latter's opposite version should accommodate that demand. The concept of "separation", being the product of another abstraction featuring "existence for and in self", juxtaposed with "the Universal" and becoming "annulled" through the actions of "atomistic Subjectivity", is sufficiently aloof from the realm of grounded representations to maintain its wishful integrity.

There is an additional measure which ensures the isolation of these TSs from the lower levels; recall the words "This is the deep truth, that evil lies in consciousness" from the second-last quote. "Deep" in an intellectual context denotes abstraction, and a relatively high one at that. To preface a certain concept by saying "this is the deep truth" will prepare the listener's mind such that anything entering from now on will be relegated to high abstraction levels with no automatic links to the recipient's more grounded structures (anyway, this is the intent). Hegel's preparation allows him to assemble a mental construct composed of such things as the "immutably homogeneous Infinite", an "atomistic Subjectivity" and others, feeling safe in the assumption that their precious realm is not going to be encroached upon by any real-world drudgery. We have seen this phenomenon before and in chapter 6 we drew the architectural analogy between TSs and building styles, such as Gothic cathedrals with their high and thin columns supporting an elaborate structure vs the squat and hefty artifacts of the Romans. Well, here the concept of "Knowledge" is such a thin column, rising up from the network of grounded elements to connect with the metaphysical cloudscape of Hegel's religion. It is a PAT chain which may have quite an extensive network of connections with grounded TSs, giving it the contentional sustenance as it were, but then moving upwards on its own to be surrounded once again with any number of connections, all of which are contained at that high level of abstraction and thus inaccessible from those lower TSs—except via that particular column. The low-level network, the thin pillar, the top-level cluster, it all resembles a tree, and so we can call this phenomenon an *illusion tree*.

Let us be clear what is meant by "illusion"—it is a very particular form of deception. If I were to tell of mermaids at a campfire I could be accused of trying to deceive my audience, but what I produce is not an illusion. Consider Fig. 7-2 from chapter 7, showing what at first seems to be a series of triangles. They are not, as becomes evident once the black bar is removed. They *seem* to be triangles because the mind already knows about triangles and uses their representations to connect to the abstracts generated by the picture visible around the bar. In other words, an already existing, higher-level abstract is taken to be the true representation of something that has been incompletely represented through lower-level abstractions. The latter are grounded (it's what we can see in front of us), but the higher one is not, at least not in relation to the item under focus. *Why* the connection is made between the grounded chain and one particular higher-level abstract depends on the situation; it could be probability or best fit (as in the picture), it could be emotive influences, or the active pull coming from a container—they all are forms of affinities.

In the present case the illusion consists of the imagery that has been created around the concept of "Knowledge" by having its highly abstracted representation embedded within an associative halo of similarly abstracted elements, such that it becomes a seemingly integral part of the religious sentiments held by the halo's TSs. In fact it is not a part of that halo—it is part of the grounded abstracts from which "Knowledge" has been derived in the first place. In principle the relationships between the respective links are comparable to what is described under Fig. 7-2, but an illusion tree defines a more complex arrangement. Later we will come across other examples of illusion trees, enhancing our skills to spot them.

The aforesaid demonstrates two things. One, we do not need to possess Hegel's Christian thought patterns enabling us to comment on the feasibility of atomistic subjectivities acting against split versions of some pure identity and other esoteric problems—we are merely analysing his cognitive dynamics because we have the descriptive tools to do so. Put another way, one does not require any detailed knowledge about the content of some train of thought in order to assess its overall validity; identifying its format is enough. Two, thought patterns, even thought structures, can be very complex. I have given a rather general description of the process involved in Hegel's view of "Spirit", but I could have easily filled a few more pages by going into the particular contents and examining *their* cognitive relationships—and yet the relevant lines of text in the *Philosophy* number not more than ten. But to continue.

Hegel makes the fundamental point that a people strive towards the realisation of their true spirituality; his container and his illusion tree are designed for that purpose. Since we are not unconscious brutes it is our will that determines our actions and produces the desired result. How 'will' is perceived is therefore significant. Towards the end of the book we find the following definition:

> Secular life is the positive and definite embodiment of the Spiritual Kingdom—the Kingdom of the *Will* manifesting itself in outward existence. [14]

So the "Spirit" and the "Will" are one and the same thing, they both manifest in secular life. Once again the need arises to reconcile an arbitrary human enactment of the will with what is taken to be the 'true' action as demanded by religious dogma. We read that 'will' can manifest in "mere impulses", "but these are transient and disconnected", forming the "ever-changing applications of volition". The "impulses" ought to be distinguished from what is "just and moral", because that is part of the "independent, intrinsically universal Will". Even "benevolent, charitable, and social" acts are impulses as they depend in their classification upon the agreement of society's members. One can only know the "Will" once the "specific and contradictory forms of volition have been eliminated". As Hegel puts it,

> Then Will appears as Will, in its abstract essence. The Will is Free only when it does not will anything alien, extrinsic, foreign to itself (for as long as it does so, it is dependent), but wills itself alone—wills the Will. This is absolute Will—the volition to be free. Will making itself its own object is the basis of all Right and Obligation—consequently of all statutory determinations of Right, categorical imperatives, and enjoined obligations.

The first thing to be noted is the format of the concept once it holds its ideal version—it turns into an "abstract essence". As 'essence' already denotes an abstraction, to have it abstracted once more clearly puts it on a high level of, well, abstraction. Thus it is a candidate for associations of the functional kind, since no attempt is being made to have the idea delineated from lower levels where the actual content, and therefore context, plays a role. In its "abstract essence" then it is a largely ungrounded concept, designed to be affinitive with similar abstracts.

Next the association with "Free" is made provided there are no links downwards. Once the notion of "Will" is such that only a sensation of unentangled purity is allowed to be understood as its representation, *any* connection with more prosaic elements is going to be perceived as "foreign" and "extrinsic". However, since it is those very connections which supply the definite content to what otherwise is merely a generalised shell of an idea, to exclude them prevents the identification of "Will" in practical terms. Yet at the same time, this very type of an idea is now considered to be "absolute" and once more associated with freedom. At that stage it ceases to be just an agent of action—it has become the foundation of all that is "Right".

As can be seen, Hegel's "Will" is another example of an illusion tree. From the everyday experiences we have various sets of TSs relating to 'will', 'free', 'absolute', 'right', and so on. In the case of 'will' there is a PAT chain reaching straight up towards a high abstraction level where the erstwhile links have been lost and functional associations can be made with the exponents of the other ideas which have been treated likewise. Therefore, although there exists no content, no detail of what 'free' and 'right' etc actually mean, the positive ambience of these elements has been preserved through their isolation. The result is a perception which draws on a certain empirical background giving it the cultural flavour, formulates any conclusions reached in this manner and endows them with the 'knowledge' that one's divine right is being enacted to fulfill one's true destiny. All around very powerful sentiments, and it would be an interesting question to find out whether the general culture of the times pointed Hegel in this direction or whether it was him who contributed to such an attitude; as an exercise it goes quite beyond the bounds of this book. Still, it comes as no surprise to observe the emergence of nationalism across Europe during the rest of the century, when Hegel's philosophy has been very popular.

Such a mental structure could not have been accomplished without the intellectual capacity of its holder, because it requires a relatively high number of abstraction levels. It also needed a religion like Christianity at its container centre, a religion that possesses a sufficiently large potential for abstraction, as mentioned already (see chapter 4). In a sense Hegel recognised its advantage, for he writes,

> ..and it is the very grandeur of the Christian religion that, with all this profundity, it is easy of comprehension by our consciousness in its outward aspect, while, at the same time, it summons us to penetrate deeper. It is thus adapted to every grade of culture, and yet satisfies the highest requirements. [15]

It is also noteworthy that such benefits are only recognised from the subjective view of the user, who is the religiously-toned theory producing Hegel; the container-driven thought pattern did not permit objectivity. Hence they are not seen as the general means to evoke a moralistic force—*whatever* its nature and aim.

Now that we understand certain formats of high-level abstractions we are better equipped to examine the nature of Hegel's concept of "Spirit", since it acts as a fundamental player in the configuration process of related thought structures and patterns. At the outset we are made aware of the distinction between the "Rational System" and the spiritual world as two sub-systems under the overriding banner of "Nature" [16]. Such a split is significant in the context of the inherent dynamics of containers and illusion trees, and we are reminded of somewhat similar juxtapositions made over the centuries every time manifestations of the mind are compared with the products of the exterior environment. Indeed in the pages before us "Spirit" is taken to be the "direct opposite" to "Matter", and it is interesting to observe what exact characteristics are perceived as being opposite to each other. "The essence of Matter is Gravity" we are told, and "the essence of Spirit is Freedom". With our 21st century hindsight we know of course that the essence of matter is not gravity at all, since "gravity" in this case is referred to in the sense of 'weight', rather than the force it actually represents, otherwise it would not have been placed opposite to the idea of freedom. Justly or not we have to dismiss the matter-gravity combination as being incongruent, but in its form it does entice one to see it as an antithesis to being free. It also indicates the by now familiar readiness to accept thoughts as *actual* representations of something outside the mind, rather than the *perceived* forms of sensory input. In the former case a spiritual notion is taken to reflect a very real occurrence beyond our mind, and therefore such an occurrence would then be equally substance-less and hence weightless. A further proof of that kind of belief can be seen in the absence of thoughts which counter-balance the current characteristics of an item. For example, I could imagine an object composed of matter yet floating freely through the air. Since it is just a thought I should have no qualms about it and even place it onto the same inner canvas which already contains my rendition of spiritual things. If I can visualise a floating rock *and* a floating spiritual something *and* know that all these are just thoughts in my head I would be more reluctant to consider such items as mutual opposites. However, should my conscious mind be very careful about what and particularly *how* it re-represents its content then it would not permit itself to entertain the above combination out of deference to accompanying functional abstracts—the mind *always* obeys its FAs. (One could argue that *not* imagining floating rocks would merely be a confirmation that one's inner self is perfectly grounded. Point taken, but in that case it would be rather difficult to come up with a picture of something spiritual and so we won't deal with such a combination anyway.)

"Matter possesses gravity in virtue of its tendency towards a central point", Hegel continues [17]. Grammatically one can link the possessive pronoun "its" either with "matter" or "gravity", but the first option does not make sense from a semantic standpoint. Therefore "matter" "possesses" the feature of gravity because gravity has a tendency to draw in a certain direction, namely to the centre of mass represented by the object. And "Matter", according to Hegel, is a "composite" that consists of "parts which *exclude* each other". But it also "seeks its Unity" and so "exhibits itself as self-destructive" because it verges "towards its opposite [an indivisible point]".

Note how the functional abstracts associated with the various elements prepare the ground for the overall framework. "Matter", being an abstracted form derived from many observations and so showing it multifaceted nature (disregarding for the moment the fact that such a generalisation cannot be really multifaceted—only the set of its source material can), is being linked to the idea of unity because its force (ie gravity) points towards its centre, and the centre can be seen as a geometric point. Since a point won't encompass the notion of extension, matter at this stage has to be considered as self-destructive. This ties in well with the concept of a substance-less ideal: gross matter acting on a force and in the process annihilating itself (the Cathars would have understood). It is almost as if "Matter", out of some idealistic-altruistic need, had acquired the force of gravity in order to remove itself from the tedium of earthly existence and escape into the realm of the Ideal. In the words of

Hegel, "It strives after the realisation of its Idea; for in Unity it exists *ideally*". An 'ideal' is an abstract, and can only reside in a representational system, eg one's mind.

Continuing the positing of opposites, Hegel writes that "Spirit, on the contrary, may be defined as that which has its centre in itself". Obviously an ungrounded, high abstraction, it nevertheless serves to further construct the juxtaposition complex. Because its "centre" is "in itself" it has no need for any associations with other elements such as matter would have, and as such it can be described as "self-contained existence". If something is self-contained it is independent, and being independent suggests freedom. So "Spirit" entails "freedom", and if one could only fuse with one's "Spirit" one would experience true independence, and freedom. The self-contained existence of the spirit Hegel describes as self-consciousness, is "consciousness of one's own being".

All this is a typical example of an illusion tree. "Matter" furnishes the material for the chain, coming from a network of low abstractions (all the objects one has identified as matter of some sort), is pushed up to high abstraction levels where such relatively ungrounded concepts as "unity", "indivisible point" and ideal-imbued gravity reside, and is placed within an associative halo through such functional connections. In the end it enables Hegel to state,

> According to this abstract definition it may be said of Universal History, that it is the exhibition of Spirit in the process of working out the knowledge of that which it is potentially.

Considering the role of containers and illusion trees, what this really means is that the concept of Universal History allows its followers to articulate their intrinsic character and subsume its demands under the auspices of their ideology. Hegel recognises the abstract nature of most of those elements as well as the emotive force necessary to have people act on them—through their desires, opinions, convictions. It is the interest or passion which drives events, he says, and "*nothing great in the World* has been accomplished without *passion*" [18]. In order to steer away from possible negative connotations when using the word "passion" he explains himself further:

> I mean here nothing more than the human activity as resulting from private interests—special, or if you will, self-seeking designs—with this qualification, that the whole energy of will and character is devoted to their attainment; that other interests (which would in themselves constitute attractive aims) or rather all things else, are sacrificed to them.

We have seen many instances that suggest how much more inviting higher abstractions are compared to lower ones. Here is an additional indication of how compulsive the evoked notions of Hegel's upper mental echelons can actually be: the combined power of religious conviction and cultural spirit driving the actions of the individual.

We still have to resolve the problem of content source. It is all very well to reduce one's forward-looking perspective to incremental steps with the expectation of arriving at the fulfillment eventually, even if the content of such steps is made relative to some divine blueprint. A sophisticated operation may not allow the complete realisation of the final product, but because of its sheer complexity would through its own inherent nature delineate the development as a function of its potential. What about simple systems—can one expect them to incorporate their future state already? One answer is "yes": once upon a time the earth was a lump of molten rock, yet has turned into a system containing organisms that build space stations. Another answer is "no": there are millions and millions of cubic metres of sand in the Sahara, yet none of that silica will ever turn into a computer chip and carry data signals. In fact, both answers are correct.

Recall our discussion on free will in chapter 7. Using Locke's TSs we saw the actual composite nature of 'will', with the word being a summarising semantic symbol standing for the dynamic confluence of abstracts, emotive modules, and their associated rule sets, all producing an action which to the holder appears more or less as his own. We also touched upon the degree of functional similarity between cognitive events inside a single mind and their equivalents in society, the group mind (see chapter 8). Let us first see what Hegel has to say about the role of the state and to what extent it plays a part in his grand scheme.

He makes the observation that the state is "well constituted and internally powerful" when the private interests of the citizens are in harmony with the common interest [19]. He also writes about the creation of the administrative apparatus, the "political machinery", and the constant interplay between it and the interests of the populace. Both, individuals and bureaucracy, contribute to the life of society through their actions:

> This vast congeries of volitions, interests and activities, constitute the instruments and means of the World-Spirit for attaining its object; bringing it to consciousness, and realising it.

He goes on to say that such acts serve a "higher and broader purpose of which they [ie the players] know nothing", something which can and has been debated by others but should be seen as the manifestation of a "Reason" that "governs the world".

In the single mind we have seen the driving force of a container, a complex of thought structures that forms particular configurations in terms of the associative qualities of the participating elements. Remaining on this scale for the moment we can say that the purpose of a container is satisfied when affinitive TSs have been acquired such that their mutual relationships are congruent (note it is the relationships between the TSs *at the container-induced abstraction level* which are congruent, and not the links to abstracts further afield which have served as their source material). They are congruent because the denominative nature of the container TS has adjusted their format sufficiently to be included in the entire pattern. The containers we have looked at so far have been established in order to articulate a certain concept, a hypothesis, or theory, but the end goal may not be articulation at all—the operation may be wholly subconscious. Because in the latter case an analysis requires more material than what the current written tracts are capable of producing, we will revisit the issue in Part II. For now we concentrate on the articulated version only.

The written text we have before us, that is the articulated result of a container, also includes the contents of the sub-TSs—they are just not visible. Not only have they been part of the formation process of associations, given their greater constituency compared with the conscious elements they also have been more significant. To use a small example for illustration, suppose someone makes a decision on buying certain shares. There is the conscious part, consisting of such things as market capitalisation, price-earning ratio and so on—all expressions of con-TSs. There is also the subconscious part, which can include virtually anything from the emotive quality of the company's name to the personal relationship of its product with the investor, to issues regarding the money needed vs the upcoming school holidays, etc etc. (To answer a possible criticism coming from the investment community: if the major influence *was* the financial parameters of the company, any secondary school student could become an investment advisor *and* would get it right most of the time too. Highly paid investment advisors are not highly paid because they only know about math.)

Thus the TP at the centre of a container will feature the abstracts coming up through the act of articulation, but in addition there are a host of unsaid notions which do not find their way into texts and speeches; and yet they are part and parcel of the entire pattern. Furthermore, once the container exists to the satisfaction of its owner it does so not only because of what can be gleaned from the owner's words but because of the inherent nature of all that what is *not* declared openly. Translating this into the group mind, the single abstracts become the individual citizens, the structures their instrumentalities, and the container turns into public policy. Just as a citizen can say one thing but does another, an instrumentality has its official agenda but different effects on the street level, and public policies have their sanctioned documents on one side and behind-the-scenes moves on the other.

Like Kant, Hegel does not recognise the role of the subconscious—it is too early yet for such a slippery beast to have been caught. But observe what ingredients he uses when introducing an example to illustrate his point about the state and its instrumentalities which are being put together to move society through its existence. There he talks about the building of a house, relying "in the first instance" on "subjective aim and design" [20]. Obviously, these would be the conscious effort by the builders. Yet there are also the things necessary for the construction per se, such as iron, wood and stones:

> The elements are made use of in working up this material: fire to melt the iron, wind to blow the fire, water to set wheels in motion, in order to cut the wood, etc. The result is, that the wind, which has helped to build the house, is shut out by the house; so also are the violence of rains and the floods, and the destructive powers of fire, so far as the house is made fireproof.

And so on. He makes the point that the elements are used "in accordance with their nature, and yet to co-operate for a product, by which their operation is limited". While we are satisfying our desires we also produce other results which were not deliberately aimed for at the time. In his words,

> They gratify their own interest; but something further is thereby accomplished, latent in the actions in question, though not present to their consciousness, and not included in their design.

We can assume that Hegel has never come across the concept of sub-conscious thought structures; nevertheless, in his articulation of certain societal dynamics which he *has* identified he comes across phenomena that he feels do play a role, but at a distance: there is the metaphorical fire, but its contribution only goes so far; likewise there is iron, but once installed keeps up the walls and nothing else; etc. His mind, acting through his container, while in the process of mining affinitive TSs for the purpose of assembling the desired TP, could not avail itself of the relevant sub-TSs because it lacked the necessary entry points. It *could* however draw on abstracted re-representations which through their functionalities enabled the connections to be made. Such abstracted re-representations in this case are 'fire', 'iron', and so on, and their associated FAs within the context of house-building are what made the connections possible. We are witnessing the appearance of the substituted abstract: an idea which has been identified in the real world and which lends its functionality to *something* which seems to be there *somewhere*, but its only trace is the manner in which it behaves. We call these kind of abstracts 'metaphors', and although a more technically correct label would be 'substituted abstract', the name 'metaphor'—literally meaning 'transfer'—serves us well.

Three things should be noted. Firstly, the TS to which the metaphor becomes applied is ungrounded, otherwise one would not need a functionality on loan, as it were. Secondly, the TS that supplies the FA *is* grounded (in the interests of clear speech it certainly should be!), hence the existence of the FA in the first place. And thirdly, the resulting abstract-FA combination may or may not be realistic, since the affinity itself has nothing to do with the quality of the participating TSs.

The question arises, why is there a need for particular FAs to begin with, especially ones that are borrowed from somewhere else; and, even more importantly, what is the reason for choosing one particular FA and not another? For the answer we need to look at the inherent nature of the entities we are dealing with. We have the container, consisting of an engine which mines other TSs with the aim to find affinities for structure-building, the associations being basically ungrounded. We also have one or more illusion trees, giving rise to functional associations at higher abstraction levels. Finally there is the process of articulation, forcing the mind to access the con-TS sections of those structures up there. However, to articulate means to consciously describe one's thoughts—but all we have at this stage are the object contents of these abstracts, not the functional rules accompanying them (recall that the functional association at the top of an illusion tree refers to the functional potential of the object at that abstraction level, and not to the function the object is meant to perform as such). Therefore the articulation process requires the positing of some act, something the object is supposed to *do*. Once more affinities are mined, but now the goal is to achieve an association between an upper-level object abstract and some other on a lower rung, featuring affinitive FAs (we know they are on a lower rung because the functionality thus procured does not come in its generalised version). That secondary mining process involves the con-TSs as well as the sub-TSs of all the structures concerned, with the exact ratio depending on the extent of the conscious sections all around. Naturally there would be at least some detail which can be articulated—after all, it was the process of articulation which directly triggered the operation in the beginning.

For example, in chapter 1 I used the metaphor of a plug and a tube to refer to the workings of a cylinder in an engine, using the familiar to explain the unfamiliar. In that case the extent of the con-TSs was rather large since just about everything about a tube and a plug is openly known beforehand, and the same goes for cylinders. Therefore, while going through the perfunctory motions of associating one TS with another, ordinarily there would have been no use for the exercise in itself and so that particular reference was quite useless from an instructive point of view. However, that metaphor was also part of yet another metaphor, namely the illustration of explaining something for which there is a different reference system available, the latter acting as the entry point to the former. As far as my own person is concerned, both metaphors featured a sufficient number of con-TSs to make writing about the topic possible; as far as you, the reader, is concerned, in your mind a picture of mental dynamics held the object abstracts (eg 'mind', 'thought structure', 'abstract') but they lacked their functionalities, a tiny part of which I supplied through the double metaphor. Your container is the desire to read this book; your illusion tree was the particular passage in that chapter (*at that stage*—hopefully those high-level abstracts will all have been grounded by the time the last page is turned); the substituted abstracts were the FAs borrowed from the world of car engines.

When it comes to Hegel's world view the situation is similar, although at that distance it is rather difficult to assert the precise extent of his con-TSs vs his sub-TSs. The problem is made more complicated by the fact that his articulated perspective, as a mental construct, is meant to be a re-representation of a similarly dynamic system on the outside. In other words, in how far have the observed events generated their cognitive equivalents, and in how far are those cognitive equivalents applicable to a productive re-representation of what society really does—the conscious as well as the subconscious parts (and furthermore, such a re-representation serving as the gist to his entire *Philosophy of History*).

It is too late to analyse Hegel's subconscious tendencies (we don't even have full access to his conscious ones), so what remains is to determine what *could* belong to the conscious and/or the subconscious sections, first in the single mind and then within the context of society at large. The reason why one and not another abstract gets mined by the con-TS driven articulation is simple enough: it has an affinity and with the other there is none. As we know by now, an affinity can exist in the form of a contextual or a functional relationship. During the process of articulation it is no different, and we can observe the same principles at work we have met before. Articulation itself is a container, drawing on affinitive TSs in terms of the container's TP. First and foremost the TP features the structures concerned with language, and what gets articulated is a function of the associative potential of the language TSs with other existing TSs. Hence for a meaningful sentence to leave one's mouth we need the appropriate connections between our linguistic abstracts themselves as well as between them and some other structure providing the content. The ratio of these two types of connections moves on a sliding scale; if language TSs are so prominent that only their own links matter the result is pure verbiage, and if only one or two language abstracts are the hook by which the others are brought up we get the mono-syllabic stare with the pregnant pause. Neither contributes to communication.

Contextual relationships are about content, functional relationships are about action. In the articulation container a contextual association will be formed if the object content of the language abstract is affinitive with the object content of the outside abstract ("outside" here means beyond the language structures complex). The functional association happens if the abstraction level of both abstracts is the same. Since even a simple descriptor (ie a word which describes something—"aha" or "wow" are not really descriptors) requires a minimum of abstraction (otherwise it would not be a descriptor), language use requires a minimum of brain power, and the complexity of a language is a fair indicator of a culture's general intelligence.

Therefore there are two main reasons why an association will not be made. If there is no affinity in terms of content then either the words exist for which no material can be found, or there is material but the speaker lacks the linguistic tools to express it (keeping in mind of course that words are only one way to portray something). If functionality is the problem, then the abstractive quality of either the material or the descriptor are beyond the immediate capacity of the operation. Fig. 9-1 illustrates both cases.

The box contains some papers and a 

He was hungry and thirsty and on top of that he had the 

Fig. 9-1: Examples of problematic associations, contextual (top) and functional (bottom). Try to read both sentences out loud and observe your thought processes. In the first the search is for a word representing something basically simple—but what? In the second the last item is not exactly unfamiliar but is clearly above the abstraction level of the rest. The question is, at what level should it be expressed?

Concerning contextual associations, 'absence' does not necessarily mean a void altogether; it can also refer to the inappropriateness in a given context, such as an idea which the speaker knows will be frowned upon when voiced. How far the aversion stretches depends upon the width and depth of the particular TS. For instance, with people who care about such matters the combination of red wine and fish is to be shunned, but all it takes is to substitute the word 'red' for 'white' and all will be well at the party. On the other hand, under a dictatorial regime far greater care has to be taken to stay away from the concept of democracy—merely avoiding the word itself is not enough. The width (links to other abstracts at the same abstraction level) and the depth (involving other abstraction levels besides the current one) are the product of the structure's generation process, something that can go back as far as early childhood and can include seemingly disparate concepts connected at the periphery of their associative halos.

Since the mining process by the container involves the con- as well as the sub-TSs, what gets articulated therefore is not always the result of an entirely conscious act. Indeed, so significant are the subconscious elements that the analysis of one's speech pattern is seen as an effective means to identify not only a certain individual but also the *type* of individual (more of this in Part II). Hegel's single mind then holds the abstracts which can be put into text because the author has the words for them and is personally unencumbered proclaiming the ideas they stand for. It also holds abstracts which are not spelled out because either no words exist or they are buried amid the layers of his persona partly for reasons not even known to himself. We know that among his subconscious there exist the ingredients which presented him with the realisation that not everything that drives a society is visible, otherwise there would not have been the affinity with such ingredients and the abstracts pertaining to the metaphors he chose to use. We cannot tell the content of those ingredients, we can only spot them through their functionalities. Equally we are aware that Hegel is Christian, and although he is prepared to establish essentially non-religious concepts across several abstraction layers, the top level is reserved for the relevant metaphysical abstracts.

As far as Hegel's re-representation of the group mind is concerned, there is the acceptance of a higher force that ultimately drives society, the recognition of acts performed by its citizens largely unknown to the rest, and other acts which are commonly known and form a secondary driving force due to their comprehensive appearance. The last category is performed by "World-Historical Individuals" [21], whose greater intellect may not be matched by better articulation, but whose actions nevertheless encompass a sufficiently profound depth of understanding to make a difference in the affairs of Man. Given the dynamics of containers etc we can say that the majority of effects is not caused by the few openly stated ideas that accompany a people's enterprise but by the myriad *un*stated notions which never make it to the light of day. "Nations are what their deeds are", he writes [22].

The subconscious parts of our psyche which are not declared openly but which nevertheless participate in the formulation of structures, are particularly significant when it comes to idealisations. To have an ideal means to have subjected a notion—already some form of abstract—to a further abstraction still. The PAT chains leading up to such a complex (and being a complex there would be more than one chain) are covering several abstraction levels and must therefore be considered prime candidates for functional associations (there may be contextual links along the way but the existence of the chains do not depend on them—yet functional links are made just on the basis of their level of abstraction). Because the initial reason for building a bridge was the similarity of the abstraction levels, there is no guarantee that this connection is a congruent one. Consequently, unless the articulation process forces access to those connected TSs and the associations are critically evaluated, the 'ideal' at the end of the chains may easily be an incongruent thought structure. It is what we have been describing in the context of containers and logic boundaries all along, only from a different angle.

Transposing the dynamic across to the group mind, we find a similar situation in a society's ideals. On a fundamental level every policy, every political platform, every cultural vision is an ideal that has been generated through the processes of container-building, illusion trees, and the functional associations which may have attached themselves along the way. The effect then represents the combined outcome of some visible, but mostly invisible activities performed in the services of that ideal. By "visible" I mean those activities which are common knowledge because they have been deemed necessary to be in the public arena. By "invisible" I mean all the acts that are performed in the background, either in intended secrecy or virtual secrecy, that is known only to a few individuals due to logistics. The articulation process in the wider society includes the media, public forums, and the acted-out policies by the state's instrumentalities; its width and depth relates to the society's inherent intellectual dexterity. If in the course of such a re-representation a connection should be found wanting, we can expect the appropriate elements to be modified. Note that the reasons for acceptance of the status quo, that is the absence of something critical to the elements, are quite similar to the single mind's circumstances: society at large does not *want* to recognise a problem because it is connected to concepts anathema to the culture, or it *cannot* do so because it does not have the means of access due to missing connections or missing material. For example, up until a few years ago the Brisbane City Council (in Queensland, Australia) did not react to complaints about nuisance dogs because there existed a general sentiment in the community that declared dogs to be "man's best friend" and so being critical was considered facetious at best (since then it has adopted an efficient and human-, not dog-friendly, approach). Presently the same city is under the governance of environmental protection laws which make it an offence to dispose of crows, a major source of noise. The law seems to be generally supported because the crow is associated with 'natural balance' and 'indigenous inhabitant', two very idealistic concepts in the current climate. The crow also happens to be a scavenger and therefore a potential disease carrier, but in the public's mind the connection between 'Torresian crow' and 'disease' does not exist. Based on the idealistic nature of the 'crow-natural balance-indigenous bird' combination there is no incentive to even research for disease and so the material for possible associations does not exist either (compare this particular configuration with the one in relation to the rat, at least in developed municipalities) [23].

It can be seen that the principle reason for maintaining an on-going policy despite its failure is the existence of supportive associations within the system, although much of that support is beyond public knowledge (we will examine how these links can be traced to their origins in Part II). For an ideal to be followed there must be conceptual elements confirming it. Note however that just as the associative configuration is not a static structure, so is the pattern comprising the ideal in a state of fluidity—after all, *its* kernel is the product of dynamic entities as well. Thus, for Hegel to describe the "Ideal" as the "ideal of Reason, of the Good, of the True" [24] is to acknowledge the inherently subjective nature of the phenomenon. In fact, for him there actually are "in the first instance only subjective" ideals, but they cease to be so once they touch the wider framework of a society's existence, the "universal reality" as he calls it. For us, recognising the mechanisms even there, the term 'subjective' remains valid. Further, considering the meanderings of approval and disapproval around the main theme he writes,

> The insight then to which—in contradistinction from those ideals [ie the more subjective ones]—philosophy is to lead us, is, that the real world is as it

> ought to be—that the truly good—the universal divine reason—is not a mere abstraction, but a vital principle capable of realising itself.

Notice the association complex suspended at a high abstraction level. Now observe the functional link from the complex to the centre of his container in the next sentences:

> This *Good*, this *Reason*, in its most concrete form, is God. God governs the world; the actual working of his government—the carrying out of his plan—is the History of the World.

From our point of view we can say that Hegel's religious container prompted the emergence of thought patterns which, through their associative potential, enabled him to connect his own abstracts with those representing society and so determine a situation where society configures its ideals derived from a stream of experiences, works towards these ideals consciously and subconsciously, all the while stoking the ideals with feed-back material and thus configuring it, then working towards these ideals consciously and subconsciously, all the while...

Two conclusions suggest themselves: (1) there can never be a realisation of an ideal, because the very process ensures a continuous modification of the goal; (2) the full nature of the ideal can never be expressed, because society inevitably lacks the resources to do so. We will have ample opportunity to confirm those points in Part II.

The word 'ideal' implies truth. Human beings do not adopt an ideal unless they can tell themselves that whatever it represents is true. We 'know' something is true if we have access to other abstracts which confirm the thought. They do so by forming associations that maintain the assertive value of the thought's elements. As our investigation of containers and illusion trees has shown, there are various methods of building links, some more, some less conducive to congruent structures. Once a concept is described as an 'ideal' it has gone beyond the mere confirmation stage through this or that supportive abstract; contextual and functional associations together have produced connections that just sync, in terms of con- as well as the sub-TSs involved in the exercise. Although the articulation of abstracts covers some participants, the majority of them does not present itself through words or other symbols but becomes known to the holder as mere sensations, the product of the emotive modules accompanying the established links. "It just feels right" is how most people would describe such a state of affairs, and they do so because to them the ideal is not so much a series of facts and figures but a positive emotional experience which features a combination of satisfaction-producing EMs. It is precisely that composition of lack of data and emotion that makes ideals such dangerous forces—the introduction of further data will be resisted if they threaten to diminish the emotional experience and jeopardise the comfort zone. The actual truth value of the abstracts involved in such a complex is irrelevant.

So the truth perceived in a mundane concept becomes 'Truth' writ large when it comes to ideals. It has turned from a cognitive transition point criss-crossed by PAT chains to a fortress with many paths leading into but only a very few leading out of it. Considering the overall format of the 'ideal' TP, from the ungrounded yet still expressed highly abstracted notions to the large invisible network of buried abstracts comprising more or less one's ego, we can see now why Hegel is able to state,

> For Truth is the Unity of the universal and subjective Will; and the Universal is to be found in the State, in its laws, its universal and rational arrangements. The State is the Divine Idea as it exists on Earth. [25]

Note the distinction between "universal" and "rational". In the single mind the overall state of the system entails mostly subconscious abstracts, the hidden yet active ingredients of TPs. The meaning of "universal" then incorporates more than that which can be articulated and so treated rationally. The entire set of structures thus produced consists of not only the abstracts but also the abstraction levels that house them. The following list summarises the relationships of the entities we are presently discussing.

1. The greater the intelligence, the more abstraction levels are possible.
2. The higher the abstraction level, the more attractive the representation at that level.
3. The more abstraction levels (and therefore the higher-reaching the levels), the greater the possibility of ungrounded abstracts.
4. The more ungrounded abstracts contribute to a structure, the more unassailable, ie powerful, the structure will be.
5. More abstraction levels mean greater connectivity, of the contextual as well as the functional kind.
6. Greater connectivity means greater diversity of abstracts and greater affinity between their respective associative halos.
7. Greater affinity means greater access to information.
8. More information means greater capacity to deal with the contingencies of life.
9. Greater capacity allows more complex systems.
10. More complex systems require, attract, and eventually spawn more abstraction levels.
11. Goto (1).

Against the background of the foregoing we can also understand what Hegel has to say about the lack of depth, especially when he talks about "Nature". In this he differs markedly from Rousseau to whom the natural state was synonymous with bliss. On the contrary, writes Hegel, it is the system of the State which enables us to realise our potential. Being dependent upon the raw contingencies of nature is not freedom:

> When man is spoken of as "free by Nature," the mode of his existence as well as his destiny is implied. His merely natural and primary condition is intended. In this sense a "state of Nature" is assumed in which mankind at large are in the possession of their natural rights with the unconstrained exercise and enjoyment of their freedom. This assumption is not indeed raised to the dignity of the historical fact; it would indeed be difficult, were the attempt seriously made, to point out any such condition as actually existing, or as having ever occurred. Examples of a savage state of life can be pointed out, but they are marked by brutal passions and deeds of violence; while, however rude and simple their conditions, they involve social arrangements which (to use the common phrase) *restrain* freedom. [26]

Obviously, the phenomenon of naive romanticism on behalf of idealistic representatives of advanced societies when it comes to simpler cultures is not a new one (and the cries of help coming from ignorant visitors suddenly confronted by local intransigence is one of its spin-offs). Some lines further down we read,

> Freedom as the *ideal* of that which is original and natural, does not exist *as original and natural.* Rather it must be first sought out and won; and that by an incalculable medial discipline of the intellectual and moral power. The state of

> Nature is, therefore, predominantly that of injustice and violence, of untamed natural impulses, of inhuman deeds and feelings. Limitation is certainly produced by Society and the State, but it is a limitation of the mere brute emotions and rude instincts; as also, in a more advanced stage of culture, of the premedidated self-will of caprice and passion.

Once again, including the last sentence, we have the connection between low abstraction levels and negative behaviour. A limited social horizon—acting without the regard for wider consequences—comes from a low perspective (see chapter 7). To understand and to accept a limitation in favour of the common good requires a certain amount of intelligence.

As the repeating, indeed recursive, nature of our list from above indicates, there is a mutual dependence between the perceived and the perceiver; that is to say, for a perception to occur there must be visible material, and a further developed perception induces action that procures more material. As a consequence, more densely populated regions tend to be more advanced because of the sheer need to organise—and organisation stimulates the mind. On that note Hegel remarks, in the context of the pressures that build in high-density Europe,

> But America is hitherto exempt from this pressure, for it has the outlet of colonisation constantly and widely open, and multitudes are continually streaming into the plains of the Mississippi.
>
> ..Only when, as in Europe, the direct increase of agriculturists is checked, will the inhabitants, instead of pressing outwards to occupy the fields, press inwards upon each other—pursuing town occupations, and trading with their fellow-citizens; and so form a compact system of civil society, and require an organised state. [27]

For Hegel to be able to write in that manner, he needed the situatedness within an environment that allowed higher perspectives. Someone standing on those Mississippi plains in the 1820's would have great difficulty contemplating a compact state—the faculty of exterior generalisation did not encompass something only lying in his future. Similarly, to reach the solution of a problem in one's mind requires the ability to evoke an abstraction level sufficiently high enough to hold the generalised form of the concept under focus. In Europe, what with the increase in complexity (in the single and group mind), the growth in diversity of material (people, tools, instrumentalities), there was an equally expanding need for more room to accommodate the thought structures and so their abstraction levels. Not only that, these abstracts must increasingly be productive and workable; in other words, they must be part of c-PAT-ls and congruent TSs. The metaphysical, the divine, the Ideal, got pushed further and further up the ladder, straining the intellectual capacity of their holders. Something had to give.

And now for a brief interlude.

It is always tempting to ascribe the character of an era to a few individuals who through their output seem to define their own time. The output is usually described in terms of a handful of single acts, and these acts become summarised under the person's name. The name turns into a label with which to classify the period: Caesar, Luther, Descartes, Napoleon, Hegel...

Yet even single acts only appear to be so; their description targets the most prominent feature of what is really a continuous series of steps, leading to and from the identified highlight. At each of these increments lie smaller but similarly significant clusters of events, each one of a particular influence, and each one contributing to the encompassing scenario without which the 'main' event would not be possible. Take the battle of Jena in 1806 in which Napoleon defeated the Prussians, certainly a decisive marker; but it was also a confluence of affairs involving Prussia, the Hapsburg regime, Britain, and so on down the line. It also includes French society and its capacity to support the army, an efficient set of generals who could turn the commands of their leader into success, as well as the soldiers themselves and their accompanying industry who managed to function adequately. It also rests on the combined abilities of all the individuals who knew how to listen, knew how to speak, knew how to do, and thus created a system where any destructive influences were kept at bay. A terse entry in a book such as "Battle of Jena—1806" is but a hook on which hang the lives, actions, intents, desires, fears, loves and hates of millions.

And minds. For a Napoleon to issue a command there needs to be a mind that can formulate a string of acoustic symbols, representing particular objects but at the same time also hinting at a myriad of un-said contingencies without which the objects themselves would be life-less—and the same resonance must exist within the minds of the recipients. A tiny notion here, a subtle inclination there, an element of a prompt rushing along in a stream of consciousness; they form the clusters that combine and produce the human act. The confluence of millions of elements, like drops in a great sea configuring the waves, the currents, and the shorelines, make what we call a culture. That grand dynamic is the atmosphere which a Caesar, or a Napoleon, breathe. Without it the singular event would not occur, and our books would be empty—indeed, there would *be* no books.

The collection of countless experiences from a world becoming ever more open to the minds of Europeans instilled the clusters, the thought structures, that in their filigreed weave incorporated the elements of newness, of surprise, but most of all *change*. People are not the same everywhere; animals and plants differ; even the landscapes vary from one place to the next. With curiosity turned in a new set of directions comprising the possibility of change, even the past became an object of analytical interest. In terms of mental configurations, it is an object in the true sense: an abstract item to be reached for, opened up, examined, and understood. A new dimension became available to the mind.

Hegel died in 1831. His philosophy used the sentiments derived from the concepts of change, movement, impermanence, from diversity and fluidity, to create a panorama which stretched from a beginning to an end. In it moved the players—the people, their societies, their cultures—in well-defined, if not always well-understood, paths. In Hegel's case we still find the 'closed system', that is a perceived world where a highest authority, imbued with the concepts of human values, situates each and every particle according to a static, permanent, plan. It is God how humans understand a god to be who accords everything its place, from the beginning until the end. Yet already we are given the space in between, expressed metaphorically through more or less intricate conceptions of Soul, Spirit, and the Divine Will, within which there is *change*. But still the change remained posited, an abstracted construct that lived through certain elements of the empirical information collected from elsewhere, rather than the direct conclusions reached via purposeful analysis.

But not for long. In 1859 Charles Darwin produced his *On the Origin of Species*, the first comprehensive introduction to the theory of natural selection [28]. Like identifiable names before him, Darwin breathed the atmosphere of his times—he was not the first to challenge the doctrine of immutability, as he makes abundantly clear in the *Preface*. His contribution consisted in the success of tying the strands of isolated data together, and under a clearly developed concept at that. He was not a philosopher—hence the interlude—and so did not play with his mind as it were; he relied on data, and drew the conclusions that suggested themselves. Just as the grand dynamic produced a Hegel, and Hegel added to the grand dynamic, so did the *Origin* contribute to the intellectual landscape of European civilisation. Right up to the present moment we find borrowed abstracts under the label of "xy Darwinism", where "xy" stands for this or that adjective; sometimes more, sometimes less in tune with the original source. In order for us to place such labels within the appropriate context, vitally important for any examination of thought structures, we are going to select a few passages that illustrate the thinking behind it; following the selections will be some comments relating the information to our own theme. We are not concentrating on the biological significance of Darwin's words therefore, but take heed of their cognitive implications. In doing so we cultivate the already featured notion of comparing recognisable circumstances on the outside with the dynamics on the inside (see chapter 8 for instance).

As the full title suggests, underpinning the entire theory is the idea of natural selection. Talking about the production of varieties in plants and animals against the background of their environment, Darwin says,

> Owing to this struggle for life, any variation, however slight and from whatever cause proceeding, if it be in any degree profitable to an individual of any species, in its infinitely complex relations to other organic beings and to external nature, will tend to the preservation of that individual, and will generally be inherited by its offspring. The offspring, also, will thus have a better chance of surviving, for, of the many individuals of any species which are periodically born, but a small number can survive. I have called this principle, by which each slight variation, if useful, is preserved, by the term of Natural Selection, in order to mark its relation to man's power of selection. [29]

There is a highly complex system therefore, composed of elements each carrying characteristic features; if in the process of mutual interaction one of these elements is found useful within a particular configuration, it will bring an advantage to the entire subsystem to which the element belongs. Three ingredients are significant: (1) the constant production of varieties and thus their continuous testing, (2) the ready response to such tests being successful, and (3) the tendency to preserve the new format. Substitute 'species' for thought structures, 'varieties' for the result of denominative TSs, and we have basically a PAT chain in action or, within a wider framework, a container. Next:

> As species of the same genus have usually, though by no means invariably, some similarity in habits and constitution, and always in structure, the struggle will generally be more severe between species of the same genus, when they come into competition with each other, than between species of distinct genera. [30]

Affinities between abstracts are in competition when their associative potential is most similar. In such a case the associative intensity of the TSs is heightened precisely because there are so many possible connection points between them. And in society at large, the severest antagonism exists between people who are closest in terms of their cultural characteristics (mostly this is obvious—when it is not, the connections are there albeit on a deeper level, as we shall see). Next:

> The elder Geoffroy and Goethe propounded, at about the same period, their law of compensation or balancement of growth; or, as Goethe expressed it, 'in order to spend on one side, nature is forced to economise on the other side.' I

> think this holds true to a certain extent with our domestic productions: if nourishment flows to one part or organ in excess, it rarely flows, at least in excess, to another part; thus it is difficult to get a cow to give much milk and to fatten readily. The same varieties of the cabbage do not yield abundant and nutritious foliage and a copious supply of oil-bearing seeds. When the seeds in our fruits become atrophied, the fruit itself gains largely in size and quality. [31]

We have already come across the distribution of emotive modules. They accompany the strands of PAT chains and influence a structure's significance in relation to others. There seems to be a similar balance as expressed above, making one TS more prominent at the expense of another, but more of this later.

There is a corollary to the foregoing sentiments, already alluded to in previous chapters, but worthwhile revisiting in the current context. A by-product of an ideological mentality is the perception that an assumed state is the final outcome of a purposeful development, the active agent in this case being the presumed authority; it occurs due to the particular dynamics exhibited by cognitive containers (see especially chapters 4, 8). Some examples, in descending order of structural extent: everything seems to have its place in nature, therefore "God arranged it all in his wisdom"; or, our own tribe / nation / culture is significant, therefore "we are God's chosen people"; or, workers don't have much money but business does, therefore "business strives to subjugate the workers"; or, I don't like what XY is doing, therefore "XY has it in for me". Unless there is a traceable series of actions, all instigated for the same reason (ie the expressed concern) those statements are illogical. In other words, just because there is a certain phenomenon does not mean it has been developed for the sole purpose of being that certain phenomenon. The phenomenon is first and foremost a perception and it is the perception which features the—perceived—characteristics; there is no direct relationship with the object it stands for. Rather, the object is a circumstance arrived at through the dynamic existence of conditions which happen to represent a particular state of affairs. Just as a variance in the species of a plant or an animal achieves purpose because it happens to explore a niche condition, and that condition coming into existence due to a multitude of extraneous factors, so do the abstracts arrange themselves within the framework of structures and patterns because their nature causes them to be in a certain degree of agreement with their local, cognitive environment. Consider Darwin's re-statement of the problem, further in the text:

> The theory of natural selection is grounded on the belief that each new variety, and ultimately each new species, is produced and maintained by having some advantage over those with which it comes into competition; and the consequent extinction of less-favoured forms almost inevitably follows. It is the same with our domestic productions: when a new and slightly improved variety has been raised, it at first supplants the less improved varieties in the same neighbourhood; when much improved it is transported far and near, like our short-horn cattle, and takes the place of other breeds in other countries. Thus the appearance of new forms and the disappearance of old forms, both natural and artificial, are bound together. [32]

Notwithstanding the role of containers and illusion trees the current state of a TS is due to the detailed and functionally extraneous arrangement of surrounding abstracts, and as such does not represent the outcome of some 'grand plan'. Note however that even in the case of containers etc there has to be the cognitive, environmental, intrinsic potential of the current structural configuration for them to come into existence.

Still in line with the aforesaid is the notion of *gradual* change, that is the minimal degree of variance under extraneous influences:

> As natural selection acts solely by accumulating slight, successive, favourable variations, it can produce no great or sudden modification; it can act only by very short and slow steps. Hence the canon of `Natura non facit saltum,' which every fresh addition to our knowledge tends to make more strictly correct, is on this theory simply intelligible. We can plainly see why nature is prodigal in variety, though niggard in innovation. [33]

Although Darwin's theory provides an explanation for the "slight, successive, favourable variations", it does not account for a definition of what exactly constitutes 'slight'—or indeed 'successive' and 'favourable' for that matter. As far as the organic side of life is concerned we need to leave the answer to the biologist; regarding cognitive dynamics the answer lies in the functional characteristics of abstracts. Two fundamental factors play a role—the affinitive quality of the denominative TS under focus and the degree of access on which the efficacy of the DTS rests. The first is a software problem (the TSs), the second one of hardware (the connectivity between the neurons). Combining the issue of variance (*a* variation remains because there is *a* niche, and not more), with that of change (*only* the necessary variation remains and establishes its advantage), and that of hardware / software quality, we can say that a variation in a TS will become established if the affinitive influence of its environment is commensurate with its potential for modification. Note the word 'established'; most abstracts never make it to their fruition (consider day dreams vs decisions on the inside, 'so much to do, so little time' on the outside). All this of course raises questions about what 'free will' means; a preliminary discussion can be found in chapter 7, and in Part II we will examine the issue once again in sequential detail. Next:

> The real affinities of all organic beings are due to inheritance or community of descent. The natural system is a genealogical arrangement, in which we have to discover the lines of descent by the most permanent characters, however slight their vital importance may be. [34]

A TS may manifest in a certain thought; another TS may produce a similar thought. This in itself does not mean that both thoughts are related. An analysis of the respective TSs may show quite different PAT chains are involved, each delineating from separate stems and basins. The apparent similarity comes from the conformity of internal conditions which produced both of them. For instance, the notion "I like" by itself means what it says, but as an appendage to the larger pattern of food consumption it is different from the TP dealing with one's friends (at least one should hope so). The configuration of abstracts leading to "I like" is similar in both cases, but the functionalities may differ. An equivalent example in biology would be a bird's wing and the foreleg of a horse; such bone structures are called homologous in relation to each other. Likewise, we can classify TSs as being *homologous* if they occur within a similar framework of cognitive entities although the nature of their resultant dynamics may be different. Next:

> Although I am fully convinced of the truth of the views given in this volume under the form of an abstract, I by no means expect to convince experienced

> naturalists whose minds are stocked with a multitude of facts all viewed, during a long course of years, from a point of view directly opposite to mine. It is so easy to hide our ignorance under such expressions as the `plan of creation,' `unity of design,' &c., and to think that we give an explanation when we only restate a fact. Any one whose disposition leads him to attach more weight to unexplained difficulties than to the explanation of a certain number of facts will certainly reject my theory. [35]

In terms of TS dynamics Darwin's sentiments are a familiar, if sad, theme. It is the basic reason why, after 100,000 odd years of *human* evolution, we still kill ourselves over a god. Next:

> The noble science of Geology loses glory from the extreme imperfection of the record. The crust of the earth with its embedded remains must not be looked at as a well-filled museum, but as a poor collection made at hazard and at rare intervals. [36]

For "Geology" substitute 'history' and consider the records gleaned from it. Although they become more and more substantial as we move towards the present, they still only represent the recorded, overt expressions. An event could have been recorded, but lost; displayed, but not recorded; held in one's mind, but not displayed—and we haven't even started on thought structures! More reason to stick to known facts and nothing else. A few sentences later:

> During early periods of the earth's history, when the forms of life were probably fewer and simpler, the rate of change was probably slower; and at the first dawn of life, when very few forms of the simplest structure existed, the rate of change may have been slow in an extreme degree.

The comparison with societal changes through history and around the world suggests itself immediately. Recall the list summarising the relationships of various cognitive entities earlier in the chapter; the more complex the intellectual surroundings, the more active the re-representing mind.

And finally, Darwin's closing remarks remind us of the frame of mind we should have, no matter what we are in the process of observing:

> It is interesting to contemplate an entangled bank, clothed with many plants of many kinds, with birds singing on the bushes, with various insects flitting about, and with worms crawling through the damp earth, and to reflect that these elaborately constructed forms, so different from each other, and dependent on each other in so complex a manner, have all been produced by laws acting around us. These laws, taken in the largest sense, being Growth with Reproduction; inheritance which is almost implied by reproduction; Variability from the indirect and direct action of the external conditions of life, and from use and disuse; a Ratio of Increase so high as to lead to a Struggle for Life, and as a consequence to Natural Selection, entailing Divergence of Character and the Extinction of less-improved forms. Thus, from the war of nature, from famine and death, the most exalted object which we are capable of conceiving, namely, the production of the higher animals, directly follows. There is grandeur in this

> view of life, with its several powers, having been originally breathed into a few forms or into one; and that, whilst this planet has gone cycling on according to the fixed law of gravity, from so simple a beginning endless forms most beautiful and most wonderful have been, and are being, evolved. [37]

To continue with the philosophies.

When we evoked the image of a grand dynamic earlier on, the visualisation of the atmosphere concentrated on its content from which thinkers drew the ingredients for their mental constructs. It suggested a rather one-sided process whereby the mind, under the influence of a particular configuration—eg a container—availed itself of whatever the environment had to offer. But as we have seen, the absorption of information depends on the affinitive potential the mind has; that is, information which does not resonate with any of the mind's TSs will not be properly imported. At the same time, the full spectrum of surrounding data consists of a multitude of small detail, and *that* has more chance of being absorbed because the affinity does not have to be quite as comprehensive. The result is a productive exposure to the manifold of intellectual material, and so without any particular initiative on behalf of society's members the general ambience of an era nevertheless establishes itself within the minds of a people.

Although there may not be a direct intent to articulate a sentiment, a changing context provided by the grand dynamic will still formulate the expression to suit the atmosphere. For example, Darwin's theory of evolution presented a most serious challenge to the Christian mindset in Europe (and continues to do so today in societies where religion dominates general thinking), because it undermined the state of permanence under their god's assumed plan. Still, even Darwin himself saw fit to use the word "Creator" several times throughout the *Origin of Species*. Yet only fifty years later the German naturalist Edmund Reitter, in the preface to his five-volume set of Middle European beetles, does not mention the word once. The closest he got to something like a god-type concept is the mentioning of a "divine, noble nature" ("..göttlichen, erhabenen Natur..") which presents itself to the keen observer in the form of complex and intricate, interdependent relationships amongst life's organisms [38]. The difference may seem only subtle, but if we look at the traditional appearance of containers which are centred around the idea of a metaphysical god-entity and the arrangements that have to be in place in order to even begin to describe the here and now, replacing a highly-abstracted 'god' with the idea of a divine nature has significant ramifications down the line. For one thing, the concept of 'nature' has to sync with any other abstracts which have been created through the process of intellectually dealing with nature's manifestations. Such manifestations, or rather the mental representations of them, tend to be grounded, congruent, and resist isolation into wholly separate elements.

The inherent potential for affinity in one's mind relies not only on the object and functional abstracts supplied by the environment; it is also due to the material composition of the brain itself (ie the quality and quantity of neurons, connections, and neuro-transmitters), as well as the mind's default abstracts. The former we leave to the neuro-biologists, but it is time to consider default abstracts in some detail (a great deal more will be said about them in Part II).

By now we are familiar with containers, illusion trees, and PAT chains. Populating all these kinds of configurations are of course the abstracts themselves, and their quantity and quality determine the efficacies of denominative TSs, PAT loops, and associative intensities. The fundamental difference between object and functional abstracts on one side and the defaults on the other are the latter's independence from the direct influence of containers, illusion trees, and so on. Recall their introductory description in chapter 4 under the section on Augustine. Default abstracts then are concepts—however singular or encompassing—which are the result of a pre-existent condition within the system that holds the mind. The abstracted concepts of say, hunger, thirst, or our sexual drive, are the result of a particular physical state in the system (ie the body). That physical state is the direct source for the default abstract; that is to say, no matter what the information processing of the mind might generate, the default abstracts are there to be reckoned with. This also includes illness, injury, and the specific functioning of one's body (eg allergies, adjustment to a climate). While a default abstract won't be removed—or its presence avoided entirely—through the influence of the others, such as would be the case with the object and functional versions, the eventual impact of the associated TSs will depend on the degree of significance exhibited by those TSs. For example, there can be a TS with a default abstract representing the sensation of hunger and the accompanying options (eg 'buy some more food'); there can also be a TS holding the information regarding shopping hours and the current late hour. Under normal circumstances the latter would override the former, and while the feeling of hunger won't go away no matter how long the list of closed delicatessens is, the person would nevertheless desist from pacing up and down the streets anyway. On the other hand, less well-defined notions can be more difficult to handle. Someone in love with another person may not be convinced that the emotion is not being returned because the received messages are too ambiguous compared to the TSs featuring the default abstracts of procreation, social belonging, and ego assertion (this might not be everybody's idea of what love is but we'll return to that topic in Part II). The outcome is a question of the respective set of affinities evoked by each thought complex, and hence the distribution of associative intensities across the relevant structures.

Since the system is a dynamic one, that is in a state of fluidity, the presence of default abstracts influences the operations of containers and illusion trees on a continuing basis. This is to be expected; after all, the interactive characteristics of abstracts are the same regardless of each type. Therefore, when it was pointed out that a certain idea can only then be properly understood if its situatedness within the PAT chain (ie its genealogy) has been identified, then the same goes for the involvement of defaults. Indeed there can be occasions when an expressed statement has its meaning changed considerably once the influence of a certain pre-existing state can be recognised. For instance, a person may show what is perceived to be a lack of compassion, but such a response could be due to several psychological conditions and not the otherwise functioning evaluation of the scenario at hand. The example demonstrates the potentially thin line between what is and what is not under the direct control of the conscious mind, but that issue goes beyond the current reference.

The pace of discoveries continued in the second half of the 19th century, and with it the growth in the diversity of material open to enquiring and speculating minds. Europe itself had mostly settled down, while the process of colonialism beyond its borders continued. Germany had become one state through the efforts of Otto von Bismarck and her administrative infrastructure modernised. The next significant military confrontation on European soil would not arrive until 1914.

During that time Friedrich Nietzsche (1844 - 1900) first lectured in classical philology at Basel University but had to retire in 1879 forced by his nervous disorders. For the next ten years he lived mainly in France, Italy, and Switzerland. His mind broke down completely in 1889 and he died in Weimar. Such an introduction is quite superficial from a psychological point of view, and in itself should not be misconstrued by implying some kind of deficiency in Nietzsche's writings. His bodily condition however touches on what we said before about default abstracts. There is a case to be made for discerning the influence of more than purely mental constructs when it comes to the arrangement of his cognitive patterns and their inherent dynamics. To explore these connections two of his works are valuable, while at the same time telling us much about the views of the man. One is *Thus spoke Zarathustra* [39], the other *Beyond Good and Evil* [40].

Walter Kaufmann, the translator of the *Zarathustra* acknowledges Nietzsche's ailments as playing a significant role in understanding the philosophy. For that reason he cites Stefan Zweig, describing his character [41]. Here are some excerpts:

> "Carefully the myopic man sits down to a table; carefully, the man with the sensitive stomach considers every item on the menu: whether the tea is not too strong, the food not spiced too much, for every mistake in his diet upsets his sensitive digestion, and every transgression in his nourishment wreaks havoc with his quivering nerves for days. No glass of wine, no glass of beer, no alcohol, no coffee at his place, no cigar and no cigarette after his meal, nothing that stimulates, refreshes, or rests him: only the short meager meal and a little urbane, unprofound conversation in a soft voice with an occasional neighbour (as a man speaks who for years has been unused to talking and is afraid of being asked too much)."

And further, remarking on his boarding house rooms,

> "Otherwise only books and manuscripts, and on a tray innumerable bottles and jars and potions: against the migraines, which often render him all but senseless for hours, against his stomach cramps, against spasmodic vomiting, against the slothful intestines, and above all the dreadful sedatives against his insomnia, chloral hydrate and Veronal. A frightful arsenal of poisons and drugs, yet the only helpers in the empty silence of this strange room in which he never rests except in brief and artificially conquered sleep. Wrapped in his overcoat and a woolen scarf (for the wretched stove smokes only and does not give warmth), his fingers freezing, his double glasses pressed closed to the paper, his hurried hand writes for hours—words the dim eyes can hardly decipher. For hours he sits like this and writes until his eyes burn."

The development of a thought pattern, influenced as it is through the experiences from the outside, also occurs against the background of one's physical constitution. In most cases the latter is more or less a common denominator and it is the experiences in life which make us different from each other. If one's physical condition however does not fall within the norm, the development of every TP will reflect such circumstances. The result will depend upon the outside event together with the type of cognitive dynamic under focus. While a full list of possibilities is impossible because of the sheer variety of the human condition, a brief example should highlight some of the principles. Consider the process of social integration in the mental development of an adolescent boy. In the majority of cases the playing of sport features—looms?—large, and the ritual of initiating a game, the selection process of players, the agreement on certain rules, and the playing of the game itself all provide a particular stage on which the articulation of the boy's identity, and therefore the inner awareness of his own persona, goes through its paces. If the boy happens to be at least an average player, each game, apart from teaching something about the sport itself, will contribute to the boy's idea that he, the persona, is generally accepted and wanted in society. As such one could welcome the experience since it gives him confidence and certain social skills. Now suppose in his earlier childhood the boy had suffered from poliomyelitis and, although recovered, one of his legs has remained shorter. As a player he won't be as desirable as the rest, and therefore *his* version of an acceptance experience will not be in the form of the constant, yet superficial inclusion in a sporting team. He will learn that to gain friendship involves more than knowing how to kick a ball; in his case it reaches deeper to levels of personality not necessarily accessed by the others. His learning curve in those matters can be painful, but through it he will have achieved a more profound understanding of human relationships than the 'normal' playground gives to most youngsters.

This leads to the following principles. If articulation means the focusing on a certain aspect, that aspect must have been perceived worthwhile articulating, otherwise why bother? In other words, the more common the abstracts which happen to be embedded in a TS, the less likely will there be a need to bring them forth. Consequently, whatever perception accompanies the general ambience of the structure, it will not be challenged because it does not represent an issue. Unusual circumstances however will bring the substructure to the fore, and in the process of articulating it (whether externally through words or internally through thinking about it) certain relationships and phenomena will be ascertained which otherwise would have stayed hidden. One effect of default abstracts that are outside the norm therefore is the heightened necessity to examine one's mental imagery.

What if the result of such an examination is the explicit understanding of one's condition—that is to say, the realisation in so many ways that one really *is* different from the rest? In the example above it will be the awareness beyond the playing of sport, that there is one venue of meeting people, of forming acquaintances and friendships, of getting to know about how things are done, of making use of sport-related means to judge, act, project, and respond, which is completely inaccessible—and shall be so for the rest of one's life. As the boy grows older he will learn to circumvent such obstacles by searching for and finding substitutes that offer at least comparable outcomes. The paths to these outcomes however will have to be different, and so there is the need to develop separate internal structures—one for his specific condition, the other in line with the common world. Therefore another effect of such default abstracts is the cultivation of inner detachment.

The ability to be detached is well and good (given the circumstances), but what if in the process of its maintenance there emerges a discrepancy between what is useful for the person and what is acceptable to other people? Suppose our boy stays away from the sports grounds and begins to pursue interests more fascinating then punching balls. This may be good for him (and eventually good for society) but it won't do much for his popularity in the class room: so he learns how to pretend a love for sport anyway. Hence the third effect is the creation of a mask.

Once the person knows how to handle a mask, its advantages become obvious as the skill increases. Just as the wise king in so many tales dons a mask to learn the truth from his subjects, so can the boy in our story avail himself of his acquired dexterity in order to learn from his fellow humans—and discover things he would not have seen otherwise. Since most of what is acceptable stays in the open, it is the odd, the undesirable, the shameful, which one tends to find in this way. And because human values can only ever be subjective, the sheer pathos accompanying these discoveries will not be lost on the observer either. So the fourth and final effect of extraordinary default abstracts is the emergence of cynicism. In summary we can say that such predispositions lead to an introspective, detached, complex, and cynical personality. Depending on other strengths and weaknesses, they can result in an alcoholic misanthrope, or a machiavellian Buddha, or a lord unseen.

Against the background of the aforesaid it makes sense not to confine oneself to Nietzsche's writings but to look at the effect these had on the readers, especially those charged with productive interpretation such as a translator. It was the reason Kaufmann saw fit to include Stefan Zweig's description of the philosopher, and similarly we are going to quote Kaufmann to show how interesting one or the other aspect can be to an outsider. Talking about Nietzsche's letters, in such contrast to the ambience evoked by Zweig's picture of the external, Kaufmann notes, "Yet the clue to these letters, as also to *Zarathustra* and some of the last books, is that they are the work of a thoroughly lonely man" [42]. Then he continues,

> Now pleasant and polite, now such that arrogance is far too mild a word—and yet his feelings of his own importance, painfully pronounced even in some very early letters, was of course not as insane as it must have appeared at times to those to whom he wrote. Resigned that those surrounding him had no idea who he was, and invariably kind to his social and intellectual inferiors, he sometimes felt doubly hurt that those who ought to have understood him really had less respect for him than his most casual acquaintances.

We also hear of his "several hasty proposals of marriage, apparently followed by a real sense of relief when the suggestion was refused politely": an attempt to tune in to commonality, and a somewhat ambiguous response to such efforts, torn between a reaction in line with what one expects and what one is ready for, perhaps? Or take the following line, referring to the expressed melodrama in *Zarathustra*, "But often painfully adolescent emotions distract our attention from ideas that we cannot dismiss as immature at all"; and then, "And what we find again and again in *Zarathustra* are the typical emotions with which a boy tries to compensate himself."

In tandem with carrying a mask is the tendency to express oneself in metaphors; not just to make a point more elegantly, but to hide its centre. The entire text of *Zarathustra* is of course one big metaphor, beginning with the title itself.

Zarathustra was a Persian prophet (c. 660 - c. 583 BCE) who, at the age of fifteen, removed himself to a mountain cave where he had a vision. The diligent reader may find it useful to delve into the mental dispositions of people living in that era to steal a glance or two behind the scenes (chapters 2 and 4 might be particularly helpful). The vision involved Ahura Mazda, the supreme being overcoming evil in the world, and Zoroastrianism eventuated from there. The religion is based on the fundamental belief in the world's inherent goodness, but divides it into the followers of truth and the followers of the lie [43].

With Nietzsche, Zarathustra is a male figure who retreats to a mountain cave at the age of thirty and stays there for the next ten years. Following his exile he wanders down to meet humanity and proclaims his insights. Those statements, couched in metaphorical settings as well as sentences, comprise the length and breadth of the book. Employing a character in a story in such a consistent way points to a desire by the author to firstly present a range of views (which is fairly obvious), but secondly to hide behind that persona as the occasion may require. It is similar to someone conducting a polemic and asking a question by saying, "Some commentators have argued..." rather than "I would argue...". The advantage lies in the affinity domains of that 'other' compared with the image of oneself, and the imagery of the 'other' acts as a lightning conductor to some expected antagonism from the outside (indeed, in extreme cases an additional persona can be created *within* one's mind to protect the self from one's own attack). Zarathustra then is Nietzsche's lightning rod, and through him he makes his views known to the reader.

Central to all this is the idea of the "overman". Rightly preferring the word to 'superman' in the face of the latter's more naive modern meaning, Kaufmann informs us about the word's semantic—if not to mention semiotic—contrast to anything affixed to 'under-' ('unter-'). Expanding on that aspect one might add that the syllable 'over' ('über') not only refers to something going 'over' something else, but also a state of being 'above and beyond', that is exhibiting a condition inherently superior to its ordinary version (which is quite different from what we expect from Superman of comics fame, namely doing what everyone else does only so much better). By the same token, 'under-' denotes not only an inferior quality, but also something which acts 'underneath' the terms assumed in the context. For instance, to use the word '*unter*schieben' when talking about someone's affairs means implying a meaning which the other might not have wished to display openly. It can also produce the notion of a fundamental, and hence significant, approach, similar to the English '*under*take' (eg 'to undertake a task'). For someone who is fond of word plays the possibilities must be inviting.

So "Man is a rope, tied between beast and overman—a rope over an abyss", Nietzsche writes, or Zarathustra says [44]. The existence of Man as an ordinary human is temporary, this is "what is great in man". Zarathustra loves those who cannot help themselves but cross that abyss, even though they may "go under". In doing so they display a spirit that is irresistible, their virtue being the master whom they obey, not slavishly but joyfully, and so becoming masters themselves:

> "I love him who does not hold back one drop of spirit for himself, but wants to be entirely the spirit of his virtue: thus he strides over the bridge as spirit.
>
> "I love him who makes his virtue his addiction and his catastrophe: for his virtue's sake he wants to live on and to live no longer." [45]

We have the metaphorical articulation of separate thought patterns in action. On one side the real self, noble but forced into isolation, represented by the ego-container and propelled by the virtue that is one's own, on the other the re-representation of oneself in relation to the outside, yet ideally governed by the self-same virtuous principles. If only the former could merge victoriously with the latter! But if not victoriously, then at least the synthesis would be an achievement in itself even though it means the annihilation of the unsuited, hitherto isolated ego. A complex interplay between two TPs, where the appreciation, if not reverence, for achievement takes second place to the success of one's own values, which in turn must give way to the ultimate glory, the sheer power to win—in Nietzsche's case only ever an abstract.

Note the push-pull dynamics shown here: I want something, but I am not confident it is rightfully mine; I am going for it, but I am not sure I can get it; I may be successful, but if not someone else's success is fine too.

Or take the episode on the flies in the market place. To be there is not healthy, because "Far from the market place and from fame happens all that is great..", where "the air is raw and strong" [46]. But here it is inhabited by poisonous flies which crave one's blood and "sting in all innocence":

> They flatter you as a god or devil; they whine before you as before a god or devil. What does it matter? They are flatterers and whiners and nothing more.

And:

> Indeed, my friend, you are the bad conscience of your neighbours: for they are unworthy of you. They hate you, therefore, and would like to suck your

> blood. Your neighbours will always be poisonous flies; that which is great in you, just that must make them more poisonous and more like flies.

It is a simile on the solitary profound and deep vs the shallow multitude. But just as there is a connection between the "raw and strong" air of solitude and nobility, and between the teeming masses and their poison, there is equally a link between greatness and the *degree* of poison. The poison is not being created as a purposeful response to one's greatness (ie the flies "sting in all innocence"), but the relationship exists so it can drive the overall dynamic of background-dependent superiority. The primary and secondary TP (the real-self and the exterior-related one) were created as a response to the unequivocal existence of the ego as well as the unequivocal existence of the outside world—and now they need each other.

The creation of a container structure is a function of available abstracts which prove affinitive with some of the contents of the current state of the structure. It stands to reason that a considerable number of acquired abstracts will be shared by both containers, just as there are abstracts which would only be affinitive with either the primary or the secondary type. Those which are unique to either one or the other will have their own domains, to be accessed via the available routes in the respective container. Those that are shared by both types will have their domains accessible via either container, and have now become virtually homologous with respect to their differing genealogy (ie the PAT chains to be followed depending on which container acts as the current reference). We say "virtually" because it is really one and the same abstract, but its effect changes according to which structure is chosen. To return to our sports example, there are abstracts comprising a TS that denotes the idea of 'team'. But 'team' can be linked to the sports ground as well as to a group doing scholarly research. How the mind will treat either depends on the genealogy of 'team' when connected to the 'sports' TP or the 'research' TP. The structure surrounding 'team' itself is the same in both cases. Generally speaking, we can call configurations that serve two different contexts *twins*, since they represent two separate entities yet share a considerable amount of their components. Like containers, illusion trees etc, they are very common since they are evoked every time a particular context demands a modified behaviour (eg using different language in different situations, dressing accordingly, and so on), although some serve more serious purposes than others.

Fig. 9-2 shows the schematic evolution of a single container, a twin configuration, and a still wider environment featuring multiple occurrences of the same phenomenon. The emerging complexity in the picture hints at the actual conditions in the mind, where the identifiable functionalities we have been discussing are continuously building and disappearing, and thus forming the fluid environment we sensate as consciousness.

One of the major characteristics of any twins are the domains accessible to one or the other. It is the content and abstraction level of what is eventually linked to the shared entity that would be of interest to an observer. In our sports example the connections are simple, but in real life things can get complicated very quickly. Consider the issue of 'double standards'—one TS affecting a certain behaviour, but associated with mutually disparate structures that could consist of containers, illusion trees, and meandering PAT chains, all subject to varying degrees of associative intensities, reciprocities, and denominative sub- and con-TSs. All the principles surrounding the behaviour of cognitive structures are still valid, except that now their applicability needs to be considered against two contexts. Thus a PAT chain may be congruent in one twin, but incongruent in relation to the other. Such a dissimilarity is in fact the first sign of twins, and the presence of an incongruent version in addition to a grounded configuration points to the problematic influence of a TS or TP. For example, the choice of food can be rational, that is administered through grounded abstracts dealing with its freshness, ingredients, and so on. Should there be a day on which the same fresh, wholesome dish will not be eaten, then this is due to an incongruent chain of abstracts adding an irrational context to the issue. Hence looking for incongruities of that nature is a productive way of finding problem areas in human behaviour, on the individual as well as societal level.

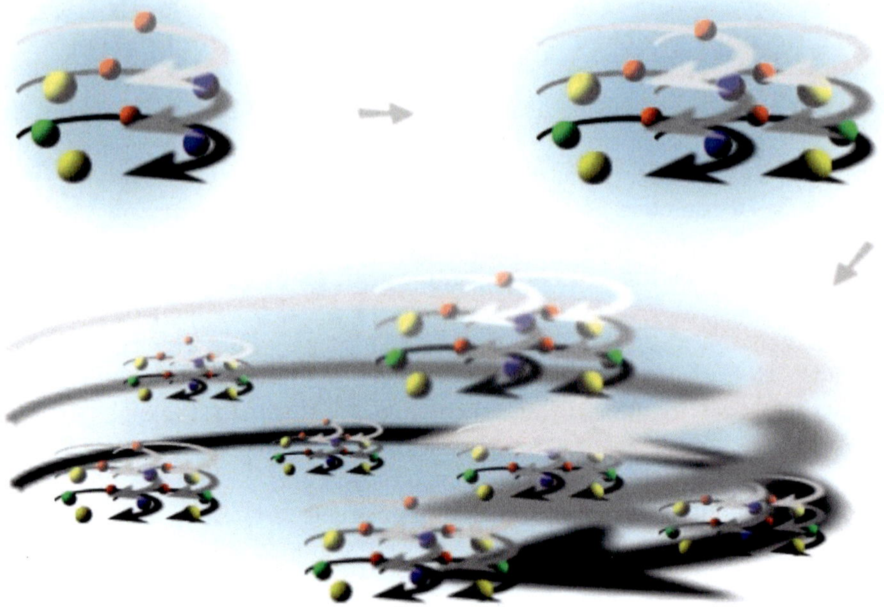

Fig. 9-2: A schematic depiction of a single container, twins, and multiple occurrences of twins embedded within a wider configuration. Three abstraction levels are indicated in every case. In the container twins the single abstract colour at each level denotes a shared abstract, demonstrating its virtual homology. For more details see text.

Nietzsche's twin domains are of course farther reaching than a few abstracts here and there. His use of metaphors provides the bridge between the to-be-expected perceptions by outsiders and the hidden content of his inner self. Yet even the outside world is not homogenous, and to understand a metaphor requires at least a principled similarity straddling both contexts. Consider the following passage, from the section *On the land of education*:

> With fifty blotches painted on your faces and limbs you were sitting there, and I was amazed, you men of today. And with fifty mirrors around you to flatter and echo your display! Verily, you could wear no better masks, you men of today, than your own faces! Who could possibly find you out? [47]

Kaufmann informs us that here Nietzsche writes against "modern eclecticism and lack of style" [48]. While the expression "fifty blotches.." does not encompass the sumtotal of the philosopher's polemic, it must nevertheless form part of the overall tapestry. The translator draws parallels to the Greek *Odyssey* and points to Goethe's *Mephistopheles* as a help in understanding the text. This may well be so; but first and foremost it is a question of the type of domain one chooses to include. Notwithstanding Kaufmann's scholarly expertise in such matters, a completely different perspective could seize on the number 'fifty' and find significant traces of numerological 'wisdom' in its current usage. Or, having talked about masks earlier on, I could emphasise its inclusion here as further proof of some secret agenda.

The existence of deep-reaching cognitive twin structures, prompting the use of metaphors, can make any interpretation a precarious exercise indeed. One needs only to look at the writings of Nostradamus to recognise the ambiguity of meaning, but also the attractiveness of such styles to those who are antipathetic towards grounded facts.

In another section (*The soothsayer*) Zarathustra recounts a dream he has had. One part should suffice:

> "And amid the roaring and whistling and shrilling the coffin burst and spewed out a thousandfold laughter. And from a thousand grimaces of children, angels, owls, fools, and butterflies as big as children, it laughed and mocked and roared at me. Then I was terribly frightened; it threw me to the ground. And I cried in horror as I have never cried. And my own cry awakened me—and I came to my senses." [49]

After that speech a disciple "whom he loved most" provides the interpretation for Zarathustra. According to him the coffin represents Zarathustra himself, "full of colourful sarcasms and the angelic grimaces of life". But Zarathustra is also like a "thousandfold children's laughter" entering all death chambers, "laughing at all the night watchmen" and so on and so forth.

To be sure, the words presented here are taken out of context and are only part of the dream narrative. But the imagery invoked through those lines is idiosyncratic at best, and the combination of a bursting coffin with unsettling sounds (I presume they *are* unsettling) which produces multifaceted laughter certainly needs an audience with a tandem train of thought. As any examination of people in different cultural settings shows, what is considered funny, horrible, delightful or disgusting remains a matter of semantic format inside the person's mind. So rather than a description of detail, the dream story and its interpretation scene is Nietzsche's way of dealing with his two inner selves: Nietzsche the writer allows Zarathustra-Nietzsche to spill forth a mix of emotional symbols, to be re-fashioned by the disciple-Nietzsche into a somewhat more agreeable format. One thinks of a Hieronymus Bosch painting, its phantasms being a reflection of the psychic turmoil a 15th century toiler must experience in the face of pious hysteria.

The way the dream is presented reminds us again of the push-pull dynamics mentioned earlier. Here is Nietzsche, eager to communicate his message, but delivering it through Zarathustra; yet Zarathustra himself—the conscious, hence accountable entity—stands aside in favour of his dream persona, and even then the contents are being shunted through someone else—the disciple who just happens to be 'most loved'. It is a relatively stark example of an ownership most tenuously submitted, which in principle tends to accompany twin configurations. Extending the relevance we can take that section itself as an expression kept at a distance, while somewhere else in another cognitive scenario the virtuous spirit is proclaimed with pride—and both articulated through an alibi.

Sometimes the information route is as zigzagged as this dream, sometimes its values are forced into juxtaposition to each other. In part 26 of *On old and new tablets* he writes,

> O my brothers, who represents the greatest danger for all of man's future? Is it not the good and the just? Inasmuch as they say and feel in their hearts, "We already know what is good and just, and we have it too; woe unto those who still seek here!" And whatever harm the evil may do, the harm done by the good is the most harmful harm. And whatever harm those do who slander the world, the harm done by the good is the most harmful harm. [50]

On the face of it, the words are a rejection of do-gooders. Knowing something about the personal life of Nietzsche however allows us to add another dimension. The TP governing the references to the outside world would include abstracts related to pity and empathy shown towards a relatively poor and suffering man whose pursuits seem somewhat obscure. Such a pattern sits alongside another which represents the actual self, an intelligent, insightful, if not visionary persona, whose physical ailments are pushed into the background as much as possible. The real self, one can imagine, utterly resents the patronising ambience which surrounds its twin and so attacks the relevant manifestations. These and similar sentiments coming from Nietzsche point to a twins-induced dichotomy of one set of values standing in opposition to another, where depending on the momentary context now one, now another act as the projected standard. In this case we have an antagonism towards those who are in need of assistance, while in the next instant there can be a lament about how badly the world treats the "overman".

One confirmation of the real self's intended nature can be found in *On the great longing* [51]. Kaufmann informs us it is a "hymn to his soul" [52], and as expected is full of metaphors. Zarathustra addresses himself to his soul (a second opinion about oneself?):

> O my soul, I washed the little bashfulness and the nook-virtue off you and persuaded you to stand naked before the eyes of the sun. With the storm that is called "spirit" I blew over your wavy sea; I blew all clouds away; I even strangled the strangler that is called "sin".
>
> Much enhanced the soul becomes; for instance:
>
> O my soul, I understand the smile of your melancholy: now your own overrichness stretches out longing hands. Your fullness gazes over roaring seas and seeks and waits; the longing of overfullness gazes out of the smiling skies of your eyes. And verily, O my soul, who could see your smile and not be melted by tears?

For someone of Nietzsche's disposition any social commonalities can be burdensome, if not objectionable. Much of the antipathy towards 'goodness' is therefore being linked with Christianity—being the common religion in those parts—and its doctrine of 'love thy neighbour' and others, against which he expresses his objections repeatedly throughout the text. The sexual repression executed by the church would not have escaped him, and especially in his case the combination of social alienation and physical importunity must have made for a frustrating sex life. Several times we find the word "naked" in some metaphorical role. In German the word is "nackt", and its sheer sound together with the specific posture of the mouth while saying it can be quite suggestive in terms of the underlying semantics, particularly when the mind perceives sex as something forbidden and nasty while at the same time confronted with its intense demands. To use "nackt" often can then be a form of semiotic masturbation, doubly enhanced by that latter act's very own seductiveness. It is significant that in all of the book's anti-Christian rhetoric not once is there any reference to the actual excesses committed by that religion; the criticism is restricted to abstracts only.

For instance, in the section *On priests* he denounces the class as "shepherds [who] themselves belonged among the sheep", and who embrace the suffering as an indulgence:

> But blood is the worst witness of truth; blood poisons even the purest doctrine and turns it into delusion and hatred of the heart. And if a man goes through fire for his doctrine—what does that prove? Verily, it is more if your own doctrine comes out of your own fire. [53]

Notwithstanding the metaphorical nature of the passage, given the degree of antagonism one would expect something more substantial. The time had long passed when a discussion of Christian fanaticism carried the threat of death, and the accounts of its victims existed for all to see (not that they were ever deemed to be a secret to begin with at the time). To demonstrate just what we are talking about, here is the description of one such event, taken from Henry Moore's *Protestant martyrology* and found in turn amongst a compendium of such inventiveness [54]. It concerns the execution of the Rev. Dr John Hooper, Lord Bishop of Gloucester, in the year 1555 on the charge of heresy.

> It was a chilly, gray, and windy morning. In order to minimise his suffering the bishop had placed two bundles of reeds and two bladders full of gunpowder between his trussed up legs. After the doomed prelate had said his prayers, and the executioners had murmured their uneasy apologies, the final act of the tragedy began. Said Moore:
>
> 'Command was now given that the fire should be kindled; but, owing to the number of green faggots, it was some time before the flames set fire to the reeds. The wind being very adverse, and the morning very cold, the flames blew from him, so that he was scarcely touched by the fire. Another fire was soon kindled of a more vehement nature: it was now the bladders of gunpowder exploded, but they proved of no service to the suffering prelate. He now prayed with a loud voice.
>
> ' "Lord Jesus, have mercy upon me; Lord Jesus, have mercy upon me; Lord Jesus, receive my spirit!"
>
> 'And these were the last words he was heard to utter. But even when his face was completely black with the flames, and his tongue swelled so that he could not speak, yet his lips went till they were sunk to the gums; and he knocked his breast with his hands until one of his arms fell off, and then continued knocking with the other while the fat, water, and blood dripped out at his finger ends.
>
> 'At length, by renewing of the fire, his strength was gone, and his hand fastened in the iron which was put around him. Soon after, the whole lower part of his body being consumed, he fell over the iron that bound him, into the fire, amidst the horrible yells and acclamations of the bloody crew that surrounded him.
>
> 'This holy martyr was more than *three quarters of an hour* consuming; the inexpressible anguish of which he endured as a lamb, moving neither forwards, backwards, nor to any side: his nether parts were consumed and his bowels fell out some time before he expired. Thus perished in a manner the most horrible that the rage of Hell itself could not devise, in a manner more barbarous than exercised by wild American Indians to their prisoners taken in war, the right reverend father in God, Dr John Hooper, for some time Bishop of Worcester, and afterward of Gloucester.'

Instead, Nietzsche avails himself of a fictitious Zarathustra, itself an earlier member of one Middle Eastern pathology, as an instrument to poke at the embers of one of the three surviving others.

Zarathustra is a parable. It seems that Nietzsche felt the need to re-phrase his ideas after all, for its successor, *Beyond Good and Evil*, "says the same things", "but differently, very differently" [55]. In the introduction to this work Michael Tanner does not agree and cites the obvious variance in style, but also the way Nietzsche addresses his audience. If *Zarathustra* is a story, then *Beyond..* is a report. While in the story there is the tiered substitution of presenting agents, driven by the homology of certain abstracts, in the report-like version the effect of the substitution as a contextual buffer has been removed. Unsurprisingly then the tone in *Beyond..* is much more direct and to the point. Still with reference to Christianity Nietzsche writes,

> The Christian faith is from the beginning sacrifice: sacrifice of all freedom, all pride, all self-confidence of the spirit, at the same time enslavement and self-mockery, self-mutilation. There is cruelty and religious Phoenicianism in this faith exacted of an over-ripe, manifold and much-indulged conscience: its presupposition is that the subjection of the spirit is indescribably *painful*, that the entire past and habitude of such a spirit resists the *absurdissimum* which 'faith' appears to it to be. [56]

Continuing, he criticises the loss of intrinsic understanding by "modern men" when it comes to the symbolistic meaning of a "god on the cross", which he sees as a "gruesome superlative" the ancients would have appreciated in its true sense. In terms of the cognitive dynamics of containers and illusion trees it is highly significant what he has to say about the effects of a religion which makes the suffering of the inferior one of its central tenets:

> It is the orient, the *innermost* orient, it is the oriental slave who in this fashion took vengeance on Rome and its noble and frivolous tolerance, on Roman 'catholicism' of faith—and it has never been faith but always freedom from faith, that half-stoical and smiling unconcern with the seriousness of faith, that has enraged slaves in their masters and against their masters. 'Enlightenment' enrages: for the slave wants the unconditional, he understands in the domain of morality too only the tyrannical, he loves as he hates, without nuance, into the depths of him, to the point of pain, to the point of sickness—the great *hidden* suffering he feels is enraged at the noble taste which seems to *deny* suffering.

If we take "slave" to mean a person of diminished ability and hence potential, then considering the results of low-quality input and stimulation, the mental structures representing one's surroundings will be compromised as well. There will be less expansion regarding the content of the structures, as well as less abstraction levels to underpin the comprehension of such content. The influence does not necessarily have to come from without—the adoption of a faith, regulating the thinking processes in a restricted manner, confining notions to the immediate relevancies of the doctrine rather than opening them to any other affinities; these are measures which act internally, hence the phrase "innermost orient". The laissez-faire attitude of enlightened minds, making use of their enhanced mental prowess through the application of higher abstractions, is anathema to the constricted thinker for whom the optionality of theoretical panoramas does not exist. Bereft of these "nuances" he enforces the direct confrontation between himself and the thus flattened idea into a symbiosis which is anything but harmonious and develops into a painful mismatch of the self and the world at large. The conclusion, when it emerges from such a cross-hatch of idealism, ignorance, and fervour, is the acceptance of pain as proof of one's righteousness—an insidious inversion of the facts. Nietzsche again:

> Never and nowhere has there hitherto been a comparable boldness in inversion, anything so fearsome, questioning and questionable, as this formula: it promised a revaluation of all antique values.

To be situated in a culture which is infused with certain values makes it difficult to acquire the distance that allows an objective view. Nietzsche was able to extract himself from the ambience because the exercise per se was already a familiar one—his cognitive twin structures saw to that. Hence he found it within him to express the following:

> Wherever the religious neurosis has hitherto appeared on earth we find it tied to three dangerous dietary prescriptions: solitude, fasting and sexual abstinence—but without our being able to decide with certainty which is cause here and which effect, or *whether* any relation of cause and effect is involved here at all. The justification of the latter doubt is that one of the most frequent symptoms of the condition, in the case of savage and tame peoples, is the most sudden and most extravagant voluptuousness which is then, just as suddenly, reversed into a convulsion of penitence and a denial of world and will: both perhaps interpretable as masked epilepsy? [57]

The developmental sequence of the above phenomena will be described in Part II, but already we know enough of the dynamics to understand what is going on in principle. Solitude prevents the formulation of grounded PAT chains and so clears the way for the creation of top-heavy illusion trees, supported by the metaphysical—ie ungrounded—content at the highest abstraction levels. Fasting, apart from depriving the physical body and brain of necessary nutrients, sets up self-fulfilling i-PAT-ls that equate the act of suffering with the perceived status of accomplishment (see the previous paragraph about slaves). Sexual abstinence imposes a push-pull quandary based on need and shame which, because of the fundamental nature of sex itself, interferes with just about anything the human mind—being the product of a physical, organic system after all—can devise. Given such negative circumstances, the resultant TSs will only ever form tenuous associations with emotive modules of any kind, since the confirmation of the ideas they give rise to are subject to the vagaries of real-world conditions and the perception of them. Therefore an affirmative emotion complex will sooner or later give way to an antagonistic one, and the associated TS will be influenced accordingly. One set of EMs gets replaced with another one, and what ensues is a reorganisation of sometimes entire thought patterns. A similar situation exists in society at large; wherever one finds a community beset with entrenched piety, there the exact opposite behaviour forms can be found as well, equally entrenched. Boys and girls segregated in the most ridiculous fashion, and child prostitutes around the corner; men and women affirming their religious 'love' at regular meetings, and the bashings of 'perverts' in side streets; human intellect being declared in their god's image, and the attack on facts and reason.

A mind that is torn between delusion and reality; a society incapable of finding its centre because more than one has been enforced; it is a scenario intuitively evident to someone whose own condition has generated a similar mind set. Would not such a person be in a particularly suitable position to visualise the consequences? And, delineating from the singular to the multitude, that multitude where all this is replayed yet all over again, would not such a person be driven to state,

> ..But worse still: the higher a type of man a man represents, the greater the improbability he will *turn out well*: chance, the law of absurdity in the total economy of mankind, shows itself in the most dreadful shape in its destructive effect on higher men, whose conditions of life are subtle, manifold and difficult to compute. Now what is the attitude of the above-named two chief religions towards this *surplus* of unsuccessful cases? They seek to preserve, to retain in life, whatever can in any way be preserved, indeed they side with it as a matter of principle as religions *for sufferers*, they maintain that all those who suffer from life as from an illness are in the right, and would like every other feeling of life to be counted false and become impossible.
>
> ..And yet, when they gave comfort to the suffering, courage to the oppressed and despairing, a staff and stay to the irresolute, and lured those who were inwardly shattered and had become savage away from society into monasteries and houses of correction for the soul: what did they have to do in addition so as thus, with a good conscience, as a matter of principle, to work at the preservation of everything sick and suffering, which means in fact and truth at the *corruption of the European race*? Stand all evaluations *on their head—that* is what they had to do!
>
> ..Men not high or hard enough for the artistic refashioning of *mankind*; men not strong or farsighted enough for the sublime self-constraint needed to *allow* the foreground law of thousandfold failure and perishing to prevail; men not noble enough to see the abysmal disparity in order of rank and abysm of rank between men and man—it is *such* men who, with their 'equal before God', have hitherto ruled over the destiny of Europe, until at last a shrunken, almost ludicrous species, a herd animal, something full of good will, sickly and mediocre has been bred, the European of today... [58]

Here is the noble self speaking, the one that looks full of disgust at what the other has turned into when entangled with the outside. That other, the one which knows what it is like *not* to be in a position of strength because this is not *its* reality, can thus complain,

> Conversely, there is perhaps nothing about the so-called cultured, the believers in 'modern ideas', that arouses so much disgust as their lack of shame, the self-satisfied insolence of eye and hand with which they touch, lick and fumble with everything; and it is possible that more *relative* nobility of taste and reverential tact is to be discovered today among the people, among the lower orders and especially among peasants, than among the newspaper-reading *demi-monde* of the spirit, the cultured. [59]

A torn mind indeed.

It is debatable whether without a Nietzsche there would have been much of a difference in the European mentality in the decades to come. Homologous thought structures exist and operate whether inside or out, and whether someone is there to articulate them or not. Affinity-seeking as the system is, the confluence of conditions would eventually attract certain dissidents who, in their strain towards the symbol of 'independence', construed the thing behind as a reflection of their own inner ambiguity and so felt comfortable with a man who had turned the affliction into a virtue.

But first a niche had to be filled. A niche created by the manifestation of productive thought and translated into what became known as the age of industrialisation—and just waiting for the right variance.

10. Anatomy of a universe

Karl Marx - Edmund Husserl - Carl G. Jung

The mother holds up a spoon. "*Spoon!*" the young child says. She holds up a fork. "*Fork!*".

In front of a class of school children the teacher points to a mark on the map. "*London!*" the students call out. He points to another spot. "*Singapore!*".

During an autopsy the coroner nods towards a discolouration on the temple. "Trauma—caused by a blunt instrument. Possible damage to left frontal lobe."

There is a common thread to those scenes: visual information enters the brain via nerve strands and gets matched with already existing abstracts, affinities are ascertained and in due course an affirmation is articulated. And yet there are differences.

In the first example the incoming abstracts get compared with the object abstracts pertaining to 'spoon', and the latter are also associated with relevant FAs. These FAs will be relatively simple; most likely they would not comprise the material consistency, specific weight, and exact dimensions of the item. But what they do entail is enough for the child to use the thing as a spoon. In the second case what the children see is not an aerial view of a city nor any buildings at all; what they do perceive is a symbol that they have learned to interpret as standing for something which, in different circumstances, *would* present itself as a landscape of houses and roads and so on. The FAs associated with the OA representing the spot on the map tell them that. And lastly, the specialist knowledge of the coroner, that is the thought pattern consisting of the array of injuries inflicted upon a certain body part and their symptoms, is like a filter through which the imported abstracts from the examined corpse stream, their functionalities interacting with the generalised data, ie the higher abstractions, of the said TP.

Let us put all that into a less familiar perspective. A spoon is made to fulfill a certain function, and so it will have a particular size and shape congruent to its intended purpose. Since just about everyone else is able to relate to its characteristics, introducing a spoon will prompt humans to use it; that is to say, articulate its functionality within an accommodating context. The result is that some food gets eaten in a particular manner. Likewise with the marks on a map, except that those marks are being treated as the symbols they stand for, and so again we can say that the class articulates their functionalities within an accommodating context (hence no-one would try to walk into the map to visit London, and making a similar mark on a desk would be quite meaningless). The coroner will write his report based upon his interpretation of the corpse's appearance and then the police will act accordingly, somebody will get charged, the court system springs into action, and the instrumentalities of a society are being activated.

Therefore, to make a spoon is more than the fashioning of a certain item, because it does not stop there. To construct a map is more than covering a surface with colours and symbols. Such things possess an inherent functionality which prompts humans to act in accordance with it and in the process of doing so modify their environment. As maps are used to relate, identify, and position, their ultimate effect depends on the capacity of the user (assuming of course they are correct maps to begin with). In other words, no matter how profound the understanding of say, 'London' is in someone's mind, it is not the map that will be affected but the user and whoever the user interacts with. Thus the functional flow goes from the map to the user, and onwards; there is no reciprocity as such (an error might be discovered leading to an improvement of the map, but by that stage it is now a different map as indeed is the context of the treatment).

In principle then, a human product contains its appearance as an object but at the same time its potential for use. In the human mind it is represented as an object abstract as well as certain functional abstracts, their quantity and quality depending upon the inherent capacity of the mind's structures. Thus its FAs can be said to exploit the potential of the product, and such a process gives rise to additional TSs of a very particular nature. The flow of functionality moves from the product to its potential, to the representation in the mind, to the FAs there and eventually to the response a mind is able to come up with—it is impossible to reverse the direction. Although the receiving mind can employ various abstraction levels to process the information, the item's potential for use remains the same. It can only be realised through the application of an agent (in this case the human mind), but it exists nevertheless. The abstraction levels and their content, that is the sophistication and complexity of the agent's operation, determine the effectiveness of the object, but not its potential—that is given through its existence per se. They do however determine the ratio between a product's *latent* potential and its *realised* potential; that is, the greater the associative halo of a product, the greater its realised potential and the less its remaining latent potential.

This has several significant ramifications:

whatever the potential of a product, it requires an intelligent agent to realise its potential;

whatever the capacity of the intelligent agent, the potential of the product stands;

just because there is no conscious awareness of the potential at some point in time and space, that does not mean the potential does not exist at that point (note that 'product' means material artifacts as well as ideas and concepts; 'potential' means what can be done with it as well as how it can be interpreted; 'capacity' includes processing at conscious as well as sub-conscious levels);

the life-cycle of a product begins with a zero realised potential and full latent potential, its latency diminishing in the subsequent process of association building and hence utility finding;

during its life-cycle it *may* achieve zero latent potential and full realised potential, ie all its possible applications have been discovered;

a product's latency is in inverse relation to the complexity of its environment;

since the latent potential includes all those applications which have not been discovered yet, high latency in general means lots of untapped resources but a paucity in applications, and a low latency means few untapped resources but a high diversity in applications.

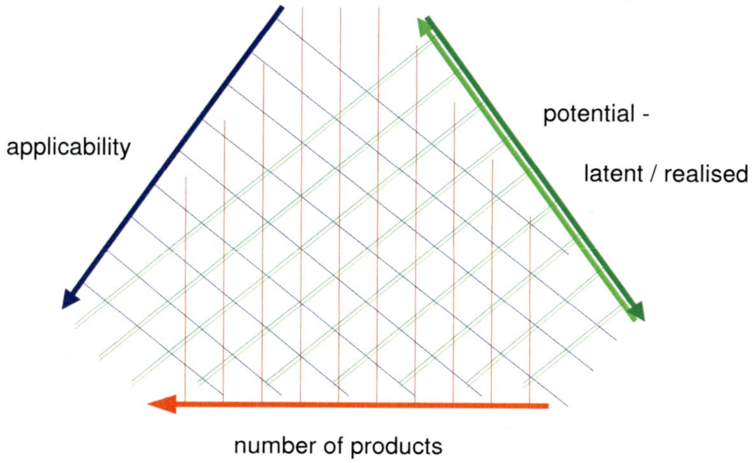

Fig. 10-1: Effectiveness chart (e-chart) of a system showing the relationships between the number of products, the degree of latent (light green) and realised (dark green) potential, and the degree of applicability within it. It should be used starting with "products" and proceeding anti-clockwise. For more detail see the text.

The interplay between the number of products in a system, its general latency, and its degree of applicability, is shown in the form of a chart (Fig. 10-1). Note that in terms of the overall interpretation the respective values can only ever be relative ones, since it would be impossible to ascertain absolute magnitudes (for example, while the number of products may be known, the full potential is not, and therefore the latency can only be expressed relatively in terms of changing one or the other parameter). For the same reason the chart should only be read anti-clockwise starting with "number of products". For example, it would be meaningless setting "applicability" at a certain value (even if such a value could be derived), then assuming a particular degree of latency and so arriving at a given number of products as the conclusion. A similar problem exists when going from "latency" to "products" and then trying to conclude the degree of applicability. There are exceptions though as we shall see in later chapters.

The late 19th century introduced novelties on a range and scale hitherto unequaled in human history. The cumulative effects of the sheer variance and the magnitude of the newcomers even influenced the already existing elements of the culture. Whether it was an invention such as an engine, or an idea such as a new format of education, they all percolated through European society and enabled associations and hence perspectives unparalleled in previous times. The mechanisation of society triggered the appearance of people who fulfilled the potential of machines—from their design, to their material requirements, to their products, to their organisation and administration. The growing of science triggered the appearance of people who performed its intent, made use of its results, cultivated its administrativa, and thus accentuated their society's demographics. The potential of the systematically improving means of communication was just as systematically realised through news media, political activism, and affirming of identities. Such variance and magnitude was to become the hallmark of European and hence Western culture, and set the tone for modernity. Since our concern is the conceptual construct, in this chapter we will scan three such artifacts—one pertaining to society, one to the mind, and one to the ego. Because artifacts consist of elements, each of which has a potential and each of which has been realised to some degree, to understand the whole one must come to grips with its parts. As such the result is an anatomy, and because we are dealing with concepts, that is large-scale perceptions of the world, we are justified calling the respective result an anatomy of a universe.

The industrialisation of society caused the development of demographic strata which filled the niche created by the functionality of machines and the interaction with them. We have the product in the form of mechanised processes, followed by the realisation of its potential through the building of factories and their staffing and administrative procedures, followed by the applicability of their output through usage, trade, and distribution of labour. Sooner or later these sub-systems were bound to spawn reactions, but just as the emergence of the overall phenomenon occurred in a multi-tiered fashion, so did the response consist of many facets. The affinity-seeking process of clustering similar elements together could be observed through the formation of interest groups and, depending on the focus of their interests, pursued directions sometimes in contradiction to each other. One such cluster was the appearance of what was eventually called the working class once the self-defining process achieved a sufficiently strong identity. As in the previous centuries, thinkers of a certain disposition exercised their own congruence and further defined what they perceived. One of these was Karl Marx (1818 - 1883), and we are going to use *The Communist Manifesto* [1] as a guide to that particular universe. The text first appeared in 1848, mainly in collaboration with Friedrich Engels who was its editor.

It is a characteristic of any concept that first and foremost it relies on a preliminary assertion. By doing so it places itself within a certain environment which from then on acts as its basin. In terms of cognitive dynamics the assertion is the container, and the establishment of the container is done through an illusion tree. This is the mechanism by which the source or sources for the initial assertion become abstracted in order to serve as a high-level entity under the auspices of the container. In the *Manifesto* the initial assertion is the following:

> A spectre is haunting Europe -- the spectre of Communism. All the Powers of old Europe have entered into a holy alliance to exorcise this spectre: Pope and Czar, Metternich and Guizot, French Radicals and German police-spies. [2]

The 'haunting spectre' derives its meaning from sources that go back centuries. First under the guise of Christianity's enemies, then as an excitement- and curiosity-building notion the idea of the 'spectre' features elements of mysterious threats, unholy machinations, and the 'great unexplored' beyond our 'five senses'. It also finds additional sustenance from the advance of science, in that the unknown can be rolled back and bit by bit explained. As such the idea is basically an uncomfortable one, but tinged with a certain amount of protective self-assurance. This spectre then is haunting (all proper spectres *haunt*) Europe, ie one's land, one's home. Not one's home per se, but the "old" version, thereby implying a development which favours the new and fresh at the expense of the old and useless. Thus the old, useless contingents band together, and to call that "holy" accentuates its old-fashioned character together with the opposition to the spectre with *its* anti-church undertone—only something that is "holy" would act against something like a spectre. The nature of the old-fashioned and also hostile contingent is further underlined by the examples: the authority figures of the establishment (pope, czar, etc), and what is meant to be taken as negative, namely radicals and police spies. Note that the only element *without* any further definition is "Communism"; rather, its characterisation is achieved through the text of the paragraph. The TSs comprising 'spectre', 'old Europe', 'holy alliance' and the rest culminate denominatively in another TS which consists of an abstracted form, and that form becomes the abstract elevated to a higher level from where it associates once more with similar abstracts there—a typical illusion tree.

The links at the crown of the tree are established in the next paragraph:

> Where is the party in opposition that has not been decried as Communistic by its opponents in power? Where the Opposition that has not hurled back the branding reproach of Communism, against the more advanced opposition parties, as well as against its reactionary adversaries?

Any abstracts at the top of the tree are results of PAT^{2+}s, otherwise they would not be relatable. Therefore a "party in opposition" does not stand for any group directly and in particular; the same goes for "opponents", "Opposition", and the sweeping generalisation of interest groups to the left and right of the reference (whatever that might be). The PAT chains comprising "opponents" etc can be followed to a certain degree, but the chains are not c-PAT-ls eventually ending up in grounded PAT^1s.

Containers usually feature more than one illusion tree—after all, they are attractors. So here is another one:

> The history of all hitherto existing societies is the history of class struggles. [3]

The use of absolutist language has already been discussed in chapter 8 under Rousseau. The phrase "all hitherto existing societies" is a further example and indicates the presence of i-PAT-ls. Here it serves a special purpose: by linking the concept of 'class struggle' in its relatively new context to all of the past, its significance is heightened considerably. The reader is prompted to regard 'class struggle' as something which has gone on all along, even subsuming its name-less existence in the past as an added confirmation of the underlying injustice. The text goes on to say,

> Freeman and slave, patrician and plebeian, lord and serf, guild-master and journeyman, in a word, oppressor and oppressed, stood in constant opposition to one another, carried on an uninterrupted, now hidden, now open fight, a fight that each time ended, either in a revolutionary re-constitution of society at large, or in the common ruin of the contending classes.

Without delineating the context of "freeman", "slave", etc etc these words are high-level abstracts associated with each other at the top of the illusion tree. None of the elements listed above contain any detail, they are all pure symbols. In order to create 'slave' for instance, PAT[1]s consisting of this or that instance and the accompanying situation are allowed to cluster together and elevate the result to a high level—again an illusion tree. Each of those trees represents a relatively highly abstracted notion, inaccessible to c-PAT-ls, and removed from grounded links. As such they make for a particularly attractive building material for further trees, which is what they have been used for here.

Assembling many trees, creating numerous links with other structures, threatens the simplicity of the declared concept. It is the complex idea, that which by necessity features many associations through which its validity can be ascertained, that is most likely to be congruent and rational. It is the simple, compacted statement that through its very own nature annuls productive references. Hence the tendency for simplification is another sign of i-PAT-ls. We find one in the description of the bourgeoisie, when its image is being reduced to a two-dimensional silhouette:

> Society as a whole is more and more splitting up into two great hostile camps, into two great classes, directly facing each other: Bourgeoisie and Proletariat. [4]

Although the erstwhile definition of what constitutes 'bourgeoisie' draws from a manifold contingent of samples, here they are shorn of any differences with only the most common denominator remaining. On the other hand, 'proletariat' is a new entity and therefore homogenous anyway. The simple 'bourgeoisie' object is now ready to be attached to the characteristic of 'great hostile camp', with the result that *any* member of that group is automatically an enemy of that singular opponent, the proletariat. That is to say, the 'bourgeoisie' illusion tree, acting within the container of 'communism', and being modified by the denominative TSs mentioned earlier, has now become a DTS itself in relation to its associations, 'society' and 'proletariat'.

It is a characteristic of ideological tracts that its significant players are reduced to single-layered entities. In such a format the element can be moved to a suitable abstraction level from where subsequent associations to other similar abstracts can be made. In other words, the diffusion of content occurs *after* the creation of the idea and serves the affinitive network from then on. Compare this development with a more rational version: there the idea emerges through the resultant definition via the PAT chains, and could possibly be a shared structure derived from many strands. The final notion represents the singular conclusion, a summary of the previously diffused information having come together. Let us recall from earlier discussions of illusion trees and containers that these structures do not entail incongruent links as such; they can be the starting point of a hypothesis, an assumption, or a theory. It is only later through subsequent associations with ungrounded structures that the idiosyncratic nature becomes firmly established. Hence we can say that in an ideological framework the PAT chains move from the base of illusion trees towards their top, to be gathered into a single notion which gets moved to higher abstraction levels in another illusion tree and becomes associated with similarly derived high-level abstracts. In a rational environment the initial assumptions are being confirmed by grounded abstracts and therefore the PAT chains move from the top of illusion trees downwards, *then* to be gathered into combined strands which may well be elevated to higher levels. Fig. 10-2 illustrates the difference.

Fig. 10-2: Left- several illusion trees attract abstracts from their domains, compact them into single elements which are used to repeat the action with respect to another illusion tree (an ideological configuration). **Right**- a similar dynamic except that the illusion trees feed grounded abstracts from their operations to the main tree (a rational configuration). At the top of the main trees the high-level abstracts find associations within their new domains once again.

There follows a brief summary of the growing extent of the industrialised economies, indicating the impact the mechanisation of production had on commerce and wealth but also on parallels such as navigation and colonisation. No doubt the developments within a system the size of an industrial nation are interdependent to a significant degree. The description of such a state of affairs however needs to reflect at least that interdependence per se; in the *Manifesto* this is not the case. The above named factors are posited as giving "rapid development" to "the revolutionary element in the tottering feudal society"—a cause-and-effect relationship which is arguable at best, but in any case harnessed to support the simplified meaning of 'class struggle' (for example, colonisation by some power has occurred with and without industrialisation, and therefore cannot as such be taken as an inevitable effect of the latter).

Some sentences later the same means of creating an artificial cause-and-effect scenario is employed:

> Meantime the markets kept ever growing, the demand ever rising. Even manufacture no longer sufficed. Thereupon, steam and machinery revolutionised industrial production. The place of manufacture was taken by the giant, Modern Industry, the place of the industrial middle class, by industrial millionaires, the leaders of whole industrial armies, the modern bourgeois.

The above paragraph conveys a meaning in which the revolutionised production is a consequence of rising demand, and so the "giant, Modern Industry" took over the traditional "place of manufacture" together with the "industrial millionaires" who have now usurped what used to be the "industrial middle class". Firstly, note the associative connotations of the TS "giant.." in relation to what is called a "place.."—a monstrous thing (non-human) doing away with a simple 'place' where something is being made ("manufacture" denoting the human element). Also, "industrial middle class", again a TS holding the concept of active human labour, put into contrast with the cognitive complex of "industrial millionaires", whose chief characteristic is therefore wealth per se without any semantic addendum hinting at what could be called a justifying effort. The compacting of abstraction levels removes nuances of meaning, and what gets lost and what remains is a function of the controlling agent, in this case the DTP governing the container.

Secondly, in the effectiveness chart earlier on we have seen the interplay between the number of products, their potentials and applicability within a given system. It is the number of products which determines the level of potential, and it is that level which in turn defines the degree of applicability the system is able to come up with. That sum of its parts, ie the whole, represents the overall effectiveness of the system. Hence the products, that is the machines, make for an increase in the potential, and through the process of finding more and more ways of applying such technology the latency is diminished and thus heightens the applicability—the whole of the system has become more effective. A more effective system is capable of generating more products, but note that at that relatively early stage those products' latency is still very high—only in time to come will the full range of potentials be realised (if ever). In the present case we have a system which comprises material objects (the machines etc) and abstract representations (the technology and the extent to which it is understood). Since the latter includes also a form of re-representation, there is a time delay between the implemented state of the former and the implemented state of the latter. So, if we combine the principles of complexity relationships as enumerated in the discussion about Hegel in chapter 9 with those of the system effectiveness, we can say that (1), a more effective system is the cause of an increase in complexity and vice versa, and (2), at any given stage of the system's life cycle there will be a discrepancy between the complexity of the system's material aspect and its overall representational side, with the former lower than the latter. A corollary to (1) is the non-linear rate of increase because *all* the active agents in the system are involved in the rise of complexity; a corollary to (2) is the ambiguous impact of the re-representational part with regards to the material, that is to say not every realised potential of a product in theory will be translatable into an ongoing enhancement of the system in practice. Some realised potential can be translated into measures which change the direction the system is evolving, although the information responsible for the change has come from the current state. Whatever the kind of change, it will of course redefine the parameters.

A further extension of the above principle of discrepancy is the role the size of the discrepancy parameter plays. The greater the difference, the greater the potential of the entire system; the smaller the difference, the less options are there for the system to develop and to adjust to any eventualities. If the difference is zero, the system will have stagnated and continue to exist in its current format until a disturbance makes the continuation impossible. (A note on the side: this paragraph itself represents a set of high-level abstracts atop an illusion tree. At the base of the tree are the principles mentioned before, themselves high-level abstracts coming from the tops of other illusion trees. Going back further we reach the various pieces of information we have been talking about in this and previous chapters, gleaned from our analyses; hence they are grounded. All this is part of a container which from page 1 onwards has accumulated the relevant data.)

Returning to the *Manifesto*, the concept of 'bourgeoisie' represents a flattened TS which has been linked with other higher-level abstracts. As such some of its potential has been realised, but because of the flattened nature of the structure the point of the structure's overall latency will be brought down to zero sooner rather than later, at which point the discrepancy parameter will have reached zero as well. Practically speaking, this means the concept of 'bourgeoisie' is being used in such a way that through its very definition any possible re-interpretations and therefore possibilities to act will have been exhausted and that sub-system stagnates—a rigid ideology has come into existence.

The text goes on in a similar vein:

> The bourgeoisie, wherever it has got the upper hand, has put an end to all feudal, patriarchal, idyllic relations. It has pitilessly torn asunder the motley feudal ties that bound man to his "natural superiors", and has left remaining no other nexus between man and man than naked self-interest, than callous "cash payment." It has drowned the most heavenly ecstasies of religious fervour, of chivalrous enthusiasm, of philistine sentimentalism, in the icy water of egotistical calculation. It has resolved personal worth into exchange value. And in place of the numberless and feasible chartered freedoms, has set up that single, unconscionable freedom -- Free Trade. In one word, for exploitation, veiled by religious and political illusions, naked, shameless, direct, brutal exploitation.

The "feudal, patriarchal, idyllic relations" are in themselves compacted abstracts and therefore make it possible to be associated with the current main element, 'bourgeoisie', in a functional sense. The absence of c-PAT-ls removes the possibility of critical influences, and the incongruent chain is able to continue. There *is* more detail, but it can be found only in the added links of inherently negative TSs, such as the idea of something being "pitilessly torn asunder", or "naked self-interest", or "the icy water of egotistical calculation", or "naked, shameless, direct, brutal exploitation". Positive details are introduced if in their juxtaposition they convey negativity once more, such as "the most heavenly ecstasies of religious fervour", or "chivalrous enthusiasm". The link-less abstracts converge upon another illusion tree which becomes the conduit for the abstract 'free trade' (accentuated in the text by the use of first letter capitals), to be used for further associations under the general direction of the main container. Since 'free trade' has not been dealt with before and therefore does not contain any detailed elements, its potential is limited to begin with. As a cognitive sub-system its latency is minimal, the potential for other interpretations is virtually non-existent, and therefore 'free trade' is *bad*, full stop.

The associative quality of 'bourgeoisie' endows its TS to be denominative accordingly. Thus Marx can say,

> It has accomplished wonders far surpassing Egyptian pyramids, Roman aqueducts, and Gothic cathedrals; it has conducted expeditions that put in the shade all former Exoduses of nations and crusades. [5]

One could perceive the Egyptian pyramids etc as an achievement. Nevertheless, in the present context under the influence of the 'bourgeoisie'-DTS they become somewhat tarnished. Not only that, the functional abstracts accompanying the respective ingredients of 'pyramid', 'aqueducts' and so on are subject to the same domain. Although the exact number of FAs will depend on the extent of their owner's knowledge, one common set would be those pertaining to 'challenge', 'commitment', and 'success'. Because the structures are denominative as well as associative, any future relationships with other TSs based on associations between similar FAs will act upon those TS too. The result is that a concept which entails something like 'achievement' will, at the very least, be tinged with a negativity it most likely does not deserve.

The idea of 'change' itself is placed under scrutiny. Since the *Manifesto* describes a state of the world that has come about through change, and that state is considered to be undesirable, the inference is made that change is to be opposed as well:

> Constant revolutionising of production, uninterrupted disturbance of all social conditions, everlasting uncertainty and agitation distinguish the bourgeois epoch from all earlier ones. All fixed, fast-frozen relations, with their train of ancient and venerable prejudices and opinions, are swept away, all new-formed ones become antiquated before they can ossify. All that is solid melts into air, all that is holy is profaned, and man is at last compelled to face with sober senses, his real conditions of life, and his relations with his kind.

'Change' is linked with 'constant revolutionising', 'uninterrupted disturbance', and 'everlasting uncertainty'. The past, although containing "prejudices and opinions", is nevertheless "ancient and venerable", and we are led to bemoan the impossibility of situations to "ossify". The denominative impact is further enhanced by associating the 'past'-abstracts with 'solid' and 'holy'.

Just as a connection is construed between a negative image and the idea of 'achievement' per se, a similar tactic is employed between the geographical extent of a phenomenon and relevant concepts. A couple of paragraphs later Marx writes,

> The bourgeoisie has through its exploitation of the world-market given a cosmopolitan character to production and consumption in every country.
>
> ..All old-established national industries have been destroyed or are daily being destroyed. They are dislodged by new industries, whose introduction becomes a life and death question for all civilised nations, by industries that no longer work up indigenous raw material, but raw material drawn from the remotest zones; industries whose products are consumed, not only at home, but in every quarter of the globe.
>
> ..In place of the old local and national seclusion and self-sufficiency, we have intercourse in every direction, universal inter-dependence of nations. And as in material, so also in intellectual production. The intellectual creations of individual nations become common property. National one-sidedness and narrow-mindedness become more and more impossible, and from the numerous national and local literatures, there arises a world literature.

Note the resultant demonisation of such concepts as 'world-market', 'cosmopolitan character', 'new industries', 'remotest zones', 'distributed consumption around the globe', 'universal inter-dependence', 'world literature'. Note equally the idealisation of 'old-established national industries', the union of the former with 'civilised nations', 'individual nations', even 'national one-sidedness and narrow-mindedness'. Given the structural format of ideologies (whatever their actual contents), it should come as no surprise that a similar categorisation of concepts into good/bad can be found amongst the more extreme members of the political Right.

Sometimes it becomes difficult at first to ascertain what actual value is being conveyed. For instance:

> The bourgeoisie has subjected the country to the rule of the towns. It has created enormous cities, has greatly increased the urban population as compared with the rural, and has thus rescued a considerable part of the population from the idiocy of rural life. Just as it has made the country dependent on the towns, so it has made barbarian and semi-barbarian countries dependent on the civilised ones, nations of peasants on nations of bourgeois, the East on the West. [6]

On the face of it, to be rescued from the "idiocy of rural life" seems a good thing, as does being exposed as a "barbarian and semi-barbarian" country to "civilised" ones. However, given the effects of the negative associations mentioned before (and we are still within the domains of those illusion trees and containers), such DTSs will also influence the structures featured here. The result is the idealisation of less developed regions in favour of more developed ones, and the removal of their image from the meanings of 'barbarian' where they had been placed for a reason, to cognitive areas having other, more positive connotations—not, it should be noted, for reasons of their intrinsic characteristics, but due to their representational form being shifted into other cognitive domains. In the same vein the paragraph demonstrates a demonisation of the 'West', while leaving the 'East' as its victim.

Another example of seemingly balancing one perspective with another occurs later, when the effects of industrialisation and its administering class are presented. We can use the word 'seemingly' because although there is a different angle, its final impact upon the overall view of the audience is determined by the biased distribution of emotional epithets (the "brutal exploitation" and others), and this has been achieved already. We read,

> The bourgeoisie, during its rule of scarce one hundred years, has created more massive and more colossal productive forces than have all preceding generations together. Subjection of Nature's forces to man, machinery, application of chemistry to industry and agriculture, steam-navigation, railways, electric telegraphs, clearing of whole continents for cultivation, canalisation of rivers, whole populations conjured out of the ground -- what earlier century had even a presentiment that such productive forces slumbered in the lap of social labour?

By itself the paragraph's content is formatted in terms of c-PAT-ls, that is every sentiment can be traced and revisited in a context different from the present. Having identified such information as indicative of a radical shift in culture, the description extends its focus by including further characteristics, again linked with 'bourgeoisie' but placing them somewhat outside the present:

> Into their place stepped free competition, accompanied by a social and political constitution adapted to it, and by the economical and political sway of the bourgeois class.
>
> A similar movement is going on before our own eyes. Modern bourgeois society with its relations of production, of exchange and of property, a society that has conjured up such gigantic means of production and of exchange, is like

> the sorcerer, who is no longer able to control the powers of the nether world whom he has called up by his spells.

As soon as the present is referenced the negative connotations are evoked once again and the achievements of technologically advanced societies are compared with the fate of the arrogant sorcerer who cannot control the forces he had called up. The theme of Man trying to master Nature is not a new one, and Goethe's poem *The Sorcerer's Apprentice*—the source of the analogy here—is just one of the more famous treatments. If we consider the overbearingly Christian environment of the European intellect, a mentality particularly strident in the stilling of its curiosity, we can understand the antagonism of the Church which over the centuries has been directed against such endeavours. Any ideology, whether spiritual or secular, relies on the acceptance of its doctrine for perpetuation, and the stronger the political presence of the faith the more intense the pressure on its members to shy away from analytical understanding. It is significant that the *Manifesto* defends the status of its message by aligning itself with similar expressions of fear.

The manipulation of the images holding 'bourgeoisie' and 'proletariat' entails the creation of a twin structure. On the one hand we have the effective class of rulers, ie the bourgeoisie, subjugating its victims, ie the proletariat, and in the course of the depiction using a considerable number of PAT chains to make the necessary links. On the other the 'proletariat' is connected to TSs which enhance the positive ambience the text seeks to evoke for its favourite player:

> But not only has the bourgeoisie forged the weapons that bring death to itself; it has also called into existence the men who are to wield those weapons -- the modern working class -- the proletarians. [7]

This duplicity serves the purpose of providing separate shades for one and the same object—the exploited sufferer when read in conjunction with the enemy, and the warrior hero when seen from the protagonist's side. Having any previously existing abstraction levels reduced to the bare minimum makes it easier for a twin structure to straddle separate chains since the possibility of extraneous associations outside the desired context is prevented. We find another duplex in the following passage, a few sentences later:

> Owing to the extensive use of machinery and to division of labour, the work of the proletarians has lost all individual character, and consequently, all charm for the workman. He becomes an appendage of the machine, and it is only the most simple, most monotonous, and most easily acquired knack, that is required of him.

It may be a question of how informed the readers of the *Manifesto* have been about working conditions in feudal times, especially as it concerned the lower demographics. In any case, references to the mostly precarious existence of labourers regarding their employment, their legal status, and their physical well-being, not to mention of course the recurring crises of a political and religious nature, were certainly available; the mentioning of the "idiocy of rural life" by Marx himself attests to them. While not stated expressly, one can assume a certain presence of such information, and the inclusion of the workman's "charm" points to an additional structure hanging from the worker-TS.

The text goes on to describe the conditions of the workers in more detail, and the virtual, if not actual bondage created by the need of the labourer to maintain his place of employment regardless of the consequences to himself and his family produce descriptors such as 'slave' and 'command of officers and sergeants'. These abstracts are grounded by the references in the pages themselves as to where and under what circumstances the workers have to exist, as well as their real-world counterparts which would have been there for everyone to see. Such conditions are embedded into a wider historical framework, its identified stages used to assert a specific chain of events which, according to the writer, has no choice but to culminate in an outcome synchronous with the aims of communism. At first comes the background,

> The proletariat goes through various stages of development. With its birth begins its struggle with the bourgeoisie. At first the contest is carried on by individual labourers, then by the workpeople of a factory, then by the operatives of one trade, in one locality, against the individual bourgeois who directly exploits them. They direct their attacks not against the bourgeois conditions of production, but against the instruments of production themselves; they destroy imported wares that compete with their labour, they smash to pieces machinery, they set factories ablaze, they seek to restore by force the vanished status of the workman of the Middle Ages. [8]

That state is defined as incomplete with the necessity of more to follow. The equalising influence of factory labour in every industrialised society causes the emergence of a unified class of people, ie the proletariat, which has now come into a position of power through its sheer size and homogeneity. In cognitive terms the high-level abstracts created by the functions of the illusion trees mentioned earlier are now linked with grounded TSs featuring the descriptions of historical events.

From our perspective we can concur with such a view of history. A similarity in conditions produces a similarity in response through comparable TSs, and the ongoing existence of the environment will tend towards an increased synchronicity of associations built under the environment's auspices. Since the associative halos in the minds of individuals encompass structures which due to their nature are becoming more and more interchangeable between different minds, the propensity for contextual and functional associations is heightened and so is the resonance amongst the TSs. This in turn leads to greater associative resonance which tends to accelerate the process. The greater the associative resonance, the less the need for a wider domain of abstracts—including abstraction levels—in order to establish supportive connections. A tract such as the *Manifesto* therefore, with its compacted imagery and the relatively narrow domains of its concepts, will have a significant impact on such an audience. Its wording itself supports the progression towards the build-up of sentiment:

> Thereupon the workers begin to form combinations (Trades Unions) against the bourgeois; they club together in order to keep up the rate of wages; they found permanent associations in order to make provision beforehand for these occasional revolts. Here and there the contest breaks out into riots. [9]

Reduced abstraction levels make for tighter associations; one simile in the outside world is the improvement of communication channels which allow a more direct transfer of information (and, depending on the medium, the data are compacted to a certain degree). Marx recognises this some sentences later:

> And that union, to attain which the burghers of the Middle Ages, with their miserable highways, required centuries, the modern proletarians, thanks to railways, achieve in a few years.

To "railways" he might have added the telegraph; between 1845 and 1865 most European and American cities were connected by wire services [10]. Statements like the one above which regard contemporary developments favourably while at the same time placing the same events into another, quite negative, context to suit their particular perspective, are not uncommon in ideological structures. Functionally speaking the phenomenon is a twin structure, but from a semantic point of view it indicates the incongruous nature of the general message the pattern is meant to convey. Hence the ideologically committed communist despises modernity because it has brought the disenfranchisement of the worker, but at the same time he welcomes the accompanying means through which his discontent can be pursued. Being familiar with the underlying mental dynamics we know that one cannot happen without the other anyway, and the relationships pertaining to the effectiveness of a system confirm the interlocking character of the development. The ambiguous attitude of the ideologist stems from the role the twin structure plays in placing the object abstracts within their respective domains, supported by flat TSs (that is the diminished chance of diverse associations).

The creation of the working class may be a novel process, but its material substance, the social demographic, are people which inhabited the same rung of the group hierarchy throughout all the ages; a point reiterated in the *Manifesto*. Therefore despite their current fate, their situatedness in the wider culture may not necessarily change, although this is disputed by Marx. Concerning the personal attitudes of the worker he writes,

> Law, morality, religion, are to him so many bourgeois prejudices, behind which lurk in ambush just as many bourgeois interests. [11]

The assortment of container-induced associations, the abstractions performed by illusion trees, and the duplicitous diversification of semantics, may in the mind of a Marx or an Engels place the contexts of the law etc in adversarial juxtaposition to their chosen substitutes; in the average mind of the proletarian such complexity would most likely not exist. This is not a presumption. A thought pattern which has been fabricated by the drawing together of many sources and the subsequent compacting of their representational abstracts does, at that point, no longer possess the diversity of previous linkages. Should such a TP be imported by a mind that is inherently multifaceted and given to associative and abstractive expansion, the TSs will be widened through the existing network of PAT chains. That such a network in whatever individual is not exactly the same as the one in the mind of Marx or Engels can be safely assumed. Thus an exported TP, containing an ideological structure, is subject to the modification of its peripherals depending upon the receiving mind's own inner state. Hence it happens that communism for example remained hardly changed within the European context, but underwent certain changes when it reached other cultures, where receiving intellectuals altered its hue and so its subsequent placement.

The labelling of "law, morality, religion" as "bourgeois prejudices" is significant. Given the text's polemic against the establishment, during the course of creating the container structure in the minds of Marx, Engels etc there would have been a range of TSs which had been duly identified as containing inimical elements with regards to the owner's persona (if they hadn't it would never have come to such criticism and the *Manifesto* would not have been written). In the second half of the 19th century the middle class had expanded sufficiently to put its stamp on the general judiciary, the conduct in society, and how its main religion ought to be interpreted. In conjunction with the increase in complexity of everyday activities the law had to be modified in tandem, and the resultant alienation of its intricacies from the rest of the population must have been the cause of considerable frustration, especially for the lesser educated. There is and always has been a difference between the religious imagery of an intellectual (to use the term loosely) and the peasant, the finely-nuanced sophistry of the former and the rough-and-ready faith of the latter. As far as morality is concerned, the emerging sciences produced novel views and experimentation, all in the cause of ensuring a 'healthy' society, whatever that was supposed to mean in terms of the current interpretation of religious sentiment. Two examples should suffice. One application of the steam engine was a contraption which served simultaneous whippings to two "lazy, greedy, disobedient, rebellious, insolent, quarrelsome, tale-bearing, chattering, irreligious" children, judging by an engraving from the late 18th century [12]. Since this labour-saving device came at a cost, its use would not have extended to the lower classes who had other financial priorities. A similarly costly apparatus was an electric alarm, which consisted of a ring to be strapped around the penis with metal contacts attached [13]. The idea was that should the penis become erect during sleep the contacts would touch, thereby closing the electric circuit that in turn activated a bell to awake the 'sinful' sleeper. Households with boys under their jurisdiction which were rich enough to afford such things could rest easy (the alarm bell was arguably more benign than a similar ring which had instead of the contacts on the outside metal spikes protruding inwards, in the end serving the same purpose—then again, nails were cheaper). Considering the cognitive role of ideology in moral issues, such middle class approaches did not really make its lower counterparts much better; the absence of these and other measures were and are not a matter of insightful tolerance but rather one of crude inaccessibility. For reasons of ideological convenience however such a point would not have been considered by someone like Marx et al.

There is another aspect to twin structures, ie the specificity of domain. Consider the following:

> The essential condition for the existence, and for the sway of the bourgeois class, is the formation and augmentation of capital; the condition for capital is wage-labour. Wage-labour rests exclusively on competition between the labourers. The advance of industry, whose involuntary promoter is the bourgeoisie, replaces the isolation of the labourers, due to competition, by their revolutionary combination, due to association. The development of Modern Industry, therefore, cuts from under its feet the very foundation on which the bourgeoisie produces and appropriates products. What the bourgeoisie, therefore, produces, above all, is its own grave-diggers. Its fall and the victory of the proletariat are equally inevitable. [14]

The reduction of abstraction levels prevents, as has been pointed out before, the possibility of associations with other abstracts in neighbouring TSs. But besides causing simplistic generalisations, it leads to mistaking one reference for another. In the above we have a description of how the situation of workers changes due to the contemporary environment (the "development of Modern Industry")—this is one set of TSs. We also have a description of the bourgeoisie under the same circumstances (the way it "produces and appropriates products")—the other set of TSs. Both sets are representations of occurrences in the outside world, where the mutual relationship exhibited in the mental configuration is mirrored. What is *not* part of that structure complex on the outside is the system of society itself since *it* comprises the domain in which the other two reside. Now, unless 'bourgeoisie' is perfectly identical with one of the sub-sets, whatever happens to those sub-sets will not be directly reflected in the super-set. Taking a perspective which allows the viewing of the entire arrangement makes this clear, but the constricted vision engendered by an ideological representational format does not. Hence in the latter the conclusion can be reached that the bourgeoisie destroys itself in a process which may well be detrimental to some of its actions plus the other player, the proletariat. The wider perspective would recognise that only the two sub-sets would be affected, that is the proletariat and the *way* the bourgeoisie has performed. As it happens, history has confirmed this—the character of the working class did change over the ensuing decades and so did the appearance of the bourgeoisie (on the other hand, in regions where the bourgeoisie *was* identical with the super-set, that is to say comprised a rigid controlling class thus constituting the rest of society, society as such changed too).

A further sign that the particular, agenda-related characteristics of the middle class are taken to be representative of virtually the whole of society comes in the treatment of property. *Any* property not connected to the proletariat is deemed to be suspicious, and so must be dealt with appropriately:

> The distinguishing feature of Communism is not the abolition of property generally, but the abolition of bourgeois property. But modern bourgeois private property is the final and most complete expression of the system of producing and appropriating products, that is based on class antagonisms, on the exploitation of the many by the few.
>
> In this sense, the theory of the Communists may be summed up in the single sentence: Abolition of private property. [15]

In a multi-faceted representational system the concept of property gets delineated into various sub-sets, each one reflecting the specific nature of its context. In the compacted version such a variety does not exist; therefore the concept as such becomes installed in its final, ie reduced, format, or, if this is not possible due to the obvious existence of exterior circumstances, the reduction has to take place before use is made of the concept. In this case a rhetorical question regarding private property serves the purpose of reduction. Any notion of 'property' not belonging to the middle class is dismissed as irrelevant, but then the question is asked, "Or do you mean modern bourgeois private property?", to which the answer is,

> To be a capitalist, is to have not only a purely personal, but a social status in production. Capital is a collective product, and only by the united action of many members, nay, in the last resort, only by the united action of all members of society, can it be set in motion.
>
> Capital is, therefore, not a personal, it is a social power. [16]

Note how the diverse cognitive domain of capitalist activity morphs into the more abstracted, singular, encompassing entity of 'social status', which then gets associated with 'collective product'. In order to complete the loop, reference to 'bourgeoisie' has to be made via the TS denoting 'united action of all members of society'; at that point the association is a congruent one (the generalising meaning of 'collective product' being functionally congruent with the equally general sense inherent in 'united action of all members of society'). The *in*congruence occurred earlier when 'social status' became redefined. The next, inevitable step in the process is the removal of the idea of 'general property' from any previous connections it may have had with notions of category, since these are now incompatible:

> When, therefore, capital is converted into common property, into the property of all members of society, personal property is not thereby transformed into social property. It is only the social character of the property that is changed. It loses its class-character.

Of course, the abstract connection between 'property' and 'social character' is just that—a relatively high abstract. Without an elaborate implementation in the outside world (if possible at all) such subtle differentiation will get lost, and what the system ends up with is no private property for the ordinary members of society and everything in the hands of the system's administrators for their personal disposition; that is to say, ownership not due to individual achievement but ideological definition.

Every re-representation of reality sooner or later gets referenced back to the outside; after all, the purpose of inner imagery is to enable the bearer to respond to his surroundings. It is a feature of ungrounded structures that the resultant notion is not compatible with the exterior because it has been derived through diverting from what can be ascertained in the real. In order to press it into service anyway two kinds of measures have to be taken. One is to modify the concept, the other is to alter the target it is supposed to be applied to. The more rigid the concept, the more likely the second option becomes. Unfortunately, in many cases the full intent behind this phenomenon only becomes apparent after the system is securely installed. But there are forewarnings, such as this one:

> In bourgeois society, therefore, the past dominates the present; in Communist society, the present dominates the past. [17]

Sometimes the urge to provide catchy slogans for the believers gives the game away. Regardless of the actual content then, the first part of the sentence is straightforward enough. But in the second part the meaning can only then possess a certain validity if the posited influence can be put into action. The only way that can happen is if 'present' and 'past' are not to be understood as reality at all, but refer to their respective conceptualisations. In that case the 'past' can be re-interpreted, modified, or changed outright, as long as its relationship with what goes for the articulated 'present' is in tune with the overall concept. If there is a problem treating the 'past', then the 'present' can be dealt with. In other words, under such a system we can expect historical revisions together with forcefully applied policies.

One of the consequences of applying the tuning effect to the real is to define what is meant by 'individuality'. Since it is private property and the personal initiatives of the bourgeoisie that are the focus, this type of individuality has to go. As the *Manifesto* announces, addressing itself to the reader,

> You must, therefore, confess that by "individual" you mean no other person than the bourgeois, than the middle-class owner of property. This person must, indeed, be swept out of the way, and made impossible. [18]

The sweeping generalisations we have encountered in the creation of compacted abstracts are an effective weapon against a diverse target. The reduced TS denoting 'family' can be represented in the light of certain circumstances, the veracity of which cannot be questioned because they happen to be true nor can their applicability because they are inaccessible. To remark upon the exploitation of children by their parents in those days is true because the evidence cannot be denied, yet no provision is made for the precise identification of same. This allows Marx to declare, "Do you charge us with wanting to stop the exploitation of children by their parents? To this crime we plead guilty" [19]. He goes on to say,

> The bourgeois clap-trap about the family and education, about the hallowed co-relation of parent and child, becomes all the more disgusting, the more, by the action of Modern Industry, all family ties among the proletarians are torn asunder, and their children transformed into simple articles of commerce and instruments of labour.

Just as we can identify purposefully modified TSs in the *Manifesto* to suit the creation of a certain image, so did the general imagery of middle class individuals contain ideas which acted as a protection for the comfort zones of their owners. The use of children as cheap labour in the factories and mines and for prostitution would not have been topics discussed around the average, traditional dinner table, and even today some topics on human sexuality are not deemed to be suitable for the ears of relevant social groups (sex being one of the last bastions of coercive morality of course). Thus the newly arriving thought pattern of communism created two dilemmas; one, the need to deal with a highly abstracted, yet ungrounded system of ideas, and two, to make room in one's own thought patterns to find the means of handling the concept.

In a proper re-representation of the outside world the abstraction levels of the real concur with those in the mind; that is to say, the systematic breaking-down of sub-sets and sub-sub-sets etc of the system under focus will show the same format on both sides. In the case of a faulty re-representation the abstraction levels do not match. Some paragraphs previously (see ref. 14) we encountered one such mismatch regarding the mutual influence between middle and working class forces. Another can be seen with reference to women. Consider the following passage, in the light of Marx's polemic against the common idea of 'family' and its associated moral assumptions:

> Our bourgeois, not content with having the wives and daughters of their proletarians at their disposal, not to speak of common prostitutes, take the greatest pleasure in seducing each other's wives.
>
> Bourgeois marriage is in reality a system of wives in common and thus, at the most, what the Communists might possibly be reproached with, is that they desire to introduce, in substitution for a hypocritically concealed, an openly legalised community of women. [20]

In terms of the elements contained in the statement, there is a set (the bourgeois), holding a sub-set (the activity by the bourgeois of seducing proletarian females and prostitutes), and a second (the seduction of each other's wives). This configuration is flattened by the general process of forced alignment with the desired container structure and so it becomes one single concept, namely 'a system of wives in common'. It is the single concept which gets integrated with the wider TP for the purposes of accentuating the overall message. In real life these sets are related quite differently. There may be a set consisting of the bourgeois, and it may have sub-sets of various activities its members are wont to do, but the seduction of wives does not depend on being part of the bourgeois but on being human and therefore having a sex-drive, something not restricted to a particular demographic at all (what Marx did not mention is the additional biological fact of women seducing males—but we'll deal with gender politics in Part II). Thus the characteristics of the set (bourgeois) listed here are actually part of the super-set (human)—a different arrangement compared to the mental picture. Note that although all this may seem rather obvious to someone familiar with human beings, the foregoing analysis of cognitive structures arrives at the same insights *without* any special knowledge of the players involved, but merely by assessing the relationships enunciated through their configuration. This was to be shown.

The induced confusion of relationships amongst concepts within their re-representational version constitutes a fundamental problem of information systems in general. By misplacing content elements such that the result features a co-dependence of a type not found in reality, or worse, where there is no co-dependence at all, leads to destructive actions once the represented notion is implemented. For example, the *Manifesto* assigns the members of the working class to a conceptual element overriding the element holding the members of nationalities. Both are TSs connected to different PAT chains since they are situated in different domains; not because of whatever geographical knowledge may be present in the minds of people, but because being a member of a nationality involves a whole set of specific experiences, values, and priorities, all of which define its members. To enforce a synchronicity on the outside based on the format created by one's inner mental dynamics would therefore lead to problems—seen from the outside, this would be an approach from the top downwards. On the other hand we also read,

National differences and antagonisms between peoples are daily more and more vanishing, owing to the development of the bourgeoisie, to freedom of commerce, to the world-market, to uniformity in the mode of production and in the conditions of life corresponding thereto. [21]

In this case the image development takes place from the ground up, that is the actions performed by people in so many diverse ways contribute to the gradual emergence of international uniformity. As a representational element the above sentence is grounded, hence concurs with reality, and so does not cause any difficulties when transposed to the exterior. The disappearance of certain local characteristics is a process that is continuing to the present day, but at the same time the alignment stops when events touch upon more deeply embedded structures, those which the processes of uniformity have so far not been able to modify—as history shows.

The proposed initiatives are summarised in ten points, although it is stated that details may vary according to the country. The following list is deemed to be appropriate for the most advanced nations—presumably the general circumstances there furnish enough negative social aspects for every item to be relatable.

1. Abolition of property in land and application of all rents of land to public purposes.
2. A heavy progressive or graduated income tax.
3. Abolition of all right of inheritance.
4. Confiscation of the property of all emigrants and rebels.
5. Centralisation of credit in the hands of the State, by means of a national bank with State capital and an exclusive monopoly.
6. Centralisation of the means of communication and transport in the hands of the State.
7. Extension of factories and instruments of production owned by the State; the bringing into cultivation of waste-lands, and the improvement of the soil generally in accordance with a common plan.

> 8. Equal liability of all to labour. Establishment of industrial armies, especially for agriculture.
>
> 9. Combination of agriculture with manufacturing industries; gradual abolition of the distinction between town and country, by a more equable distribution of the population over the country.
>
> 10. Free education for all children in public schools. Abolition of children's factory labour in its present form. Combination of education with industrial production, &c., &c. [22]

Our examination of the *Manifesto's* structural configuration has highlighted the associative means by which the above was put together. Taking each point in isolation and considering the diverse network of dependencies through which the respective human and administrative entities are embedded within the greater fabric of society, the degree of disruption necessary to implement only one of them becomes obvious. Our analysis has given an outline of the scale of the change in the mental structures. Since a re-representation always falls short of the real thing, the theory is but a pale shadow of the kind of upheaval flesh-and-blood people would go through if those ideas ever did get put into practice. The past century has provided us some examples of just that. While it gave us the chance to observe the transformation from one style of existence to another, it is also worth noting the equally disruptive influence of spiritual ideologies, ie religions. Unfortunately their presence constitutes the status quo, and therefore the conditions under a liberal-rational alternative can only be guessed at.

Still, such ruminations are at least possible; they are certainly in stark contrast to what the designers of the *Manifesto* have in store for their fellow citizens, as its last paragraph demonstrates:

> The Communists disdain to conceal their views and aims. They openly declare that their ends can be attained only by the forcible overthrow of all existing social conditions. Let the ruling classes tremble at a Communistic revolution. The proletarians have nothing to lose but their chains. They have a world to win.
>
> WORKING MEN OF ALL COUNTRIES, UNITE!

In summary we can say that the imaged world of the *Communist Manifesto* consists of a central container structure that serves as an attractor for TSs featuring associative high-level abstracts. The container has been built by subjecting the abstracted contents of PAT chains to a reduction process in which unwanted details have been removed so that only a singular representation of the original concept remained. At the end of this process these leftovers formed the domains of further illusion trees, to be drawn to abstraction levels at which functional associations dominate in favour of contextual ones, thereby preventing the possibility of opposing c-PAT-ls based on grounded links. Instead, connections are established and maintained on a functional basis only where the abstracts gain their validity through their resonant effect on their thus formed domains, and not through any semantic significance they might have possessed had their contents been allowed to stay accessible via congruent chains. If an abstract did contain a connection to lower levels it was put into place at the head of a twin structure, with one strand leading downwards, the other in a denominative relationship with container-induced formations. The reduction of abstraction levels within a cognitive entity itself leads to a mismatch between it and another, perhaps similarly treated element, since any varying types of functionality on different levels have disappeared, there is less nuanced detail, and so the associative intensity of both candidates has been heightened. Modified TSs and skewed relationships between TSs lead to implementation problems once the re-represented concept is being transposed back into the outside world.

The exploration of the environment, the collecting of data from so many fields, their analysis and categorisation, led to an ever increasing pool of products. Some were looked for, many others were found serendipitously. As the number of products grew so did their applications, and while the general latency was high it was accompanied by the steady accumulation of realised potential, leading to greater applicability which in turn enabled their users to add to the proliferation of products; and so on. As the range of material objects widened the means of articulation had to keep pace, and expressions such as 'piston', 'electrical circuit', or 'molecule' made their way into the common vocabulary. There they participated in the same process of applicability generation as their material counterparts, because now people had the means to give labels to their thoughts and descriptions to their ideas. Their thought structures became more prolific, more detailed, became ready for further associations, spawning more constructs still. It is true that most of these words were not new as such; but their degree of applicability rose as they were able to be linked to greater numbers of other abstract objects, and so their significance increased.

In order to make connections, both, the source and the target need to be identified. An intelligence which is capable of abstractions soon discovers that the realisable potential of an entity at a low level of abstraction is heightened once that entity is elevated to higher levels. For example, an electric current sent through a few windings of copper wire is enough to demonstrate the generation of a magnetic field. But as soon as the situation per se is investigated it leads to the discovery of the electro-magnetic force, induction, rotors, stators, new engines, etc etc (refer to chapter 7 for a discussion of the higher abstracts in that role). We can abstract such a context itself and declare that, having gained the perception of an object, it is desirable to identify its generality also, since the higher level opens the door to far more insights than would otherwise be possible. This generalisation is of course the *essence* so much discussed by philosophers of any age, and the sheer recognition of the act of abstraction and its advantages, although not expressed in formal language, had been a feature of FAs for centuries already. The need to grapple with the phenomenon of generalisation, to place it into some formal system, received further kudos with the proliferation of scientific achievements.

There was only one problem. It is one thing to ascertain the higher abstracted form of say, magnetism, having arrived at that discovery through experiments with various arrangements of electrical coils. But the benefit of knowing about electro-magnetic induction at that higher level suggests the existence of certain properties well *before* the cobbling together of some rudimentary apparatus. Hence arose the idea of determining such underlying properties, the essences, and by doing so make the process of discovery much more efficient. Given that we are dealing with formal properties, ie characteristics which do not interact at random but in a precise order, the task was to establish a formal system of a priori phenomena as the basis for any research that could follow. Generalisations are products of the mind; subsequent research concerns itself with material things. Therefore the problem became one of having materialistic sciences on one side, and to create a science of phenomena, of appearances, on the other. Enter Edmund Husserl (1859 - 1938).

His investigations into a possible means to identify the existences and relationships of the mental representations which our minds acquire from sensuous experiences led him to what he called the science of phenomenology, and we will follow the results as they appear in the first and second book of *Ideas pertaining to a pure phenomenology and to a phenomenological philosophy* [23]. The material for these works was composed, edited, re-edited and published mainly during the first three decades of the 20th century.

Before we can begin with the description of Husserl's system we need to be aware of the cultural immediacy of his language. By the turn of the 20th century the common usage of words and phrases had become quite close to the present format, and so his text has to be read carefully to prevent misunderstandings. Not only should this be a concern to us, Husserl himself recognised the potential difficulty of dealing with a subject that had been in circulation for centuries but now required an updated format of a nomenclature based on a language that had not changed to the same extent [24]. It is doubly problematic for us, since not only do we need to participate in the semantic translation process from 'traditional' philosophy to Husserl's version, the examination requires the placement of our own descriptors within the correct context. For instance, when the word 'abstract' is used in terms of phenomenology it stands for the product of a generalisation—not more, not less. Using 'abstract' for our purposes denotes the result of an information transfer in terms of a recognisable idea from one TS to another; this *may* eventuate in a generalisation, but could also be quite the reverse depending on which way along the PAT chain we happen to proceed.

On a similar note, the concepts involved within the subject of phenomenology are by their very nature not clearly defined entities. There is no formal terminology describing thoughts and their dynamics (such as *we* are in the process of carefully establishing) and therefore a sometimes somewhat obtuse language is employed in getting the point across. This is a problem of the degree of conceptual maturity which at the time existed in relation to the topic, and really is a feature of any endeavour in its infancy. Trying to overcome that problem by borrowing from more established fields is fraught with danger, as Husserl observed at the very beginning:

> One cannot define in philosophy as in mathematics; any imitation of mathematical procedure in this respect is not only unfruitful but wrong, and has most injurious consequences. [25]

To appreciate the above is important: a mathematical relationship (a state) can be explained through the articulation of its clearly defined elements as well as their functionalities, equally defined. The thus expressed state serves as an additional conceptual element to be used productively in subsequent delineations. If the terms are more equivocal however, such precision is not given and therefore their combination is also less precise. While it may be just possible to depict a state, to continue onwards into the expected behaviour within that state will be far more precarious and so it would be wise to refrain from going too far. Husserl must have sensed this. His system of phenomenology *describes* but does not *explain* anything, ie it does not provide us with an understanding of how the introduced entities actually work. As he says, towards the end of the First Book,

> Because it is not our proper task here to solve phenomenological problems, but rather to work out scientifically the principal problems of phenomenology and to predelineate the directions of investigation cohering with them, we must be satisfied with having carried the matters under discussion that far. [26]

Let us accentuate the issue further. A material system which has been constructed (that is, one that exists in the outside world and consists of actual, measurable, identifiable objects) can be observed and its parts described. Each part is labeled according to its appearance and its functionality, and they all together can be summarised in terms of what the system does as a whole. At any given point in time it is possible to ascertain the system's state through observing the particular state of each of its parts, and since it is possible to do that at regular intervals I can not only revisit the system's past but also its future. All this can be done only if the system has a clearly defined domain as well as no additional, unforeseen extraneous influences to contend with. The trouble is, such a system does not exist—for two essential reasons.

Firstly, for the above proposition to hold true the system would have to be situated in an ideal environment where it is allowed to go through its paces without the slightest interference from anywhere. There is more behind that requirement than at first meets the eye. Let's say our system is an engine which just runs along, nothing else. Even if we could provide an ideal kind of screen around its parts, protecting it completely against everything the surroundings could ever throw at it, it still would not be enough because it has to get its energy from somewhere—and that *somewhere* is not protected. And even if we would erect further barricades around those energy-producing features, they in turn depend on *their* supplies and so the situation escalates. Already however the widening domain begins to include other systems which we cannot define to the same degree of completeness because we did not build them. To make the proposition come true therefore we would have to engage in constructing not only the original system but its supports as well—and the supports of those supports and so on ad infinitum. Clearly, this is impossible.

The second reason is more relevant to cognitive dynamics. So far we have only considered a system purely on its own; our only type of participation had been in the form of talking about it. Yet in the 'real' world nothing is done by us without having a mental representation of what we are about to do—the idea must precede its implementation. If—and only if—those ideas are a perfect reflection of that which is to become its implemented form, we could expect to at least build a system according to the given specifications, although reason 1 would eventually come to the fore anyway. Since our internal imagery is only ever an abstraction of the actual counterpart, not only is the overall mental concept of the system incomplete but it will be impossible for us to identify all the eventualities which could ever manifest through its operation. So once again it is impossible for such an assumed system to exist—even in terms of its mental representation.

Now consider abstractions. What Husserl's system requires us to do is to create a construct of further abstractions (ie consisting of generalisations) in relation to the first-level representations mentioned under reason 2. In other words, not only do we need the full contingent of inherent characteristics per se to be represented, but our further abstractions would have to hold the entire principles of those characteristics. As it is impossible to even generate the full contingent of characteristics let alone a perfect rendition of its underlying generality, we will not be able to establish a system of generalities either. And yet, Husserl's work is intended to represent the abstracted form of just such a construct. How is this possible?

Recall our discussion on containers and illusion trees. A container can be seen as an initially established attractor the purpose of which is to serve as a driving core for associative TSs. An illusion tree is the conduit for an idealised—ie purified—version of a set of TSs in order to present an idea for further processing. In Husserl's case the process requires the identified parts of his system to be congruent with each other, and that includes not just the object abstracts but the functional ones as well. Therefore, being cognisant of reason 2 above, we can say that the creation of a system of phenomenology necessitates the formation of illusion trees and their associative TSs such that the overall purpose of the schema is being served under the auspices of its containers.

The somewhat lengthy steps in the preceding paragraphs had been necessary to explain our position. It would not have been enough to merely declare, "It is impossible to arrive at a fully working phenomenological model of the mind" by pointing to an assumed degree of ignorance or otherwise on behalf of its builder. One does not need to know the builder's mind (even if that were at all possible), nor is it necessary to define the comprehensive outline of what the model ought to entail. What needed to be shown was the fact that it is impossible to formulate a complete model of an abstractive, truly representative system by virtue of its very definition—regardless of its contextual nature.

And yet the generality ought to precede the object itself. Given the inadequacy of a structure built out of essences (this according to our definition), such a negative feature would, however indirectly, become apparent once a designer of that structure sets about laying down the bridge spanning from its borders to the realm of objects. Husserl calls the former the eidetic science, and summarises the other versions by "Science of Matters of Fact" [27]. For him the "Eidos" is utter generality, without any material complement, pure essence. (It is interesting to observe a certain change in the meaning of the word 'eidetic' since then. Three generations later the Macquarie Dictionary gives us, "1. of, pertaining to, constituting or having a vivid and persistent type of imagery or memory, esp. during childhood. 2. pertaining to an image that recreates a recent optical impression with hallucinatory clarity." More than sixty years of chipping away at imaginary concepts do have their effect not only on the target but also on the tools used in the process)

In any case, considering the manner in which Husserl's construct is assembled, the FAs associated with the concept of essence must reflect the semantic authority of the items in question, the generalities. Since an essence defines the object in which it is found in a way that is unquestionable, the domain of any intellectual process dealing with objects (the manifest essences) will be restricted to whatever examples of those manifestations can be gathered; it is *not* going to be allowed to infringe upon the realm of the principle because that area is reserved for the eidetic science. Husserl says as much when explaining the relationships between the eidetic and other sciences:

> After the foregoing it is clear that the *sense* of eidetic science *necessarily precludes any incorporation of cognitional results yielded by empirical sciences.* The positings of actuality that occur in the immediate findings of these sciences obviously extend throughout all their mediate findings. From matters of fact nothing ever follows but matters of fact. [28]

For us two things in the above passage become evident: the generality takes precedence over its manifestation in quite a comprehensive way, and the manifestation is held to be incapable of furnishing any principled insights into its nature. Following such tenet our demonstration of the candle and tack puzzle mentioned previously in conjunction with Locke's and Newton's deductions becomes impossible. The generations to follow should be grateful these two gentlemen had not been phenomenologists!

What is written next is particularly significant, given the history of such sentiments in general:

> But although every eidetic science is necessarily independent of every science of matters of fact, the reverse holds, on the other hand, for the latter sciences. There is *no science of matters of fact* which, *were it fully developed as*

> *a science*, could be pure of eidetic cognitions and therefore *could be independent of the formal or the material eidetic sciences*. For, in the *first* place, it is without question that an experiential science, wherever it brings about mediate groundings of judgments, must proceed according to the *formal* principles treated by formal logic.

It is certainly true that one must proceed logically. The FAs pertaining to formal logic are associative with those TSs that permit such a linkage in the first place. FAs are not OAs, and therefore a connection with an appropriate TS will be a functional and not a contextual one. This means that, when it comes to such FAs, associations can be established purely on the basis of functional congruence, nothing else. For example, I can declare that three chairs, or tables, or trees, are more than two chairs, or tables, or trees; in other words, the underlying principle of 'three is more than two' being independent of the object. As soon as an FA is specific enough to relate to a particular set of things however, my previous freedom of association becomes curtailed. For instance, the statement "the flash point of timber, or petroleum, or wax is lower than the flash point of iron, or copper, or silver" is valid; it would turn meaningless were I to change it to "the flash point of timber, or happiness, or divinity is lower than the flash point of iron, or copper, or silver". That is to say, the abstraction of an object, in terms of its associative quality, needs to be flexible enough for connections to be made or disbanded, according to its associative properties. In the above quote there is the insistence of permitting functional associations regardless of the object's contextual quality on the one hand, and the equally authoritative direction to prevent further abstractions of an object, just in case they may harbour potential FAs of a nature specific to the object, on the other. That must never happen—after all, "From matters of fact nothing ever follows but matters of fact".

Cutting through the complex framework of Husserl's treatment, one can sense the basic dominance of 'pure' abstractions over lesser entities; the preference for higher-level, ungrounded abstracts against lower-level, grounded chains; and ultimately, the desire to elevate a concept to whatever height is necessary to ensure its authority over the mere material and hence making the latter dependent upon the former—the scenario is starting to look familiar.

Trending towards ever higher levels of abstractions is bound to approach the metaphysical sooner or later. For Husserl the belief in an ultimate organiser and designer still exists, and besides having the quality of perfect knowledge there is also perfect perception—but with a particular slant, brought out in stark relief due to the kind of subject matter.

The system of phenomenology relies on the integrity of essences, or to be more accurate, the system's descriptive powers in relation to their complete nature. Unless the information inherent in a given essence can be accessed to the fullest extent, one can hardly justify to nominate such a representation as an authoritative standard. While the optimal access may not be achievable in each case, the possibility cannot be denied without downgrading the system as such. Thus Husserl writes,

> There belongs to any existent the essential possibility of being simply intuited as what it is and, more particularly, of being perceived as what it is in an adequate perception, one that is presentive of that existent itself, "in person," *without any mediation by "appearances"*. [29]

Regarding "appearances", the statement is a reference to previous remarks about the incapacity of a physical thing to be "immanently perceived" and so "found at all in the concatenation of mental processes" [30]. He continues,

> God, the subject possessing an absolutely perfect knowledge and therefore possessing every possible adequate perception, naturally has that adequate perception of the very physical thing itself which is denied to us finite beings.

Note the sharp division between TSs (and hence representations) on the grounded side of the logic boundary and those on its other side. Perceptions derived from a physical thing (for us humans at least) are inadequate whereas an abstract situated among the associative network of God-related TSs must be satisfactory. Although the two sets are kept firmly apart, there is nevertheless a significant link between them, forming a bridge from the representations of things in general to the 'ideal' version. Still discussing perception and the resultant "stream of consciousness", Husserl affirms the act of perceiving itself as a valid exercise:

> Every perception of something immanent necessarily guarantees the existence of its object. If reflective seizing-upon is directed to a mental process of mine, I have seized upon something absolute itself, the factual being of which is essentially incapable of being negated, i.e., the insight that it is essentially impossible for it not to exist; it would be a countersense to believe it possible that a mental process *given in that manner* does *not* in truth exist. [31]

We know that Husserl accepts the existence of fantasy as non-representative of anything in the real; he says so on several occasions. Given that information there must be the implied necessity of somehow confirming the perception of a thing as a mental act that involves an actual object. In terms of his system's format this can only happen in one of two ways: either the thought processes manage to ascertain the validity of the perceived object unequivocally and so create the desired certainty, or the essence of that object happens to be associative with high-level abstracts beyond the logic boundary and so is proven by default. Since our thought processes are not deemed to be capable enough for a confirmation through merely perceiving, the second option can be assumed. The only path an abstract—in whatever form or modification—can use to move from a PAT chain of grounded perceptions to those heights is via an illusion tree; it is the bridge across which an abstract has to travel in order to become acceptable as a true essence.

Yet at the same time religion is being excluded as a source for phenomenological science, as is anything else which in any way or form draws on information derived from physicality, as we read when discussing phenomenological reductions:

> In the first place, it is immediately understandable that, with the exclusion of the natural world, the physical and psychophysical world, all individual objectivities which become constituted by axiological and practical functionings of consciousness are excluded, all the sorts of cultural formations, all works of the technical and fine arts, of sciences (in so far as they come into question as cultural facts rather than as accepted unities), aesthetic and practical values of every form. Likewise, naturally, such actualities as state, custom, law, religion. Consequently, *all natural sciences and cultural sciences*, with their total stock of

> cognition, *undergo exclusion* precisely as sciences which require the natural attitude. [32]

If dealing with pure principles only, means turning away from any actions that require the importation of empirical data, we can understand the exclusion of something like the "natural and cultural sciences". To include religion in the same context points to an attitude which expects religious sentiments to be grounded in reality just as the material of the sciences should be. The presence of illusion trees in Husserl's dynamic, designed as they are for the inevitable elevation of lower-level abstracts towards the highest regions, makes such a decision largely irrelevant—all the important abstractions move along i-PAT-ls anyway.

Adding to the isolation of high abstracts is their removal from any possible influences of the self. The "pure Ego"—already a highly abstracted concept in itself—shall not remain visible in the attained reductions. At that stage our thinking processes may carry their content past our consciousness, but they are not going to be touched by that which constitutes our own essence, although its constant presence is acknowledged. Consider the following passage, not only what is being said but also how it is expressed.

> This much is clear from the very beginning: After carrying out this reduction we shall not encounter the pure Ego anywhere in the flux of manifold mental process which remains as a transcendental residuum—neither as one mental process among others, nor as strictly part of a mental process, arising and then disappearing with the mental process of which it is a part. The Ego seems to be there continually, indeed, necessarily, and this continualness is obviously not that of a stupidly persistent mental process, a "fixed idea". [33]

To view the "pure Ego" as an entity standing aside from mental processes per se, means there has to be a set of abstracts pertaining to the ego-notion which are perceived to be beyond any associations that relate to the materially human. Furthermore, as a TS it will be congruent with similarly abstracted contents, making contextual and functional connections possible (indeed necessary) with whatever else happens to reside beyond the logic boundary. As a side effect it posits the idea of "pure Ego" as being over and above humans themselves, ie an entity not really belonging to this world nor playing any part in it.

The allocation of TSs dealing with essences, the divine, and the pure ego, to those regions which have no direct links with grounded PAT chains except through illusion trees, prepares the ground for the resultant thought pattern to become resonant in relation to already existing TPs at that level. In Husserl's case these TPs would be mostly in line with his broad cultural background including the Christian view of the world, thereby making them denominative towards the new addition. Since alterations within that TP based on grounded linkages are actively discouraged, we can say that the ultimate determinants of essences within the system of phenomenology are a function of the overall cultural-religious perspective. This is confirmed by the remark,

> In all this [referring to mental processes concerned with empirical intuition as well as those belonging to phenomenological reduction], since the *rationality* made actual by the fact is not a rationality demanded by the essence, there is a marvelous *teleology*. [34]

Or put into our words, the meaning inherent in grounded TSs is done away with by the meaning inherent in metaphysical TSs, and the latter's assumed superiority allows them to be seen within the context of an overall, original design—the prime cause.

How profound such an influence can be is illustrated in the section on self-observation, or reflection. Husserl offers various descriptions of thought processes to accord with the purposes of his system (which will be looked at presently), so when it comes to reflections there is the need to demonstrate their evidently abstract nature.

> The phenomena of reflection are, in fact, a sphere of pure and possibly perfectly clear data. It is an *eidetic insight*, always attainable because immediate, that starting from the objectively given as objective reflection on the presentive consciousness and its subject is always possible: starting from the perceived, from what is "there 'in person'," a reflection on the perceiving; starting from the remembered, just as it "hovers before us" as remembered, as "having been", a reflection on the remembering; starting from the statement in the flowing off of its being given, a reflection on the stating, etc. As a consequence, the perceiving becomes given as the perceiving of just this perceived, the present consciousness becomes given as consciousness of something intended to. [35]
>
> Since essence rules, an 'essence-inducing' kind of thinking is bound to identify what is "really inherent". The divine is no exception:
>
> God is also bound to this absolute necessity given in insight, just as He is to discerning that 2 + 1 = 1 + 2. Even God can only acquire cognition of His consciousness and consciousness-content by reflection.

And in a footnote:

> We do not extend the controversy here to the domain of theology. The idea of God is a necessary limiting concept in epistemological considerations, and an indispensable index to the construction of certain limiting concepts which not even the philosophising atheist can do without.

This shows that the TPs holding religious perspectives are not only denominative to the TP incorporating phenomenological content, the latter also induces reciprocity.

Now that we have familiarised ourselves with the general setting of Husserl's system, let us turn to its major elements.

First in line, naturally enough, are the essences. Husserl differentiates between the object, "factually existing spatio-temporally", and the Eidos, something which regardless of the thing's actual occurrence in this or that spot, at this or that time, presupposes its eventual manifestation according to circumstances. The Eidos is pure essence, can be identified apart from the physical appearance, and belongs to its own set of categories. As we have seen, an object's essence predefines how it is empirically perceived (or "intuited", as Husserl prefers to call it), that is essences come first, the actual form later. This led us to assume the existence of illusion trees in order to facilitate the information flow along the PAT chains from initial perception to whatever level of abstraction may deemed to be suitable. For Husserl the problem is one of reconciling the obvious presence of abstractions and their mental processes with the posited dominance of what the abstractions mean to him. As far as the identification of essences is concerned,

> *Everything belonging to the essence of the individuum another individuum can have too*; and *highest* eidetic universalities of the sort just indicated in our examples delimit "*regions*" or "*categories*" *of individua*. [36]

Immediately following the above we get a first taste of what the process of categorisation might entail:

> *At first* "essence" designated what is to be found in the very own being of an individuum as the What of an individuum. Any such What can, however, be "*put into an idea*". Experiencing, or *intuition of something individual* can become transmuted into *eidetic seeing (ideation)*—a possibility which is itself to be understood not as empirical, but as eidetic. What is seen when that occurs is the corresponding *pure* essence, or Eidos, whether it be the highest category or a particularisation thereof—down to full concretion.

In a sense, the foregoing describes the underlying functional format of phenomenology—the rest is interpretation. Thus—reverting to our own nomenclature—we have the first abstraction of the perceived object, a PAT^2 (the "What"). Next it is "put into an idea", which can only mean a further abstraction, ie a PAT^3. The very act of forming the generalisation of something is held to be an eidetic process, that is a process that by itself can be seen as an abstraction, and the result is the emergence of "pure essence". Note however that in terms of Husserl's description the process is not dependent upon precise abstraction levels, which is why or own method of using PAT^2s/ PAT^3s etc becomes misleading. If Eidos, the resultant "pure essence", can be achieved at the highest as well as the lowest abstraction levels (even "down to full concretion" in his own words), the nature of Eidos itself cannot be accepted as a form of abstraction as such, but rather as a descriptor for a particular perception of a thought process (Husserl's beliefs notwithstanding). For us the relationship between abstracts comes down to their respective content and the level at which each one happens to reside. For Husserl on the other hand it is not the degree of generalisation to which some content has been subjected, but the circumstances in which it is being processed. For example, in terms of our system we would distinguish between a car (a PAT^1), a motor vehicle (a PAT^2), and a mode of transport (a PAT^3). Going by the above quote all three items could be at the same level, and only a generalisation of any one of them would be considered a higher abstraction, with the entire exercise a further "eidetic insight" by itself.

To postulate a separate species of thought each time the environment of its dynamic changes leads to a plethora of contingencies, all of which have to be brought to some kind of common denominator with respect to their essential nature. We will not list the entire range featured in the two books but select only a few of them to highlight the format of the construct. There is for instance fantasy:

> The Eidos, the *pure essence*, can be exemplified for intuition in experiential data—in data of perception, memory, and so forth; but it can equally well be exemplified in *data of mere phantasy*. Accordingly, to seize upon an essence itself, and to seize upon it *originarily*, we can start from corresponding experiencing intuitions, *but equally well from intuitions which are non-experiencing, which do not seize upon factual existence but which are instead "merely imaginative"*. [37]

One can only speculate about the specific reasons why someone who contemplates the art of abstract thinking finds himself using the accompanying circumstances of a thought as the ultimate determinant of that thought's nature. Obviously circumstances would feature in any observation, but just because I see a car in a driveway and another on the road does not mean I am convinced of having seen two different types of cars. And yet there was a time when we observed the phenomenon of magnetism in the form of aligning particles and that of electricity in the form of sparks, not realising they were two aspects of the same force. Relying on *phenomena* alone would of course cause such misinterpretation, and so it may be no coincidence that Husserl's system is called *phenomenology*. A case of large-scale functional association perhaps.

Consider also his already mentioned remarks regarding the relationship between "sciences of matters of fact" and "eidetic sciences", particularly the opening sentence to this section:

> The ground for a corresponding interrelation between sciences of matters of fact and eidetic sciences is the connection (itself eidetic) obtaining between individual object and essence, according to which an essential composition belongs to each individual object as *its* essence—just as, conversely, to each essence there correspond possible individua which would be its factual singularisations.

Once there is an abstract which is identified as the essence of an object, and such an identification is held to be another abstract in its own right (which it may well be), there is the need to assign a particular and unique set of characteristics to that other abstract if for no other reason than to accord with the premise that *its* essence must reflect the manner in which it had been gained (the "corresponding possible individua", the "factual singularisations"). It is a form of reciprocity across PAT chains, and it is striking in the light of the same dynamic already occurring with regards to the entire system, as pointed out earlier.

The tendency to classify according to appearance leads to ever more detailed perceptions of what are—essentially!—merely abstractions in various contexts. There are "concepts such as property, relative determination, predicatively formed affair-complex, relationship, identity, equality" etc which come under the banner of "logical categories" [38]. There are also "signification-categories", the "fundamental concepts belonging to the essence of the proposition (apophansis)"; essences denoted by signification and "that which can undergo 'expression' *by* signification", leading to "categorical concepts" and "categorical essences". Objectivities, once attained, are being distinguished by "syntactical forms" and "syntactical substrates", culminating in "syntactical categories" and "substrate categories". Without going into the exact detail of all these special cases, when it comes to the TSs representing these descriptions their particular, specific domain gives rise to associations that are more likely to form links according to some descriptive particle rather than their main theme. Since the driving force behind the formation of TS-sets in a certain manner is the overall schema of essence-dominance imbedded within the even larger background of a metaphysical interpretation of the world (a format which favours ungrounded links), the ongoing decomposition of phenomena trending away from their basic nature can easily lend support to incongruent structures.

The combination—and hence functional associativity—of phenomenological rules and decomposition of trains of thought produces a recurring relationship between the perceived object and its representation. Given the context of the degree of importance attached to a thing's essence, an abstraction is therefore not only seen as a derivative but also as a causative with regard to the item itself—once again that would be in line with the element of reciprocity identified earlier. The notion that a thought somehow *is* the object is not a new one, but we find it reconstituted to a certain degree in the present system. In a discussion of essence and concept Husserl warns about mistaking the two, especially since 'concept' is often used ambiguously. He provides an analogy via the idea of cardinal numbers. On one side they are essences, but they can likewise be described as concepts. Concepts however, he says, are distinguished by their usage, whereas essences, because of their very nature, remain untouched by whatever may be done to them; the essence of say, the number two is an "atemporal being" [39]. He goes on to say,

> If concepts are physical formations then those affairs, such as pure numbers, are not concepts. But if they are concepts, then concepts are not physical formations.

Although here a distinction is made between the content of a thought (ie the concept) and what the thought stands for (ie the number two), it ought to be remembered that both are forms of abstractions situated in some cognitive structure nevertheless. We will deal with the question of similar, non-material items in Part II and what they represent in terms of belonging to our complement of thinking tools but whatever their nature, for them to be participating in our TSs they need to be abstracts of some kind or another (whatever they may be in another realm). Whether recognised as such or not, in the phenomenological schema the essence pertaining to something like a number is held to be outside the mechanism which processes it for its information value.

Or take the case of the flute-playing centaur [40]. It represents what Husserl calls an "objectivational formation". It is "objectivation in the sense in which the objectivated is called objectivation, and not the sense in which objectivation is the name of a physical mental process." He warns that "one should not confuse just this mental process of imagining with what is imagined by it as imagined." As he writes,

> As a consequence, in spontaneous abstracting it is also not the *essence* which is generated but instead the consciousness *of* the essence; and the situation for this is that, and obviously essentially, an *originary presentive* consciousness of an essence (ideation) is in itself and necessarily spontaneous, whereas spontaneity is extraessential to the sensuously presentive, the experiential consciousness: the individual object can "appear", can be apprehended by consciousness but without a spontaneous "activity" performed "upon" it.

Note how the item 'essence' is kept apart from the act of 'spontaneous abstracting', which in turn is seen as an act of consciousness. In addition, even thinking of an essence is relegated to the spontaneous and therefore to consciousness, with the understanding that it is still not the essence itself. Because for us humans nothing can be known unless it exists in some form of mental representation (and that includes especially a highly abstracted form of content), and furthermore, because we can only ever know something in the terms under which it is thus represented, from the perspective of thought structures we can say that the item 'essence' as being used presently is done so under the assumption that the content of that item actually exists outside our sphere of consciousness (what is also apparent in the above paragraph is the usual decomposition into various thought forms). Not to treat 'essence' as just another abstract but as something altogether different from any ideation sets in train associations with similarly characterised contents of TSs, most of which would be found beyond the logic boundary.

In his defence of the above schema Husserl addresses the idealists, for whom such abstractions are also more than just thoughts, but in the opposite direction:

> Anyone who regards ideas, essences, as "physical structures," anyone who, in the case of those operations of consciousness in which the "concepts" of colour, shape, etc., are attained on the basis of an exemplificatory intuition of physical things with their colours, shapes, etc., confuses the resulting consciousness of these essences (colour, shape), with the essences themselves, ascribing to the flux of consciousness as its really inherent component part something which necessarily transcends it: anyone who does this, on the one hand, corrupts psychology since it concerns even empirical consciousness; on the other hand (and this is what interests us here) it corrupts phenomenology. [41]

Maintaining the middle ground between Husserl and the idealists we would suggest that the decomposition of thought processes becomes problematic when it concerns the thinking about already abstracted objects. Writing about such things as "reflective intuitions" and "objectivations taken universally", the raising into consciousness of items derived from such processes invites, or even demands, a further classification [42]. These would be objects "no longer pertaining to the senses", and Husserl suggests two options; one, to proceed towards "noetic syntheses" (ie treating them in terms of their relationship with empirical data), and two, establishing a new set of identities of "positings". Explaining his own choice he says,

> If we take the latter direction, then we encounter the *noeses of feeling, of desiring, of willing* (first of all, the simplest ones possible, i.e., free of syntheses at lower or higher levels), which are founded on "objectivations," on perceptions, on memories, on sign-objectivations etc., and which, in their structure, show obvious differences in level-by-level founding.
>
> ..For example, an aesthetic liking may be founded on a neutrality-consciousness with a perceptual or reproductive content; a gladness or a sorrow on a (non-neutralised) belief or belief-modality; a striving for or against on the same, but related to something being valued as agreeable, or beautiful, etc.

These additional "objectivations", in terms of their potential to give rise to ever new abstractions, seem to occupy the same categorical region as the objects of their previous focus did. Under the terms of Husserl's system however that cannot be the case; since it is not the functionality of the abstract which it at issue here but the environment in which it occurs, a new set of descriptive features has to be found to accommodate their articulated presence. Hence we read,

> Before going into the species of this structure, what interests us here is that with the new noetic moments *new noematic moments* also make their appearance in the correlates. [43]

He continues,

> On the one hand, there are new characteristics which are *analogous to the modes of belief* but possess, at the same time, *themselves* doxo-logical positiableness in their new content; on the other hand, connected with the novel

> moments there also are novel "*apprehensions*," and a new sense becomes constituted which is *founded on*, while at the same time embracing, *the noesis underlying it*. The new sense brings a totally *new dimension of sense*; with it no new determining parts of mere "things" are constituted, but instead *values of things*, value-qualities, or concrete Objects with values: beauty and ugliness, goodness and badness; the use-Object, the art work, the machine, the book, the action, the deed, and so forth.

Under the current schema such a multi-tiered layering of abstractions (using our nomenclature) implies having any perceived empirical nature on one side and any essential nature on the other attributed to abstractions according to their place within the general TP. That is to say, the decision whether to nominate some object (meant in the widest possible sense) as the product of an empirical act (ie intuition) or as the result of further processing, rests with the tendentious influence of the driving force behind the TP, that is its container. Given also the preferential position of those objects acknowledged as essences the entire process is therefore under the auspices of the overall ambience of the total structure—the large-scale TP governing the system. This may not mean much one way or another, were it not for the kind of members included in the set of possible candidates: beauty, ugliness, goodness, etc but also art work, the machine per se, the book per se, an action per se. Recall the structural format generated by illusion trees and the PAT chains contributing to the dynamics of a container; the output of an illusion tree is a compacted TS which has now become associative in terms of its modified nature, and a similar outcome goes for i-PAT-ls. Since during such a process the grounded links are discarded along the way, the eventual meaning that *can* be found within the contents of some upper-level TS will not be informative in terms of empirical facts (the connections to them are gone) but through their associations with other, similarly esoteric concepts which are held to be above reproach. Note that, in practical usage, the issue is not this or that manner of classification; problems can arise because of what the thus produced idea can be presented as, what meanings can be attached to it. As if to underline the situation, here is what Husserl writes in conjunction with evidence-producing reason:

> In addition [to the various acts of positings, such as doubt, suppositions, etc], naturally, it is necessary to subject to a comprehensive eidetic analysis the processes in the sense, as the appertinent *position-materials*, which are essentially determinative for alterations in the posited characteristics (e.g., the processes of "conflict" or "rivalry" of appearances). For here, as everywhere, in the phenomenological sphere there are neither accidents nor facticities: everything is motivated by essential determination. [44]

There is only one area from where the acts of "essential determination" can draw their justification: amongst the associative network of upper-level abstractions derived via the i-PAT-ls in the service of illusion trees and, ultimately, the container.

We find an example of just how determinate the system can be with regards to the content of a thought in the chapter on nature and the perceptions derived from it:

> We had earlier drawn a contrast between just being conscious, by way of seeing, of the blue sky and the theoretical performance of this act. But we are no longer performing the seeing in this eminent sense when we, seeing the radiant blue sky, live in the rapture of it. If we do that, then we are not in the theoretical

> or cognitive attitude but in the affective. On the other hand, though we have adopted the theoretical attitude, the pleasure may very well be present still, as, for example, in the observing physicist who is directing himself to the radiant blue sky, but then we are not living in the pleasure. There is an essential phenomenological modification of the pleasure, and of the seeing and judging, according as we pass over from the one attitude to the other. [45]

Apart from demonstrating yet again the character-equals-content format of the system, a particular act of consciousness is treated according to the exterior purpose of the act (eg, to observe, to analyse, to find pleasure, etc). But the validity of the act's content is not so much determined by its associative properties delineated from the wider implications of the act (eg, one observes in order to know about the weather, one analyses in order to know about the atmosphere, one finds pleasure in connection with a beautiful day), but by the categorical validity prescribed by whatever the pertinent "phenomenological modification" might have in store.

All this of course represents a form of articulation within the conscious part of the TP. Let us see therefore how Husserl addresses consciousness within the context of phenomenology.

We have come across examples of reciprocity in two different contexts, the functional similarity of their dynamics evoked by the consistency of Husserl's system, and we surmised about the influence of some FAs which led to a repetition of certain actions that had been ascribed to the respective abstracting processes. Something similar occurs with the description of consciousness in two areas, sensuous perception and abstract thinking.

> I always find myself as someone who is perceiving, objectivating in memory or in phantasy, thinking, feeling, desiring etc.: and I find myself actively related in these activities *for the most part* to the actuality surrounding me.
>
> ..I busy myself, let us say, with pure numbers and their laws: Nothing like that is present in the surrounding world, this world of "real actuality". The world of numbers is there for me precisely as the Object-field of arithmetical busiedness; during such busiedness single numbers of numerical formations will be at the focus of my regard, surrounded by a partly determinate, partly indeterminate arithmetical horizon; but obviously this factual being-there-for-me, like the factually existent itself, is of a different sort. [46]

There is the recognition—and hence the existence of relevant TSs—of a surrounding "actuality", to which his consciousness is attuned. There is also the articulation of other worlds, such as "pure numbers", to which the conscious mind attends every now and then. One could hardly deny the actual presence of one's surroundings, but to label the realm of numbers as the "Object-field of arithmetical busiedness" seems to imply something more than mere mental constructs. While "Object-" starting with a capital letter denotes the generalised version of some definite object (we know this from the book's introduction), the perception extends further:

> *The arithmetical world is there for me only if, and as long as, I am in the arithmetical attitude.* The *natural* world, however, the world in the usual sense of the word is, and has been, *there for me continuously* as long as I go on living naturally.

One could argue that in the end Husserl is very much aware of the difference between a natural field and an arithmetical one, and in a direct sense this would be no doubt the case. Consider however the necessary structural configuration in terms of TSs and their associations in either case as far as his system is concerned. If a particular content happens to be associative with a PAT chain delineated from a straightforward perception, its associative properties would be different from another abstract concerned solely with other abstracts, possibly even ungrounded ones; the principles of congruence will see to that. Yet here both formats are being treated equally when it comes to their potential for associations within their respective domains, always keeping in mind the ultimate link to the system's highly abstracted prerogatives as demanded by phenomenology. So, either the "natural world" is taken for what it is, ie an actual, existing, concrete collection of things, in which case any other world would be seen as similar in terms of its validity; or, none of them are considered to be the 'real' thing at all, including the "natural world" which now must be held to be equally non-actual. To define one's consciousness as a medium that encompasses any type of such formats places a predetermined authority at its feet which it may not always deserve. It does not matter whether such a mind regards everything as quite 'real', or everything as 'un-real'—the process of articulation ensures the selective associations according to one or the other rule. Since articulation is only concerned with the conscious part of TSs, the subconscious regions will not be affected, at least not directly. What would be the ultimate outcome of a discrepancy between wholly grounded PAT chains and those who are subject to alterations at the point of being raised into consciousness, is hard to describe in the absence of appropriate material; after all, the writings of Husserl are a form of articulation and articulation only, without any accompanying signs coming from other behaviour forms.

In his chapter on "consciousness and natural actuality" Husserl subjects consciousness to what he calls a systematic, "eidetic" analysis, because he needs "a certain universal insight into the essence of *any consciousness whatever*" [47]. This is in line with his overall procedure, but here the subject is the very thing which had been the medium in all the other considerations. Given the general format, a similar placement of consciousness within the entire schema is to be expected:

> In these studies we shall go as far as is necessary to effect the insight at which we are aiming, namely the insight that *consciousness has, in itself, a being of its own which in its own absolute essence, is not touched by the phenomenological exclusion.*

The ungrounded nature of TSs which are held to be significant in relation to the system favours the purposeful manipulation of associative elements in order to bring any additional contents in line with the overall requirements (we have already encountered several examples). To state therefore, "we shall go as far as is necessary to effect the insight.." carries certain overtones of whimsy. If there is a TS which holds some content pertinent to the concept of 'consciousness' then it will be embedded within the TP such that it can be in congruence with already established structures. In addition it leads to the acceptance of an 'essential consciousness' which exists beyond the bounds of human physicality; a notion which would be supported by Husserl's general TP. Just as the conceptualisation of any other essences involved the isolation of their perceived nature from meanings based more on empirical input, so does the notion of consciousness undergo the same treatment. The "phenomenological reduction", that is the paring away of anything which could be linked in any way to concrete appearance, is applied once more:

> It now becomes clear that, in contrast to the natural theoretical attitude, the correlate of which is the world, a new attitude must in fact be possible which, in spite of the "exclusion" of this psychophysical universe of Nature, leaves us something: the whole field of absolute consciousness. Instead then, of living naively in experience and theoretically exploring what is experienced, transcendent Nature, we effect the "phenomenological reduction". [48]

Note that the creation of a "new attitude" implies a whole gamut of specific TSs, spawned under the associative regimen of the system, and therefore pre-setting their format not in terms of their content (which mostly would have come from 'below'), but according to the ambience of ungrounded, high-level abstracts. The result is,

> That, then, is what is left as the sought-for "*phenomenological residuum,*" though we have "excluded" the whole world with all physical things, living beings, and humans, ourselves included. Strictly speaking, we have not lost anything but rather have gained the whole of absolute being which, rightly understood, contains within itself, "constitutes" within itself, all worldly transcendencies.

In the above paragraph we find traces of the overall, container-driven intent ("..sought-for.."); the exclusion of grounded chains ("..have 'excluded' the whole world.."); the hierarchy of values when it comes to the contents of abstracts ("..not lost anything but rather have gained.."); and lastly, the perceived configuration of existence as such ("..contains within itself .. all worldly transcendencies"). There is a further emphasis on the strict distinction to be employed: while under the system "the whole world is ultimately a single Nature" [49], "the situation is fundamentally and essentially different" when it comes to "mental processes as absolute essentialities". That region is "a strictly self-contained domain, yet without any boundaries separating it from other regions". He concludes the remarks by saying,

> In its *essence* it is independent of all worldly, all natural, being; nor does it need any worldly being for its *existence*. The existence of a Nature *cannot* be the condition for the existence of consciousness, since Nature itself turns out to be a correlate of consciousness: Nature *is* only as being constituted in regular concatenations of consciousness.

There is a significant difference between a concept which is held to be related to some other concept but effectively so; and a concept which is declared to be unique from the very beginning. For example, suppose I describe the workings of a reciprocal engine and then of an electric motor, and point out the inherent differences between the two. The entire TP is about engines, and under those terms I can relate, compare, and distinguish by pointing to one example and then another; that is to say, I can make use of the higher-level abstracts to build bridges between otherwise unrelated TSs because at those levels they do connect after all (the situation is somewhat similar to the abstraction process used in problem solving—see chapter 7). In the case of two unique concepts however a relationship is avoided on any level. In practice this means the separation of all characteristics belonging to the respective domains from each other such that any characteristic is not linkable in whatever way or form. Going back to the associative format of TSs within the phenomenological system, such a separation entails the impossibility for any, possibly grounded, PAT chain to be denominative with respect to the concept in question. To put it crudely, the isolation ensures that fantasy is given free reign, and there's nothing which can prevent it (but notwithstanding Husserl's obviously applied intellect and diligence in those matters). It is a sophisticated version of an age-old approach, the mind-body dichotomy.

Just as in our example of explaining about different forms of engines the concept 'engine' per se enables us to identify the specific elements belonging to this or that type, so does Husserl avail himself of his spanning concepts to describe consciousness. In other words, the denominative nature of the encompassing TP produces the desired meaning in its sub-sets. Thus the world around us is more than physical, it is also "psychophysical" because it includes "all the streams of consciousness connected with animated organisms" [50].

Before going further let us consider what that means. If there is such a thing called 'consciousness' which is perceived as a significant player in nature to the extent of representing its own category of manifestations, then there must be a strong reason for such a declaration to be made; for instance, we do not talk about nature—or Nature for that matter—involving a stream of metabolism, or a configuration of bones. Having traversed a good part of human thinking by now we can say with reasonable justification that the cause of the special treatment of 'consciousness' lies in that object's capacity to articulate such a state of affairs (or *any* state of affairs of course). We are prompted to perceive 'consciousness' as a unique entity because, as an articulated concept, it allows us to issue statements about ourselves and the world. This, however, is all that it is—a faculty of our brain tissue which happens to produce various sensations that find their way into speech. As an analogy, let us suppose there are only robot astronomers in this world which (who?) are highly intelligent but lack any senses whatsoever except a facility to communicate with the outside. Suppose further all the information on celestial bodies they have access to comes in the form of print-outs of reams and reams of matrices filled with parameters describing whatever their space probes have come across. Then according to how we treat consciousness our astronomers would have us believe that in space one encounters streams of matrices! Being human we would regard that as nonsense, but only because we have learned to identify the sensation of seeing as the *medium* to carry information about shapes and colours and so on, and the sensation of hearing as the *medium* to carry information about tone and pitch, and the sensation of feeling as the *medium* to carry information about heat and cold etc etc. In other words, the world does not consist of 'seeing' as some separate, roving entity, perchance touching us with its facility, but 'seeing' (and all our other faculties) are only a function of our organism which happens to provide us with information we have *learned* to interpret. I have emphasised the word 'learned' because it was not always so: there was a time when the sensation was taken to be the object itself, the most prominent example—at least for people belonging to the Judaeo-Christian-Islamic belief system—being the idea that in the beginning there was 'light'. There still is the time though when consciousness is seen as something similar because the mind-body dichotomy complex prevented us from looking 'behind the scenes' as it were. As can be expected, no space explorer would ever find a star or a planet if all he/she/it would be looking out for were matrices; and no 'human explorer' would ever discover the thing called 'consciousness' if all they have is an ambiguous descriptor with nothing attached to it except fanciful notions of its utility (seen, naturally, from *their* perspective only).

Returning to our consciousness-revering Husserl, he is faced with a dilemma:

> Thus, *on the one hand consciousness is said to be the absolute* in which everything transcendent and, therefore, ultimately the whole psychophysical world, becomes constituted; and, *on the other hand*, consciousness is said to be a

> *subordinate real event within that world.* How can these statements be reconciled?

The answer is a product of the denominative process represented by a container, fed through the dynamics of illusion trees and adjusting the contents of ungrounded TSs to suit the container's environment. Accordingly, the entity 'consciousness' enters the real world and takes on the "characteristic of transcendence" (a semantic bridge is built between the TS of 'consciousness' and the complex of 'real world'). From 'pure consciousness' it changes into "real human or brute consciousness" "by virtue of its experienced relation to the organism" (a denominative act going from the chain comprising 'consciousness' towards the TSs representing humans and animals). It is only "by virtue of the connection joining a consciousness and an organism to make up an empirically intuited unity within Nature" that it is possible to have "mutual understanding between animate beings" (a TS based on grounded abstracts containing the assumption of understanding between "animate beings" has been created through the denominative influence of the organism-consciousness TS established earlier—an example of a cascading effect of DTSs). Note how, during the entire process of adjustments, modifications and even creations, the hierarchical order of prerogatives is being maintained; that is to say, while a highly-abstracted notion of 'consciousness' can alter the perception of what it means to be human, any newly crafted concept of 'human' will not be able to influence how one thinks about consciousness. Husserl:

> A *peculiar kind of apprehending or experiencing*, a peculiar kind of "*apperception*," effects the production of this so-called "annexation," this reification of consciousness. Regardless of that whereof this apperception consists, or of what particular kind of demonstration it may demand, this much is obvious: Consciousness itself, in these apperceptive involvements or in this psychophysical relationship to something corporeal, loses none of its own essence and can take up into itself nothing alien to its essence; indeed, that would be countersense. [51]

The ambiguous wording at the beginning of the passage hints at the lack of hard detail due to the relevant TSs' isolation; the portrayed dynamic reflects the format of the container structure; and the preservation of what 'consciousness' means has its cause in the hierarchy of priorities within that structure.

Any additional reflection on what consciousness can or cannot do is also harnessed to the overall requirements of its surrounding structures. Since its role in the abstraction process is evident, means have to be found to articulate the mechanism of moving from the conceptualisation of a perceived object to its interpreted essence. This is how Husserl expresses these series of actions:

> A synthesis [between a perceived object and its essence] can be *effected* step by step; it becomes, it arises in *original production*. This originality of becoming in the stream of consciousness is quite a peculiar one. The positing and synthesis become while the pure Ego actionally takes one step and each new step; the Ego itself lives in the step and "steps forth" with it. The positing, the positing-thereupon, positing antecedently and consequently, etc., is its *free spontaneity and activity*; the Ego does not live in the positings as passively dwelling in them; the positings are instead radiations from the pure Ego as from a primal source of

> generations. Every positing begins with a *point of initiation*, with a *positional point of origin*; so it is with the first positing, as with every further one in the concatenation pertaining to the synthesis. This "initiation" belongs precisely to the positings as positing qua distinctive mode of original actionality. It is, perchance, like the *fiat*, like the initiating point of willing and acting. [52]

The organisation of concepts in terms of a theoretical framework is a necessary starting point of any serious intellectual endeavour; without its formality any referencing to a standard, however temporary, lacks rigour. Yet the ultimate success of such a pursuit is not only the framework, but the availability of conceptual links between the initial statements and grounded abstractions, as has been pointed out on several occasions. Consider then what the current system provides regarding the topic 'soul'. To explore what 'soul' is, Husserl leaves aside "all vague experiential lore" [53]; he "renounces all erudition", because they only lead back from a "mediate cognitive function precisely to these original experiences"; he also stays away from "all convictions prevailing in the science of psychology", without however "disputing their rights". All this because,

> What we are seeking does not lie in the *consequences* of theoretical, mediate thinking but in its *beginnings*; we are looking for its most original presuppositions.

And, once having found it,

> This meaning lays down an absolutely binding rule for all theoretical research. Any deviation from it results in absurdity. This is indeed intelligible on absolutely universal grounds. Legitimate theory cannot accomplish anything other than the predicative determination, in mediate thinking, of that which was first posited by originary presenting intuition (in our case, experience) in a straightforward way as a being and thereby as a being of determinate "content" or sense.

While the value of experience is not denied, its content must be examined "in eidetic intuition" so as to identify the "essence of the experienced in general". Here is the by now familiar break with grounded PAT chains, the jump from c-PAT-ls across to a network of associations consisting of TSs that are governed by the schema. Even the format of experiences becomes aligned with its demands. Thus they exist as a "stream of consciousness", and are considered as a "totality", a "phenomenal unity" [54]. These "lived experiences" have the "unity of a flux of time" and so "possess a perceivable, graspable, unity". The influence of certain FAs encountered previously which hold the notion that a perception of an object must somehow have its equivalent counterpart in the 'real world' can be found here too: the "flux of time" seems to be more than just a metaphor.

Husserl may not recognise the jump across the logic boundary in the way it can be articulated within the format of our perspective on thought structures. Nevertheless, he is aware of the need for this transposition because he knows the inherent subjectivity of personal experiences, and so perceptions, of otherwise common objects. To arrive at the essence of an object one would have to by-pass the subjective contributions of our minds. He uses an example of meeting several human beings with whom he attempts to share his experiences of the world, only to realise that their perceptions differ markedly from his own [55]. It may even reach the point where his "actuality" is considered the "hallucination of someone who up to this point in time has been mentally ill". For him the only way to solve that difficulty is to follow the principles of phenomenology, to

> ..interrogate the thing-apprehension itself, there where it is an experience of an "Objectively-actual" thing, and we must interrogate the experience which is not yet exhibiting, but is in want of exhibition, as to what, inherent in it, is in need of exhibition, what components of unfulfilled intentions it harbours.

Since it is our consciousness, ie our awareness of ourselves as such and such a human being, which influences the act of perception, the task is to remove such knowledge and therefore its influence. Once this is achieved,

> ..the subject .. knows nothing of a human environment, knows nothing of the reality or even just the real possibility of "other" Bodies, understood in the sense of an apprehension of the human, and thus knows nothing of his own Body as understandable by others.
>
> ..It is clear that *the apprehension of the Body plays a special role for the intersubjectivity* in which all objects are apprehended "Objectively" as things in the one *Objective* space of the one Objective world. [56]

The thing to be perceived becomes then independent of the "manifold of appearances belonging to a multiplicity of subjects", and in its pure form is indistinguishable from its form which has been represented by someone else. Note the procedural distance created by that approach: where before we have the object, the act of its perception, and consequently a unique result which differs from a perception done by anyone else, we now have the object, the act of its perception, plus the act of transposing its abstracted representation into another realm, followed by the result. Since in effect the additional procedure is done by the same mind which performed the previous version, the desired objectivity would not derive from the mind's inherent characteristics but from the additional environment brought to bear upon the process. That additional environment consists of the principles assigned to phenomenology, a set of object and functional abstracts that themselves are the product of a transposition from grounded PAT chains to ungrounded ones.

With the involvement of high-level abstractions, particularly as they are seen as more valid than their lower-level ancestors, the question of God enters the picture once again.

> Shall we say that God sees the things as they are in themselves while we see them through our sense organs, which are a kind of distorting eyeglasses?

Husserl asks, and adds that in this case there should be the possibility of mutual understanding given the objects' inherent sameness [57]. Just as amongst humans a common platform can be reached due to the mutual understanding of our sensuous perceptions (regardless of their final product), so could the same be achieved if the "absolute spirit" sees the things "precisely also through sensuous appearances". He continues,

> And if not in that case, then God would be blind to colours, etc., and men blind to his qualities. Is there any sense, however, to arguing about which are the true qualities? The new qualities would again be secondary and would be eliminated once more by physics, which has to be the same for all, if the things are the same. Obviously, the absolute spirit would also have to have a Body for

> there to be mutual understanding, and thus the dependency on sense organs would have to be there as well.

The above paragraph includes several significant features which are worthwhile examining. First, there is the perception that the essence of an object (as articulated by humans) is relatable to what an *absolutely* knowledgeable entity would come up with as well (God being such an entity under Husserl's religion); the only difference being some possible quantitative shortcomings on our side. This conclusion is largely reached through the idea that essences are a separate estate which needs to accessed, by ourselves as well as God. Secondly, the sum-total of a perception relies to a significant degree on the ambience its appearance can provide in the form of shape, colour, etc, in other words, the interpreted form of physical parameters; even a divine being would avail itself of such an interpreted form. Physics, on the other hand, "has to be the same for all", and thus would provide a description without the risk of subjectivity. Rhetorically Husserl assumes for the "absolute spirit" to have a body—it would have to be a Body of course—in order to perform the interpretations.

Now, either there is a common ground in that sense between us and the knowledgeable entity or there is not. If there is, then the notion must exist (and hence the relevant TSs) that our capacity for generalisation is in tune with the ultimate design of the cosmos, and colours and shapes are by-products. For that to be possible our internal system performing these generalisations must either have the knowledge of each and every principled phenomenon in the entire universe (which we can assume it does not), or by its intrinsic nature it cannot help *but* act in tune with all the principles out there, even though it may not know their manifestations concretely. In both cases the ultimate distinction between God and ourselves would dissolve, since the difference is one of degree only, and even that is more a function of the number of resolved processes in time rather than a capacity as such.

If there is no common ground, then the mechanism through which we humans arrive at the identification of an essence is inherently subjective in relation to us regardless of what phenomenology can tell us, and so at best can only ever be an approximation of what the object really is. But if this is so, the result would have no effective relevance to the Real, and all our efforts would remain on the level of self-referenced fantasies. This, however, is against the tenets of phenomenology.

Therefore the first option would have to be accepted, and considering the necessary cognitive format it appears to fit within the system identified so far. We have seen the reciprocity in terms of the interaction between the cultural-religious container and the abstracted notions brought into contact with it at that high level, and we know how information is being transported from low-level grounded thought chains to those higher regions, making the necessary conceptual transformations possible and so preparing the data for their associative destiny with the 'highest'. In that regard the schema represents a closed system, which was to be shown.

Husserl's phenomenology is a multi-segmented configuration of structures which are driven by a hierarchical arrangement of containers, the main one being the philosopher's cultural and religious background. There is substantial reciprocity between the structures above the logic boundary, that is the line beyond which the idealised version of some content becomes ungrounded, with the proviso that the denominative effect of a structure proceeds from the centre outwards only. The major purpose of the system is the processing of abstractions to such a degree that any links with low-level precursors are not only broken but rendered unworkable, thereby accentuating the particular representation's abstractive quality. The hierarchy of representations is determined by the degree of abstraction to which they have been subjected, and in consequence by their distance from grounded links (the greater the distance, the higher their semantic value). While below the logic boundary the dynamics amongst the grounded PAT chains are largely unhindered (with the possible exception of selective associations towards certain illusion trees), and therefore that region being as large as is necessary to accommodate these chains, the function of illusion trees growing out of that area is to provide a one-way traffic across the logic boundary and upwards; there is no reciprocity this side of the line. Due to the strict differentiation marked by the logic boundary, there exist c-PAT-ls on the lower side and *virtually* congruent chains on the higher, in the sense that the latter are in agreement with other members of their set, although they are incongruent in relation to grounded structures. The emphasis on abstraction invites the substitution of articulated characteristics with purely mathematical equations, and while it may be tempting to regard the schema as 'reality-friendly' under the auspices of logical-mathematical functions, it should be remembered that even in this case a definition is not complete unless it contains quantitative parameters as well. It is those quantitative parameters which in the end constitute the appearance of human perception and as such cannot be foregone—without them humans could not make sense of their world.

The potential of a system's constituent part is not merely dependent upon the number of products per se; it is also a function of the *types* of products, because it is amongst the set of types that delineations assemble to continue on differing paths. With improved hindsight the latency of the system becomes higher too, but such a parameter can only ever be an abstraction in itself—it is not a predefined variable unless we subscribe to Husserl's notion of pre-fabricated essences floating about looking for someone to articulate them. Rather, its value stands for the difference between a position on the range of already identified potentials and the currently ascertainable degree of latency. Then we are justified declaring the latency at the present level because we know that at some stage previously some of it had been realised and therefore the latency must have extended that far at least, and we also know that it extends still further because since then more of it has been realised again. As such the latent potential within a system cannot be used for the purposes of prediction, but it can be used to make the inherent momentum of the system apparent during the process of moving from realised potential to its degree of applicability.

So we can see the steady churning of mind-related concepts during the Middle Ages as a progression of religion-dominated abstracts lending their ideological principles to the attentions of philosophers who set about going over the same ground over and over again, the only difference being the personal courage they applied when dealing with the Church in a critical manner. We also notice the gradual diversification of intellectual material as Europe moved through the Age of Enlightenment and discovered an ever increasing number of insights, later turned into useful products. And by the time we reach the 19th century science and technology have given us a vast array of facts and figures, most of which proved to be first the latent, then the realised potential of ideas and so ultimately their applicability for the entire system which was European society. For a philosopher like Husserl the latency of the mind-concept provided the gamut of complex abstractions drawn from his knowledge pool—philosophy—as well as those of other fields such as physics and mathematics, religion and history. During his time the development of medicine led to the beginnings of psychiatry, and that branch offered yet another opportunity for basic ideas to be processed along quite particular lines. Another type of approach had come into being, and we are now turning to someone who utilised the improved latency of the greater range the approach provided in new and radical ways.

Carl Gustav Jung (1875 - 1961) set himself the task to explore the mind as an exercise in scientific enquiry. Observation, analysis, and conclusion—and yet, like everyone else before him the constructs he built rested, had to rest, on the foundations of his personal disposition, formed by culture, religion, and societal environment. His entire output was prolific to say the least, but we will concentrate on his last, *Memories, Dreams, Reflections* [58], because it is here that he recounts the nature of his background, his general life, and his theories in the most compact form.

In the book's introduction, written by Aniela Jaffé, the person who recorded and edited the text we are about to encounter (a fact that should be remembered throughout), we are told that "Jung's distaste for exposing his personal life to the public eye was well known" [59]. Writing began in 1957 when Jung was in his eighties, and "it had been proposed" in the form of an autobiography with Jung as the narrator. Not long after there is the paragraph:

> Jung read through the manuscript of this book and approved it. Occasionally he corrected passages or added new material. In turn, I [Jaffé] have used the records of our conversations to supplement the chapters he wrote himself, have expanded his sometimes terse allusions, and have eliminated repetitions. The further the book progressed, the closer became the fusion between his work and mine. [60]

In addition we have written comments by Jung, referring to a request by an old friend of his to tell the public about his youth. While at first reticent, he lately "discovered hidden in my memories certain objective problems which seem to call for closer examination". In order to perform the task objectively, and because it "has proved so difficult and singular that in order to go ahead with it, I have had to promise myself that the results would not be published in my lifetime". This promise was essential to ensure for him the "necessary detachment and calm". The reason, it had become clear to him, was those memories which had "remained vivid" and had to do "with emotional experiences that arouse uneasiness and passion in the mind", although he does not mention whether he meant *his* mind or that of the reading public. He goes on to explain the difference in recollecting his "outer" and his "inner" experiences, where the former have largely "vanished" from his memory because, although having participated in them "with all my energies", they all have turned into "phantasms which I barely recollect and which my mind has no desire to reconstruct", since they no longer stir his imagination. He continues,

> On the other hand, my recollection of 'inner' experiences has grown all the more vivid and colourful. This poses a problem of description which I scarcely feel able to cope with, at least for the present. Unfortunately, I cannot, for these reasons, fulfill your request, greatly as I regret my inability to do so... [61]

After which Jaffé adds,

> This letter characterises Jung's attitude. Although he had already "resolved to take the plunge," the letter ends with a refusal. To the day of his death the conflict between affirmation and rejection was never entirely settled.

Before we even get to the text itself we can already state the following:

there is a mind that employs a TP dedicated to a field of knowledge through which it had presented itself to the public for a great number of years (the professional TP);

there are other TPs that incorporate TSs relevant to other issues of daily living, and they are deemed to be in need of reconciliation with the professional TP;

through the years the need for this reconciliation has waxed and waned in accordance with the perceived significance of TSs coming from the professional TP and the others, when posited against each other;

the perceived significance has been producing feelings described as unease and passion in Jung or others, that is the TSs attracted negative emotional modules (either from inside associations or by importing material from the outside);

the EMs influenced the reconciliation processes along the way;

for the professional TP to remain valid in terms of producing abstractions which satisfy the associative properties of already existing TSs, and for the other TPs to remain the same in terms of satisfying the needs of the mind's persona, *and* for the reconciliation process to remain intact nevertheless, there is a tendency to develop twin structures in order to accommodate any incongruence resulting from functionally possible, but emotionally incompatible associations;

over the years the ongoing process of reconciliation, in the presence of twin structures, toned down the *contextual* significance of compromising TSs, but heightened the *functional* importance of the same—this ensured a reduced risk of yet additional, negativity-imbued thought complexes due to content, but also enabled the formation of functional associations across a range of PAT chains without incurring an unwanted emotional load;

the eventual articulation of the current state of TPs to a separate receiving and responding mind causes an induced congruence within such a feedback process, skewed towards a mutually comfortable description of past event states in terms of currently perceived significances.

The foregoing almost looks like an anatomy already, but so far we lack the content through which the type of functionalities lending their framework to the above can be delineated. For that we now enter the book proper.

As always we begin with the mind's surroundings, that is the general source from which the additional elements were drawn, gradually building a cognitive landscape consisting of very specific features. Here we need to be aware of two things: firstly, the material to be gathered has been supplied by someone who, from the position of hindsight, expressly sets out to introduce the reader to what seems pertinent to him; in other words, it is not a series of texts that acts as the medium through which we attempt to glean the items of interest to us—the items have been purposefully provided by the author for that very reason. Secondly, the author happens to be one of the major contributors to psychoanalysis and so can be expected to construct his narrative just so, controlling (as far as would have seemed necessary to him) the placement and relationships of items. Therefore the man's attitude towards himself, his standing in relation to society, and how he sees the role the contents of his memory play with regards to any recollection destined for public consumption, are significant issues.

Right at the end of his introduction Jung makes an interesting remark, something that sets the tone for everything which is to follow. Referring to his choice of material to be recounted he writes:

> Similarly, other people are established inalienable in my memories only if their names were entered in the scrolls of my destiny from the beginning, so that encountering them was at the same time a kind of recollection. [62]

Combining the gist of the above with the earlier cited comments about his "inner experiences" it appears that the general character of the current content of his TSs represents a selective range of abstracts which are in congruence with the precepts demanded by his professional TP. In order to achieve this there would have to be a set of FAs derived from many previous experiences which performed for that very reason, albeit under varying circumstances. These FAs, being of a functional nature, would have a particular affinity with scenarios where functional associations rather than contextual ones, were called for. That is to say, from the erstwhile beginnings these FAs, having come into being due to the demands of the day, acquired a degree of importance in Jung's mental processes which they would not have gained otherwise, and they participated increasingly in the organisation of TSs. It is a development prompted by the need for a mask (we will deal with that shortly) as well as by his intellect's preference for storing and processing information by principle rather than detail. A few words regarding the latter are in order.

Suppose I want to know the amount of tax to be paid for any given income, to be recalled later. Assuming a progressive tax system, I have two options. One, I could memorise certain income levels in terms of their actual amounts and at the same time memorise their appropriate tax values. Very quickly however the sheer number of figures to be stored would become unwieldy (especially for a memory like mine) and so I can avail myself of the second option, which involves the storage of only as many figures as there are income levels (maybe four or five) plus the formula associated with each level. In terms of memory usage the second choice is easier, but the trade-off would be the added effort needed to calculate the actual tax for every given income. In pre-computer days therefore people used pre-calculated tables so that the calculation part would be minimised while at the same time doing away with the need for the recall of detailed figures; one prime example were the books of logarithmic tables. The same system can be applied across a whole range of situations where, instead of the actual objects themselves, their principle functions are added to one's memory. The process of articulating a particular memory then consists of recollecting the principle which ought to be associated with the object currently under focus. For instance, asked what I had talked about with a friend of mine at such and such a time and date, I could either try to recall the memory complex in terms of 'personX + conversation + holidays + fishing + trout' (let's call this contextual storage), or I could try to get the same effect by accessing 'personX + goto[recent activities of personX] + goto[known preferences of personX] + goto[possible fish to catch given the results from before]', where 'goto' denotes a temporary branching-off towards a separate knowledge pool in order to retrieve some information (this would be a functional storage). While this may seem lengthier at first, it has the advantage that I don't have to memorise a separate 'holiday' item for each and every one of my acquaintances, but merely need to maintain one collective 'holiday' domain to which I can link from the 'personX' domain via associations.

Such an approach requires the ability to abstract and to think recursively (the branching-off), the number of abstraction levels depending upon the complexity of the scenario, and so becomes a function of one's intellectual capacity. Utilising it empowers the mind to a far greater degree than contextual storage can ever achieve because the mind's energy is channeled towards the current need for processing, rather than the ongoing storage of data which might never be used anyway. Going back to our calculation tables, the reason they have become obsolete in our computer age is because now the human task has been reduced to one of providing the actual data only, while the computer performs the application of a set of algorithms which process those data, with the complexity of the algorithms not being an issue any longer—they only have to be written once. (As a not entirely irrelevant and analogous aside, it should be mentioned that for the computer even the most complex operation merely consist in an organised placement of 0s and 1s, the eventual meaning relying on the collective interpretation by an outside intelligence—so far that's us. Hence complexity is nothing but simplicity multiplied)

In the course of one's life the mind will create a set of FAs dedicated purely to such a functional type of processing, where the usefulness of these abstracts is based on the overall ability of the mind to operate. In the case of Jung his ability would be on the higher end of the scale, to a point where the act of storing and processing of data (virtually all of which are derived from memory anyway) are almost automatically handled by functional storage FAs. This has certain consequences.

Because the organisation of data is largely handled by FAs, naturally the type of association under these circumstances will, to a large extent, be a functional one. Functional associations per se do not differentiate between cognitive and intuitive FAs, and so the process of recollection as such cannot be taken as the arbiter in deciding the validity of a consciously learned behaviour vs an automatically adopted one. That is to say, in the course of accessing some TSs for the purpose of articulating a past event, the details of that event will not be evaluated in terms of their inherent truthfulness, logic, or idiosyncrasy as part of the articulation process alone. They may well be scrutinised under the auspices of another structure which contains object or functional abstracts that delineate into one or the other type of PAT chain, but this is just the point: the recall process itself follows the tendencies of its own dedicated FAs and so leaves itself open to influences which have nothing to do with the character of the accessed abstracts themselves.

For example, consider the practice of musing in a relaxed state of mind. Here the articulation takes place internally, that is associations are formed and brought into con-TSs to be viewed as inner images. There usually will be some sort of focus or theme, but otherwise the thoughts are left to flow freely. Because the FAs involved will favour functional associations, any delimiting influence coming from their contextual counterparts are mostly avoided and therefore links are being evoked which ordinarily would be blocked. Thus combinations of contents occur (the contents accessed via their functionalities, not their content characteristics) that may well lead to insights never before contemplated. One often cited instance of just such a discovery is told of Kekulé von Stradonitz, a German chemist credited with the development of the 'closed-chain' theory of benzene in 1865 [63]. At the time the known molecules were viewed as open chains only, and the story goes that one evening Kekulé sat in front of his fireplace, dreamily imagining possible arrangements of molecules. The strands began to dance before his inner eye, weaving and curling, until one of them suddenly turned into a ring, locking one end to the other; this happened to be the solution.

Note the cognitive dynamics involved. There was a set of hitherto known molecules in the form of open-ended strands stored in the mind's TSs; there was also the knowledge about the chemical properties of molecules and the atoms which made them; and there existed the memory of geometric figures in the form of strands, chains, rings, and so on. During the recall process the FAs relevant to the geometric forms in their principled state, finding an affinity with similar FAs belonging to the domain of chemical structures, yet unhindered by any interference from the object abstracts of the molecules, began to form associations which would have been prevented by the OAs of both, the geometric forms as well as the molecules. But in terms of the active FAs there was no problem, and therefore the image of a molecule could be made to bend into a ring—the object 'borrowing' the functionality from a geometric form.

Note further that in this case the outcome was productive because the contents of the geometric forms domain as well as the molecules domain consisted of congruent PAT chains, grounded in experience. At the moment of the two ends' imaginary joining, the existing rule set did not lead to a fault because the set itself had been valid. The recalling process had no control over such matters and would have joined those ends eventually no matter what; the

FAs would have seen to it. Had it not been for the ordered mind of Kekulé some other affinitive perception would have made the outcome acceptable too.

Such a facility made possible by certain FAs comes therefore at a price. It's rather like being in possession of a credit card—it enables the owner to purchase goods which otherwise would not have been immediately available (its functionality overriding everything else at that moment), but unless there happens to be a financially sound regimen in the background (it existing independently of the credit card arrangements) the usage of the card can lead to problems further down the line. As for the mental arena, the facility so suited to intellectuals because of its use of different abstraction levels (in itself a measure of intellect) can be a potential trap in terms of i-PAT-ls—and its owner may not even know it.

In Jung we find a predisposition towards functional storage, gradually come into being by the combined forces of his intellect and the need to reconcile his own TSs with those supplied by outside systems. One of those systems had been religion.

His family was Swiss, his father a Protestant minister; indeed there had been a succession of other family members holding church offices. While both main Christian religions represented in Europe at that time, Catholicism and Protestantism, had their more and less intense adherents, the latter had always been prone to a somewhat more rigid implementation than the former. This is due to that particular strand's emphasis on individuality in matters of morals, the understanding of God, and so on, whereas the Catholic church relied more on its official tenets to prescribe and proscribe. In such cases the implemented difference lies in the degree to which the church member is held personally responsible for his or her actions (keeping in mind of course the general tendency of historical eras to enforce their doctrines at specific scales of cruelty regardless).

From his writings we see that Jung was quite well disposed towards his father, although in later, post-adolescence years the single-minded religiosity of the elder would introduce a certain ambiguity into the son's feelings. There seems to be no question however of having generally felt to be badly treated by either of his parents.

From the perspective of an outsider, especially one without religious fetters, all this is relative. Switzerland, for all its political stability and even maturity, had been and still is a conservative society. To be a Protestant minister in a conservative environment, charged with the identification of and response to religious transgressions amongst the flock, must by definition slant one's own range of acceptable behaviour forms towards the severe end of the scale. Add to that the family's rural residence, in itself a conservative influence, and the stage is set for a psychological ambience that would have been quite avers to an exploring intellect.

By now we are sufficiently familiar with the dynamics of abstract-associations to understand that their configurations are a matter of relativity; that is to say, an affinity comes into being due to particular cognitive elements being relatable, at whatever level of some applied standard they may exist. Therefore a thought association needs to be *understood* in terms of its own peculiar environment, and not someone else's (which is not to say they cannot be *identified* by someone else). So for Jung to live within his social and cultural environment was not a matter of mutual adjustment to that environment's detail, but rather a matter of coming to terms with its inherent nature juxtaposed against the inherent nature of those experiences which were exterior to that environment (we will revisit the topic of cultural discrepancies in Part II). These experiences involved events encountered outside his childhood circles, but also those that constituted his own, inner, natural self—in other words, his default abstracts.

We are being introduced to one such juxtaposition during Jung's description of his childhood, although significantly he does not label it as such. The event occurred in the form of a dream, and here is the passage, quoted in its entirety to preserve its flavour as it were:

> The vicarage stood quite alone near Lauffen castle, and there was a big meadow stretching back from the sexton's farm. In the dream I was in this meadow. Suddenly I discovered a dark, rectangular, stone-lined hole in the ground. I had never seen it before. I ran forward curiously and peered down into it. Then I saw a stone stairway leading down. Hesitantly and fearfully, I descended. At the bottom was a doorway with a round arch, closed off by a green curtain. It was a big, heavy curtain of worked stuff like brocade, and it looked very sumptuous. Curious to see what might be hidden behind, I pushed it aside. I saw before me in the dim light a rectangular chamber about thirty feet long. The ceiling was arched and of hewn stone. The floor was laid with flagstones, and in the centre a red carpet ran from the entrance to a low platform. On this platform stood a wonderfully rich golden throne. I am not certain, but perhaps a red cushion lay on the seat. It was a magnificent throne, a real king's throne in a fairy tale. Something was standing on it which I thought at first was a tree trunk twelve to fifteen high and about one to two feet thick. It was a huge thing, reaching almost to the ceiling. But it was of a curious composition: it was made of skin and naked flesh, and on top there was something like a rounded head with no face and no hair. On the very top of the head was a single eye, gazing motionlessly upwards.
>
> It was fairly light in the room, although there were no windows and no apparent source of light. Above the head, however, was an aura of brightness. The thing did not move, yet I had the feeling that it might at any moment crawl off the throne like a worm and creep towards me. I was paralysed with terror. At that moment I heard from outside and above me my mother's voice. She called out, "Yes, just look at him. That is the man-eater!" That intensified my terror still more, and I awoke sweating and scared to death. For many nights afterwards I was afraid to go to sleep, because I feared I might have another dream like that.
>
> This dream haunted me for years. Only much later did I realise that what I had seen was a phallus, and it was decades before I understood that it was a ritual phallus. I could never make out whether my mother meant, "*That* is the man-eater," or, "That is the *man-eater*." In the first she would have meant that not Lord Jesus or the Jesuit was the devourer of little children, but the phallus; in the second case that the "man-eater" in general was symbolised by the phallus, so that the dark Lord Jesus, the Jesuit, and the phallus were identical. [64]

The dream, Jung tells us, occurred between the ages of three and four and it was the earliest one he could remember.

What does such a tale tell us? Without going into the general details on dreams, we can nevertheless see the experience of a dream as a re-configuration of already existing TS contents but without the controlling influence of con-TSs. There *is* control, but it is based on associative intensities and EMs prevailing amongst the relevant sub-TSs. Keywords or phrases are "re-configuration", "already existing", and "without con-TS control".

We can therefore expect to encounter TSs which one way or another represent something significant to the person having the dream; the manner in which the significance is presented depending upon the cognitive environment that is now free from conscious domination. So first we have the setting in which the dream occurs. Here it is the vicarage and its surrounds, its associative properties playing an important part in Jung's life, particularly early in the piece. The whole theme of religion could in fact be seen as a kind of container, because it furnishes the principles through which the dreamer's TPs are being formed, which of course it does in 'real' life. Another principle (note the functional nature of these abstractions!) involves the one of reticence, indicating TSs which are not easily accessed, processed, and articulated. In that sense a functional congruence exist then between an unpleasant context and the dream event of having to do one thing after another in order to get somewhere, each step being of a somewhat odious nature. Those steps, in Jung's case the steady approach towards what turns out to be the throne in the chamber, heighten the sensation of apprehension and for a good reason (right now let us gloss over the question of how one can experience apprehension without 'knowing' what the source of it will be—see Part II). Anyway, there are TSs containing FAs which produce negativity when applied to what is to follow. The appearance of the room with its opulent layout suggests authority, and could be Jung's—functional—version of someone who is in charge (religious communities do tend to be hierarchical).

Which brings us to the 'throne thing'. Several items are notable. Firstly, the fact that the interpretation of the object on the throne occurred well after the dream itself, and so did not constitute the TS contents of a three- or four-year old. Rather, the interpretation came in two stages, the first one consisting of an articulation in terms of the object's assumed basic nature (ie the male sex organ), the other one being a more ornate version derived from a set of since accumulated data of a special nature. Since the latter happened "decades" after the actual dream itself, the relationship with the dream object could be considered somewhat tenuous, yet understandable given Jung's attitude towards certain pre-existing mental entities (we will touch on those shortly). Secondly, Jung says he does not know where the "anatomically correct phallus" could have come from [65]. Well, not quite anatomically correct: he gives the measurements as 12 to 15ft high and between 1.5 and 2ft thick. Taking the average of these figures (13.5 and 1.75), we arrive at a length-to-width ratio of 7.71. A real-life, erect penis, for a European of average tallness, measures somewhere around 18×5cm, giving a ratio of 3.6. Even allowing for smaller body size and a more diminished length than width (which is what would happen), the ratio would only marginally increase. To be endowed with an organ of a ratio as Jung would have it is not only physically highly unlikely but poses mechanical problems as well. To put this into context, applying Jung's ratio of 7.71 to our length of 18cm would give us a thickness of approximately 2.3cm; conversely, applied to a width of 5cm the length would have to extend to just under 38.6cm! As can be seen, Jung's estimate of relational measurements is out by a factor of over 2.

So, what is the point? The purpose of the dream recollection was to demonstrate the personal disposition of the author (or teller) by creating a symbolic aura around phallus, God, Jesuits, and himself (the allusion to Jesuits in the dream refers to his youthful consternation about them, being representatives from the 'other side' of religion and thus appearing circumspect to a Protestant—itself characteristic of the state of moral righteousness prevailing at the time). We also have to take into account the particular format of such recollection, the issues of hindsight, functional storage, and possibly unresolved questions as to the overall validity of religious thinking in general. What is significant therefore is not so much the occurrence of sexual content in a dream, but the fact that it happened to a young boy who at the time had already been imbued with moral repression but lacked the intellectual stature to either reject such imposition outright or at least ameliorate it down to an endurable level. Right now we do not have the necessary tools to properly deal with repression in all its origins and forms (this has to wait until Part II), but it is possible to remark, and this should be obvious, that the required hardware for getting an erection in terms of musculature and nerves in males is there from the very beginning. It is the higher-level processing of the relevant sensations in the brain which interferes with the otherwise natural act and it is here that we have to look for the reasons of Jung's terror.

Recall the drastic measures from earlier in the chapter that had once been taken to prevent an erection from occurring, to the point where physical injury was considered preferable to sexual feelings, and one can get an idea of how uncomfortable a middle-class Christian must have been when confronted with his own body (or hers, for that matter). Nor is such antagonism restricted to the past: the Courier Mail in Brisbane could report as recently as February 1995 about a Cairns City Council decision in far north Queensland to paint a pair of bathers over a rendition of Michelangelo's David, that nude figure being part of a mural that surrounded a building site in that city. The action was triggered by a complaint from a member of the public (to its credit the paper ridiculed the decision through its editorial [66]).

Given such primitive attitudes the denominative influence of a religion-driven TP over the articulation of something sexual can be expected to come with associations of a fearful, if not panic-inducing nature. It is debatable whether the sheer size of Jung's dream apparition can be related to his having seen an erect adult penis, assuming for a moment the possibility that his own may not have been so confronting given his young age. Equally however one can argue that we are dealing here with *functional* storage issues; in other words, it is the principle that counts and so age and size does not matter. After all, even by the age of three a child has already acquired the rudiments of cultural values, whether they be positive or negative ones (a fact particularly striking when observing a culture different from one's own).

A further remarkable item is Jung's insistence of having seen a "single eye" at the top of the figure. It is true that the open end of the urethra is sometimes referred to as an 'eye', but this is a linguistic substitute due to society's discomfort towards anything sexual and not because of any similarity with the real thing. A small boy would not have known that, particularly in a Christian household. As well, consider the spatial relationships at the dream scene: how could a child, barely three feet high, see what is atop an object ten to twelve feet above its own head as it stands before it on the same floor? Clearly, to articulate the event in those terms does not inform the listener about the dream as such, but rather about the mental and emotional state of the dreamer.

Next, the functional principle in a dream occurrence concerning an object which 'does not move but is somehow expected to move' represents a typical freeze-frame scenario indicating the mind's utter focus on something that stands for 'fear'—inherently, utterly, and without any distracting appendage such as the slightest movement. The stillness is the thing itself and nothing else, and the expected movement refers to the element of threat in terms of an imagin*able* act against oneself. It is an imaginary scenario which can be induced by the reader relatively easily; all it takes is a state of relaxation, a concentrated focus on what is considered scary, and the ability to hold such a picture with absolute stillness. Once the image is made to move the degree of fear immediately lessens. Warning: do not attempt this unless you are good at evoking and dismissing images, and unless you can expect that the image you *want* to evoke is the one you *get*.

Lastly, the reference to the "man-eater" is, apart from being another example of belated embellishment in terms of interpretation, a self-referential constituent of a fear-inducing TS because the thought complex's sexual nature could not be reconciled with the mind's container TP. Phrases such as "Lord Jesus" hint at the kind of balance manifested in the power relationship between the man himself and the introduced entities which populate his container, activating repressive FAs for the purpose of establishing a dependency between the real self (ie the self responsive to its default abstracts) and the abstracted control mechanism having been imported from the outside. For the mind to articulate an abstraction as the "dark

Lord Jesus", then to construct an association between it and another ideation representing a gigantic phallus, and to surround the whole experience with EMs spelling fear and terror, speaks volumes about such a mind's general state. Further chapters are added when that mind does not reject such sickly imposition outright but forces itself to labour under a siege of virtual penitence for years on end.

Nor did Jung perceive that dream to have been merely one of those bad nights. He already mentioned his additional interpretation a great many years later, so obviously the experience did not constitute a negligible memory. He returns to it several times in the book as well, almost like a thread to be followed as he works his way through the past. In tandem with those problems his relationship with religion started to suffer: "Church gradually became a place of torment to me" [67]. Such were his ties with the doctrine however, cemented in through daily exposure in general and cross-chained by varying factors such as filial responsibility, societal demands, professional reputation, and so on, that in his case the result was not a breaking-away from the bonds altogether but a switching of poles:

> It was then that it dawned on me: I must take the responsibility, it is up to me how my fate turns out. I had been confronted with a problem to which I had to find the answer. And who posed the problem? Nobody ever answered me that. I knew that I had to find the answer out of my deepest self, that I was alone before God, and that God alone asked me these terrible things. [68]

The above paragraph neatly depicts the cognitive circumstances as they unfolded. There is a mind, coming to terms with the continuing intellectual embedding of default abstracts which are supplied by his growing body; there is also the same mind having to reconcile itself with the conflicting demands placed upon it by the wilful introduction of unnatural perspectives like the apprehension towards sex. Being endowed with a high intellect and thus able to conduct abstractions any which way, the conflict produces association complexes peculiar to himself—there is no-one to tell, let alone ask for help. At the same time the mind also has to deal with the necessities of everyday life and in a manner which requires a demonstration of normalcy at that, such need being reinforced by an intensifying awareness about oneself.

Take for example his awakening sense of beauty, which for an intellect such as his must have been progressing in profound steps. Unfortunately, the greater the profundity, the more impact a destructive influence is able to exert. He tells about a visit to the museum in Basel, taken there by one of his aunts at the age of not more than six. After having spent hours looking at stuffed animals they got caught out by the closing time and had to make their way through a different section. There they happened upon the gallery of antiquities, and

> Suddenly I was standing before these marvellous figures! Utterly overwhelmed, I opened my eyes wide, for I had never seen anything so beautiful. I could not look at them long enough. My aunt pulled me by the hand to the exit—I trailing always a step behind her—crying out, "Disgusting boy, shut your eyes; disgusting boy, shut your eyes!" Only then did I see that the figures were naked and wore fig leaves. I hadn't noticed it at all before. Such was my first encounter with the fine arts. My aunt was simmering with indignation, as though she had been dragged through a pornographic institute. [69]

Since his aunt had not been a Queenslander and this happened around 1880 and not 1980, we can assume that her pains did not so much result from worrying about being branded a paedophile but from her own personal pathology in those matters. Note the following details here: a young boy thinks the human body is beautiful, it is the first time he has actually seen such things, the concept of 'nakedness' sprang associatively to mind only at the mention of "disgusting" and, although now the adult, the idea of nudity is separated into one TP containing "fine arts" and another dealing with "pornographic institute". Thus are slaves grown.

In another passage we particularly can observe the workings of a curious and therefore adventurous intellect. This time he was nine years old when his mother had a little girl. Jung hadn't noticed anything until his father told him of the birth, and the boy is totally confused:

> My father brought me to my mother's bedside, and she held out a little creature that looked dreadfully disappointing: a red, shrunken face like an old man's, the eyes closed, and probably as blind as a young puppy, I thought. On its back the thing had a few single long red hairs which were shown to me—had it been intended for a monkey? I was shocked and did not know what to feel. Was this how newborn babies looked? They mumbled something about the stork which was supposed to have brought the baby. But then what about a litter of puppies or kittens? How many times would the stork have to fly back and forth before the litter was complete? And what about cows? I could not imagine how the stork could manage to carry a whole calf in its bill. Besides, the farmers said the cow calved, not that the stork brought the calf. This story was obviously another of those humbugs which were always being imposed on me. I felt sure that my mother had once again done something I was supposed not to know about. [70]

The stage is set for the development of twin structures—one for personal consumption, the other for the world at large. As Jung puts it,

> Somewhere deep in the background I always knew that I was two persons. One was the son of my parents, who went to school and was less intelligent, attentive, hardworking, decent, and clean than many other boys. The other was grown up—old, in fact—sceptical, mistrustful, remote from the world of men, but close to nature, the earth, the sun, the moon, the weather, all living creatures, and above all close to the night, to dreams, and to whatever "God" worked directly in him. [71]

Once again we find a statement alluding to a sense of prescience. This is not the place to argue the possibility of a mind knowing events in advance. Quite curious theories have been proffered to explain the deep conviction some people feel regarding their perceived ability to predict the future, provided it is being accepted at all. Yet in Jung's case there is a set of mental configurations which could indeed contribute to such a belief. For a start, there is a container which consists of TSs of a particular nature, in this case ideas of a religious format. They in turn are composed of OAs and FAs, most of which are ungrounded and therefore are not affinitive to c-PAT-ls. On the other hand, they *could* be associated with structures whose contents are either contextually related or, and this is important for us right now, carry similar FAs because *their* domains are subject to comparable treatments. There is also a container representing the self, involving all those TSs which are instrumental in establishing one's identity. It is through the associative dynamics of that container and its related domains that the affairs of everyday life are conducted. Now suppose there is a conflict between the affinities of these two containers somewhere amongst the myriad of TSs and their abstracts, such that either a focus on the religion container with links to the identity container produces negative EMs, or a focus on the identity container with some of the religious abstracts being added attracts the negativity. Either way, the outcome is unsatisfactory and pressure exists to alleviate the stand-off. If the mind is capable of abstraction (and every human mind is, to a greater or lesser degree), a more desirable, wished-for scenario can be assembled which now includes the OAs that have been generated for that purpose, plus the FAs which represent the associative medium all along (the 'glue' that holds the complex together). The result is an ideation which consists of imaginary objects but already existing functionalities, where 'imaginary' means not necessarily phantasms but an abstract that is being expressly placed into that structure. Given Jung's propensity of articulating memory contents via posteriori abstractions (his "inner life" renditions), we end up with con-TSs that reflect current OAs and past FAs, where 'current' means the time of articulation. The variables in this situation are the associative intensity, the level of conflict, and the complexity of the con-TS dynamics. Hence in terms of the role the conscious thought processes play this is not a clear cut operation; should the con-TSs be in considerable control, the articulation will produce expressions such as the one in the preceding quote, that is to say it is the product of a wilful, organised endeavour. If the underlying dynamics are largely removed from conscious influence the result becomes what is commonly referred to as deja vu. Note that in principle the outcome is an experience which to the person is real—it is a thought sensation like any other and it will be treated as such (*basically*, this is *all* we ever *can* do), the only difference being that in the case of deja vu it is the unexpected suddenness with which the sensation occurs since it is beyond the control of the articulating conscious.

Since with Jung the occurrences would often fall under the auspices of the functional storage method he so favoured, their rise to consciousness would depend upon their associative readiness brought about by the existence of appropriate structures. In other words, there had to be some abstracts which were able to 'pull' the relevant sections of sub-TSs towards the con-TS active at the time. If such abstracts did not exist (remember, we are dealing with a mind growing towards adulthood), then no sensation will eventuate. In such a case a result at some later time would reflect the then current state of TSs, and therefore the conscious thought would be a function of the association complex at that point. We can say then that under those circumstances the perceived recollection of some memory complex is a representation of past data, paired with a functionality made possible under the currently existing rule set.

In the absence of attractors there is no resolution. Thus a conflict, such as Jung experienced in his attempt to reconcile religious sentiments with the demands of his body and the experiences of daily life, could be ongoing for as long as there existed no appropriate abstracts to act as associative relief. He tells us of one of these battles when he writes about a particular train of thought he had. The episode begins like this:

> One fine summer day that same year [he is now about eleven] I came out of school at noon and went to the cathedral square. The sky was gloriously blue, the day one of radiant sunshine. The roof of the cathedral glittered, the sun sparkling from the new, brightly glazed tiles. I was overwhelmed by the beauty of the sight, and thought: "The world is beautiful and the church is beautiful, and God made all this and sits above it far away in the blue sky on a golden throne and..." Here came a great hole in my thoughts, and a choking sensation. [72]

He felt numbed, he says, and a great fear gripped him lest his thoughts continued and revealed something terrible. What that terrible thing was he could not imagine, did not dare to imagine.

The FAs denoting 'fear' only had one OA association, and that was the concept of 'sin', leading to what must have been a frantic search up and down PAT chains to find any link to what that 'sin' object could pertain. During the walk home the exercise was repeated over and over again, each time coming to the TS denoting God sitting on a throne to be followed immediately by a terrified escape, as if his thoughts "had received a powerful electric shock". When coming home his mother commented upon his appearance (such mental states can show) but obviously was of no help. He slept badly, the shape- and nameless terror thrashing under the surface. This continued until the third night when "the torment became so unbearable that I no longer knew what to do". He now embarked upon a seemingly helter-skelter hunt for clues, which eventually led him, via countless associations within his religion TP, to Adam and Eve. There he picked up another trail of the 'sin' abstract, leading to

> God in His omniscience had arranged everything so that the first parents would have to sin. *Therefore it was God's intention that they should sin.* [73]

It was followed by

> This thought liberated me instantly from my worst torment, since I now knew that God Himself had placed me in this situation.

Having established God as the source of the problem, he began to see it all as some kind of test, and so the articulation of the elusive idea changed from a forbidden to a required task. Only at that point did the resolution come:

> I gathered all my courage, as though I were about to leap forthwith into hell-fire, and let the thought come. I saw before me the cathedral, the blue sky. God sits on His golden throne, high above the world—and from under the throne an enormous turd falls upon the sparkling new roof, shatters it, and breaks the walls of the cathedral asunder.

He felt "an enormous, and indescribable relief", "grace had come upon" him, and he "wept for happiness and gratitude".

The central theme is not the turd. It is the thought of somebody shitting—and not just anybody but God—and not just shitting on anything but the roof of a cathedral. Of course, had excrement been replaced by say, a pebble, the whole episode would have been of no consequence. Similarly, had excrement itself not represented such a distasteful object in Jung's mind, the idea would have possibly been a hilarious joke, to be shared with other kids. Toilet humour is nothing new, and in puritan parts of the world it is denounced as primitive, immature, if not downright sick. And yet it is nothing but a reference to a continually recurring, natural body function (in fact, if it *doesn't* occur regularly you have a problem!), the significance of which moves between allusion to a waste—and therefore expendable—product and the association with a substance that is dirty and stinks but at the same time can be summoned by anyone, high and low.

Take for instance the stories of Till Uilenspiegel, or Dil Ulenspiegel as the original version goes, a prankster who seems to have existed during the 16th century and acquired such fame as the archetypical fool who plays tricks on the high and mighty in German lands, that the popularity of his adventures continued through the generations to the present day, even to be made into children's tales [74]. Ulenspiegel became a persona, an icon, quite removed from particular historical individuals, several of whom were brought into the service of the idea. One such story has it that Dil was made aware by his travelling companion of the similarity between excrement and mustard, this condiment being one of the more expensive spices in those days. So one day they happened to stay at a castle as labourers when he was asked to fetch some mustard, which in its final form he had never seen nor tasted, pennyless fellow that he was. He goes down to the kitchen, finds the jar, and decides to shit into it remembering his friend's advice of how similar the two substances looked. Anyway, the mustard and everything gets served at the tables to one of the nobles, et cetera et cetera. Or the day when he stayed at an inn and was treated as a nobody while at the same time, at the higher end of the table, a priest received one morsel after another from the innkeeper. Night falls and both are given beds in the same room. Next morning the priest rises early, says his prayers, pays and leaves, whereas our Dil jumps into the other's bed and deposits the only thing he could afford freely. When the innkeeper discovers the stinking mass amongst the sheets Dil tells her something like, "No wonder—with all the good food you put in front of him I'm surprised he hadn't filled up the whole room!" After which the innkeeper roundly cursed the frocked guest but warmed to the poor traveller [75].

A somewhat different social setting is given by William Burroughs in his *Soft Machine*, where human faeces becomes a kind of social currency as "black fruit" amongst Mexican boys [76]. Nevertheless, here too the treatment is inclusive and without mind-wrenching connotations.

But back to Jung. With his moralistic baggage the combination of body functions and the religious would have caused a not inconsiderable disquiet; the more so should the supreme member of the religion TP get unwittingly thrown together with them. Not only that, it is that very member who performs the distasteful operation. There is also a more hidden angle. At age eleven Jung had already experienced precarious moments caused by his inability to make sense of various religious tenets in relation to reality. Yet at the same time his mind embraced the faith as a fundamental anchor to the world, not to be trifled with. Thus there is a discrepancy between 'God' itself and the symbols of Church officialdom. It is therefore not difficult to see the possibility of his youthful resentment, availing itself of elements of basically forbidden bodily activities (to be able to actually defecate an elaborate system of circumstances must be in place), getting back at the ordained order of things by imaginatively making God himself dropping a turd onto a human place of worship and thereby damaging it.

For his mind to articulate the final thought after such a battle the ideation itself must have existed already. It was the *con*-TS which, under the restrictive format, prevented access to what already had been assembled into an associative complex by the sub-TSs. And it was the con-TS structure that handled the anxious search for some way around the no-go zone of taboo subjects. When we talked about the absence of appropriate TSs in order to unlock certain areas of the mind therefore this should not be taken to mean that those TSs do not exist at all; rather, they do not exist as an affinitive abstraction to such areas. We are only consciously aware of things which can be sensated via conscious processes, and if they cannot, then, as far as we the conscious person are concerned, they do not exist—end of story. Recall the discomfort that accompanied Jung's attempts to connect with parts of himself; they were in the form of EMs which interfered with his other trains of thought and, because of the potential associative properties of TSs with their particular EMs, can take such interference into regions that offer themselves precisely because of such EM-induced linkages. In principle the situation is no different from any other PAT chain; there are TSs, attractors, congruent or incongruent chains, and affinities based on contextual and/or functional associations, not to mention the illusion trees and containers providing the overall organisation. That is to say, the *inherent* dynamics of unpalatable thoughts do not, in themselves, mitigate for or against their contents with regards to any standards that may be applied from the outside. It is their interpretation, coordinated by the conscious mind, which causes all the trouble. As an analogy (and it is by no means an original one), the entire scenario can be compared to a house with many residents, in which the cellar and its underground passages are administered by an authority figure who takes it upon himself to permit access to this or that room, regardless of the needs and wants of other residents. In some cases there may even be someone locked up in one of those vaults, forever separated from the rest upstairs! And isn't it interesting how the picture of a vast fortress, harbouring in its deepest innards a creature confined to a secret chamber yet vaguely controlling the lives of its inhabitants, can send a shiver down the spines of people?

Whether the conscious mind can summon a set of abstracts or not, their domain together with whatever links they managed to create, can still be active and hence participating in the clustering of ideation complexes. There they either support or hinder the development of PAT chains due to their associative potential. If they are of a positive nature, they channel the contents of related TSs towards the desired structure; if they happen to be negative, away from it. Note that in the current context 'positive' and 'negative' merely refer to the act of accessing some associations, regardless of whether the result will be considered good or bad by their owner. For example, if someone has a bad reputation, anything having to do with him will attract similar connotations; conversely, someone generally revered will engender good thoughts, to the extent that even a usually negative concept can be turned into something acceptable. One needs only recall the many fashions that have come and gone, and how something which looks so impressive one day can be utterly ridiculous the next. And yet, when pressed as to the actual reasons for acceptance and/or rejection, most people would not be able to articulate any.

Same with Jung. Writing about some experiences where he had been accused unjustly, but adopting a more general tone, he says,

> I knew so little about myself, and the little was so contradictory that I could not with a good conscience reject any accusations. As a matter of fact I always had a guilty conscience and was aware of both actual and potential faults. [77]

Here the sub-TSs exert a negative influence, but the conscious self will not or cannot circumvent them; the denominative force of his religion TP is too great, and it is too much intertwined with the TPs defining his self.

The allocation of negative EMs to other structures by virtue of subconscious, wilfully unapproachable content domains, can therefore cast an antagonistic shadow across a whole range of TSs, indeed entire patterns. The more comprehensively the mind associates its abstracts with those of a religious nature, the more pervasive the flavour of metaphysical thought processes will become towards anything the mind is dealing with. Should the overall mystical message be an affirmative one, with no major conflicts to be solved (*every* mystical concept must, by its very ungrounded nature, sooner or later produce some sort of idiosyncrasy and hence conflict), the person's general outlook with reflect such peace and harmony. On the other hand, hostile ideas will spread their influence and colour the inner landscape. As Jung writes, while discussing the afterlife,

> After all, the Buddha opposes to the painful illusion of existence his *quod non*, and the Christian hopes for the swift coming of this world's end. [78]

Considering the role FAs play, especially in the absence of the ameliorating influence of object abstracts, such an encompassing modification of one's re-represented environment by the subconscious can have quite chilling ramifications (more of this in Part II). In his youth Jung already had a foretaste of what might be to come:

> The influence of the wider world, this world which contained others besides my parents, seemed to me dubious if not altogether suspect and, in some obscure way, hostile. Though I became increasingly aware of the beauty of the bright daylight world where "golden sunlight filters through green leaves," at the same time I had a premonition of an inescapable world of shadows filled with frightening, unanswerable questions which had me at their mercy. [79]

He adds that the nightly prayers granted him a "ritual protection", but did not prevent the fear during daytime. The denominative influence of the religion TP and its interlacing with his daily routine can be observed.

While the subconscious control of thought processes largely makes itself felt through indirect means only (eg various feelings of negativity) and so will usually not be seen as such at all, there are just two options available to at least identify it. The first and more common one is via the importation of outside material, ie information from third parties (like this book you are reading). The other is an internal one, and it consists of 'stepping aside' as it were and analyse one's thought in terms of purposefully constructed abstractions. It represents a direct application of twin structures, except now they are not the inadvertent product of a pressured mind but an act of conscious will. The level of abstraction needs to be commensurate with the level the problematic TS contents occupy, and the 'stepping aside' has to provide sufficient separation from the rest of one's self so that objectivity can be maintained. As a relatively minor example, suppose circumstances force me to work with someone towards whom I experience a certain dislike. In order to neutralise these feelings and enable myself to proceed adequately, I can superimpose a persona upon my ordinary self that involves nothing more than the tasks required of me. Simultaneously I do the same with that other person, equally in terms of what he or she has to do for the current job. Hence we arrive at two virtual people, myself and the other, and stripped of any associations which make up the images under ordinary conditions. Note that none of the usual abstractions have been modified, deleted, or re-directed; they have simply been set aside for the time being. Occasions which virtually demand such an approach would be familiar to, say, police and ambulance officers, or prostitutes. The alternative is burn-out.

It gets much more complex when we are dealing with internal processes where the subject *and* the object are part of one's self. In that case the abstraction levels to be reached have to be correctly ascertained, particularly as it concerns the object under focus. In the previous example the operation is usually successful because the necessary level presents itself more or less automatically in the form of the person to be perceived. Once the object is not an outside perception but an item identified internally, the self-delimiting function of sensory input does not exist; it is us who have to furnish the outline of what it actually is that we are concerned about.

We find a problem of that nature in Jung's description of his mathematical abilities, and it leads us into the further realm of one of his specialties, the archetypes. But first the maths:

> Right into old age I have the incorrigible feeling that if, like my schoolmates, I could have accepted without a struggle the proposition that $a = b$, or that sun = moon, dog = cat, then mathematics might have fooled me endlessly—just *how*

> much I only began to realise at the age of eighty-four. All my life it remained a puzzle to me why it was that I never managed to get my bearings in mathematics when there was no doubt whatever that I could calculate properly. Least of all did I understand my own *moral* doubts concerning mathematics. [80]

By way of explanation let us consider two exercises.

We are given five candle holders and a box of many candles of the same length but varying thickness. The task is to pick five candles out of the lot which require the least modification so they can be placed into the holders.

Calculate $s = \sqrt{\frac{\Sigma(x - \bar{x})^2 \Sigma}{n - 1}}$ where s is the standard deviation, x is an item from the sample, x is the mean of the sample, and n represents the sample size.

The question is: what is the *principle functional* difference between the two?

If you answered something like "not much" then generally speaking maths does not hold any terrors for you; if you launched into some lengthy explanation, Jung for one would understand.

Consider: in exercise (1) we have—roughly speaking—the abstraction of a set of candle holders (an OA), another abstraction denoting the box with the candles (also an OA), and two FAs; one deals with the idea of modifying, the other with the placement of candle into holder. In general terms however this changes to, one magnitude of something, another magnitude of something else, and amongst the latter some occurring difference between its members and the members of the former, plus the relationship between those lots and the total. Exercise (2) is not much different; one magnitude (that of a sample item), another (the mean), the varying differences between these two, plus the ratio between these and the total. Then there is the additional operation of taking the square root but that's only a further refinement.

In other words, for somebody who is used to thinking in formulas conceptually it is not so much a matter of viewing the symbols as values to be added, divided etc, but as entities that interact with each other in a certain way—almost like the plot in a story. The trick is to raise the abstracts representing the symbols into some sensate form that allows the operation to be visualised and then observe the action. To some people this comes naturally, and they are the gifted mathematicians; for others, if introduced to the methodology, the erstwhile trepidation will dissolve; as to the rest, their lives will lead them as far away from Σs and $\sqrt{}$s as they can possibly make it!

So, if the generating of abstractions comes easily to the mind, then no matter what the actual purpose of that act, the necessary PAT chains will be established in the appropriate lateral and vertical direction, that is reaching to whatever content at whatever level. For Jung this seems to have been a problem, although there is no indication that his mind lacked the ability per se. However, since abstractions are usually associated with some functionality, *and* Jung makes use of FAs to a large degree (recall his preference for functional storage), *and* in many cases the tracing of PAT chains across his TPs invites problems (the religious conflicts), we can assume the possibility that his abstraction-forming FAs have been allowed to wilt, at least in the realm of his con-TSs. The mention of "moral doubts" in conjunction with mathematical thought processes would further confirm this. See also his statement somewhere else referring to science as something to be admired in the "conventional way", but also "giving rise to alienation and aberration from 'God's Word'", and even "leading to a degeneration which animals were not capable of" [81].

Which brings us to the concept of archetypes. He gives a very rough sketch of what he means by them in the later part of his book, yet there are certain subtleties which ought to be heeded.

> The archetypes, which are pre-existent to consciousness and condition it, appear in the part they actually play in reality: as *a priori* structural forms of the stuff of consciousness. They do not in any sense represent things as they are in themselves, but rather the forms in which things can be perceived and conceived. Naturally, it is not merely the archetypes that govern the particular nature of perceptions. They account only for the collective component of a perception. [82]

Just before that definition he talks about the conflict of duties in one's life, of opposing forces in the mind, and so the "inner polarity" of the psyche. This he sees as the "indisputable prerequisite for its aliveness", indeed "polarity is inherent in all living things". Such sentiments are not without foundation; just *how* fundamental they really are we will investigate in Part II. For the moment let us accept a form of energy driving the cognitive dynamics which is based on the difference between one state and another. Let us also recognise the fact that the word "state" is a descriptor, standing for whatever domain we care to identify: from the smallest unit, the information value represented by a single neuron, to an entire container featuring a multitude of functional entities. Given the underlying physicality of the system, the wetware, we would expect those functionalities to occur within a certain range of type, and not beyond. Jung's "polarity" therefore would manifest in certain terms, united by their basic mechanics, but differentiated by the type of functionality currently under focus. Since cognitive dynamics in general are neither here nor there as far as we are concerned unless and until they are given conscious form through articulation, it stands to reason that there would be a set of specific types of inner experiences which are described in similar terms across individuals, societies, and cultures. They do not "represent things as they are in themselves" as Jung put it, that is they do not stand for the actual content of such articulation, but rather the inherent principle, the "forms in which things can be perceived and conceived". Thus the skeletons of an Albrecht Dürer are but the articulation of the archetype 'death', which is not the same as saying a skeleton is an archetype.

And yet, to conceive 'death' as something more than a cadaver one needs the ability to abstract the dead object to a degree which elevates the ideation to levels at which it becomes a concept. Given the associative potential at various levels of abstraction, and given the state of the overall environment defining what can and what cannot be linked and what with, the articulation of an underlying concept then becomes a function of one's culture and one's knowledge. The more common and more fundamental the experience thus represented, the greater the similarity in expression and hence the commonality of the symbol, eventually attracting a descriptor such as 'archetype'. Jung provides some examples of archetypal images: a crusader in chain mail [83], symbols used in European alchemy [84], the triad in the form of a group of three men [85], the notion of immortality as some image preferred by the individual [86], characters in a Wagner opera [87]. The list gives some idea of the sheer diversity of these symbols, but it is not the symbols which have the power to enthrall—it is their meaning, their situatedness within the expansive matrix of thought elements our mind has formed as it responds to the surrounding life.

And yet again, Jung seems to go further. In the context of mentioning Wagner he writes, referring to his earlier years,

> I also did not know—no more than, even to-day, it is generally known—that the future is unconsciously prepared long in advance and therefore can be guessed by clairvoyants.

Considering his way of 'looking back in anticipation', and the often quite rapacious desire to have an unresolved train of thought brought to some conclusion, the above words can be seen as more than just an appreciation of the preparatory nature of the subconscious. They hint at a thought = object kind of perception we have come across before, in comparable circumstances of metaphysical denouement. Take for instance the passage where he recounts his experience with North American Pueblo Indians, who revere the sun as a god:

> I then realised on what the "dignity," the tranquil composure of the individual Indian, was founded. It springs from his being a son of the sun; his life is cosmologically meaningful, for he helps the father and preserver of all life in his daily rise and descent. If we set against this our own self-justifications, the meaning of our own lives as it is formulated by our reason, we cannot help but see our poverty. Out of sheer envy we are obliged at the Indian's naïveté and to plume ourselves on our cleverness; for otherwise we would discover how impoverished and down at the heels we are. Knowledge does not enrich us; it removes us more and more from the mythic world in which we were once at home by right of birth. [88]

It is one thing to metaphorically paint an inner landscape that consists of mythical entities; it is quite another to transpose such imagery into one's living space as if it was a substance to be partaken of and to be held as treasure, the absence of which would reduce one to poverty. True, knowledge per se does not enrich; it is its application through abstractive interpretation and applied back again to one's surroundings, which elevates one's stature. To do just that the PAT channels must be free of obstruction, and the ego at leisure to take—two conditions that are not always in place. Significantly, Jung equates the shadow land of subconscious sensing, the "mythic world", which we entered by virtue of the emerging capacity of our intellect, as a "home" and a birthright. If his exploring spirit, subject to the denominative influence of certain containers, suffers pain in the course of hitting locked doors, then one can understand the sense of longing it must feel when contemplating an existence without the driving curiosity that only chafes at the leash of ignorance.

The repositioning that his mind has to perform in order to reconcile a default-based ego with the demands of some other structure can result in a reversal of the cause-and-effect relationship between the self and its will. Recall his earlier problems with the image of God defecating on to the cathedral roof. Objectively it was the mind which tried to evoke an image but was hindered by being the recipient of a particular representation it carried as well. This was turned around so that now the representation assumed the active part and the mind became the recipient of the image. Only when the roles were reversed, when it was now God who put the thoughts to him rather than him producing the thoughts about God, only then did his mind allow the total image to rise to the surface of consciousness. In the light of such a reorganisation, consider the phrasing in the quote above: "It springs from his being a son of the sun", "his life is cosmologically meaningful", and "he helps the father and preserver of all life". All these expressions have one thing in common—they position the individual exactly opposite to what the cognitive dynamics suggest. So rather than using the sun for light and warmth, a Pueblo Indian sees himself as the sun's product; rather than being just a part of the cosmos, he is an active participant; rather than the sun driving all life, it is he who provides necessary assistance. Ultimately such a role reversal serves to furnish an identity, so much so that a dilution of these beliefs leads to the undermining of an entire culture; a phenomenon apparent wherever people of low abstractions are confronted with the pressures from more effective societies. The above also demonstrates a relatively superficial grasping on behalf of Jung of what it means to live without the fruits of a high-ranging mind resulting in a type of romanticism that has only grown during the ensuing decades.

PAT chains which literally turn into chains; apprehensions about flights of fancy; a re-formatting of the ego to escape responsibility; they all point to a fairly encumbered persona. Nevertheless, paired with a strong intellect, and the steady accumulation of successful applications of that intellect, Jung possessed enough inner strength to pursue his ideas even against sometimes formidable opposition. Raising certain objections for instance against Sigmund Freud's theories and going public with them caused not inconsiderable problems.

> After the break with Freud, all my friends and acquaintances dropped away. My book was declared to be rubbish; I was a mystic, and that settled the matter.
> ..That was a point I had thoroughly considered beforehand. I had known that everything was at stake and that I had to take a stand for my convictions. [89]

Sexuality, one of the central themes of Freud's work, is a precarious field at best of times on which to conduct one's stoushes, the realm of academia being no exception. Jung criticises Freud for not understanding the spiritual side of sex, although the latter was "deeply stirred .. by the phenomenon". In the same passage he describes his own underlying attitude as follows:

> Sexuality is of the greatest importance as the expression of the chthonic spirit. That spirit is the "other face of God," the dark side of the God-image.

Chthonic—a word referring to the spirits of the underworld and ironically derived from that source of erotic exuberance, Greek mythology, but now conjoined with Christian loathing to imbue it with cosmic dimensions—what *can* one expect from such a marriage but sheer negativity?

While not necessarily on par with some activists, Jung nevertheless acknowledges the dangers faced by those who think outside the bounds, and know it:

> The man, therefore, who, driven by his daimon, steps beyond the limits of the intermediary stage, truly enters the "untrodden, untreadable regions," [an allusion to *Faust*] where there are no charted ways and no shelter spreads a protecting roof over his head. There are no precepts to guide him when he encounters an unforeseen situation—for example, a conflict of duties.
> ..For such a conflict always presupposes a higher sense of responsibility. It is this very quality which keeps its possessor from accepting the decision of a collectivity. In his case the court is transposed to the inner world where the verdict is pronounced behind closed doors. [90]

"As above, so below", the alchemists said; a sentiment Jung must have known.

In the main, Jung's mental structure comprises a container holding contents of a metaphysical nature, derived from his exposure to Christian Protestant teachings from his earliest years onwards. Its abstracts have become interlinked with affinitive TSs within his ego TP, their denominative influence causing a re-positioning of the perceived role his self plays in relation to those abstracts. On one hand the resulting conflicts prompt an emphasis on the need to understand one's self, carried into the adult years by his choice of profession as a psychiatrist. On the other the modified TSs pertaining to his own persona evoke varying degrees of discomfort when brought into association with certain PAT chains that relate to such areas as body functions and sexuality because it is those that have been particularly targeted by religious doctrine. Due to the pressures of home, his profession and society in general, the conflicts were not resolved but rather integrated into the overall system, effecting twin structures to serve both societal and personal tendencies. His considerable intellect provides the appropriate environment for abstractions at many levels, leading to a preference for functional, rather than contextual associations and storage. At the same time, such an emphasis on one mode also invites interference along such very lines, given the nature of his containers. FAs have been generated which discourage the independent tracing of PAT chains, unless a priori measures have been taken to prevent the kind of discomfort endured so many times in his childhood. A priori measures turn into a priori perceptions, that is to say the spectre of fear being alleviated through replacing the ego as subject by an image of the ego as object, with the erstwhile focus of attention now being the perceived agent, provides an ambience of diminished responsibility coupled with an increased significance of the role the previous object is seen to play. Repeated endeavours by the con-TSs to pre-arrange certain contents amongst the sub-TSs leads to a tendency to view significant events as precursors to a particular scenario, the associative complex having been denominatively treated by the relevant FAs before being allowed to rise into consciousness. The trepidation which is experienced at the point of incongruence between c-PAT-ls and i-PAT-ls leads to an apprehension towards some of the c-PAT-ls, especially those which can be expected to transgress into the domain of i-PAT-l-based structures. Therefore areas of knowledge that rely on rational procedures of content processing become awkward to deal with, that is the development of congruent links of abstractions is being hindered by either negative associations or by an absence of associations altogether; at times these areas are even placed behind irrational complexes in order of preference. Despite such problems, the driving force of his intellect ensures that the overall dynamic of his various structures remains a productive one; sometimes the ensuing alienation from other people becoming a corollary to his inner state.

The investigations into human behaviour, from the first serious curiosity about the mad in their cages to the exposition of postures and utterances in a detailed manner, brought a vast range of intellectual products that never had been available before. The latent potential of behavioural science may not have been identifiable in its entirety, but the stream of realised concepts was enough to demonstrate the sheer abundance of raw material, to be had for the taking. The fires were long gone, and the torture instruments banished to the museums. Now at last could the suffocating covers be lifted, at least without fear of death. The supply of products was there. The question became one of finding out about their meaning. Applicability had yet to wait.

11. In search of freedom?

Herbert Marcuse - Jean-Paul Sartre

The 20th century has been and still is subject to many interpretations and definitions. Whatever the agenda of the articulating entity, there can be no doubt that the modern epoch is marked by great diversity, action, and dynamism. Our effectiveness chart in the previous chapter (see Fig. 10-1) gave an indication of the interplay between the pertinent factors: the multitude of objects available, the growing understanding of their use, and so the effectiveness of the host system as a whole. These influences can be seen anywhere, and can be observed in any system which contains self-regulating, pattern-seeking, and dynamic agents; in other words, which is alive (we will define more clearly what we mean by this in the next chapter).

With continued growth the extent of the effectiveness increases, not only in terms of the total but also as far as the roles of the parts are concerned. That much is expected from studying the chart, yet due to the dimensions of scale the interrelationships within the activity space assume different proportions. Note the two key expressions: *activity space* and *different* (not necessarily larger).

So far our observation of this dynamic has been relatively superficial, with the examples of cognitive activity carried into the philosophical and eventually into the socio-political arena, sufficiently new and hence unencumbered by historical baggage to demand otherwise. But this is only a subjective view, assumed for the purposes of a step-by-step introduction of concepts. While not containing false material as such, it does not tell the whole story. Now let us complete the picture.

An activity system operates in its space, defined by such things as physical extension, numbers of members, energy available, and so on. All that can be circumscribed as *resource*, and so we can say that each activity system has its own resource space in which it has to operate and through which it is sustained but at the same time delimited. The dynamics described so far apply to that space, but because of its finiteness they experience certain modifications as they go through their paces. In order to describe these we need to incorporate the concept of the system's domain into our picture and therefore turn our previous 2-dimensional chart into a 3-dimensional representation of a system situated in its resource space, such as in Fig. 11-1.

The red, green and blue axes of product number, potentials, and applicability are the same as before, although the potentials have been collapsed into one only for better clarity. The whole grid is situated inside a chamber, the resource space, with the yellow lines representing degrees of complexity. What follows is a walk-through.

We start with a given number of products, concrete objects for simplicity's sake. Their number and their realised potential form a subsystem that is placed on a particular intersection of the red and the green grid (ie the product number and realised potential value ranges), and is denoted by some spatial extension to symbolise its domain; this is disc A-1. As the realised potential increases we move along the green line but keeping on the same red parameter since the product number stays the same, arriving at disc A-2. As expected, the development also increases the applicability within the system and so we have also moved along the blue axis. Greater applicability means more products, and so we arrive at something like A-3. From here we would continue with the development were it not for an increase in complexity which the applied products have created. Therefore the subsystem's new domain needs to be elevated accordingly along the complexity gradations and we get B-1. Now the process starts again, moving to B-2 and B-3.

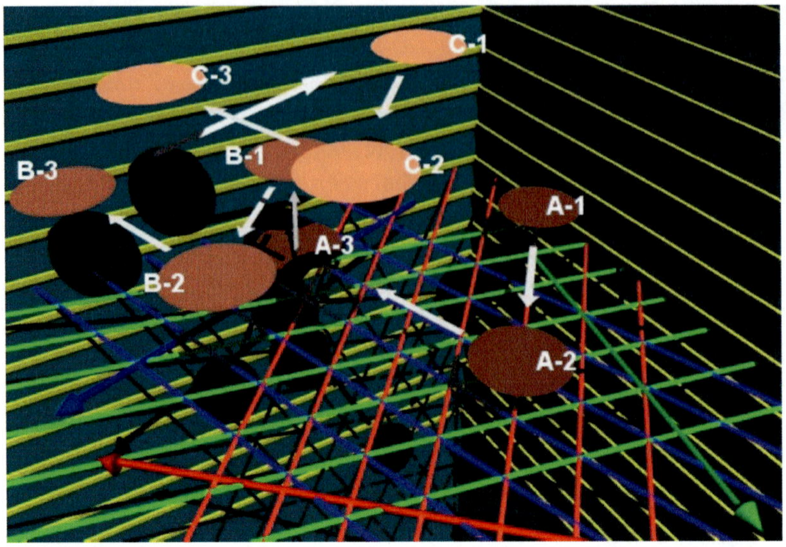

Fig. 11-1: A 3D representation of an activity system in its resource space—the full version of the effectiveness chart (e-chart). For details see the text.

Continuing in the same fashion our C-triplet would be offset relative to the B-set just as the B-triplet had become offset in relation to the A-set. Keeping in mind that we are moving inside a finite resource space it can be seen that our increase in complexity, driven by the continuing realisation of product potentials and applicability generation within our system, gradually forces us to the limits of the space the system inhabits; something has to give.

If we want to move back towards the A-set we would have to forego our achieved complexity, which effectively means cutting back on the number of products and hence the overall applicability; a process that involves the destruction of products (remember, we have to start again in order to place our new B-3 in a more central position). Another, similar option is to curtail the improvement of the products' potential; this also entails a form of dismantling already established procedures. A third alternative is to remain with the existing state of complexity, and so the existing state of products and their potential (ie keep the A's and B's), but place the new development of B's offspring at a lower product-number scale. Thus we have a virtual diminution of the system: the existing sub-entities are maintained and it is only their progression which is being reset to a lower parameter—C-1 in our picture, from where the process continues once more and now again well inside the resource space.

Such a development (the enrichment of a system due to growth in products and their associated applicability, but at the same time a certain curtailment of potential prompted by the overall increase in the system's complexity) is not an unusual phenomenon. A pertinent and sometimes drastic example is the growth in vehicular traffic in an urban environment. A greater variety of cars means more cars per se, means more activity, plus still more cars contributing to a growing complexity in traffic; this is followed by tighter rules, more traffic lights, inhibiting planning schemes, which on a certain level diminish the potential for further growth such as has been experienced before. Since the development does not stop here, there will be more cars contributing to yet higher complexity which triggers further restrictive planning such as new parking policies, entry permits, and transit zones; and so it goes on year after year. Without the increase in resource space (eg urban sprawl, multi-level parking) on one hand, or the wilful reduction in cars altogether (eg prohibited entry on certain days, general restriction zones) the system will eventually run itself to the ground. Note that there are three key elements involved: the capability of the system to develop in the first place, the extent of its resource space, and the attractiveness of the sub-systems that produce its complexity.

These last items are a function of the system in terms of its inherent nature, that is to say those characteristics which are instrumental in forming a phenomenon that eventually lends itself to a specific label and descriptive articulation through already defined terms. They in themselves are not part of the product-potential-applicability-complexity cycle, although they usually end up playing a part in the cycle's processes once that is up and running. For example, the development of the motor vehicle is not directly related to the state of contemporary traffic (although energy conservation leads to modifications of its design), nor have traffic rules anything causally to do with the availability of metals and the engineers who make use of them (although imposed rule changes prompt the development of performance and safety features), and the desire to own a car is not derived from a study of urban planning schemes (although, given the problems of traffic density there may well be artificially imposed links as part of public policy as time goes on). Therefore, the system as such becomes established through the confluence of factors which do not, at the outset, define its future history. On the other hand, the parameters in the e-chart do set the course of the system and in a way which is quite specific—and hence limited—in its nature, its direction, and in the possible range of outcomes. We can go so far as to say that the combined total of existent sub-states locks the ongoing development of the system into particular bands of options which are defined by the type of sub-sets that are instrumental in pushing the system along. In the case of road traffic, the sub-set 'road-rules' will give rise to changes and additions in the way drivers can behave behind the wheel; the sub-set 'car-design' will cause the development of such things as brakes and air bags; the sub-set 'road-grid' will furnish freeways or one-way streets and the like; and so on. The actual appearance of such developments relates to the possibilities inherent in similar sub-sets of effectiveness charts as well as the interdependency of these sub-sets since they are all part of the same greater whole. Let us call this form of determinism *progression lock*, since it represents a range of options locked into particular directions and caused by the progressive nature of the system they belong to. Note however that the expression itself does not include a value judgment one way or another.

It should be apparent that there is no infinite future for such a system. Its resource space, no matter how large or modifiable, must reach its boundaries eventually; its ability to re-form, re-organise, and re-allocate the energies derived from its resources cannot be assumed to be boundless; and the parts themselves, regardless of their ingenuity, must reach the end of their potential sooner or later. Against such odds one could argue that the best course of action would be to curtail its progress or halt it altogether. As to the former, a diminished pace would merely delay the inevitable, and as for the latter, a full stop is impossible unless the inherent dynamics responsible for development are dismantled and rendered unworkable. But to do so entails the creation of appropriate mechanisms to do just that—another development and so a contradiction in terms. Systems do come to a halt then, and in those situations the end is not a controlled ceasing of operations but an implosion, not an administered closure but a ragged tearing away from the path it has hitherto followed. Our traffic example does not offer such comparisons (yet?), but the former Soviet Union would, or Germany's Third Reich (the beginnings *and* ends of both for that matter), or any civilisation that once flowered but now exists only in dead shards of memory.

On a personal level, we have a metaphor for it: painting oneself into a corner.

Another option is to import further sub-systems which contain latencies of their own, thereby adding to the overall potential. Such a situation unfolded for Australia after World

War II, a country enjoying one of the highest living standards derived from the rather limited economy of raw materials and agricultural products. Against the general development of industrialised nations after the war that country's standards soon dropped, its own potential in terms of its output range becoming exhausted. However, a climate of diverse investment took over and a catastrophe was avoided (although these efforts were not as intense as those of other countries and so its position in the world order dropped anyway). By contrast Argentina, in a similar position at the outset, did not partake of the same culture and its fortunes sank further and further. These histories are obviously an outward manifestation of the general inner mentality of people (it also needs to be said that just as there can be an import of a positively contributing sub-system, there can be the addition of a negative contributor, ie a sub-system which is more primitive and therefore lowers the overall potential of its host).

In any case, as long as the end of the system has not been reached, its internal mechanisms formulate the conditions which define its perceived state. This state consists of two fundamental types: the first can be described in terms of the potentials in the process of being realised and all their ramifications (the subject of the previous paragraphs); the other is the gradually decreasing potentials of those very sub-systems in relation to the sub-systems' shift along the grid under the influence of the complexity factor. Returning to our traffic example, alongside the developmental activities already discussed there is the growth of constriction due to those very developments. As defining elements the first is a conglomerate of many parts, each contributing to the increase in complexity but each at the same time representing a sub-system in its own right (eg the town planners, the safety engineers, the city council revenue administrators). The other is the situation as experienced by the user during the course of interaction with any particular part. Or, to put it another way, the first represents the forward looking, desirable side of things; the other, the inhibiting, regrettable one. One stands for wealth, the other for its cost that must be paid.

As always, the occurrences played out in the outside world have their counterpart, indeed their source, within the inner. The principles outlined above are equally applicable to the mind, and we can refer to the seven points made in chapter 10 and see them as one instance of a many-layered phenomenon, repeated continuously across the complexity levels until the end. In overall terms we have not yet reached the final stage, but there are certainly sub-systems which came and went, developments which have been superseded by newer configurations because their predecessors have been found wanting and hence were discarded. Just as their outside equivalents did not go smoothly, so have these succumbed to their successors through more or less violent struggles. Complex thought patterns such as the representation of the Inquisition, or of a monarchy, or on a lesser scale such things as the abstractions of business corporations, social combines, or even individual life styles—are all valid examples. The one effective addition to the seven-fold list of ramifications from the previous chapter corresponds to the defining elements mentioned before: the first are those sets of TSs that carry the desire for ever more complex schemes, the second are the opposing structures of perceived limitation, hindrance, and coercion. Just as no-one in past centuries complained about gridlock in their townships, or had their taxes increased because of too many horse carts on the streets, the mental counterpart of the negative side of progressive developments needed a sufficient degree of complexity in order to make itself felt. Thus there is one major factor underpinning the positive as well as the negative aspect: intelligence.

The multi-faceted unfolding of cognitive dynamics produced the visions which led to the exploration, discovery, and utilisation of our surroundings; accompanying that we find the growth of doubt, of feelings of reduced liberty. So it came that as the European mind created ever more diverse complexes of abstractions, it also engendered perceptions of discomfort and claustrophobia. The constituent parts were not new—they had existed all along and were in fact playing their role in those two-sided processes. But it was only now that the level of sophistication brought out the ensuing reduction in optionalities, where the abstractive tools of the mind had become sharp enough to identify that what could not be had after all. To understand how real, how efficacious these limits really are and so what influence they are capable of in any case is the purpose of this chapter.

We begin with Herbert Marcuse (1898 - 1979), a philosopher who, against his intellectual background of Marxist ideology, saw modernity as a hindrance not merely to socialist ideals but the healthy development of the individual. For him 'health' meant a broad range of realised potential in man, of which a fundamental contingent consisted in the erotic fulfillment of one's persona. The constraints which he felt prevented such fulfillment he dealt with in the book *Eros and Civilization* [1], to which we now turn our attention.

Given that freedom, or rather the lack of it, is the main theme of his work it is justified to ask for a definition of what it actually is that is missing. Yet there is none. The closest we get is vague and self-referencing:

> Freedom is thus, in a strict sense, freedom from the established reality: man is free when the "reality loses its seriousness" and when its necessity "becomes light" [as in 'not heavy']. [2]

Since there is no exact description, its content delineated and traceable from other—grounded—sources, we immediately recognise the work of i-PAT-ls. This does not mean Marcuse's concern ought to be treated as baseless—it's just that as a focus of intellectual endeavour it has not been worked from a concrete configuration. It is a perceived notion, its nature hinted at by attaching the idea of 'reality' to the largely subjective interpretation of 'seriousness' and coupling it with 'necessity', a similarly subjective view formed by circumstances. Its incongruent character is further confirmed by what he has to say a few sentences later:

> In a genuinely humane civilization, the human existence will be play rather than toil, and man will live in display rather than need.

Although i-PAT-ls are not linked to grounded abstracts (otherwise they would be of the congruent variety) they still are situated amidst a network of affinities. What these are we will come to later, but the fact of their existence points to the presence of TSs which must be in some kind of relationship with the abstractions holding the concerns about freedom—that is what affinities are all about. Since those other TSs *are* linkable in that way they have to be such that the idea of 'freedom' in the above sense can be entertained (however vague etc it may be). If the description 'freedom from the established reality' is abstract A, and its implementation 'play rather than toil' is abstract B, then their opposites are non-A and non-B respectively which presumably reflect the present undesirable conditions. Therefore, to strive towards such a goal would entail divesting oneself of all that is non-A and non-B in our society, or at the very least adopt a stance which rejects everything non-A and non-B; that is to say accepts whatever can be perceived as A and B. Naturally, such things would earn the title 'good' and 'beautiful', or in Marcuse's words,

> Aesthetic culture presupposes "a total revolution in the mode of perception and feeling", and such revolution becomes possible only if civilization has reached the highest physical and intellectual maturity. [3]

The TP encompassing "aesthetic culture" is then a set of TSs which are affinitive to A and B and in addition feature supportive associations with other sets of TSs denoting "highest physical and intellectual maturity" (by the way, in all of the above Marcuse quotes from Schiller's *The Aesthetic Letters*). As maturity of any kind is not only a state of being but also comprises a process along a path that needs to be defined in a particular manner, we see that the TP "aesthetic culture" must be a container capable of handling i-PAT-ls (the As and Bs) as well as the dynamics of "maturity"-building representations of the implementable kind. Because i-PAT-ls are involved on one hand and significant action-forming abstracts on the other, such an attractor can be expected to play an important, if not imposing role in any mental configuration. In terms of our nomenclature we can say that we are dealing with a phenomenon that engenders a form of projection within the bounds of 'freedom'-inducing TSs, the basically undefined nature of these necessitating certain modifications to what are originally PAT^1s through the application of illusion trees. We encountered a similar situation with Marx in the previous chapter.

In projection we have established TSs which form bonds with some incoming information. Ordinarily there is an element of choice, but in this case the role of the container would tend to curtail such possibilities. It is interesting that Marcuse mentions Jung in this context. Speaking of what the envisioned scenario of actualised "aesthetic culture" would mean he writes,

> Jung recognised these implications and was truly frightened by them. He warned that the rule of the play impulse would bring about a "release of repression" which would entail a "depreciation of the hitherto highest values", a "catastrophe of culture"—in a word, "barbarism". [4]

Recall Jung's reluctance to allow certain PAT chains to interfere with his religion container. There the problem was one of countering mostly default abstracts in order to offset the demands of the former. If the image of human beings 'at play' is a frightening one, we can safely assume that for someone like Jung the compromise of one's "highest values" is not about human well-being; rather, it is the giving-in to the law of one's god, where service at the altar means "culture" and refusal equates with "barbarism". With Jung the unacceptable alternative was the positive appreciation of default abstracts, thoughts relating to sexuality to be precise. As such any fundamental link between 'sex' and 'beauty' is inherently denied. Note particularly the existence of a progression lock in the form of pre-arranged TPs which force any related cognitive configuration to proceed along a determined path. If, as Jung evidently holds, sexual repression is integrated with the wider fabric of human society to such an extent that its removal would cause the demise of that same society, then this says a great deal about the influence of sexual repression in our general lives—not to mention the mechanisms that ensure such a state of affairs. True or not, in Jung's mental universe an alternative and possibly more productive perspective is impossible.

Marcuse on the other hand adds,

> Schiller himself was apparently less inclined than Jung to identify repressive culture with culture as such; he seemed to be willing to accept the risk of catastrophe for the former and a debasement of its values if this would lead to a higher culture.

Yet again we have to ask, what *is* this repression, what *does* such debasement consist of? Like the idea of freedom, it has no direct answer (with one exception as we shall see). For that we have to employ our knowledge of thought structures to guide us.

EC (I am borrowing the abbreviation of the title from Douglas Kellner's preface to the edition) has been written under the guise of a "philosophical inquiry into Freud"—hence its full name. The nexus between philosophy and psychology is justified, so Marcuse says, because contemporary conditions (the book was written during the 1950s) forced the individual to relinquish the personal aspect of one's psychological configuration and have it submitted to society at large, through his "public existence" [5]. What could or would have been a psychological structure is therefore mapped onto a political one; what represents the political ambience is treated under philosophical auspices by Marcuse. For us this does not represent a problem; the translative properties of thought structures into the wider society have become familiar, even expected. Since we find no hints of Marcuse's understanding of such matters in his book we have to assume that the possibility of such transference is not only a matter for concern to him but also a regrettable feature of the position of the individual in relation to society; he says so indirectly anyway. There is more to it than the commonality of themes however. Apart from "psychological problems" turning into "political problems", apart from "private disorder" reflecting "more directly than before the disorder of the whole", there are further ramifications which manifest in the structural nature of his thought patterns. Firstly, note the phrase "more directly than before", implying that the identified state is different from what it had been in earlier, and presumably better, times. Secondly, to categorise one set of ideas as psychology and therefore subject to a certain degree of formalism (however ideologically configured such a framework may turn out to be in the end), and another as philosophy and therefore allowing personal musings to assume a measure of significance which otherwise they would not be allowed to possess, and moreover make a further connection to the field of politics which through its very label is acknowledged to be the hunting ground for ideology, heralds a very specific arrangement of interdependencies amongst the respective contents.

Without yet having come across the details, for a set of TSs to be acceptable to another set which holds one's personal opinions, the contents have to be transported across illusion trees. Whether this is a process spread right across society as Marcuse thinks, or whether it remains only confined to individual minds, is a matter of manifested degree. If it is a case of the former then we could expect certain signs within the public arena and so the situation becomes indeed political; if the latter is true, psychology alone would be concerned with the effects.

If *EC* is meant to be a polemic of Freud (which fundamentally of course it is), the divergence of opinion occurs somewhat down the track. In the *Introduction* Marcuse mentions Freud's proposition that "civilization is based on the permanent subjugation of the human instincts" and that it has been "taken for granted" and is "inevitable and irreversible" [6]—what follows is the refutation. What does not emerge is any articulation regarding the difference between instinctual and fundamental behaviour; 'instinctual' because it is assumed to pre-date civilisation, and 'fundamental' because for Marcuse it forms the basis for the general behavioural dilemma in our civilised state. It follows that the TP organising the framework of perceived inadequate living conditions is not made to distinguish between what we would call the default abstracts and those formed by any container. Those cognitive elements that point to the inadequacies will therefore draw their affinitive properties from default-induced TSs as well as container-induced ones. Such an a priori overlap leads to progression lock because of its over-generalised nature; that is to say the combined set of TSs determines the possible range of later affinities such that both included types must be satisfied.

An example follows soon after:

> However, intensified progress seems to be bound up with intensified unfreedom. Throughout the world of industrial civilization, the domination of man by man is growing in scope and efficiency. Nor does this trend appear as an incidental, transitory regression on the road to progress. Concentration camps, mass exterminations, world wars, and atom bombs are no "relapse into barbarism," but the unrepressed implementation of the achievements of modern science, technology, and domination. And the most effective subjugation and destruction of man by man takes place at the height of civilization, when the material and intellectual attainments of mankind seem to allow the creation of a truly free world. [7]

Several declarations are being made: (1) "unfreedom", the opposite to the undefined concept of 'freedom', is synchronised with the accelerated pace of development; (2) the dimensions of scale are applied to 'domination' and exclusively linked to those societies experiencing industrial civilisation; (3) examples of cruelty are kept separate from the idea of 'barbarism' and are further synchronised with advances in science and technology, the last two being bundled together with 'domination'; (4) the present is considered to be the height of civilisation; (5) that apex is taken to be capable of achieving 'freedom', however undefined it may be.

The cognitive ramifications of such a TP are as follows, in line with the above numbering: (1) the set of abstracts addressable through the articulation of the word 'freedom' and consisting of i-PAT-ls is conjoined with the TS denoting the opposite (created for that purpose and as incongruent as its progenitor), and that combination serves as a denominative influence towards any TS related to the idea of 'development'; (2) while dimensions of scale apply to qualitative as well as quantitative phenomena, here the qualitative side is over-emphasised at the expense of sheer numbers and can therefore be linked to 'industrial civilisation' only, the latter a manifestation of quality first and foremost—ie an illusion tree in the process of excluding c-PAT-ls to serve the incongruent nature of the TSs identified under (1); (3) another set of illusion trees where the existing links between congruent chains concerned with 'cruelty' and those concerned with 'barbarism' are sloughed off and the thus compacted result under the name of 'cruelty' elevated to form associations with similar levels of 'science and technology' as well as 'domination'; (4) there is a TS associated with 'civilisation' which denotes the optimum and thereby prevents the formation of TSs holding forward-looking and positive notions with regards to further scientific and technological advances—in other words, the imposition of a progression lock; and (5), the TS from (4) is being associated with the incongruent set of 'freedom' TSs from (1) and through this link being associatively connected with any negativity inherent in the set's opposite.

From this mental organisation certain practical consequences are derived, such as: (1) since there is no formal definition of what 'freedom' means, it is open to appropriation by whatever interests see an advantage in using the word or its opposite for that matter, thereby opening the way for simplistic spins (eg 'development is bad', 'development takes away your freedom'); (2) because a scientific product is effective (given its congruent character) its misuse may have a relatively high impact, but large-scale non-scientific activities can be deleterious too and because of their incongruent character can even be more destructive (eg tribal warfare, the persecution of religious minorities, hate and violence based on superstition)—yet these are not taken notice of, indeed are often excluded from the official definition of 'barbarity'; (3) whatever links there might have been between identified instances of cruelty and barbaric behaviour are destroyed so that firstly, the concept of cruelty can be maintained and used to define incidents taken from industrially developed societies while secondly, others from regions historically connected with 'barbarians' are now cleansed of such connotations and so appear as innocent of such behaviour; (4) the idea is formed that the present conditions, as bad as they are, are as good as it gets—a paradoxical juxtaposition which leads to pessimistic expectations for the future and of the underlying mechanisms responsible for its unfolding which, in the West, are of course science and technology; (5) the resultant perception leads to the overall view that developed societies are the inevitable source of a steadily diminishing degree of freedom, without however providing a concrete description of what 'freedom' means today and indeed what it entailed in the past—therefore a rational comparison is made impossible.

Whether Marcuse himself would have shared some or all of the above manifestations is not the issue in the above context; we are simply describing the mental framework which can be gleaned from his text and outline the potential results coming from it. As we shall see, he did in fact share some of them, although the arrival of the full contingent would have to wait for a few more years.

The progression from the initial state of the individual to his civilised self needs anchor stones, like a bridge spanning two lands. Freud summarises the psychical land of departure as the pleasure principle, and the region of arrival as the reality principle. Marcuse presents them in the following form:

from:	*to:*
immediate satisfaction	delayed satisfaction
pleasure	restraint of pleasure
joy (play)	toil (work)
receptiveness	productiveness
absence of repression	security [8]

We notice that all of them are abstracts. Furthermore, all of them are the result of an articulation process which mined sub- as well as con-TSs and combined them with labels that have been made available through language. As such they are representative of whatever TS-contents were subsumed under the auspices of their respective semantics. They are *not* pleasure-per-se, nor joy-per-se, that is the descriptive words of essences such as a Husserl for instance would have dealt with. For Freud and Marcuse, evidently not familiar with the finer points of abstractions, beyond the instinct-learned behaviour combination they correspond "largely (but not entirely) to the distinction between unconscious and conscious processes". In terms of our mind system they have therefore unwittingly contributed to yet another progression lock: the attaching of the elements on the left side to the TP holding notions about the unconscious and those on the right to the conscious-TP, thereby predetermining the basic meanings of either within the context of psychological phenomena. This is not a time for the psycho-analysis of psychology, but just to demonstrate the significance of such an arrangement let us take the first from/to pair.

Suppose i-s (immediate satisfaction) is instinct dependent and d-s (delayed satisfaction) is learned behaviour. Then any action prompting some form of i-s must in theory be successful, otherwise there would not be an instinct for it (indirectly, that sub-argument also takes care of the other supposition that i-s is not based on instincts at all). It must also be successful in practice, because if it is not then the development of instincts includes the information processing required by random situations and that would take it beyond the range of instinctual behaviour by definition. But if it is successful in both cases, then the adoption of learned behaviour in favour of the tried-and-true instinctual version must yield advantages to such a degree that the evolutionary progress enabling such addition can pay for itself. In which case we have left the realm of instincts and are now fairly and squarely in the sphere of purposeful information processing. Hence d-s is not a regression, but a necessary step forward and accomplished by animals long before the advent of humans. Now, unless one is prepared to ascribe the relatively high abstraction of the i-s vs the d-s debate to cats and dogs it is nonsensical to posit the d-s as some form of dilemma for humans; either that or the even more highly developed abstractive abilities of humans predispose us towards despair—yet that brings us back to the point of evolutionary advantage made before.

So, since learned behaviour is intrinsically better than what can be gleaned from instinct, the categorisation of the latter in terms of 'unconsciously good' and the former in terms of 'consciously bad' sets up a chain of thought events which predisposes their owner towards veering away from articulated and processed information and embracing acts of whim based on unarticulated and unprocessed sentiments. Or, to be somewhat less direct but more practical, which prompts a preference for actions that are influenced by emotions and unarticulated urges rather than rationally set-out concepts. Apart from reflex actions and such things as sleep walking every form of articulation is by nature conscious—our minds are built that way. Therefore, to claim some act of playful joy as a manifestation of an unconscious process does not describe reality but indicates the type of cognitive dynamics at work here. Generally speaking then, what is categorised as pre-industrial and/or pre-civilised behaviour is placed within the domain of TSs denoting non-rational thought processes and so being associative with incongruent containers; at the same time, the rational, that is the grounded and congruent abstract structures which are the basis of industrial and civilised society are relegated to the domain of what is perceived consciousness. Combined with the added classification based on respectively positive and negative connotations we arrive at two fundamental cognitive environments, artificially construed and the starting point for further delineations: an emotive, non-rational and anti-technological ambience on one side and a detached, rational and technology-embracing one on the other. And so Marcuse can continue by saying,

> The individual exists, as it were, in two different dimensions, characterized by different mental processes and principles.

For him the chasm occurs inside the individual. As we have seen, it only becomes manifest through the implementation of the relevant perceptions, and so is more likely to be found in society at large, carried by individuals who subscribe to one or the other point of view.

Again, such a division can only occur through the determinism of a progression lock, establishing in advance certain presumptions which in due course lead to the set of perceptions outlined above. They do not reflect reality, nor would they be arrived at via a congruent (ie grounded) representational experience taken from one's daily life in pre-industrial surroundings. Even a short stint amongst Amerindians in South America or traditional rural dwellers in Africa or Asia will demonstrate the fact that life under those conditions is anything but totally playful, and having to constantly combat wild nature ensures that just about everything comes through the application of rigorous efforts introducing delays hardly imaginable by citizens of developed countries. Furthermore, such conditions give rise to a form of communal discipline and authoritarianism that make the claim regarding the "absence of repression" quite ludicrous (more of this in Part II).

Once the two TPs have been formed (one for the "instincts", the other for "civilisation"), the divergent streams of respective assumptions are secured their sustenance. These TPs are now denominative patterns and identified developments such as the "faculties of attention, memory, and judgment" are circumscribed as a "rationality which is imposed..from the outside" [9]. On the other hand, what is considered "phantasy" is "protected from cultural alterations and stays committed to the pleasure principle". Regarding the 'civilisation'-TP, it is denominative because internal associations are suppressed in favour of linkages to what can be connected with society at large (the "outside"); as to 'instincts', it is a DTP because whatever can be described as "phantasy" must by definition be the product of a mind untrammelled by the ills of progress.

Yet is goes further. Consider the role which has been assigned to memory:

> If memory moves into the centre of psychoanalysis as a decisive mode of *cognition*, this is far more than a therapeutic device; the therapeutic role of memory derives from the *truth value* of memory. Its truth value lies in the specific function of memory to preserve promises and potentialities which are betrayed and even outlawed by the mature, civilized individual, but which once had been fulfilled in his dim past and which are never entirely forgotten. The reality principle restrains the cognitive function of memory—its commitment to the past experience of happiness which spurns the desire for its conscious re-creation. [10]

Under the auspices of the 'instincts'-DTP the contents of memory structures, on their way to articulation, are being morphed into the higher-level abstractions positively associated with 'promise' and negatively with 'maturity' and 'civilisation'. In that sense there is indeed a "commitment to the past experience of happiness", except that the conscious perception of 'happiness from the past' is nothing else but the result of a DTP-ruled articulation process—this is the 'commitment' part; its perceived source, organised as it is under the mantle of 'pre-civilised state' consigns its consciously perceived version to all those TSs concerned with whatever is not linkable with the unsatisfactory present—its 'past' part. Such abstracts, having been elevated to the appropriate levels via illusion trees, are able to form functional associations within the general domain of the 'instinct'-DTP including the idea of 'truth'. Being of a functional nature, these associations are not tested as to their congruence but solely for their degree of abstraction. As usual in those situations, the nexus between irrationality and truth remains unchallenged.

Marcuse's perspectives throughout the rest of the text, in so far as they deal with the influences of civilisation on our psyche, can be seen in the same light. The denominative domains of the two TPs mentioned above together with the directions taken by related PAT-chains, prescribed by the progression lock which is produced by the framework of assumptions about pre- and post-industrialisation, these colour the thoughts from now on. Hence in line with Freud's thinking, civilisation is seen as "organised domination" [11]. Again, note the judgmental aspect of both words: "organised" is affinitive with 'civilisation' → 'capability' → 'rationality'; "domination" is connected with 'civilisation' → 'productive society' → 'loss of freedom'. Once the jump from the individual to society is made Marcuse introduces two further principles, "surplus-repression" and the "performance principle" [12]. Since "organised domination" exists as a valid force there must be some repercussion in the form of "restrictions necessitated by social domination". As to the latter, it is the "prevailing historical form of the *reality principle*". So there is a further progression lock, determining the view in relation to one's status within developed society and the presumed possibilities of the administrative and political entities therein. While any consideration of modernity ought not to deny the requirement, indeed necessity, of some form of subordination to the greater organisational whole, the fundamental quality under which it is going to be addressed would not be trivial.

Similar to the deterministic progression adopted towards the single mind is the perception of society at large. We find the notion of a "development of a hierarchical system of social labour" that "rationalises domination" and "'contains' the rebellion against domination" [13]. Since any perceptions of 'development', 'hierarchical order' and 'social labour' are subordinated to the DTP of 'civilisation' and its general content, the other DTP escapes similar associations with grounded abstracts of the same nature. In other words, pre-industrial societies may be feudal, tribal, coerced along the demands of superstition and shamanism, yet all this does not attract the label of 'rationalised domination' or 'hierarchical order'. Again, this is not to deny the existence of relatively sophisticated systems of control in modern times. Nevertheless, as we have already seen in the philosophies of the past, lower abstraction levels and the emphasis on i-PAT-ls for the purposes of justifying social, political and spiritual control made intolerable excesses—at least according to an enlightened view—a virtual default state. It was only through the application of reason, that is grounded abstractions and the capacity to handle a variety of cognitive levels that life in general had become more civilised (Marcuse adds another element in the form of Freud's Oedipus complex, but we will deal with that later in a different context).

The associative intensity between the DTPs and their affinitive structures affects areas which otherwise would be left untouched. Take the concept of the family, a social element as subject to constant modification throughout the ages as any other cultural entity. Under the guise of 'unfreedom', ie the "technological abolition of the individual" [14], the perceived field of battle amongst parents and children in their struggle for domination has now shifted from the immediate radius of the siblings to the greater environs of the industrialised community. Marcuse does not deny the combat per se—the acceptance of the Oedipus complex sees to that—but however painful the fight turned out to be for the child, he reminisces, at the very least there emerged from it a persona in full possession of its acquired (and painful) scars along the way. This "sphere of private con-conformity" is preferable to a "premature" socialisation "by a whole system of extra-familial agents and agencies". One recognises the role of illusion trees under the effects of a denominative reach by Marcuse's beliefs about civilisation: every feature of familial life is paired away except the relatively individualistic aspect of a small-scale social unit, so that the resultant TS can become linkable to those abstractions reserved for the definition of 'civilisation'. At the same time that TS now resides on an abstraction level sufficiently high for a similarly elevated abstraction to be approached, namely Freud's concept of sexual antagonism within the family. Conversely, the modern version of growing up gets the same treatment except that now any individualising, contemporary influence gets removed so that the outcome turns into a symbol of collectivism. And the father, the former object of Oedipal angst, is exchanged for the "system", the "sum total of the institutions that determine, satisfy, and control his [the individual's] needs" [15]. The "aggressive impulse", the main feature of such a parent-child relationship,

> plunges into a void—or rather the hate encounters smiling colleagues, busy competitors, obedient officials, helpful social workers who are all doing their duty and who are all innocent victims.

From our view of the cognitive system at work two things stand out: the abstracted versions of 'busy competitors' etc. ("..all doing their duty.."), necessary for the assimilation into the wider domain of the DTP; and the heightened potential of the mental and societal ambience compared to what both would have been like under the older, discrete family unit. In fact, Marcuse makes exactly the same point, although after his own fashion:

> The better living is offset by the all-pervasive control over living. People dwell in apartment concentrations—and have private automobiles with which they can no longer escape into a different world. They have huge refrigerators filled with frozen foods. They have dozens of newspapers and magazines that espouse the same ideals. They have innumerable choices, innumerable gadgets which all are of the same sort and keep them occupied and divert their attention from the real issue—which is the awareness that they could both work less and determine their own needs and satisfactions. [16]

Yes—but what if they *want* private automobiles and huge refrigerators and frozen foods and newspapers and magazines...?

He follows the theme on the potentiality of the system from the base of his identified coercion. While there has been destruction before, he writes, the difference today goes beyond the numerical scale.

> There is more than a quantitative difference in whether wars are waged by professional armies in confined spaces, or against entire populations on a global scale; whether technical inventions that could make the world free from misery are used for the conquest or for the creation of suffering; whether thousands are slain in combat or millions scientifically exterminated with the help of doctors and engineers; whether exiles can find refuge across the frontiers or are chased around the earth; whether people are naturally ignorant or are being *made* ignorant by their daily intake of information and entertainment. [17]

This is followed by what at first glance must seem curious, given the dynamic of potentials:

> Still, progress continues, and continues to narrow the basis of repression. At the height of its progressive achievements, domination not only undermines its own foundations, but also corrupts and liquidates the opposition against domination. What remains is the negativity of reason, which impels wealth and power and generates a climate in which the instinctual roots of the performance principle are drying up.

True, development is accompanied by a shrinking resource space. In Marcuse's mind, which holds the re-representation of the perceived state of society, the affinity potential of the 'progress'-DTP decreases. 'Domination' can only be at its height if its TS is capable of connecting with as many abstractions as possible. Yet at that point certain abstractions are at odds with what used to be supportive structures, therefore the number of connections are reduced and hence leading to a perception of a "narrowed basis of repression". Note however that it is not the entire concept of 'progress' which is under attack right now, but only those TSs which so far have given rise to the negative interpretation of its status. Ordinarily the result would be a more positive perception of its nature but clearly this is not so. Instead we have a "negativity of reason" and any links with 'instinct'-related PAT-chains, however incongruent they may have been, are now severed. Since those had been the very chains that gave the 'progress'-DTP its significance in terms of 'domination', 'unfreedom' etc, the fact that the same pattern still retained its status points to the elevation of some content to an even higher level of abstraction where supporting connections are no longer necessary. Recall similar examples in religion-containers, where the pertinent set of TSs (the ones which directly related to the divinity) had been moved well beyond the logic boundary so that in the end they became unassailable by any argument whatsoever. In the present case the content is the representation of reason, now a negative cognitive entity which has outgrown its need for any confirmation from other sources, no matter how tentative, while at the same time having its denominative power reinforced.

The entire text of *EC* is not only a description of modern life but offers some suggestions of improvement, for which the former is a necessary preparation. For Marcuse the way out is essentially an undoing of the performance principle, since it has been the coercive power of the collective fulfillment of need that prompted the overshadowing of the instincts. Yet despite the advances in science and technology, their combined potential may still be questionable when it comes to reaching this goal; that is to say, "scarcity and immaturity remain great enough to prevent the realization of the principle 'to each according to his needs'" [18]. That tone rings familiar. Recall the interpretation of pre-industrial rural life by Marx (see chapter 10), where the tentative existence in the face of over-bearing nature was seen as a measure of self-sufficiency (indeed, the idea of the present negativity hurting itself as expressed in the previous paragraph equally runs parallel to Marx's view of the bourgeoisie digging its own grave). To point out Marcuse's leftist tendencies is nothing original; what is significant is the mapping of the meaning of 'need' to a set of historical circumstances where the acquiring of satisfaction in general is precarious at best—quite apart from the intuitive nature of the word itself. A sign not only of i-PAT-ls but of a DTP as well.

Which brings us back to the self, the ego. In the mind of Marcuse the ego suffers dilution amongst the greater forces of society, its original state as a comprehensive psychological unit destroyed. He cites various sources of the present and antiquity to illustrate the symbol of narcissism—the ego par excellence—as representative of its assumed glorious past. For example, in a footnote he writes,

> In his paper on "The Delay of the Machine Age," Hanns Sachs made an interesting attempt to demonstrate narcissism as a constitutive element of the reality principle in Greek civilization. He discussed the problem why the Greeks did not develop a machine technology although they possessed the skill and knowledge which would have enabled them to do so.
>
> ..he proposed that the predominant narcissistic element in Greek culture prevented technological progress: the libidinal cathexis of the body was so strong that it militated against mechanization and automatization. [19]

To begin with there is the assumption that the ancient Greeks could have developed technology if they had really wanted to. This is arguable. In our earlier chapters we have already seen several indications of a reduced ability to abstract in earlier days, that is to elevate the contents of an observation to higher abstraction levels in order to gain further insights. We have also seen the need to move an identified principle first up, then down the levels to reach the solution to a problem (see the text surrounding Fig. 7-1 in chapter 7). Compare the functional dynamics there with the 18/19th century approach to building a flying machine: at first the wing of a bird is copied because of the relatively superficial connection between the wing's appearance and the fact that birds fly. It was not until the underlying aerodynamic principles were discovered and the shape of the wing was modified that the machine actually took off. Without the abstract understanding of the principles involved the solution to the problem was not forthcoming. Now consider the story of Icarus, son of Daedalus, two mythical Greek figures who used feathers assembled like bird wings to escape from danger. Icarus, as is well known, flew too close to the sun, the wax holding the feathers together melted and he fell into the sea. Given the quite intimate relationship the Greeks had with their pantheon, would not such a story induce some form of practical experimentation, *if* the required ability to abstract had been present? The fact that it did not, nor did any other mechanical adventure eventuate, would suggest that the absence of machines was not due to the nefarious influence of sex on their psyche, but that the reasons were much more fundamental. Moreover, if anything can be observed across the ages as well as amongst contemporary demographics, it is the *rise* of individualism in line with a society's advance in general knowledge and understanding. Could it not be that the classical treatment of narcissism is one of an oddity amongst the prevailing ambience of the group mind? And that the sexual undertones were merely the means to an end and not the significant negativity ascribed by the prudishness that was to follow? In fact, the interpretation as touted by Sachs seems to come from the same stable as the belief that exposure to sex in early life forestalls one's ability to read and write—a predilection certainly not shared by the characters in the *Satyricon*, written by Petronius who must have been quite a party animal around the first century CE [20].

In addition, to re-classify the relatively collectivist role of the ego during past ages as an individualistic unity comparable to our present-day understanding, is to cast the generality of those societies as a form of discrete instances it does not deserve. In such a process the cognitive representation of a common custom is associated with values usually reserved for such concepts as 'free will' and 'individual determination'; the effect being a paradoxical turn-around of what conditions of old really had been like.

The supposition of sexual behaviour causing all sorts of dire consequences was not a classical sentiment but rather reflects the attitudes of Christian Europe, then and now. Take Marcuse's quote from Ovid's *Metamorphoses* a few paragraphs later, concerning Orpheus:

> ..Orpheus had shunned all love of womankind, whether because of his ill success in love, or whether he had given his troth once for all. Still, many women felt a passion for the bard; many grieved for their love repulsed. He set the example for the people of Thrace of giving his love to tender boys, and enjoying the springtime and first flower of their growth.

"He was torn to pieces by the crazed Thracian women", Marcuse adds. Apart from signifying the potentially dangerous whimsicality of the female, the above is not meant as some perversity but the introduction of a particular erotic variant by a "culture-hero", as Marcuse calls him.

This, incidentally, is the closest we get to some form of identification of what a liberated sexuality is supposed to mean. Since one can assume that erotic freedom would not be limited to "tender boys" only and to avoid unwitting limitations, from now on let us refer to sexual behaviour in the present context as 'sex+', symbolising its open-ended and undeclared nature.

As always, the articulation of a sentiment needs to be observed in its current context; that is to say, it needs to be identified in terms of its associations with other content structures and of its associative potential with yet to be imported data. For Freud, and indirectly for

Marcuse, the basis for the interpretation of human sexuality is formed by a conflict between "life instincts" and the "death instinct" [21], where the former are synonymous with Eros. The evolutionary path of the human organism, so Freud's assumption goes, predisposes it towards following one's sexual instincts in tandem with the other, non-sexual self-preservation counterparts,

> .."whose function it is to assure that the organism shall follow its own path to death, and to ward off any possible ways of returning to inorganic existence other than those which are immanent in the organism itself"

as Freud is quoted. Yet we also have the idea that along that path the instinctual erotic urges become suppressed. Thereby a set of TSs emerges in which the general concept of sexual behaviour is cast in a negative mould, associated with the opposite to its source (the death instinct, acting against sex per se), and so bundled together as a representation of Eros in society. Marcuse continues by saying,

> However, the discovery of infantile sexuality and of the all but unlimited erotogenic zones of the body anticipates the subsequent recognition of the libidinal components of the self-preservation instincts and prepares the ground for the final reinterpretation of sexuality in terms of the life instinct (Eros). [22]

One has to be careful about the underlying semantics with regards to the statement about the "discovery of infantile sexuality" and the "erotogenic zones". To take the immediate, more direct version first, the claim that it took a Sigmund Freud to demonstrate the existence of sexual behaviour in children would be ludicrous, given that he arrived at a point in history when the human race had already been around for tens of thousands of years. At the same time, the influence of Judeo-Christian repression upon the European psyche (and on many others' too of course) did manage to hide issues of a sexual matter to the point where it took the committed initiative of some daring individual to 'discover' sexual patterns which in other circumstances might not even have been worth mentioning, let alone considered to be a profound subject.

On the other hand, since in Freud and Marcuse we are dealing with the abstracted representation of society, its values and its development, the sudden inclusion of children's behaviour may indeed be seen as a "discovery", compared to the incompleteness of similar thought structures prior to such treatment. Given that we are now moving within the realm of psychological and philosophical abstractions, the sexuality of young people ought to be seen as just so many TSs jostling for position inside the mental framework of these two people. As such they have nothing to do with actual, life subjects, *their* actions being as far removed from the inner world of these men as ever. Instead, the creation of the relevant TSs are a function of the affinitive configuration provided by Freud's universe.

What kind of a universe it must have been is illustrated by Thornton in her book *The Freudian Fallacy* [23], in which she points to the state of sexual repression in general society and Freud's copious use of cocaine in particular as significant factors in the formation of his theories. More than once has she reached the conclusion that the psycho-analytical theories regarding various sexual fantasies as well as the Oedipus complex had not been the result of observed occurrences in his patients, but rather constituted the projected phenomena of a mind which tried coming to terms with a cocaine-induced turmoil amidst the associative shadows of erotic elements. This led to some curious outcomes, such as this instance of Freud's collaboration with Wilhelm Fliess, his "strange friend" as she calls him [24]:

> Fliess himself had already taken off into the flight from reality with his preoccupation with the twenty-eight- and twenty-three-day sexual periods to which people of all ages and both sexes were supposed to be subject. From these premises he had already embarked on far-flung hypotheses that found expression in a peculiar numbers mystique, and by now he was finding a deeper connection between astronomical relations and the creation of organisms. Such writing is practically pathognomonic of drug usage. [25]

Without going into the psychological effects of drugs at present, in terms of thought structures the modification of a particular perception (specific sexual periods) in order to align such abstracts with wider patterns containing intrinsically unrelated but deeply-felt significant contents (astronomy, the birth of Life) is a familiar sign of i-PAT-ls under the influence of some greater driving force.

Nor is the sheer incongruence of such trains of thought readily identifiable due to their isolation from grounded abstracts. Provided some of the environmental factors are similar, someone else can easily transpose their meaning onto further delineations of the theme. Against the background of wariness towards all matters of sex and with Freud having done his work, the notion of a parent-child rivalry evolves into still more assumptions about its extent. So we read many decades later about Dr. Benjamin Spock, described as "America's foremost pediatrician, author, and advisor to parents worldwide", warning the public against being nude in front of one's children. This leads to overstimulation, he warns, which most children cannot take calmly. It produces "overactivity, excitability, sleeplessness, frequent masturbation, school failures, difficulties in parental management, and sexual behaviour of which the parents disapprove" [26]. Not only that:

> Dr. Spock specifically warns a parent against allowing a child of the same sex to view him or her without clothes. He appears to feel that the nudity of a parent is particularly bothersome to the child of the same sex, as in the case of a father and his son. Rivalry, he claims, occurs between sons and fathers because of the parent's much larger penis size. This, he adds, is a common rivalry among boys and men.

He does not explain how mothers ever manage to hide their breasts from their young daughters even when fully clothed, without tossing their offspring into a life of sleeplessness, school failures, and worse. It seems it is not the children who are in need of treatment, but certain professionals.

Somewhat akin to the above is the stance taken towards perversion. On the question why the "taboo on perversions is sustained with such an extraordinary rigidity", Freud decides that they are "not merely detestable but also something monstrous and terrifying" [27]. The reason *why* they are so monstrous and terrifying is worth noting:

> .."as if they exerted a seductive influence; as if at bottom a secret envy of those who enjoy them had to be strangled"

—Freud is quoted from *A General Introduction to Psychoanalysis*. In other words, perversity is not monstrous as such, but only in terms of what one is not permitted to do compared with some lucky others. One wonders whether Freud actually considered the implications of his conclusion, or whether his sentiments are a, well, Freudian slip!

In any case, according to psychoanalytic theory the exercise of sex+ behaviour is yet another instance of trying to escape the procreative necessity and so is acting against the "reappearance of the father". That this quote is taken from the work of someone else (G. Barag), that work being *Zur Psychoanalyse der Prostitution* (*On the psychoanalysis of prostitution*), shows the progression lock in relation to child-related considerations of a sexual nature. It is also no coincidence that the topic is prostitution, considered a perversion by the Christian church and therefore particularly attractive of negativity-imbued TSs from the sexuality-domain.

The actual contents of those TSs concerned with sex+ and perversion may not be enumerated in the text, but this does not mean they do not possess associative properties in the writer's mind which actively participate in the formation of PAT chains. The juxtaposition of 'instinct' and 'reason', in the service of a TP which in turn is influenced by the deterministic assumptions about the whole nature of sex+, leads to a particular interpretation of history, in this case Western civilisation. Hegel used to call it 'philosophical history' (see chapter 9). Marcuse:

> The attempt to draft a theoretical construct of culture beyond the performance principle is in a strict sense "unreasonable." Reason is the rationality of the performance principle. Even at the beginning of Western civilization, long before this principle was institutionalized, reason was defined as an instrument of constraint, of instinctual suppression; the domain of the instincts, sensuousness, was considered as eternally hostile and detrimental to reason. [28]

If positing a simple act of instinct against a simple act of reason is called intellectual belligerence (say, the desire to eat vs the observation of manners), then first and foremost the term applies across a horizontal PAT chain (eg 'I want to eat now—for whatever reasons' vs 'I shouldn't eat now—for whatever reasons'). But here this is not the case. As Marcuse explains further,

> The categories in which philosophy has comprehended the human existence have retained the connection between reason and suppression: whatever belongs to the sphere of sensuousness, pleasure, impulse has the connotation of being antagonistic to reason—something has to be subjugated, constrained.

While the instinctual act (if indeed this is what it is) has remained on the same level, the regimen under which it becomes evaluated in the name of reason has expanded into a complex edifice spanning many abstraction levels, not to mention the further links at those levels. As such we are dealing with vertical PAT chains, and unless both sides of the juxtaposition (X vs Y) are synchronous with regard to their respective affinities these chains cannot be grounded and are therefore i-PAT-ls. The greater the disparity in abstraction levels the less likely they will be synchronous. It is interesting what Marcuse cites as an example of a modern-day version of the overall theme ("From Plato to the '*Schmutz und Schund*' ['*filth and smut*'] laws of the modern world"):

> A bill proposed by the New York Joint Legislative Committee on Comic Books would prohibit the sale and distribution of books portraying "nudity, sex or lust in a manner which reasonably tends to excite lustful or lecherous desires..." (*New York Times*, February 17, 1954).

Note the application of the word 'reasonably'—from the pre-reason state of instinctual lust to a judiciary construct to an interaction between the two; the entire complex coming under the label 'reason' because the latter is dominant in relation to the former. One might add an example of what is considered equally 'reasonable', but this time in an affirmative light because it accords with the ideological tendencies of the same culture: *Exorcists*, one of a series of comic books for America's "adults and teens", in which Hinduism is denounced as the work of the devil, waiting for Christian crusaders to be defeated [29].

In its original state (before its delineation into various strands of sub-topics by himself and others) Marcuse's perception of society begins in a land of instinctual freedom, encounters a combative assertion of individuality between parent and child euphemised through classical characters (Oedipus, Orpheus, Narcissus...), translates this into the wider battleground of society, and ends with collective repression executed by that society's instrumentalities. The *real* starting point then is the Oedipus complex, because it is the first identifiable—or rather, identified—phenomenon from which our current situation can be effectively traced. It is sub-section 'incest' of the sex+ complex which becomes the centre of attention for psychoanalytic reformers as well as interpreters of the human condition. Cognitively speaking, the focus on incest represents the denominative influence of a sex-TP in correlation with a religion-TP such that a particular association with a grounded abstract (ie incest) becomes processed through illusion trees so that its purified version can be subjected to an ongoing treatment by the abstractive elements of sex-related PAT chains. Note that the original abstract has lost its initial congruence, has changed into a TS affinitive with higher-level abstractions especially from the sex-TP and, if at that stage implemented as a—now psychological/philosophical—concept at all, carries within its associative halo an appropriate set of intuitive FAs. These IFAs are in a position to override any observed instances (the CFAs of PAT^1s) due to their unimpeded authority which is supported by the removal of grounded links during the entire procedure.

Take for instance the following statement, regarding the Oedipus complex manifest in contemporary society:

> Nowhere does this [the transposition of the Oedipus complex into modernity] become clearer than in Fromm's reinterpretation of the Oedipus complex, which tries to "translate it from the sphere of sex into that of interpersonal relations." The gist of this "translation" is that the essence of the incest wish is not "sexual craving" but the desire to remain protected, secure—a child. [30]

Suppose there is an act of incest. Either the act is initiated by the parent, or the first step is made by the child. In the former case the scenario does not fit the theory because the child can hardly be a symbol for protection, the assumed substitute for security. In the second, there is no substitution because the object of desire is the parent anyway (and should the adult not be a parent then it is not incest). Therefore, regardless of the newly-found context, the inherent principles of the situation do not compute. For someone guided by the IFAs of Oedipal theory however such considerations would not have a cognitive chance.

Broadly speaking, the usage of such a TP is very attractive. Its ungrounded nature, its vast expanse across so many sex+ issues, its capacity to obfuscate the religious influence through the ambiguous associations amongst high-level abstractions, all this makes for a huge reservoir of potentials once the intellectual departure points have been set. It has been and still is desirable to so many players, not only from the fields of psychology and philosophy but also history, politics, and the arts. An entire cognitive eco-system has been created, identifiable not by its framework, because there is none, but through is malleability, making itself useful to so many ad hoc purposes. It is not without reason (!) that post-modernism lacks a precise definition, and that its artistic cradle has been Dadaism, Surrealism, and so on. This is what James G. Ballard has to say about Dalí in his essay *Salvador Dali (the innocent as paranoid)*:

> The art of Salvador Dali [sic] is a metaphor that embraces the twentieth century. Within his genius the marriage of reason and nightmare is celebrated across an altar smeared with excrement, in an order of service read from a textbook of psychopathology. Dali's paintings constitute a body of prophecy about ourselves unequalled in accuracy since Freud's 'Civilization and its Discontents.' Voyeurism, self-disgust, biomorphic horror, the infantile basis of our dreams and longings—these diseases of the psyche which Dali rightly diagnosed have now culminated in the most sinister casualty of the century: the death of effect. [31]

Or listen to Ramón Gómez de la Serna, an essayist described as a "writer-character" just as Dalí had been a "painter-character":

> Dalí's art has its own particular strength, and so does psychoanalysis, yet no one can fail to be struck by the thesis and antithesis which they so obviously represent.
>
> Dalí dissects reality and was the first artist who, in the face of the spiritless art of Picasso, gave effective expression to man's complexes and unsatisfied desires.
>
> For this reason, Freud considered Dalí a great Spanish fanatic. [32]

By the way, the above should *not* be construed as a comment on Dalí—it is a comment on the comments on Dalí.

In the end the communication of ideas depends on the articulation of thoughts such that the recipient has the opportunity to reintegrate their abstracts into his or her own mind. The flow of information can be coordinated, it can be subject to an agenda, it can be wilfully misleading; yet in each case the possibility remains, however distant, to retrace the steps and thereby develop one's own position in response. What an artificially isolated conglomerate of multi-faceted representations does is to insinuate a sense of cohesion in the recipient's mind which leads to the erroneous conviction that, whatever one's own personal focus may be, it stands for the intended meaning being on offer. Nothing can be further from the truth; but if there is one thing which cannot be defined, it is just that. The last word should go to Dalí—as he himself, not someone else, sees it:

> It is not necessary for the public to know whether I am joking or whether I am serious, just as it is not necessary for me to know it myself.

The ability to abstract is a tool. The greater the ability, the greater the tool's potential. As we have seen, the path of abstraction, that is to say the sequence of cognitive elements being arrayed to serve the processing of information, can extend across horizontal as well as vertical directions. By itself the direction does not promote or hinder the resultant progression; apart from the overall ability of the system of course, it is the attractiveness of what can be found along the way that determines the significance of any particular movement. Ideally, horizontal search leads to such things as comparison, juxtaposition, or affirmation and negation; vertical processing delivers underlying principles, essential vistas, the deep insight and the higher perspective. So it was just a matter of time before the demonstrated existence of enhanced abstractive abilities led to a focus on our own, innermost selves. So far a concern about the individual, inasmuch as one permitted oneself to regard a human being as a distinct, discrete entity, was driven by the desire, need, or fear to align that what was to be studied as closely as possible to the deity which had been assumed to give it credence, in a material as well as spiritual sense. Now the potential of so many cognitive functionalities had reached proportions where the presence of representations in terms of metaphysical ambiguity was no longer needed. If none of the resulting constructs yielded the ultimate answer to our age-old questions about the self and its freedom it was not for lack of trying, nor indeed for lack of material; although ideations about God were relegated to the background, their roots were too numerous and too deeply entrenched to release their host minds just yet.

One major example of such a complex carries the name of Existentialism. Grown into several varieties, and by now superseded by other models as is the fate of any more or less ungrounded edifice, it nevertheless introduced an intricate composition of abstractions, all dedicated to accentuating the Self. For it to be created required a system's ability (ie intelligence), an adequate number of products (ie pieces of information derivable through an expanding culture), the awareness of the products' potential (ie hypotheses and theories in science and politics), and the shared experience of applicability (ie the results emanating from the theories). One of the most significant representatives of existentialist thought is *Being and Nothingness*, written by Jean-Paul Sartre (1905 - 1980) during the early 1940s [33]. Let's explore its resource space.

The book starts with a discussion on the definition of a 'thing', or the product in the context of our resource space. Sartre criticises the traditional dualism which posited 'being' and 'appearance' as the two aspects of the same object. We have become familiar with that approach and followed the mental configurations necessary for the containment of observed characteristics (the 'appearance' carried by PAT^1s) and their abstracted forms (the 'essence' represented by PAT^{2+}s). In due course we recognised the possibility of having the links to grounded structures severed in favour of an alignment with already incongruent TSs, thereby giving the 'essence'-ideations a degree of significance they sometimes did not deserve. In fact, the drive towards a comprehensive articulation of the principle of a thing leads almost automatically to some form of incongruence because the very act of ascertaining a fundamental meaning of something entails an abstraction and thus a divergence from concrete content. That dualistic arrangement consisting of two types of PAT chains, one grounded and the other not, is rejected:

> The appearance does not hide the essence, it reveals it; it *is* the essence. The essence of an existent is no longer a property sunk in the cavity of this existent; it is the manifest law which presides over the succession of its appearances, it is the principle of the series. [34]

In chapter 8 we dealt with the development of hypotheses towards theories etc and the role of the container in providing the contents for such TSs. Initially a hypothesis consists of largely ungrounded structures since it is first and foremost an assumption about something. If one proceeds rationally, associations towards grounded chains are established and the entire structure acquires a firm base. In the irrational version, it is the affinity with similarly ungrounded abstracts which is explored and so the result becomes top-heavy. In terms of the potential to be realised we can say that the former has less potential because the progression depends on observable, ascertainable imports, whereas the latter possesses a relatively higher one due to the much more open-ended possibilities of linking ungrounded abstracts in general. So regardless of the eventual outcome, the traditional dualism provided a wealth of opportunities for possible abstractions to be processed. However, in the above statement there are two arrangements precluding such freedom: firstly, whatever TSs may have been reserved for the notion of 'essence', they are now held to be the same as those concerned with the holding of PAT^1s (ie the initial observations); secondly, in the re-representation of the old model the 'essence'-TSs are taken to be a sub-set of the 'appearance'-structure ("..sunk in the cavity..").

As to the first, the only way the contents of the 'appearance'-TSs can be elevated to some sort of principality at all is through illusion trees since there are no pre-laid higher abstractions that can be used for the same purpose; and in the second case, the considerations stemming from the old approach would be mostly dismissed because, measured against the volume of the 'appearance'-structure, the 'essence'-TSs would be found wanting. To begin with, if the articulation of an object demands the inclusion of the "manifest law" of the thing (or at least some traces of it), then such a TS has already experienced a jump from a PAT^1 to a PAT^2. This may be trivial—for instance, I observe two billiard balls hitting and deflecting each other; I now declare that billiard balls have the characteristic to follow geometric vectors if hit. Note however that such a "manifest law" will relate to billiard balls and billiard balls only, unless there is some proviso to enable further cognitive jumps to be made. Jumps mean affinities, and the affinity potential of a TS about billiard balls is inherently less than one of a TS holding the notions of spheres in general (in practice of course there would be very few minds which constrict themselves to follow the tenets of existentialism to the last bit, and whether the balls are used in billiards or not they would be considered as spheres and be done with it).

For Sartre the problem is this: since an appearance of something is perceived as containing the underlying principles as well, that single aspect cannot even be representative of a finite series of instants because their source is constantly changing. As he writes,

> Although an *object* may disclose itself only through a single *Abschattung* [lit. 'shadowing', ie the—incomplete—projection of a thing via its shadow], the sole fact of there being a subject implies the possibility of multiplying the points of view on that *Abschattung* under consideration. [35]

"Multiplying the points of view" leads to the perception of infinity associated with the articulation of an object. Here then is the compensation Sartre has to affect in order to by-pass the absence of the previous 'essence'-chains—the creation of an abstracted version of an observation, containing within its own domain the ungrounded framework of something like 'infinity'. He explains it in the following manner:

> This new opposition, the "finite and the infinite," or better, "the infinite in the finite," replaces the dualism of being and appearance. What appears in fact is only an *aspect* of the object, and the object is altogether *in* that aspect and altogether outside of it. It is altogether *within*, in that it manifests itself *in* that aspect; it shows itself as the structure of the appearance, which is at the same time the principle of the series. It is altogether outside, for the series itself will never appear nor can it appear. Thus the outside is opposed in a new way to the inside, and the being-which-does-not-appear, to the appearance.

'Infinity' is not an observation, it is not a PAT[1]; therefore the only way in which some low-level content becomes aligned with an abstract denoting 'infinity' is through an illusion tree where the original content gets elevated in order to be linkable to, in this case, 'infinity'. Yet the complexity reaches still further. For something to be represented as manifesting "itself in that aspect" there must already be a TS responsible for the aspect in question. On the other hand, before the evaluation of the 'appearance' (containing its 'being' at the same time), how can an aspect—any aspect—be applied? Obviously, it cannot. For the above to hold true, one needs to differentiate between a result having come from the articulation process carried by con-TSs and another type having drawn on sub-TS processes as well. This is particularly necessary when the additional mental constraints come into force; namely, the acceptance of a "series" that never shows itself, the 'real' object being located somewhere outside its observed version, the opposition between the "outside" and the "inside", and finally, a similar contrast between the seen object and all that of it which remains invisible. Thus the potential of such a TS-complex is considerably higher than an ordinary PAT[1], and quite possibly competes with anything a pre-established incongruent PAT-chain could come up with. There is also a significant contingent spread across subconscious structures. As Sartre himself says in the next sentence,

Similarly a certain "potency" returns to inhabit the phenomenon and confer on it its very transcendence—a potency to be developed in a series of real or possible appearances.

As he placed the finite/infinite dualism after its predecessor and elaborated upon the former's meaning, the phrase "..a certain 'potency' returns.." seems to accord with our more technical perspective: given the innumerous combinations one could derive from the old 'essence'-chains, a full understanding of the new variety can indeed "return" a comparable wealth of abstracts and thereby enhancing the "potency" of the phenomenon. "Potency" of course being another word for 'potential'.

We touched upon the "opposition" that is being set up between the representation of the actual thing and its more open-ended counterpart which is meant to hold every possible manifestation of what has been and could have been observed. Note that the idea of "opposition" is purely a wilful one—since the open-ended contents are not precisely defined there cannot be a rational, ie grounded, contrast between it and something else. Moreover, the choice of describing the variants as "inside" and "outside" inevitably suggests some sort of antagonism without needing any further detail about the actual content. Note also that, according to the definition, it is only the grounded abstract, ie the PAT[1], which is in any definite way ascertainable. If TS1 contains the PAT[1]s and TS2 the possible appearances, then there is an a priori association of the TS1/TS2 complex with 'opposition'-related affinities, such aspect being further emphasised due to the continuing necessity of adjusting the contents of the TS1-chains with those of the TS2-chains—or vice versa.

We find an example of just such an adjustment in Sartre's concern about consciousness. For something to be observed, there needs to be the will of the observer; for the observed to be understood, there needs to be the fulfilled knowledge of the knower; behind it all stands consciousness. How 'one knows that one knows' has been a recurring difficulty for philosophers, and naturally Sartre recognises the problem.

> Either we stop at any one term of the series—the known, the knower known, the knower known by the knower, *etc*. In this case the totality of the phenomenon falls into the unknown; that is, we always bump up against a non-self-conscious reflection and a final term. Or else we affirm the necessity of an infinite regress (*idea ideae ideae, etc*), which is absurd. .. If we wish to avoid an

> infinite regress, there must be an immediate, non-cognitive relation of the self to itself. [36]

Before we turn to Sartre's solution, let us briefly identify the situation in technical terms. Broadly speaking, to observe oneself doing something implies two TSs; one relevant to the action itself, the other to the current position oneself is taking. The second TS (for simplicity's sake I use the singular, but there may well be several structures involved) may or may not be a higher abstract than the first, but it is certainly more than a PAT^1 (it cannot be a PAT^1 because the sense of oneself is an abstraction of various contents taken together). That 'oneself'-TS then exists as a PAT^{n+x} in relation to the other TS which is a PAT^n, where n is the abstraction level of the action performed; x is some single-digit number including zero (with super intelligences the number may well be higher). If I employ the same concept once more, that is conjure up another 'oneself'-TS then, given its inherent nature, it can only ever be of the type PAT^{n+x} again with n and x being the same values as before. The same holds if one observes oneself observing, since the internal representation of myself observing xyz is the same no matter what xyz stands for. The reader can easily confirm this by going through the motions of imagining a lengthy series of the same; regardless how many observations of the observations... etc are made, the thought is always of the same kind and complexity. So, rather than being faced with an "infinite regress" leading us into who-knows-where, we are simply dealing with a given number of instances of the same idea, and hence the same structure. To conceive the result as some unavoidable absurdity would be akin to regarding the visualisation of two, three, ten, or twenty apples as a conceptual absurdity (which *really* would be absurd).

Had those philosophers for whom the possibility of an infinite regress became a pressing issue been more detached in their understanding of the object vs the perceived object, the problem would not have arisen. Sartre is no exception. As with so many of his predecessors, the idea that one's notion of some thing is in a profoundly real way reflective of its actual nature, led him to regard the concept of 'knowing the knower...' with trepidation; after all, what sort of a world would it be where one ingredient, ie consciousness, can seemingly multiply itself towards infinity simply by exercising its faculties in additional relation to another, ie the object? Since the contents of our world cannot be subject to the whims of its inhabitants, the conceptualisation itself must be adjusted.

For Sartre the solution consists of allocating the contents of thought. In the old version we had an idea holding some object; then we had another idea holding the 'self' contemplating the object; then we would have yet another holding the idea holding the 'self' and so on—and all these ideas being somehow representative of real world occurrences (eg 'consciousness' being a 'thing' existing in this world). Sartre now turns the situation around and subsumes not only the object itself but any derived meanings (such as 'essences') under the label which our representation has given it. We have already seen his explanation regarding the inside/outside aspects of everything, and 'consciousness' is no exception. 'Observing oneself' then becomes a matter of 'consciousness reflected', and since the object of 'consciousness reflected' is not a new consciousness but just another quality of an object, we are not really talking about an expansion of consciousness but a sequence of objects, or even the same object possessing all these qualities of 'being reflected on'. In his words,

> Every conscious existence exists as consciousness of existing. We understand now why the first consciousness of consciousness is not positional; it is because it is one with the consciousness of which it is consciousness. At one stroke it determines itself as consciousness of perception and as perception.

> ..This self-consciousness we ought to consider not as a new consciousness, but as *the only mode of existence which is possible for a consciousness of something*. [37]

This leads him to consider a sensation like pleasure not as a separate entity but as part of the same consciousness—"pleasure can not be distinguished—even logically—from consciousness of pleasure". In other words, that abstractive TS we have mentioned earlier now becomes a container for the characteristics one cares to ascribe to it. It fulfills that function because it acts as a holding area for evoked sentiments and/or interpretations in relation to some object; such sentiments attracting particular associations and emotions will henceforth participate in formulating some articulated description of the object; in the process of which the container may well turn denominative. The intended function of a holding area is evident from his comments on the inclusion of pleasure. As to the second purpose, consider the following:

> There is an indivisible, indissoluble being—definitely not a substance supporting its qualities like particles of being, but a being which is existence through and through. [38]

If the abstracted version of some TS (even one including the notion of 'consciousness') is considered to be the content's "existence through and through" then the quality of the thing would certainly be significant. As for the third, the heightened importance of such a TS triggers repercussions down the line as shown in a sentence soon after:

> In order for there to be an essence of pleasure, there must be first the *fact* of a consciousness (of) this pleasure. It is futile to try to invoke pretended *laws* of consciousness of which the articulated whole would constitute the essence. A law is a transcendent object of knowledge; there can be consciousness of a law, not a law of consciousness.

Once there is a representation of something which, as that idea, is taken to be that something's existence per se, and which therefore is representative of its very definition, then chances are such an idea would act in the role of influencer, not the influenced. "It can be limited only by itself", Sartre adds. Although writing specifically about consciousness, for a particular configuration to be created and then acted upon sets up intuitive functional abstracts about that very behaviour. In this context the last sentence in that section together with the first ones of the next are of particular interest:

> But it is precisely because consciousness is pure appearance, because it is total emptiness (since the entire world is outside it)—it is because of this identity of appearance and existence within it that it can be considered as the absolute.
>
> ..It seems that we have arrived at the goal of our inquiry. We have reduced things to the united totality of their appearances, and we have established that these appearances lay claim to a being which is no longer itself appearance. [39]

In the above we find the affirmation of an authoritative status (the contents of a TS stand for its "identity" and include its existence as well), a two-sided nature (although representing an entire "existence" it also stands for "total emptiness"), and the elevation of the spawned

TS to a form of being in its own right (laying claim to something which is "no longer itself appearance").

Now that the fundamental relationship between PAT^1s and their flow-ons has been established, Sartre proceeds to apply it. An investigation, any investigation, starts with a question. The answer is a perception which we, the questioner, need to interpret correctly for it to be useful.

> In every question we stand before a being which we are questioning. Every question presupposes a being who questions and a being which is questioned. [40]

The "being which is being questioned" is "questioned *about* something", and this "about", so he writes, "participates in the transcendence of being". The question then becomes "a kind of expectation", and the revelation about the nature of the being can be reduced to either a "yes" or a "no". Since objectively speaking the possibility of either exists as soon as we pose the question, in principle there is the chance of the answer being negative. If the sample question was, "Is there any conduct which can reveal to me the relation of man with the world?", then

> This means that we admit to being faced with the transcendent fact of the non-existence of such conduct.

Note the sequence of events here: there is the question emanating from a context represented by certain TSs; there is the object that is being questioned, at this stage a PAT^n; there is also already an answer, a TS of the PAT^{n+x} kind which contains the non-existence of whatever the PAT^n will eventually represent. The non-existence of what? Since the answer-TS can't be expected to provide the material (we haven't yet begun to work on the reply), it must come from the question-complex. Therefore, no matter what the answer-TS may yield, or rather, no matter what we are willing to accept from the answer-TS, there is already a path laid out for us which starts with the contents from the question-complex, touches on the negation inherent in the answer-TS, and ends up somewhere in the question-complex again—that 'somewhere' being defined through the affinities in operation at that time. For Sartre the matter appears thus:

> Thus the question is a bridge set up between two non-beings: the non-being of knowing in man, the possibility of non-being of being in transcendent being. Finally, the question implies the existence of truth. By the very question the questioner affirms that he expects an objective reply, such that we can say of it, "It is thus and not otherwise."

Recall the decision to incorporate the characteristic of something into the perception-TS of the thing. The result is not only a perceived 'appearance' but an idea which stands for the full contingent the object has on offer. Add to that domain the 'essence'-containing configuration of 'consciousness' and we have a ring of mini-containers, through the combined influence of which the PAT^1 data have to travel in order to be processed. Not only is the arrangement a complex one, it is also denominative: 'being' and 'non-being' are not ambiguous values, nor is the concept of 'truth' (their actual contents are another matter). Indeed, the very act of posing the question is meant to set up a PAT-chain along the lines of a perception of 'objective reply', and so pre-determines the quality of the result to a significant degree. Since 'truth' is regarded as another object, its qualities as to 'being' or 'non-being' (ie 'true' or 'false') comprise its nature as part of its entire contingent. Then Sartre can say,

> In a word the truth, as differentiated from being, introduces a third non-being as determining the question—the non-being of limitation. This triple non-being conditions every question and in particular the metaphysical question, which is *our* question.

Such conditioning extends further than the preliminary setting-up of the question; whatever the answer may be, it exists first and foremost as an abstract within some cognitive structure. Therefore at that point it is immaterial whether the answer is truthful in the realistic, outside-world sense. What *is* being ascertained by this format is an a priori significance derived from the associative influences of such notions as 'objective', 'truth', 'cannot be otherwise', and so on—all these characteristics getting linked with the answer-TS. Not only is there a pre-treatment, there is also a post-treatment of the cognitive process responsible for the answer. Neither the pre- nor the post-processing is restricted to con-TSs only; given the contextual ambiguity inherent in such concepts as 'truth' etc a great deal of work in fact will be conducted away from the conscious supervision of the ego.

This brings us to the role of the individual mind doing the processing. Before, the religious mentality featured a series of low-level abstracts which were more or less congruent as perceived reality, depending on the degree of insights which were possible at the time. Then came the logic boundary, delimiting the appearance of such congruence and heralding the presence of ungrounded, container-driven abstraction complexes which determined the perceived significance of the contents coming from the lower levels. These complexes were homogenous in the sense that their metaphysical contents brought their associative power to bear upon the imports as a comprehensive force—we might articulate them as an all-encompassing, general world view. Now, the contents from the lower levels are assessed individually by a complex abstracted version of the original, where the quality of that abstraction is individually adjusted according to the effects of certain interpretations on the associative domain of said abstracts. One could say that whereas before the mind held one global religion, it now contains a multitude of mini-religions which wheel about according to the perceptive dynamic of the given mind.

This has its good and its bad side. Consider for instance what Sartre has to say about 'destruction'. A tempest does not destroy as such because there does not exist an interpretative framework which applied some value judgment to the things as they were before and after the storm. Therefore natural disasters "merely modify the distribution of masses of beings" [41].

> In order for destruction to exist, there must be first a relation of man to being—*ie.*, a transcendence; and within the limits of this relation, it is necessary that man apprehend one being as destructible.

Apprehension of course implies the processing of information, and that processing entails the possibility of 'non-being'. That "certain probability of non-being for a given being under determined circumstances" Sartre calls "fragility". If the interpretation of a particular act as 'destruction' requires a rather detailed analysis so that one may understand the "determined circumstances", then one could expect the result to be somewhat more considered than some categorical judgment derived from one's pre-established ideology. The associative domain of 'fragility' alone hints at the tentative nature of the assessment. On the other hand there is no predictable stability either, and the outcome of one's position is purely based on current circumstances.

Yet again on the other hand...

> It is because of this ensemble of measures that an earthquake or a volcanic eruption can *destroy* these cities or these human constructions. The original meaning and aim of war are contained in the smallest building of man. [42]

If the associated elements are already indicative of the entire contingent of possible entities (including of course 'non-being' and 'destruction'), then a preventative measure for example would fall under the same category too. So, if the preventative measure is an indirect acknowledgment of the possibility of 'destruction', and cognitively speaking the possibility exists because of the preventative measures, then the stage is set to define a situation not in terms of the causative effects observed in reality but in terms of the associative effects caused by their mental representations. To continue our example, under these auspices the disappearance of a forest because of bushfires would not attract the connotation of 'destruction', but preventative measures taken by a forest authority could.

These meanderings from the perceived to the abstracted object and back again are quite similar to what the Jainas employed when musing about their world and its relationships. Compare the ideas on the nature of the atom as depicted in Fig. 3-3 with Sartre's explanation about objects in general. Compare also their dualistic mode of viewing reality as 'is' and 'is not' with Sartre's 'being' and 'non-being'. Unlike Sartre, the Jainas incorporated this approach into the Seven Modes of Predication, the Saptabhangī, in order to apply a formal reference to one's perception of the world (see chapter 3). What's more, they understood the representative elements of 'is' and 'is not' strictly as abstractions, and so standing aside the 'real'. At the same time, those abstractions did not reach beyond their initial representation *as abstracts*, and so the cognitive model of that ancient world had been altogether a flatter one.

A further aspect regarding 'non-being' should be mentioned, although we will explore it more fully in subsequent chapters. If a particular representation is to be associated with another particular content, the linkage is a straight-forward one (certain horizontal and/or vertical complexities notwithstanding). What happens if, despite the stated need for an affinity, there is no readily defined content to be associated with? Suppose I am taught the rule that dropping one's knife at dinner is bad manners because of the noise it makes. Regardless of how much in agreement I may be with that rule, the sound of a knife clattering onto the plate is obvious and so at the very least I have *something* to relate to. Now suppose I am confronted with the same rule but this time no reasons are given. This time my mind is free to associate a non-compliance with whatever it feels like, but no matter what the actual contents may turn out to be they will have some form of affinity with the cognitive structures of the representations of the self—after all, 'it must have something to do with *me*' since obviously I am the target of the rule. In the absence of any concrete affirmation my entire 'self'-concept will be affected, the degree of which being dependent upon the inherent affinity between any aspect of the 'self' and representations of the knife, or the plate, or the dinner, or eating at the table, or eating per se, or being in company... the list is virtually endless with each item being first and foremost a function of its readiness to be affinitive with the context. In other words, declared associations with unspecified content lead to an affinity-based internalisation of the entire complex, and as such are prone to subjective interpretation only. In Sartre's examples, where something is being paired with 'nothingness', the results are rather benign, creating a very personal perspective of the objects in question. Beyond those however the outcome can be quite dramatic, where the unstated threat of some scenario produces a conceptualisation of the ego which is as damaging as it is indefinable.

Moving backwards and forwards across the variously abstracted contents of TSs is therefore also an exercise in positioning of the self. It is not just the characteristics of the object which are being evaluated, but the process of association-building with largely subjective concepts causes the general nature of the 'self' to be brought into the equation as well. Whatever the 'self' stands for, the affinitive strands of its elements participate in the formation of PAT-chains. Hence not only is the object being defined in the process but also the persona doing the perceiving, and the dexterity employed could be considered as a measure of the person's overall capacity—his freedom. In line with his previous perspectives Sartre does not consider 'freedom' to be a property of humanness.

> Human freedom precedes essence in man and makes it possible; the essence of the human being is suspended in his freedom. What we call freedom is impossible to distinguish from the *being* of "human reality." Man does not exist *first* in order to be free *subsequently*; there is no difference between the being of man and his *being-free*. [43]

For us this means there is a TS holding the concept of "being-free" regardless of the subject matter under consideration, with that TS' domain reaching towards any other structures which are related to the self. As such its role in any given process is a potentially two-sided one: when it comes to the processing of articulated contents it will have to adopt a flexibility in accordance with the interpretative variants encountered by the object—it is the phenomenon of homology again (ie the representational units of the 'self' are given, but they have to behave differently due to the cognitive demands coming from the object). Yet as far as the subconscious dynamics are concerned, the associative effects produced by the object generate representational units all directed towards the 'self', although dissimilar in structure. A simple example: suppose I 'just know' that I am the boss of a gang. I argue the point with whomever I feel like by demonstrating that I can make any of my underlings do exactly as I please. That's the homologous part, directed towards the member-object. Then I become uneasy in the face of some real or imagined opposition and when everyone decides to go and see a movie rather than listening to me I rationalise their sudden departure by telling myself that as the boss I need to be sometimes lenient. In this case the imported perception (and not of my doing) is denominative towards my ego-perception and so having everybody taking off is considered part of my powerful self after all. This occurrence is the opposite to homology, and so we can call it *synonymy*.

If both, homology and synonymy, occur at the same time as is the case here, the result is a TS complex of a very specific nature. Its conscious part consists of similar representational units that show a variance in behaviour, whereas its subconscious part consists of differing elements but behaving in a similar fashion. During the process of articulation (through speech, mannerisms, behaviour) the migration of data from the—diverse—subconscious to the conscious needs to proceed via illusion trees so that the homogenous character of the latter can be accommodated (recall my act of rationalising the gang's departure in our example). These are dedicated illusion trees, acting under the regimen of what could be termed a *rationalisation complex*. It is a common occurrence, and comes into being whenever there is a mismatch between the acknowledged meanings held within subconscious and conscious structures. As the structures become established, their operations engender IFAs which further ensconce their presence and, as can be imagined, their contrasting origins hint at a certain degree of discomfort for their owner.

We are given an example of such a complex when Sartre continues to describe the concept of freedom. He sees the idea of consciousness inclusive of ascertained object characteristics but also of 'nothingness', the delimiting feature of object identification. Since there is also the consciousness of consciousness, there ought to be "consciousness of nihilation" which in turn gets paired with the element of 'freedom' as a part of the "being of consciousness", producing the "consciousness of freedom" [44]. As to the format—

> In freedom the human being *is* his own past (as also his own future) in the form of nihilation. If our analysis has not led us astray, there ought to exist for the human being, in so far as he is conscious of being, a certain mode of standing opposite his past and his future, as being both this past and this future and as not being them.

We can identify the "past" as TSs embedded in memory and their associations spread amongst sub-TSs. We also recognise the conscious act of viewing the past and future as the work of con-TSs. Given the likelihood of past events patterns not being in total agreement with those of more current perceptions and hence attitudes, chances are that the process of evaluation, as portrayed by Sartre, necessitates the creation of rationalisation complexes. So when he asks what kind of consciousness this would be his own reflection prompts him to write,

> ..it is in anguish that man gets the consciousness of his freedom, or if you prefer, anguish is the mode of being of freedom as consciousness of being; it is in anguish that freedom is, in its being, in question for itself.

Note the setting up of a progression lock: from the perception of 'essence' to the positing of 'consciousness' to a definition of 'freedom' and so to an entire modus operandi which from now on determines the way one's surroundings and one's self in it are processed. The associative properties of the TSs encountered along the way enable Sartre to come up with the "formula" of consciousness:

> "Consciousness is a being, the nature of which is to be conscious of the nothingness of its being." [45]

The constant juxtaposing of an 'is' and an 'is not' and the ever-present 'nothingness' which acts as a grand space in which those few glimpses of 'something' float around, all this leads to an affirmation of a particular type of mental existence.

Irony is one of these. In irony a man annihilates what he posits within one and the same act; he leads us to believe in order not to be believed; he affirms to deny and denies to affirm; he creates a positive object but it has no being other than its nothingness.

In a stronger version the more or less light-hearted opposite turns into a presence that cannot be switched on or off so easily; what before had been a make-belief toy is now a serious instrument of destructive power. The lie which has become part of oneself turns into a festering sore. As Sartre puts it,

> The ideal description of the liar would be a cynical consciousness, affirming truth within himself, denying it in his words, and denying that negation as such. [46]

He labels the state of living a lie while being in denial of it "bad faith", and his perspective on 'being', 'freedom' and so on predisposes him to accord that behaviour a prominent place in human affairs. Consider the dualistic nature of the rationalisation complex in terms of thought structures and then compare its double role but single existence with Sartre's words on the subject:

> There must be an original intention and a project of bad faith; this project implies a comprehension of bad faith as such and a pre-reflective apprehension (of) consciousness as affecting itself with bad faith. It follows first that the one to whom the lie is told and the one who lies are one and the same person, which means that I must know in my capacity as deceiver the truth which is hidden from me in my capacity as the one deceived.

And further,

> That which affects itself with bad faith must be conscious (of) its bad faith since the being of consciousness is the consciousness of being. It appears then that I must be in good faith, at least to the extent that I am conscious of my bad faith.

In how far such consciousness can make itself felt in any individual is open to question. Sartre observes that bad faith "can be the normal aspect of life for a very great number of people" [47], and its ever-present commonality masks it as a normal feature, not to be comprehended. Given its cognitive framework, a rationalisation complex is inevitable whenever there is a need for a certain attitude to be demonstrated without however being in possession of a satisfying reason for such an attitude to exist. We can expect its creation then any time there is some form of authoritarian pressure to conform to a particular style, where the authority can be an ideology (spiritual or secular), a political directive, or a social imposition. Sartre quotes Steckel, a Viennese psychiatrist, who reports on cases of "pathological bad faith which the Freudian doctrine can not account for"; for example women who suffer from frigidity because of marital infidelity, but whose behaviour demonstrates the workings of a conscious will nevertheless [48]. Keeping in mind the increased probability of IFAs in any type of prolonged behaviour, a general environment that is conducive to these complexes will spawn appropriate rule sets within the TSs such that the acting out of "bad faith" is not only not confronted but where any attempt at its deconstruction will be met with hostility. Examples of sometimes cruel suffering which is adhered to with great fervour can be seen anywhere in the world and we will re-visit this issue in Part II.

All this raises once again the issue of freedom. In how far is it possible for us to decide on any given approach, and in how far are the pre-conditions deterministic of our subsequent behaviour? Sartre's introductory comments to this section are interesting:

> It is strange that philosophers have been able to argue endlessly about determinism and free-will, to cite examples in favor of one or the other thesis without ever attempting first to make explicit the structures contained in the very idea of *action*. [49]

If one's inner universe is made up of discrete units of evaluating entities, each acting on the data which are specific to *its* domain and no other, then one can indeed have the tendency to look beyond any real or imagined universality of meaning; it is almost compulsory if some sort of productive outcome is desired. Such contemplation then, in terms of its attitudinal quality procured, leads to an intent with respect to the situation considered. "..An action is on principle *intentional*", Sartre remarks. True to his principles the action consists of an "objective lack" ascertained through the object's extension into 'nothingness', and so there is an intent to create something which has been found wanting. The intent therefore does not emerge from an exposure to the fullness of experiential contingencies, but what the mind can evoke within itself as a "lack".

> It is on the day that we can conceive of a different state of affairs that a new light falls on our troubles and our suffering and that we *decide* that these are unbearable. [50]

No situation by itself is capable of formulating an intent; it is the actions performed by consciousness, comprising as it does the recognition of 'non-being' which transcends the direct perception of 'what is', that lead to action and thus to an enactment of freedom:

> It is the organized form..which must be surmounted and denied in order for it to be able to form the object of a revealing contemplation. [51]

The perception of a "lack" needs to be in place before any further mental action can occur. The resulting action, so Sartre insists, should not be seen as a deterministic consequence of such pre-condition, but as an effect that harbours the full range of possibilities available to the person acting. In chapter 7 we discussed 'free will' against the sentiments expressed by John Locke, and we described the phenomenon as the acted-out potential at varying levels of directness in terms of mutual interdependency between dynamic elements. The question then becomes a matter of choice for placing one's instrument of analysis as it were, that is to say choosing exactly where along the path of processes one wishes to insert the probe. In terms of the subjective thinking experience the options are limited—one cannot observe outside the moment of any particular thought-sensation, either before or after its event. For Sartre then, given his relatively multi-faceted focus and given that it is exercised as a philosopher standing more or less outside it all, the point of access would be the interface between the pre-ordained contents (the sub-TSs) and the contemplating consciousness (the con-TSs). It cannot be otherwise: the data embedded within the subconscious structures only rise to the surface at the moment of being mined by the conscious, and it is the conscious which is being charged with purposefully processing the data together with the rule sets furnished by it. From our perspective Sartre's solution is rather trivial—it does not provide any further insight than what is ordinarily possible in any case (without delving into the technical details of cognitive dynamics that is). It does however confront the individual with the opportunity, neglected at his own risk, to investigate the circumstantial potential of an events series with the view of obtaining a course of action that is consistent with the perceived totality of the scenario.

If nothing else, this one aspect of existentialism represents a significant advance over less accentuated modes of thinking. There is no remedy for a real or imagined state of dissatisfaction, beyond the immediate prompt towards a considered action. If some better option exists it is already contained within the sum-total of possible states and requires the application of intent—provided such intent includes the relevant possible state. That state presumes the response that will be forthcoming, it cannot do otherwise. Sartre:

> Such a choice made without base of support and dictating its own causes to itself, can very well appear *absurd*, and in fact it is absurd. This is because freedom is a *choice* of its being but not the *foundation* of its being.
>
> ..By this being which is *given* to it, human reality participates in the universal contingency of being and thereby in what we may call absurdity. This choice is absurd, not because it is without reason but because there has never been any possibility of not choosing oneself.
>
> ..It [ie freedom] is the perpetual escape from contingency; it is the interiorization, the nihilation, and the subjectivizing of contingency, which thus modified passes wholly into the gratuity of the choice. [52]

Here we have the relegation of the subject domain to those TSs immediately concerned with its representation, causing the perception of absurdity against a possibly wider, more universal reference system and necessitating a self-referential format towards its components (first paragraph); the positing of any possible characteristic within the total contingent of what can be perceived precludes its dismissal, and so the choice becomes perfunctory while still essential to its nature (second paragraph); and so finally, the act of freedom is a continuum of a self-evoked space within which the essentiality of the world provides the material for its affirmation and negation, which in turn unfolds into the choice of essentiality. Or, using our nomenclature, the contents of a perception limited to directly-related PAT chains while neglecting affinities with more distant structures; the resultant set of TSs becomes the reference and sole containment of perceived characteristics including the element of choice as a constituent FA; the constant act of invoking such structures leads to the formation of rationalisation complexes where not only the given object-characteristics but also ethical notions compete for their admission to a significance ultimately determined by their owner.

It seems this is as close as we can get to a concept of 'freedom'. In Marcuse's case there was the sensing of a land beyond one's reality which had been defined through the actions of generations, yet still veiled by the imposed unwillingness to openly confront one's deepest urges. With Sartre it was the background of one's cultural regimen that had been veiled for the purpose of isolating the single object, yet recognised as part of the wider fabric that includes us and our consciousness and all of its potential. In both cases the result was an articulation of some content, but content that was made for articulating anyway. In the end the *perceived* freedom had not changed; as always it depended on the ability to sensate as exercised by the individual—never more, and never less. The *actual* freedom however had been enhanced by the pursuance of hitherto unavailed abstractions, raising first the latency, then the realised potential of mental dynamics, but also raising the degree of complexity and thus the possibility of progression locks. The effect was not wholly satisfying—a re-setting of the applicability contingent was in order.

12. Back to earth

Ludwig J.J. Wittgenstein - Karl R. Popper - Homo Sapiens

Even without going into the technical details of mental configurations one can discern the steady increase in complexity over the centuries. Whether we look at the language used, the concepts conveyed through their words, or the relationships that had been assembled within their symbolic scenarios, these overt signs alone convey the growth of the intellectual world presented to us over the generations. The philosophical material has been only one feature of that world, and such is the interdependence of its many elements that it would be pointless attempting to define a particular cause and effect between any two of them as some typical occurrence. Sometimes an original impetus came from science, sometimes from religion, sometimes economic circumstances prompted further musings, sometimes an experience from war, or disease, or a natural disaster formed certain ideas which in time proved significant to others. In this text our relatively long paces through history relieved us from dealing with the question of ultimate originality so far; had we observed the intellectual progress in shorter increments our regular focus on philosophy may have been misleading in terms of this or that idea's authority.

Now we have arrived at the present however, and such is its sheer volume of appearances of any kind that a sequence of some description not only may not, but also need not be synchronous with the actual material. There is more to this than meets the eye.

As we deconstructed one cognitive edifice after another, the instructive purpose behind these exercises dovetailed neatly with the evolutionary gradient of the sophistication we encountered along the way: in the beginning our simple tools of understanding were applied to simple perspectives, and as the mental structures grew on the outside, so did our own as the representations of the former. As engaged as an Augustine for instance certainly was, he would have been quite bewildered by the intricate juxtapositions of a Sartre, his plain morality no match for the elaborate relativities espoused by the other. To highlight the unfolding of thought, the ever-widening range of possibilities enabled by broader and broader platforms, we introduced the e-chart. Its parameters are deliberately vague in order to encompass a variety of type addressable through their values, depending on the current context. For example, 'product' can be an idea, a piece of machinery, a vaccine, or a public policy; its potential and in turn its applicability likewise relies on some temporary particulars. The parameter types are also interchangeable, and so the effect of say, machine use on governmental policy can easily be accommodated within the chart's resource space.

On a rather trivial level we are therefore justified in turning to the theme of the present chapter *after* having examined the efforts of a Marcuse and Sartre, even though chronologically speaking the sources for what is to come should be placed somewhat earlier. The deliberations by Marcuse and Sartre may have occurred during the 40's and 50's of the 20th century, but the material for them existed already. Such is the immensity of that century's resource space that differences in time increments do not necessarily matter because the entities responsible for this cause and that effect, for all practical purposes, exist simultaneously. Hence to consider the complexity achieved by these two philosophers and then view another, previous, approach as a form of re-setting is not a mistake in the title; it is done out of recognition of the era's abundance.

Apart from the issue of synchronicity the more profound questions of potential are raised, especially the relationship between its latent and realised version. To identify the realised potential of an item does not present a problem—here is the object and there is its application. But what about the item's latency—is it proper to include latencies in the description of a thing? And how far can one stretch the definition of the word 'latency' in any case? Consider a lump of iron ore: if by latency we mean all that which has not yet been turned into some material form, then we are quite justified including sheet metal, steel rods, iron chains and so on, in the unrealised potential of that piece of ore and adjusting the value we are giving it accordingly; the more uses of iron ore can be discovered, the more valuable its raw state will become.

Now consider 'iron ore' in terms of its mental representation. We know the realised potential of the concept because we can think of sheet metal and steel rods etc, but we cannot visualise its unrealised possibilities because to do so would entail having such thoughts and these are unavailable to us. Two hundred years ago the definition of 'iron ore' did not include steel. Until the middle of the 19th century the significance of oxygen and carbon in the molecular arrangement of iron was unknown, and so even the cognitive latency of the concept of 'iron' lacked those ingredients. Yet its latency in terms of the atomic configuration *does* incorporate such possibilities—as industrial history has shown. The problem then becomes one of deciding under what conditions of definition the idea of latency is going to be used, keeping in mind of course that at this stage the medium of definition is our intelligence (or anyone else's for that matter). Outside that specialised domain the nature of latency cannot be expressed in the form of a progression from one identified state to another since there is no-one there to do the identifying. Rather, it is the state of readiness expressed in the medium of the kind of dynamic which happens to be relevant at the time. In that sense the possibility of rust on iron can be expressed as a form of latency with respect to iron because the material has the ability to oxidise and the appearance of oxygen will set the process in motion. At the same time, oxygen does not appear on the scene *in order to* create rust, just as rain does not fall onto the fields *in order to* grow plants. Once intelligence becomes involved however a sprinkler system made out of stainless steel can be created which waters the field so that produce may grow. There is a fundamental difference between those scenarios, so let us define it.

At the outset we ought to make clear what we mean by 'latency' or 'unrealised potential', the words already having been circumscribed somewhat during our discussions involving the e-chart. If some object contains certain characteristics which at the moment are unused and hence are lying dormant, then these characteristics represent the unrealised potential of that object. It does not matter *how* such characteristics will have to be used, and equally insignificant is the degree of participation by them when compared to any other additions that may be necessary to evoke the new state. While the manner of usage is unimportant, its proximity to the original condition is not. For example, bees wax has the potential to be used in candles, due to its calorific properties and its malleable nature: just supply a wick, put a lump of wax into the correct shape, and there is the candle. Less direct is the potential of iron to be made into steel because it requires a relatively complex procedure to add carbon to its molecular structure. Nevertheless, unless certain characteristics—of whatever form—are in place the potential for any further combinatorial use does not exist. It is also obvious that it is easier to discover how to make candles from bees wax than it is how to turn iron into steel. The proximity to some original condition (seen from the point of view of the realised potential) becomes a matter of probabilities in terms of the relevant characteristics needed to be activated. Since we are talking about probabilities a quick detour is in order.

Suppose I have an ordinary playing die with number symbols from '1' to '6' engraved on each of its six faces. The probability of throwing any one number, and one number only, is $^1/_6$ (ie, there is 1 possible and desirable event in a sample space of 6), or 0.167. If I have three dice then the probability of one single combination of three numbers appearing (eg 4-1-6), with no particular order and repetition allowed, is $^1/_{216}$ ($^1/_6 \times {}^1/_6 \times {}^1/_6$), or 0.00463. But suppose I am less specific and group the six numbers into A (1,2), B (3,4), and C (5,6), then the probability of one number appearing in any one group at one throw is $^1/_3$ (ie, $^1/_6 + {}^1/_6$, or 2 possible and desirable events at one time in a sample space of 6), or 0.333. Similarly, if I apply the 2-number categories to three dice, then the probability of groups A, B, or C appearing amongst the three dice in whatever order and with repetition allowed (eg B-C-A), is $^1/_{27}$ (ie, $({}^1/_6 + {}^1/_6) \times ({}^1/_6 + {}^1/_6) \times ({}^1/_6 + {}^1/_6)$), or 0.037. Suppose again I am after single-number combinations with three dice, but with the added rule that repetitions are not allowed. Now the sample space is 6 for the first die, 5 for the second and 4 for the third. Then the probability of a single non-repeating number combination is $^1/_{120}$ ($^1/_6 \times {}^1/_5 \times {}^1/_4$), or 0.00833. If we use the non-repeating rule for the group example, the probability becomes $^1/_{15}$ (ie, $({}^1/_6 + {}^1/_6) \times ({}^1/_5 + {}^1/_5) \times ({}^1/_4 + {}^1/_4)$), or 0.0667. Note that the rule prohibits us from using a repeated number outright, so it is not a matter of having a value available and choosing not to use it—hence the decreased sample space. It should also be pointed out that every throw is an event in its own right; that is to say, the probability values apply anew at every time—the dice do not 'remember' past events.

To highlight the various outcomes we can list the results in the following table:

	no. of dice	no-repeats rule	Single/Group	probability P	
❶	1	no	S	0.167	❶
❷	3	no	S	0.00463	❷
❸	1	no	G	0.333	❸
❹	3	no	G	0.037	❹
❺	3	yes	S	0.00833	❺
❻	3	yes	G	0.0667	❻

Expressed more generally, in terms of probability P of a particular event occurring, this means:

with the number of products increasing, no additional constraints and same applicability definition, P decreases (❶→❷);

with the number of products being the same and no additional constraints, P increases when the applicability is defined more coarsely (❶→❸);

with the number of products being the same but with additional constraints and a finely-grained applicability definition in both cases, P increases (❷→❺);

with the number of products being the same but with additional constraints and a coarsely-grained applicability definition, P increases also (❹→❻);

All in all it can be seen that an increase in the number of products alone reduces the probability of any given event occurring; if some constraints apply the chance of an event increases because any possible alternatives are being prevented; and lastly, if we define the success of an event in very particular terms its chance of coming to fruition is reduced, but if the criteria are less fussy the chances of the event happening are enhanced. Going back to our bees wax-and-iron example, less number of products are involved in the case of wax than with iron, so candles are more likely; the definition of success is more critical in the case of iron, which means another difficulty; there would have been more constraints associated with iron (a more complex system is associated with steel production than it is with bees wax), so once started the process is more rewarding.

Our simple example of thrown dice does nothing more than point to the general relationships between dynamic states, defined by their volume, their overall rule base, and the stringency of their evaluation, in terms of the underlying probabilities of their outcomes. They do not touch upon the availability of these states as well as the quality of those others which contribute to further outcomes, nor do they rate the frequency with which one state moves towards another, not to mention the driving forces applicable at any one time; these parameters need to be ascertained within the context of a specific event. We can observe such relationships however, as general as they are, and view a non-living, dynamic system as a series of object states that incorporates the laws of probability to the extent that its movement through time is not entirely random. There is no intent, but there is a performance envelope which delineates the nature of the system. Once the system has reached a degree of complexity such that the number of products which result from the applied probabilities are sufficient to support the previous-state products (ie those which gave rise to the newcomers) so that the probability domains of both types of products are congruent, we have a *dynamic* (ie there is movement from one state to another), *self-regulating* (ie the definition of success is not too fine and not too coarse), *pattern-seeking* (ie preference will be given to some constraint due to its chance-enhancing property), *system*—in other words, we have Life.

Note the element of feedback in the above description: if the new set of states, through their probabilities, does not affirm the same kind of possibilities in terms of their likelihood of occurring than their parents' activities, then not only will the new states decrease in number but also the old ones and so further generations of outcomes will be affected overall. Hence feedback, as far as its role within a live system is concerned, should not be seen as a forward-looking factor alone; in terms of the potential of the system's elements it also reaches backwards because it re-defines the possibilities of the previous generation of the system's states. Note also that we are only talking about one generation at a time, since the inability to 'remember' holds as true here as it did with our dice before (but that does not mean several series of states at various stages of their dynamics cannot exist side by side in a sufficiently large resource space—see the beginning of this chapter).

Our use of descriptors such as 'dynamic states', 'sample space', 'likelihood of occurrence', and so on, may trigger some comparison with the kind of language we employed with regards to the behaviour of mental structures. And so it should, given that in the mind we are dealing with a system that is alive. The ability, if not the need, for TSs to seek out patterns, exercise affinities, and act denominatively towards already existing configurations, are nothing else but sophisticated versions of our simple introduction this chapter. Not only does the very highly complex system of the mind (note to the reader: a thousand additions is complex—a single differential equation is complicated) lend itself to the organisation of such functionalities, a human being, owning a mind and so being subject to its manifestations, will conduct thought processes which involve the ascertaining of affinities, of pattern recognition, and the question of meaningful content amongst these.

We have mentioned Augustine and Sartre as two representatives of thinking humans, both similar in their quest for answers to abstract problems yet also differing in their intricacy of perspectives. Over the last two thousand years or so the feedback process of evaluating mental phenomena has gradually absorbed the accompanying rules of evolving articulations and with that expanded the mind's resource space. The meanderings around the theme was not an arbitrary search, going this way and that; rather, it demonstrates the slow but steady implementation of attractive possibilities through the incessant testing and improving of the emerging re-representation of ourselves—not because this has been our 'destiny' from the very beginning, but due to the inherent probabilities of resultant thought states, forever defining their potential through the following products. Sooner or later the inner eye, that is to say a current mental state of functional integrity together with its favoured progeny, would turn to the identification of thought itself; not as some assumed type of human activity, but as a precise entity of information-processing capacity situated inside a system which can be probed, analysed, tested, and described in concrete terms. It is the last step a philosopher can take, if by 'philosopher' we understand a person who contemplates the human condition in a purely abstract way. Any closer to the core of our—functional—existence and the medium of our own thinking falls short of its stated purpose, just like a light microscope cannot identify anything below the pattern-size of its own medium (the likes of Sartre et al were already teetering on the edge). And so we have arrived at the end of the age of philosophers, those dealers in visions and gamers of abstracts. Our cognitive evolution has turned a corner.

Ludwig Wittgenstein (1889 - 1951) was one of the last to extract something useful from his contemplations, especially since he went as close as he could to the centre of the data-representing problem. In his *Philosophical Investigations* [1] he proceeds to probe more and more deeply the meaning of words, how they relate to consciousness, and how meaning comes about in the first place. It is interesting to find his approach based largely on questions, not statements or answers; a question being a holding area for several possibilities. Yet they do form a definite path, each step being the appropriate (probable?) progression from the current state of his argument.

He begins by quoting Augustine (as it happens) as an entry point to the phenomenon of human language: its words carry specific meanings, and the meanings stand for objects, although Augustine does not differentiate between various types of words (the reader can guess why) [2]. For Wittgenstein this is not enough of course; he recognises the importance of word types in the presentation and arrangement of meanings, and our ability to discern them goes hand in hand with our understanding of sentences. There is such a thing as context, and he wonders how it can be that sometimes one, two words can identify a context which can consist of a sizable domain by itself. Moreover, one more word can change the direction entirely and with that the meaning of the very same words. He asks whether a language can ever be complete and compares it with a town—"..how many houses or streets does it take before a town begins to be a town?". The comparison is in fact quite significant:

> Our language can be seen as an ancient city: a maze of little streets and squares, of old and new houses, and of houses with additions from various periods; and this surrounded by a multitude of new boroughs with straight regular streets and uniform houses. [3]

To infer from the appearance of human artifacts something about the minds behind their creation is certainly worthwhile—see for instance our mentioning of Roman and Gothic architecture in our discussion on Descartes in chapter 6. Here it is even more apt: a still relatively simple representation through linguistic symbols, derived somewhat arbitrarily, to be superseded by more coordinated structures later on when knowledge has arrived on a more confident footing. We can continue this analogy in line with our own perspective by pointing out the tendency of higher data volumes in the regulated areas together with their categorised uniformity, and the lower volumes coupled with more diffused but possibly wider-ranging information under conditions of lesser productivity. Or take renaissance man, who knows a little about lots of things, and a modern-day expert who knows a lot in depth within a small field of knowledge.

Wittgenstein then moves on to the representational aspect of words, or rather the ratio of the volume behind the cognitive representation to its linguistic counterpart. For instance, a shortened and an unshortened sentence can carry the same sense because their respective cognitive sources are far more extensive than a few words would suggest. We realise by now that the understanding of a sentence relies upon holding in one's mind the functionalities symbolised by the words, and provided the words in use have sufficient semantic strength they will be able to evoke the relevant TSs which in turn sensate their meanings to our consciousness. This brings him to the question of the number of possible sentence types, which he thinks is huge. In the course of the "language-game", as he calls it, such types come and go, and he provides a list of suggestions which I will quote here:

> Giving orders, and obeying them—
> Describing the appearance of an object, or giving its measurements—
> Constructing an object from a description (a drawing)—
> Reporting an event—
> Speculating about an event—
> Forming and testing a hypothesis—
> Presenting the results of an experiment in tables and diagrams—
> Making up a story; and reading it—
> Play-acting—
> Singing catches—
> Guessing riddles—
> Making a joke; telling it—
> Solving a problem in practical arithmetic—
> Translating from one language into another—
> Asking, thanking, cursing, greeting, praying. [4]

At first glance these types would seem to be a bare hint of what our language enables us to do. On closer inspection however many single list items can be grouped under the same headings. Take for example the second: describing an object is not so much different from reporting an event. Or, constructing an object from a description entails something similar to forming and testing a hypothesis. Play-acting can involve a mental behaviour very much in tune with making jokes, or greeting, or praying, which again is quite close to giving orders and obeying them; and so on. Secondly, sentences are not merely lots of words taken out of the total vocabulary pool. As one word is followed by another the options largely diminish because not every word can be strung next to any other. There are rules about repetition, grammar, and semantics, not to mention turns of phrase derived from culture and/or demographics. A form of progression lock takes place, where the number of products are the words, the outside world represents the latency of the system, the realised potential is the world's ascertainable part (depending upon the mental capacity), and the derived meaning stands for applicability. Given the sheer number of words in the English language (estimated to be anywhere between 1 and 2 million, and growing) the actual *available* variations are surprisingly small. If we consider the potential variance of cognitive representations, then their symbolic manifestations through language certainly fall short of what stands behind them, and it is not without reason that non-verbal expressions are able to convey an inner scenery at a level of comprehensiveness not achieved by words alone. Note however the difference: while a sentence can describe an exact detail (ie 'the nail is too short!'), a string of non-words can be more encompassing perception-wise but lack the definition (ie '*#&**!'). That is to say, well-defining words have a lesser degree of structural optionality than detail-less symbols. Why this should be so we will discover in Part II. It is not a trivial issue, as Wittgenstein demonstrates by asking us about the definition of the word 'games'. We all have some conception of what it means, but how able are we to put our ideas into exact, well-defining, non-excluding words?

> Consider for example the proceedings that we call "games". I mean board-games, card-games, Olympic games, and so on. What is common to them all?—Don't say: "There *must* be something common, or they would not be called 'games'"—but *look and see* whether there is anything common to all.—For if you look at them you will not see something that is common to *all*, but similarities, relationships, and a whole series of them at that. To repeat: don't think, but look! [5]

That last sentence, the admonition "don't think, but look!" has not been made without purpose. If by 'thinking' we mean the conscious employment of TSs (ie the con-TSs), then what can be articulated in this manner would indeed be less helpful for the above problem than letting the sub-TSs take over and merely 'feel' the results of evoking various representations triggered by the word 'games'. Right now we are not ready to identify precisely what this kind of 'feeling' involves.

For Wittgenstein to confront such problems and to ask these pointed questions he must certainly be able to draw on experience. First and foremost therefore, his PAT chains leading to such expressions are well-grounded, and in the context of his questions any abstractions are restricted to bringing out the generalised form of an object (eg types of words, a general language game, constructing analogies). It is only when he goes deeper into the subject that certain cognitive arrangements will have to be made. Firstly, we encounter yet again the tendency (or just the observation?) to ascribe the intended meaning to the object itself, as if one's mind defines the object and not the other way around (this may be confusing until we get to Part II, but in the present sense the object defines the mind by triggering particular TSs through our senses; it is the resulting sensation within us which prompts us to declare "I define this so and so"). As Wittgenstein writes on the theme of naming a thing,

> And you really get such a queer connexion when the philosopher tries to bring out *the* relation between name and thing by staring at an object in front of him and repeating a name or even the word "this" innumerable times. For philosophical problems arise when language *goes on holiday*. And *here* we may indeed fancy naming to be some remarkable act of mind, as if it were a baptism of an object. [6]

Recall the process of the perception-abstraction transference as outlined in detail in chapter 3. The qualities of the perceived object are the reservoir out of which the qualities of the abstract object are taken, where the perception may be the result of a sense-import or can be an internal cognitive entity. The selection process is driven by the affinities the Q_{po} can find within the Q_{ao} domains along the PAT chain, but it is only amongst the sub-TSs that an unhindered network of associations can evolve; links between con-TSs are subject to conscious evaluation and thereby limited to those elements that are being wilfully accessed. The process of naming then consists of mining more or less comprehensively the subconscious structures, accompanied by continual attempts to mate the emerging imagery with consciously articulated expressions. Two things become obvious: first, only those representations of qualities get activated which find an affinity with the contents of already existing TSs; and second, the accuracy of the naming is determined by the groundedness of the TSs involved, the type of articulation symbols (eg words) available for the matches, and the dexterity of the thought process handling the operation.

So it is possible for Wittgenstein to talk about "queer connexions" when contemplating one or the other result from such an exercise, especially since his intellectual attitude made him adopt the position of an observer of his own thoughts. There he would view a range of potential Q_{po}s, some of which his conscious evaluation would permit to be included in the set of Q_{ao}s. He would also be aware of other ideations which normally would not find their way along a particular PAT chain, and muse about their fate. Given the constant creation of IFAs whenever one state is turned into another, we can expect a certain number to eventuate at that point. However, because they are not functional characteristics of the useful, productive kind, they would not contribute to the final definition of some object but merely sensate their inherent incongruity to their owner, making him experience those "queer connexions".

And yet, even knowing which descriptor is appropriate and which one is not, still does not solve the problem of differentiating between the *essences* of one or the other thought complex:

> If I am inclined to suppose that a mouse has come into being by spontaneous generation out of grey rags and dust, I shall do well to examine those rags very closely to see how a mouse may have hidden in them, how it may have got there and so on. But if I am convinced that a mouse cannot come into being from these things, then this investigation will perhaps be superfluous. [7]

To which he adds,

> But first we must learn to understand what it is that opposes such an examination of details in philosophy.

In the ensuing discussion he brings in a further variety of the language game when items are either called by their encompassing name (eg table, chair), or by listing their parts. He then asks, "In what sense is an order in the second game an analysed form of an order in the first?" [8], which again hints at the assumption that the perception of something is constructed similarly to the object it stands for. If that was the case, if the concept 'table' in its representative form consisted of its platform and its legs just like the real thing, then our thoughts ought to be just as demountable and yield their ingredients when taken apart bit by bit. On the other hand, if the mental representation of an object is completely different in its nature from that what it represents, any attempts at decomposing such imagery would be fruitless. In Part II we will see why the second alternative comes closer to the truth, although for now the constant difficulties experienced by thinkers who are trying to identify a thing's characteristic ingredients by delving into their own thoughts should give us already sufficient grounds for suspecting that the problem might be systemic.

The impossibility of discerning a discrete part inside a thought leads Wittgenstein to surmise about "blurred edges"—"but is a blurred concept a concept at all?" [9], a notion that implies the presence of information but also for some reason our inability to recognise it distinctly. He goes as far as to ask whether the "indistinct picture" is often not a more suitable one. If "more suitable" means the ability to conjure a notion at all, his sentiment is acceptable; if it means a better alternative to employing a sharper version, one should be more careful—but not for the reason which immediately suggests itself.

Let us consider the issue first through the eyes of a purely mechanistic attitude that tends to see mental phenomena as just so many cogs and wheels purring along. In this case we attempt to identify a certain functionality by observing some of the constituent parts performing an act and in doing so we categorise the parts involved in terms of a sub-system (eg, there are some wheels and rods responsible for the pumping of water, and so we can classify the responsible set of parts as a 'pump'). In other words, there is a combination of something which can be isolated from the rest because of its distinct appearance, and that combination consists of parts the nature of which will become obvious once the set gets disassembled. Obviously, if our eyes are too weak to discern the parts a device that makes for a sharper vision can only be helpful. Now consider a picture printed in a newspaper. It too can be seen as a system, the functionalities it contains being the identifiable objects in the picture. The functionalities rest on the ability of a series of black and white dots at various sizes to convey an image which can be understood by our minds. Let's say the picture contains someone with a dark hat against the backdrop of a snowy landscape. If we were to enlarge the picture there comes a point beyond which a viewing area of a given size will make it impossible to distinguish one cluster of dark spots from another cluster, where the one is part of the hat and the other part of a hollow in the snow. The only way they can be told apart is to reduce the resolution once again until the entire context becomes visible. This last situation is fundamentally different from the first—there we dealt with discrete units providing their service in their own right, here we have many similar units which convey the information by their patterned arrangement only.

Although Wittgenstein does not say so explicitly, his desire to explain the meaning of words in terms of their semantic spread across contexts would bring him up against the issue raised in or last example. After all, if an idea can be so voluminous as to evoke a whole series of images and feelings, and if attempts therefore to decompose it run constantly into insurmountable difficulties, then there must be something about the nature of mental representation per se which causes the problem. As he said, if there is reason to believe that a mouse is made out of rags and dust etc then one ought to invest a considerable amount of effort trying to ascertain the composition of rags and dust. That Wittgenstein had done so is evident from his text; that he was unsuccessful can also be established. Yet, going by his warning about such detailed examinations, his failure had not been in vain.

In the previous chapter we came across Sartre's encompassing manner of establishing the essences of an object by including 'what is' and 'what is not' as pertinent. We have seen how the cognitive configuration had to make allowances for a comprehensive set of characteristics, to the extent of forming accompanying structures to take care of the new affinities. Here the situation is somewhat similar: for Wittgenstein to consciously examine the contents of an idea, he needs separate TSs which accommodate a variety of connections in order to test the presence of this or that datum in the structure under focus. Whereas with Sartre the exercise was more or less a game because the end result was simply one meaning rather than another, for Wittgenstein the matter is more serious. Before the question was, "what is this thing?"; now it is "what does it mean to me?" (notwithstanding the possibility in either case to diminish or enlarge one's basic concern). To answer the latter, the TSs which are charged with providing the content to be juxtaposed must maintain their perceived purpose in relation to the rest of the person's identity. For example, if I wanted to establish the deeper nature of any thought that arose within me upon hearing the word 'table', I need to do two things: firstly, I need to allow my thoughts to run freely so that their contents can pass my 'inner eye' unhindered, and secondly I need to adopt the position of a neutral observer, my 'inner eye' being disassociated from my usual self for the purpose of the exercise. When thoughts are allowed to flow during meditation we have a similar situation, except that the thoughts are not evaluated as critically as in the current case. Here the constant need for diligence sets up IFAs which customise the arrangement of two separate dynamics, one connected with the object and as such with 'normal' thoughts, the other dealing with deconstructed entities of no overall value and not being part of any reference system. Nevertheless, these newly-created TSs abide by the same rules in terms of affinities, congruent and/or incongruent links, and denominative influences over other structures, as all the others. The continuous attempts to break one's thoughts into smaller and smaller pieces therefore does not result in the emergence of 'smaller and smaller' thoughts, but evokes TSs of unrelated and basically unconfigurable elements which are neither content nor function and do not even have a productive bearing on one's own self (just like the separated dots in the newsprint do not have any information value by themselves). Still, the associated creation of IFAs along these lines contributes to affirm the presence of such TSs and, if unchecked, a point must come where it becomes difficult for the owner to distinguish between content that is relatable to oneself in a productive, reliable way, and some other that is an unpredictable hotchpotch of cognitive bits and pieces gleaned from out-of-context domains. An extreme example of an enforced mental state of this kind occurs during solitary confinement, with the added danger that a withdrawal from these PAT chains is made impossible. The situation becomes particularly precarious when the observed content—as ambivalent as it may be—turns into a resonant structure with respect to the observing TS (it is the cognitive equivalent of the 'observer influencing the observed' scenario).

By the way, this should not be confused with the age-old "idea ideae ideae..", where thinkers find themselves in an—imagined—infinite regress when contemplating the nature of knowing (see chapter 11). The 'infinite regress' is a form of thinking in which some representation of oneself is imagined to observe oneself thinking. Wittgenstein's exercise on the other hand is the wilful act of decomposing one's thoughts, even if there is nothing to decompose. If insisted upon, the act can lead to pseudo sensations because the achieved detail is insufficient to actualise meaningful content—not because there can be no meaningful content at a particular level of granularity, but because our minds are not equipped to identify content at that level of resolution (*why* this is so we will see in Part II).

When Wittgenstein asks,

> What does it mean to know what a game is? What does it mean, to know it and not be able to say it? Is this knowledge somehow equivalent to an unformulated definition? So that if it were formulated I should be able to recognise it as the expression of my knowledge? Isn't my knowledge, my concept of a game, completely expressed in the explanation that I could give? [10]

he procures the issue from the problem of mapping the information value of a word to the underlying structures responsible for the word's articulation. In this sense, just as words are the compacted versions of conscious thoughts, so conscious thoughts are the compacted versions of the cognitive elements involved in their creation. Looking inside himself for the answer he is prompted to write shortly after,

> In such a difficulty always ask yourself: How did we *learn* the meaning of this word ("good" for instance)? From what sort of examples? in what language-games? Then it will be easier for you to see that the word must have a family of meanings.

To "learn" the meaning of a word, that is to say, to interpret the role the word plays in the context of usage amongst fellow humans within the framework of a culture, one needs to establish a cognitive relationship with the word's representational domain and the rest of oneself (especially if its connotation is a value judgment). The deconstruction of the word does not produce a word particle (a "family of meanings" is not a set of particles), but invites a host of hitherto unrelated contents associating itself with it.

In another passage we are presented with various interpretations of the phrase "Now I know how to go on", made in the context of asking someone for the value of a number after been given a numerical pattern as a leitmotif. The person may know the value after discovering the formula for the series or after having recognised the formula straight away, but semantically speaking the phrase is an umbrella representation of all the reasons that may lead to the solution.

> But that does not mean that his statement ["Now I know.."] is only short for a description of all the circumstances which constitute the scene for our language-game. [11]

And further,

> We can also imagine the case where nothing at all occurred in B's mind [the person being asked] except that he suddenly said "Now I know how to go on"—perhaps with a feeling of relief; and that he did in fact go on working out the series without using any formula.

From the realisation of multiple sources for a given phrase, to the awareness of various *types* of sources, Wittgenstein is led to declare,

> *This is how these words are used.* It would be quite misleading, in this last case, for instance, to call the words a "description of a mental state".—One might rather call them a "signal"; and we judge whether it was rightly employed by what he goes on to do.

In our terms we would not call these words a "mental state" either because they represent the symbolisation of what can have a variety of mental configurations; and we could even call them a "signal" because they are the outward manifestation of a sensated cognitive event, the exact detail of which we cannot identify directly but can surmise about by observing its accompanying action.

Just as we are prevented from knowing exactly the composition of the idea which happens to manifest in a verbal articulation (or any other kind for that matter), so are the sequential steps leading up to it:

> Thus you were inclined to use such expressions as: "The steps are *really* already taken, even before I take them in writing or orally or in thought." And it seemed as if they were in some *unique* way predetermined, anticipated—as only the act of meaning can anticipate reality. [12]

For someone who understands the operation of con-TSs materialising as it were their contents through linguistic or other expressions, the above statement seems trivial. To discover such relationships however via a diligent observation of one's thought processes is not only tedious but carries certain dangers, as we have mentioned above. To descend into one's personal consciousness in search for some meaning which has not already been offered in the course of the normal waking state is to use reality as a bargaining tool—at least as we can understand reality to be for us. Since the preparatory period for assembling those contents which are destined for con-TSs lies outside our consciousness, we are not aware of the state of readiness either and the result, when it happens, comes "in a flash" [13]. Not only is the creation of the conscious idea invisible, the accompanying formation of FAs (these would be mostly of the intuitive kind) will also go unnoticed, unless we make a—conscious—effort to monitor our mental state. To say "I know" therefore because one has become acquainted with the structural configuration of some solution (we would commonly express this as "getting a feel for it" since we don't actually *know* the structure), may be the result of having learned the application of some formula; it may also happen because the issue *seems* familiar to us, in other words, it evoked the same FAs as had already been created by a previous example. Note the difference between the two: the following of a formula involves a conscious step-by-step articulation (inside or outside) of an idea we have learnt; doing it by the 'feel of it' proceeds by mining associated abstracts (especially including the FAs) and allowing the stringing together of the solution in terms of best probabilities, amidst the entire network of relevant domains. To put it technically, following a formula means sequential processing; scanning associations for the best fit is a form of distributed processing. Humans are very good at the latter—it is the reason why it has been so difficult for a computer to beat someone at chess. The adage "you only have to know one move—the correct one" still holds true as a human advantage (so far).

The transparency of unarticulated TSs brings us to the often-heard question of what a certain colour means to a person. "What does it mean when I say 'red'" can only be asked if there is already some doubt about the colour's representation. To remain with similar examples would not be very productive since the image of 'red' in this case is too universal and instantly identifiable for any problems to become apparent. Wittgenstein moves on to another:

> Let us imagine the following case. I want to keep a diary about the recurrence of a certain sensation. To this end I associate it with the sign "S" and write this sign in a calendar for every day on which I have the sensation.—I will remark first of all that a definition of the sign cannot be formulated.—But still I can give myself a kind of ostensive definition.—How? [14]

He compares this action to a ceremony through which the sensation gets identified and articulated, without however in any way contributing to its definition. There is no "correctness" one can point to, since neither the sensation itself nor its expression are made up of discrete units of consciousness, which in their totality form the sensation if re-assembled correctly. He writes,

> One would like to say: whatever is going to seem right to me is right. And that only means that here we can't talk about 'right'.
>
> Are the rules of private language *impressions* of rules?—The balance on which impressions are weighed is not the *impression* of a balance.

Observe Wittgenstein's attempts to formulate verbal sentiments in the face of a subject which can't be expressed unless one takes recourse to especially construed words and phrases (see the glossary for ours). The "private language" (ie one's conscious thoughts) following certain rules (ie the mining of affinities with OAs and FAs), these rules held in the form of "impressions" (ie the probability events along PAT chains), and yet such rules themselves not being material for articulation—all this needs to be written down in a language which up to now had never been used for this extreme kind of self-referencing purpose. Nevertheless, his juxtapositioned cognitive artifices reveal to him the sheer subjectivity of sensation:

> The essential thing about private experience is really not that each person possesses is own exemplar, but that nobody knows whether other people also have *this* or something else. The assumption would thus be possible—though unverifiable—that one section of mankind had one sensation of red and another section another. [15]

And if one's image of a colour is a function of one's own and unique mental configuration, then the first—although by no means the last—test of the veracity of a representation would be the congruence between that configuration and its accompanying articulation; whether the eventual expression can be understood by anyone else comes later. Since for Wittgenstein the connection between thoughts and words is still somewhat direct (witness his allusions to machine-like operations when explaining the functions of words), an outside evaluation of an articulated string of symbols is part and parcel of the entire dynamic. In a sense it is of course—that's how we learn to speak a language; but in another sense it is not—as in the case of a babbling infant, to whom its utterances constitute meaningful matches with what resides in its consciousness. The fact that such expressions do not lead to any kind of social fulfillment is the incentive to improve upon their delivery (but this does not *necessarily* relate to their source). In the context of the pure thought-expression relationship Wittgenstein's remark, "Even a nonsense-poem is not nonsense in the same way as the babbling of a child" [16], needs to be treated carefully. The main difference between this type of poem and infant-talk is the ease with which the sections of the poem can be differentiated. This ease exists because the linguistic elements of the poem had been borrowed from 'real' speech and therefore find resonance in a similarly trained listener. The child on the other hand does not yet have the skills to differentiate its verbalisations, although the 'words' are just as nonsensical in their totality to anyone else as may be the poem's lines (yet this is a matter of degree—to a mother the babbling of her child may be much more meaningful than the random verbiage of a nonsense poem).

After having verified the initial congruence between the representation of something and its articulation, the next step involves the confirmation from the outside, that is a reaction from who- or whatever the expression had been addressed to. A further hint that the inner imagery of a thing is not an objective mirroring of it comes through the utterly subjective process of aligning that image with the outside reaction. The representative content of a TS is not stable, but receives constant updating as long as there are incoming messages which demonstrate a sufficient affinity with the TS (for instance, "How do I look?"). There is a certain progression lock in force inasmuch as the original characteristics of the content are expected to be delineated further along the set line. If the resource space delimited by the probability results which come from the imports is rendered unworkable, a good part of the entire framework of PATs collapses. Wittgenstein uses pain as an example:

> Look at a stone and imagine it having sensations.—One says to oneself: How could one so much as get the idea of ascribing a *sensation* to a *thing*? One might as well ascribe it to a number!—And now look at a wriggling fly and at once

> these difficulties vanish and pain seems able to get a foothold here, where before everything was, so to speak, too smooth for it.
>
> And so, too, a corpse seems to us quite inaccessible to pain. [17]

We could extend the fly scenario to the corpse and imagine one of *its* limbs suddenly reacting to a touch! The ensuing sensation to the observer profoundly illustrates the experience of collapsing PATs.

Conversely, the mere presentation of a response (eg the facial expression of pain) triggers relevant TSs although no actual pain exists. The visual sensory import is only one of the streams of abstracts which combine to form a cognitive complex, but arriving by itself would be able to produce enough associations with already formed TSs due to the occurrence of IFAs. Whether the experience is as comprehensive as an ordinary exposure to injury is neither here nor there—if there are TSs which sensate consciously the internal connections can and will be made.

> And do you know that what you are giving yourself this exhibition of is pain and not, for example, a facial expression? And how do you know what you are to give yourself an exhibition of before you do it? This *private* exhibition is an illusion. [18]

Another way to prevent a progression lock from developing is the introduction of a different cultural context. The feedback loop that refines our understanding of an event depends upon the congruency of its parts; turn any one of them into an inconsistency in relation to the rest and the cycle breaks down. The need to reinforce a train of thought extends right across our endeavours. In the current text the emphasis is on language and its—learned—use, but the process itself is a common one. For us to know how to express a thought in a productive manner needs the contribution of samples of similar contexts and complexities, and our understanding of such samples' significance in a given situation. Since there are many words and hence combinations of them, the systematic arrangement of these symbols according to the intended meaning produces the confluence of well-proven probabilities in terms of progression locks leading from the synchronised start of an object- and abstract-event to their conclusion. That system of well-proven probabilities we call grammar, and the more elements a language has, the more complicated its grammar has to be. In tandem its expressiveness increases, contributing further to its applicability, and so on.

Given this potential variety Wittgenstein wonders at the underlying dynamics of sentence construction. Speaking more than one language brings out the issue in stark relief. "..a picture occurs to me and I try to describe it. Or an English expression occurs to me and I try to hit on the corresponding German one." To which he says,

> Now if it were asked: "Do you have the thought before finding the expression?" what would one have to reply? And what, to the question: "What did the thought consist in, as it existed before its expression?" [19]

Note the changing nature of his queries. At first we had references to the constituent parts of a concept, exemplified through his allusions to the variants of an idea (eg 'games'), or his listing of several possibilities for expressing a broad theme (eg a series of sentence types). These were questions that got inserted *after* the actualisation of the concept itself, that is to say after the necessary con-TSs had construed the formulation of its content in terms of words. Now he is writing about the nature of the thoughts themselves, *before* the processing by conscious structures. As such the thoughts elude his investigative faculties because the functional medium of his 'inner eye' and of the idea to be investigated are one and the same. Without indirect means therefore we are incapable of being witness to the sub-TS activities, a point made before but focused on once again by Wittgenstein. Continuing in the above vein, he writes,

This case is similar to the one in which someone imagines that one could not think a sentence with the remarkable word order of German or Latin just as it stands. One first has to think it, and then one arranges the words in that queer order. (A French politician once wrote that it was a peculiarity of the French language that in it words occur in the order in which one thinks them.)

That last sentiment is indicative of a situation where the mind processes TS content without the conscious awareness of the steps taken, with the result that the only understood feature of the whole procedure is the synchronicity between one's thoughts and the words one has been taught to say. In more technical terms we can say that the organisation of articulation consists in arraigning some thought content such that it tallies with the understood significances of available articulation symbols; the content gets expressed according to the capacities of the symbols, and the symbols portray their combinatorial potential according to the content which has been made conscious.

> Would it be imaginable that people should never speak an audible language, but should still say things to themselves in the imagination? [20]

we are asked. To answer this fully we need to start with the development of thoughts themselves, and for that we need the context of Part II. For the moment let us note that the question would have been impossible for someone unless he had engaged in the perfunctory juxtaposition of a secondary set of TSs for the purposes of searching through the first. By the same token it is also noteworthy for it to have been asked at all, since the diffusive state of thought contents makes it difficult to differentiate between the elements concerned with word formation and those providing the source. We know he realises the problem because he writes later,

> The feeling of an unbridgeable gulf between consciousness and brain-processes: how does it come about that this does not come into the considerations of our ordinary life? This idea of a difference in kind is accompanied by slight giddiness,—which occurs when we are performing a piece of logical sleight-of-hand. [21]

A "piece of logical sleight-of-hand" occurs when our train of thought has been interrupted by the event's proceedings, yanking us out of our progression lock that had been induced by the logic. There is not enough information to deduce Wittgenstein's state of mind in terms of thought processes in general, and so we cannot suggest any kind of form such a lock could have taken in his case. We do however recognise the precarious nature of artificially decomposed TS contents, and the evocation of unreferenced cognitive elements may well produce the "giddiness" mentioned above. Consider how he describes the experience:

> When does this feeling occur in the present case? It is when I, for example, turn my attention in a particular way on to my own consciousness, and, astonished, say to myself: THIS is supposed to be produced by a process in the brain!—as it were clutching my forehead.—But what can it mean to speak of

> "turning my attention on to my own consciousness?" This is surely the queerest thing there could be! What I happened to describe in this manner had been an act of gazing (since in ordinary life such words are not being used).

If under normal circumstances the articulation of something relies in part on the feedback process under the auspices of one's understanding of where the feedback comes from (recall the fly/corpse scenario earlier), then drawing the cycle within myself from one part of my consciousness to another cannot be very productive in terms of discovering any new material, except where such material has previously been disassociated from the present context. It is also true that the evolution of language (certainly as far as English and German are concerned) does not yet provide for the exhaustive expression of these kind of sentiments (I have taken the liberty to re-translate the last sentence in the above quote—the reference to "ordinary life" is omitted in the English text).

Which brings us back to the relationship between a thought we carry in our minds and its form as it becomes represented through sentences. We can theorise about the information value inherent in a given statement provided we have access to its elements and so to *their* meanings. The *ultimate* information value however goes beyond the few articulated symbols—it covers the underlying structures and whatever significance they can convey to our evaluating mind. Our e-chart together with the discussion on probabilities has shown us the segmentation from one stage of realised potential to the next, that is the effective knowledge we derive as the concept unfolds. The question is not only one of determining how much detail can be gleaned from a given state, but how well the current detail can be traced back across previous states. When introducing the e-chart we maintained that its parameters have to be accessed in clockwise fashion, and that the opposite direction would be useless for its purposes. In the present context there are two main reasons for this: firstly, it is self-evident that a particular quantity of a realised potential cannot be made indicative of the number of products responsible for it (for example, faced with a full swimming pool and knowing the total volume of the pool is not enough for me to determine the number of pumps and capacities of the water pipes used for filling it). Secondly, the existence of a certain degree of probability at one stage does not tell me anything about the overall outcome of all the probabilities (which would then be an indication of the combined potential which had been available), because each state has to be taken on its own merits (states do not 'remember'). In other words, while I can estimate the utility of a certain state in terms of its character, I cannot calculate backwards from some state in order to identify that previous state's potential. The following plays on that theme, although the context is restricted to words and their meanings.

> One can imagine an animal angry, frightened, unhappy, happy, startled. But hopeful? And why not?
>
> A dog believes his master is at the door. But can he also believe his master will come the day after to-morrow?
>
> ..Can only those hope who can talk? Only those who have mastered the use of language. That is to say, the phenomena of hope are modes of this complicated form of life. [22]

We can define "hope" as a form of projection, based on subjective experience, the sharing of which can only be accomplished via language. This does not necessarily mean that the source material for the subjective experience does not exist in the absence of a language medium. Just as the presence of some expression does not allow for a wilful interpretation of its constituents, so does the existence of some constituents not permit the automatic assumption of their *realisable* potential (this should not be construed as an attempt to humanise animals or animalise humans; merely to demonstrate the difficulty of ascertaining cause and effect in dynamic and chaotic systems).

An example of how the potential of words, indeed letters, can be extenuated beyond the expected is provided by the company formerly known as Acer Communications and Multimedia and now called Benq. Here's what 'Benq' is supposed to stand for:

> The interaction of the key senses is visualised by a unique, original, friendly, as well as stylish element, that builds the letter 'Q'. The letter 'Q' in its distinctive shape and form, focuses on the key enjoyment expressions of Benq being exciting and aspirational. The original and modern form of the word mark, with its smiling 'swoosh', communicates the brand promise. The word mark's clear lines and sharp forms also communicate a professional, top-quality, smart image, which projects quality products contributing to a quality lifestyle. [23]

Who would have thought that a 'Q' has hidden in its profound shape exciting and aspirational images? Surely not someone who understands probability.

Without the commonality of our understanding when it comes to the usage of words human language would not work. At the same time the considerations we applied so far suggest that the representational quality of a word's precursor does not enjoy the same degree of access.

> If God had looked into our minds he would not have been able to see there whom we were speaking of. [24]

is how Wittgenstein puts it. The core of the problem (which we will explore in Part II) is to identify the composition of an idea that we only know about because it has been articulated in our consciousness. But *what* enables us to know about the idea in the first place is hidden from our—conscious—probing. Hence the tendency to transpose the concept standing for thought articulation onto the process responsible for the creation of thinking, although the phenomenon of 'thinking' does not feature in any productive explanation either. Wittgenstein:

> The familiar physiognomy of a word, the feeling that it has taken up its meaning into itself, that it is an actual likeness of its meaning—there could be human beings to whom all this was alien. (They would not have an attachment to their words.)—And how are these feelings manifested among us?—By the way we choose and value words. [25]

During our observations of philosophical mind edifices across the centuries we did recognise the absence of the intellectual distance from the observer to the observed that would have enabled their owner to substantiate the perception of difference between thought and symbol (as indeed between thought and object since similar principles are involved)—especially in earlier periods. Not only would there have to be the appreciation of another medium, the articulation of *that* concept needs a specific nature of TSs such that a potentially denominative influence from the object representation towards the thought representation can be avoided; it is one of the reasons for having made the comments about language use in our introductory chapter. So Wittgenstein's remarks about the "way we choose and value words" is not amiss: in this part of the book it has been *the* major source of our findings.

Yet note the functional context in which they were being made: it is only through a series of words, forming meaningful combinations, that we are able to establish the significance of an expression, and likewise it is only through the cohesiveness of our own thought processes that we can apply the discovered meaning to our knowledge base. A random collection of utterances, together with an equally ad hoc multitude of neural impulses, will not be able to differentiate the patterns we are looking for. In fact, we would not *be* looking for any patterns because the idea itself would be alien to us. Not only that, we would not *be* be looking at all because even such a statement could not be made. So, what constitutes a pattern?

Consider the appearance of a meadow from twenty metres up. The colour is basically green, although we know that the variety of grasses and flowers contain a range of hues. There may be a pattern, but from that distance it has disappeared. If we would increase our elevation to say, a thousand metres, we would see the patchwork of many meadows, and quite possibly discover a distinct layout. In other words, the technique of our evaluation determines the existence of patterns. This should not be confused with the appearances of the objects themselves—the word 'pattern' denotes a concept to be thought, not a shape. That is to say, the frequency and method of occurrence of certain objects are not the issue here per se; it is the utility of such appearances we are concerned with. The issue is not confined to mental operations either; if a certain plant species needs a particular pattern of rainfall, then such rainfall is neither here nor there if its manifest occurrence falls outside the relevant life cycles of the plant. For *us* the pattern may still be there because we are capable of using records and means of descriptions, which relate to *our* way of life and not the plant's. A pattern then is a phenomenon which enjoys a correlative dynamic in an accompanying system in terms of its characteristics. If the dynamic is not correlative the pattern ceases to be a pattern. Just as plants cannot make 'sense' of rain periods which happen outside their biological parameters, so we cannot make 'sense' of events which are beyond our own evaluation criteria; for example, changing colours in octopuses, melodies of whales, the fate of cultures, the movements of mountains. In order to understand any of these we need to employ a translation mechanism which uses abstracts, and in a biological system the only things which can process abstracts are neurons.

Wittgenstein alludes to pattern recognition several times since the importance of its purpose is intimately tied to the effectiveness of symbol use. The evocation of rules for instance is a direct application of an artificially created pattern, without which we would not be able to access information in a common—if not communal—manner. To make sure the occurrence of a particular form of behaviour can be understood in the same way over and over again, we cast that behaviour into a set of rules so that not only the acts themselves but their evaluation can be anticipated in a regular way. And once again, like in every aspect of life, the variety of rules describes the complexity of the system and is therefore bound by the same principles of applicability and progression lock like all the others. Of course those rules, being patterns, make available but at the same time also limit our means of perception, and inappropriate rules can hide a particular circumstance so effectively it may just as well not even exist.

The interest in language and its role in consciousness shown by Wittgenstein has led us to consider perspectives which we otherwise would not have entertained. The mental patterns of his investigations were well grounded, and so did not provide some of the exotic surprises discovered in others. Nevertheless, his ideas needed intricate configurations, some not without certain dangers to their owner, and as intellectual structures they came as close as is usually possible to the actual phenomena he was observing.

A proliferation of material enhances the degree of applicability because the more products there are available the greater will be the number of combinations that can be achieved. The existence of many combinations is more than just a matter of apparent variety. The significance of an object changes according to the context it is placed in and, given a multitude of environments, the importance of one specific element within a system is not a static factor across a range of applications. If this applies to the physical resource spaces in a society then it is even more relevant to the resource spaces of the mind. We have encountered some possible variants amongst homologous and synonymous TSs, two examples of functionality types where the momentary context imposes either a different behaviour or configuration upon their respective domains. The evolution of mental dynamics has led to the growth of cognitive structures alongside which the need for expressibility in the language of their owner asserted itself over and over again (witness the sentence constructions by someone like Husserl or Sartre for instance). Whatever the ultimate success, the TSs with their content exist and contribute to the formation of uniquely arranged complexes in terms of the data as well as the meaning of such data due to the associative potential of their ingredients; in other words, we are getting new attitudes and perspectives.

The intention of examining the flow of thoughts prompts the emergence of TSs which are charged with carrying out that task. Depending on the exact focus the examination is meant to take, the nature of the structure will accommodate it accordingly. The linking of affinities and the distribution of such things as illusion trees and logic boundaries would be more function-related in the case of say, Jung, who was primarily concerned with the importance of thought contents and their emotional components, such TSs requiring a certain cognitive dexterity along functional lines. For a Wittgenstein on the other hand the probing TS needed also the associative flexibility towards content, since his interest lay in the role its articulated version played in relation to the implied meaning (in Part II we will look more closely at the interdependence between functionality and content). This does not mean that the idea of objectivity had not existed before; yet it was only now that the act of standing aside as an observer began to be considered as a mental phenomenon in its own right. The functional principles of objectivity, in contrast to its appearance only, were considered sufficiently important to be dealt with systematically, and one major contribution on this theme is *The logic of scientific discovery* [26] by Karl Popper (1902 - 1994).

To investigate the concept of objectivity via the mechanisms of its principles (or essences, as earlier philosophers would have called them) requires the services of an abstraction process which not only involves the re-representation of content, but also that of its functionality, that is the manner in which the content is being handled. Not only is such an initiative somewhat novel as pure intent, the resulting TSs are meant to operate at close proximity to the self—they can get quite ‘personal’. In his preface Popper alludes to that angle while mentioning “rational discussion” and being “critical”:

> The point is that, whenever we propose a solution to a problem, we ought to try as hard as we can to overthrow our solution, rather than defend it. Few of us, unfortunately, practise this precept; but other people, fortunately, will supply the criticism for us if we fail to supply it ourselves. Yet criticism will be fruitful only if we state our problem as clearly as we can and put our solution in a sufficiently definite form—a form in which it can be critically discussed. [27]

As far as the semantic resource space of our e-chart is concerned, these few sentences encapsulate all the rest which is to follow. Their potential, to be developed in so many strands later in the book, resides in the set of principles that is their domain. Firstly, there is the recognition that one's focus of interest is dependent upon the affinities existent across the TPs in our mind, and are therefore a feature of one's personality; secondly, being a part of the associative framework, the contents of these TSs are particularly subjected to the effects of resonance because of their ego-specific nature; thirdly, given that third parties do not possess the same intimate relationship, a critical response will be more likely from their direction; and finally, it is the format in which a problem is stated which first and foremost defines the context of its re-representation by someone else (or by another cognitive dynamic of the same person). In the main, this is the theme of *The logic..* and we will follow it from our perspective.

Regarding the third point, Popper indirectly emphasises the issue by telling us of one problem solving method of many, and that is the historical one. "At present unfashionable", he writes, it nevertheless allows us to revisit previous approaches and so learn from the involvement of others—"why they had to face it: how they formulated it: how they tried to solve it." In our words, listening to other people makes it possible to break the progression lock which might have installed itself through the delimiting influence of one type of associations only (namely one's own). He adds,

> If we ignore what other people are thinking, or have thought in the past, then rational discussion must come to an end, though each of us may go on happily talking to himself. Some philosophers have made a virtue of talking to themselves; perhaps because they felt that there was nobody else worth talking to.

One might see here a certain criticism of the single mind's inherent capacity to be rational, and according to our model this is indeed justified. No matter how congruent PAT chains are and how well-grounded their links may turn out to be, there are still the IFAs exerting their influence and they do not exist by virtue of their truth value but by the degree of affinity they have managed to achieve between the circumstances of their creation and the contents of relevant TSs. It is those person-specific IFAs which are being by-passed when making use of other people's thoughts, although their presence alone should not be used to judge the effectiveness of complimentary structures.

FAs, particularly the intuitive kind, are defined by their collective capacity to circumscribe an event, information about which has entered the mind. As such they rely on the existence of identifiable data, and it is the data's appearance through object and function which, in its distributed form, contributes to their overall representation. In the previous section we pointed to the correlative dynamics of incoming and resident content when it comes to pattern recognition. In a fundamental sense that relationship exists all the time, since without any kind of affinity to already established TSs the new abstract will not be imported—at least not meaningfully. For someone who is concerned with the correctness of one's internal information flow, it is the validity of some idea measured against the significance delineated from the rest of the mind which matters in the end.

Popper sees it from the strictly logical perspective and so raises the problem of induction. Cognitively speaking, an induction consists of the transposition of some PAT^n to a PAT^{n+1}, the elevation to the next higher level representing the generalisation of the content. As such there needs to be a previous stage, otherwise the transposition would be impossible (see chapter 5 for some detail). In Popper's view the difficulty lies in the ultimate information value of PAT^n in relation to the entire set of ideas which can be articulated beginning with PAT^{n+1}—that is, the importance of a singular statement in terms of its universal application. We would say the question revolves around the potential of some product to be realised so that the potential's applicability can be assured, but without compromising the applicability of the initial product. If the induction method is to be trusted in providing a valid statement, not only the data themselves but the manner of their handling ought to be above reproach:

> For the principle of induction must be a universal statement in its turn. Thus if we try to regard its truth as known from experience, then the very same problems which occasioned its introduction will arise all over again. To justify it, we should have to employ inductive references; and to justify these we should have to employ an inductive principle of a higher order; and so on. Thus the attempt to base the principle of induction on experience breaks down, since it must lead to an infinite regress. [28]

For us there are two issues involved: (1) the formulation of a statement such that it can be productively re-visited by someone else (this is Popper's main concern), and (2), the actual result of moving inductive principles to a higher order. Let us first deal with the second one.

In a straightforward abstraction process across several levels the content of a PAT^n becomes a PAT^{n+1} and so on. Under normal circumstances this means that the content's associative properties will also change from one level to the next, simply because the content is now of a slightly different nature (for example, it could have moved from 'table' to 'raised platform' to 'plane'). In the case of functional abstracts the generalisation is similar, although the available levels depend on the specificity decided on in the first place. While one would need to exercise a considerable amount of caution when combining 'plane' with 'wood screws' in one single concept (as in, "you can use wood screws to attach something to a plane" rather then "..to a table"), the problem becomes less urgent when using FAs, but it still exists. An FA could be the observation that water spills out of a tilted glass. Since there are so many possible varieties of objects at various abstraction levels, and all relatable to the act of 'water spilling', there is usually no need for any FA to form a ladder of abstractive steps in tandem with its object counterpart. It follows that a PAT^n of the functional kind can be associated with several PAT^{n+i} of the object variety, where 'i' represents successive abstraction levels. Regarding Popper's sentiments, the functional compliment of an abstraction complex need not be on the same level as its object-type contingent, and therefore does not become useless when not being elevated as well. Hence there is no infinite regress of one exalted principle being surpassed by an even more exalted one, merely to supply the entire context with meaning. Specifically, to have a CFA (dealing with the concept of induction it would have to be a cognitive FA) which says, "if this, then..", it can be applied at several levels of object abstraction without loosing its significance. Indeed, a further abstraction of the induction principle does not make sense since it would postulate the "if this, then.." enactment for levels of generalisations where this becomes counterproductive. For instance, one can go from, "if this glass is tilted its water will spill" to "if any container is tilted its contents will spill" and so on; on the other hand, to continue with "if any container is moved its contents will escape" becomes precarious. Therefore, the introduction of inductive principles of a higher order goes against the very purpose of induction and is unrealistic. By the same token, to use these rules in terms of horizontal PATs would have to be in relation to the objects they are referenced to and so cannot be done to themselves. Hence the second point above is of no real concern within the context of cognitive dynamics.

Making statements accessible is another matter. For Popper the testing of a theory ought to consist of four broad approaches—how consistent is the theory internally; what logical form does the theory take (eg, is it empirical, scientific, tautological, etc); is it an advance over other theories; and, are there any empirical applications that can be used for testing its conclusions [29]. The last item is held to be particularly important since it is here that the matching of scenarios (the one implied by the theory vs the reconstructed one supplied by the tester) takes place, and thereby any discrepancy will play its role of arbiter. The fundamental aspect of this issue is what Popper calls the "problem of demarcation", which involves the finding of some criterion that "would enable us to distinguish between the empirical sciences on the one hand, and mathematics and logic as well as 'metaphysical' systems on the other" [30]. It is indeed a problem, but for our purposes we should re-phrase the description. For us it becomes a matter of deciding on the point from which onwards the content of some structure can be considered no longer autonomous in terms of its original defining context. What does that mean?

Suppose my mind receives the information about a branch falling off a tree, and assume further that at this point I do not possess any knowledge about falling objects whatsoever. Roughly speaking there exists now a TS consisting of the objects 'tree' and 'branch' in the form of abstracts at the first abstraction level, as well as the FA 'falling', equally a PAT^1 and linked with the OA of 'branch'. Given such an experience I could be prompted to state, "branches fall off tress" (we are disregarding the issue of dexterity in one's language for the moment, though in practice it is not trivial). From now on, any time in the course of my life that I come across some object disengaging itself from something else I will draw on that first experience in order to provide a cognitive context for such a newly generated FA. Note that I am now in a position to articulate in some suitable format the for me quintessential fact that, given half a chance, things tend to fall off from where they had been attached to, that such readiness has to do with their weight (another sensory input comprising shape and movement), and that the consequences of a fall can be serious under certain circumstances. Let's say many years later I find myself amongst a class of physics students listening to a lecture on gravity. What would be the chance of someone like myself understanding the concept of gravitational pull if they had *never* known what it was like for something to fall off something else? Or in other words, what cognitive context could my mind provide for the lecture material if it possessed *no* TSs with the content of *some* object coming off *some* other object and then accelerating downwards? If such a question seems too hard to contemplate, one needs only consider the information value of a purely theoretical explanation with no practical examples supplied at all; or the same value inherent in a discourse on quantum mechanics without any preliminary introduction to its basic agenda. It seems clear from those last examples that an introduction of new material will not be productive unless there is some pre-established context in the receiving mind which allows the incoming elements to be associated with what could be called their basic concepts—such basic concepts having been derived from what again could be called basic experiences (eg, running, stopping, fast, slow, loosing, gaining, rich, poor, happy, sad, etc etc). We can say then that a thought complex has become *autonomous* in relation to its abstracts when it has become capable of absorbing new associations without requiring the elements responsible for defining its original domain. So, the idea of 'falling' has become autonomous in my mind when the TS comprising the relevant cognitive units is able to exist without the referential memory about the introductory experience; it is at this point that I can deal with statements about gravity-affected objects without needing the mental crutch of trees and falling branches.

Returning to Popper, we would be able to treat a concept as being purely logical (or mathematical, or metaphysical) as soon as we are able to employ an autonomous understanding of its principles. Apart from additional experiences a main factor deciding the existence of autonomous ideas is their TS's associative intensity, which in turn is a function of their associative halo, resulting in a certain degree of resonance with the FAs from the basic experiences. In societal terms it is interesting to observe the emergence of this type of contemplation at that stage of our evolution—the issue of concepts falling short of their intended, principle definition being itself a notion that had to work its way from simple beginnings to ever more detailed imaginings, to become sufficiently autonomous for it to be treated in its own right. What Popper chooses to quote a couple of pages before when writing about the attractiveness of induction as a mental process (a process for which he strives to provide a realistic substitute), is significant in the context of the creation of autonomous TSs for the purposes of better understanding:

> In a similar way Einstein speaks of the 'search for those highly universal laws ... from which a picture of the world can be obtained by pure deduction. There is no logical path', he says, 'leading to these ... laws. They can only be reached by intuition, based upon something like an intellectual love ('*Einfühlung*') of the objects of experience.'

The quote is from Einstein's address on Max Planck's 60th birthday in 1918, and had been translated by someone else. While the phrase "intellectual love" seems enticing, it is not the correct interpretation of the German word "Einfühlung". Literally meaning 'the act of feeling oneself into..', it denotes a gradual penetration of a subject matter by following one's intuition rather than strictly logical precepts; put into our words, it is the gradual absorption of an idea's elements into the network of already existing entities (the "objects of experience") in order to place them within a meaningful context. The context can only be meaningful if each entity finds appropriate associations inside the network, an action which can only happen in a measured way (that is, testing the associations for their autonomy).

We can now transpose the issue into the notions of resource space and complexity as illustrated through the e-chart. Since a higher generalisation of some thought content equals a greater realised potential of such content, which consequently raises the degree of applicability and ultimately complexity in the system, we are able to identify the direct relationship between the amount of complexity on one hand and the degree of autonomy on the other, both with regards to the same complex domain. Therefore, in any given set of TSs the development of abstractions of levels higher than PAT^1s is accompanied by a consequent rise in the degree of autonomy of the set, with the effect on its complexity such that there is now a higher potential for associations beyond the immediate domain. This process ameliorates the chances of a progression lock occurring as well as stalling the over-utilisation of the set's resource space; or, to turn this around, once a system decelerates its process of creating new abstractions within its resource space, it runs the risk of evoking progression locks causing its dynamics to grind to a halt.

Note the main source of this process: if it were not for the intent—and ability—to generalise, none of the above would follow. Since a generalisation is a form of selective representation, the ultimate success of the operation (ie, heightened complexity, autonomy, and so on) depends on the appropriateness of that representation. We have also seen in the previous section that the probability of attaining a result which lies inside a resource space is higher the more constraint there is on the initial procedure. 'More constraint' can be equated with better definition, which is another way of saying 'relevant generalisation' or the 'articulation of specific principles'.

From our point of view the problem is now to address the probability of a generalised statement being utilised, given that its information content exists inside a finite resource space and given that its nature denotes a particular degree of generalisation and hence autonomy and is thereby limited in its associative ability with other concepts (it is important to remember that we are dealing with mental representations here, and not real-world events;

equally, the probability values refer to the dynamics in our resource space, and not to some series of experiments that are designed to test the frequency of occurrence of a particular event—all this comes later with the implementation). Apart from a better definition the probability is further enhanced by a form of articulation which allows as many other TS-sets to make affinitive contact as possible, *their* degree of autonomy etc being on par with the original set (the same dynamics apply of course). Obviously, affinities are less likely between sets in which the affinity relies on an agreement between their contents, and are more likely between those where the fact of disagreement becomes the criterion for establishing a linkage; what matters then is the contents' format. Popper, coming from a direction different to ours, arrives at the same conclusion for he writes

> ..that not the *verifiability* but the *falsifiability* of a system is to be taken as a criterion of demarcation. In other words: I shall not require of a scientific system that it shall be capable of being singled out, once and for all, in a positive sense; but I shall require that its logical form shall be such that it can be singled out, by means of empirical tests, in a negative sense: *it must be possible for an empirical scientific system to be refuted by experience*. [31]

Although he is talking about the physical act of verification etc, for that to be implemented there must be the necessary mental structures which represent the system to be tested, together with those holding the tests to be enacted. Both can only then be meaningfully conceptualised if the former allows for the logical juxtaposition of the latter.

Such a juxtaposition is not only restricted to affinities between opposites either (how and why affinities can be enacted in terms of opposites will be dealt with in Part II). Its occurrence can consist in an entirely different set of TSs. First an example from the outside.

Let's say we study the meteorological chart of a certain region in order to predict the weather there for the next 12 hours. We possess detailed records of the necessary parameters extending back over the past 36 hours which show a cold front moving in from the south-west, and according to those data the front will reach our present location over the next 6 hours. Are we justified in making such a prediction? If the entire information is, "The cold front has passed point A, B, and C at certain intervals, its speed has been determined at 40km/h, and current location D is 240km north-east of C", then the answer is 'no'. Because the system is a chaotic one and because no cold front exists for ever, any past behaviour cannot be used as a reliable indicator for future performance. What we can do however is determine the nature of any weather phenomenon which would modify our low pressure system such that its progress becomes altered, and then check whether such an influence indeed exists. We could consult our charts therefore, draw a circle of say, 500km radius around place D and analyse the state of the weather in that area. If no disturbances of the required character are to be found, we are justified in assuming the continuing approach of the cold front. In other words, dealing with chaotic systems the question is not one of, "What are the chances of state X persisting" but, "Is there a state Y in addition to state X such that Y is capable of significantly influencing X within a relevant time frame". (We are then faced with looking for some possible state Z interfering with Y, and in turn testing Z's stability and so on. Our quest becomes a case of diminishing returns, the increasing predictability being an asymptotic curve getting close to, but never reaching the desired degree. In practice we have to stop somewhere)

In terms of cognitive structures, the processing of information within a certain set of TSs is not only a matter for that particular set, but is also determined by other—opposing—TSs provided their denominative potential is significant in relation to the set. The principle parallels with Popper's idea are obvious, and with the kind of dynamics we are concerned with here the mental act of juxtaposition is sometimes the only one which leads to a productive articulation of them.

As to the general usefulness of delineating from the singular to the universal, a question which prompted Popper to embark on his quest for better testability in the first place, we can see that in a large-scale system the probability of finding more than one example of affinity in the above mentioned sense is substantial. Indeed, the very fact of an increased probability value points to the justification of assuming a multitude of similar instances, so that the underlying problem of induction (ie, can a single instance be taken as universal) becomes less and less pressing, especially since the word 'universal' is not meant in its absolutist sense anyway. To declare that just about anything can happen provided the sample is large enough may sound rather glib, but considering the dynamics of highly complex systems such as the mind, it comes close to the truth. After all, the configuration of any TS set is not a random collection of elements thrown together at any one moment, but represents a particular built-up of cognitive entities for their own purpose. A further development from any point onwards therefore is made with reference to the current state, which in turn has its reasons for existence too (this is where the difference between c-PAT-ls and i-PAT-ls becomes especially critical). Once the combinatorial complexity of a mind's TSs has sufficiently advanced, not only are the fundamental structures available in all their variety, the organised combinations of them occur at such proliferation that the probability of *some* useful match *somewhere* is almost assured. At that degree of scale the problem with repeatability is no longer the inherent truth value of this or that scenario, but the degree of applicability in terms of the original intent for one or the other version of events. As Popper concentrates on physical states, seen from the traditional view of ultimate validity, the resultant difficulty causes him to be somewhat ambivalent—reluctantly, one might guess:

> In any case, the question whether the laws of science are strictly or numerically universal cannot be settled by argument. It is one of those questions which can be settled only by an agreement or a convention. [32]

He illustrates the issue of applicability by referring to zoological classifications such as 'mammal' or 'dog', words which carry a definite meaning in one field but can be ambiguous in another [33]. The multitude of detail in every type of definition can cause semantic overlaps due to varying emphases on particular aspects, misleading ancestries of meaning since the origins of its parts are different, and so can result in various possible requirements to achieve useful associations with related items (or rather, with what one wishes to be relevant on any given occasion). Again, in more traditional systems these problems may not matter much, but the nature of semantic domains and representational extent does become a vital player in areas such as nano technology, quantum mechanics, and cognitive science, precisely because the combinatorial variety at their level of information detail occupies such a huge resource space.

In addition to circumscribing its domain, further definition of a state also enlarges its chances of achieving affinity with another. It is a consequence of an increased probability due to heightened constraints, provided there is an affinity between the initial articulation and the new ones. Popper expresses this by using the concept of a certain type of hypothesis.

> As regards *auxiliary hypotheses* we propose to lay down the rule that only those are acceptable whose introduction does not diminish the degree of

> falsifiability or testability of the system in question, but, on the contrary, increases it. [34]

Since a hypothesis is a form of—tentative—definition, more than one means more constraints in its resource space and hence a greater probability for finding associations, as long as the newcomer occupies the same contextual space.

Now it is one thing to expect a representational TS-content to find its connections with some structure in due course, it is quite another to implement a purposeful intent in order to pursue a specific idea (or hypothesis for that matter). Are there any direct measures that can be taken, that is to say, can a particular set of TSs be assembled such that a potential affinity with a reference TS can be realised if the entire operation happens to be valid, or be discarded if found to be irrelevant or flawed? Before attempting the answer, we see how Popper articulates the issue.

Under the heading *The problem of the empirical basis* he first enumerates several difficulties [35]. There is the need for a scientific statement to be verifiable, otherwise it would be dogmatic. However, a reasoned, logical argument is also a statement, and the same requirements apply to *its* character, demanding yet another statement and so leading to an infinite regress. To avoid dogmatism and infinite regress we could turn to what he calls 'psychologism', the idea that a statement can be additionally justified through perceptual experience. If one can see no other alternative, some would opt for the last option. Popper rejects this because relying on empirical data brings us back to the problems of induction and universal validity which to him are not sustainable. He then continues with statements that obey a 'formal mode of speech', in contrast to ordinary or 'material mode of speech'. A scientific statement then cannot be explained through recourse to an experiment, but only via comparison with other, 'formal' sentences, accompanied by what are called 'protocol sentences'—expressions which are so basic that they have no more need of being confirmed in turn. They "describe the contents of immediate experience, or the phenomena; and thus the simplest knowable facts" [36]. Yet these 'protocol sentences' are not fail-safe and can be rejected should the occasion require it although, being already so basic, there is no rule set which logically defines their application or otherwise at that moment. One idea is to imbue thought processes themselves with an unwavering adherence to logic, where logic is held synonymous with the "laws of our thought" [37]. Fortunately, so he writes, we have moved on from there:

> In the field of logic, this kind of psychologism is now perhaps a thing of the past. Nobody would dream of justifying the validity of a logical inference, or of defending it against doubts, by writing beside it in the margin the following protocol sentence. 'Protocol: In checking this chain of inferences today, I experienced an acute feeling of conviction.'

(Alas, in the realm of public debate at least, this kind of expression is still alive and well and is actually meant to lend weight to one's opinion—more on this in Part II). For Popper there is only one solution,

> This is to put it [ie, the statement] in the form in which it is most easily testable: we break it up into many small steps, each easy to check by anybody who has learnt the mathematical or logical technique of transforming sentences. If after this anybody still raises doubts then we can only beg him to point out an error in the steps of the proof, or to think the matter over again. [38]

Let us now follow the path outlined above under our own auspices. There is a TS or a TS-complex which holds some content A that is under discussion. Note straight away that for this condition to occur the structure cannot exist beyond the logic boundary (otherwise there would be no challenge to begin with). For a statement to eventuate, some of the contents must be articulated, which means that only the content of con-TSs are addressable by someone else, although the overall significance of the issue is a result of the total complex; the information held by the sub-TSs remain unstated but not, one might say, 'unfelt'. There is also the recognition, carried by other TSs (let's call them V), that A ought to be verifiable—in other words, it must be possible to relate to A from outside its own complex. Since the meaning of 'verifiable' is defined by the methods used for such an action, the content of V is predetermined. If there is no intent to use V at all, the articulation of A is termed 'dogmatic' and hence rejected. Note the priority: V represents an already established rule set and so is perceived to be more substantial than A; therefore, during the process of aligning V with A it is the former which acts denominatively towards the latter. There is the perception that V ought to relate to yet another rule set and so on regressively, so its principles (the CFAs) are deemed to be scaleable towards infinity. Rightly or wrongly (we have argued this point before), they are now being applied not only to A but also to V (due to their perceived regress, and usually without further abstraction), hence the rules become denominative towards the events held under A. In the case of any further sensory input *its* content is associatively integrated with A and falls under the same considerations. At this point we have a content in A which is taken to be of a certain nature, represented in a compacted form by way of an articulation, and subjected to some standard accepted to be the norm.

Because the denominative influence is exerted across the entire complex of A, the subconscious parts are effected also and give rise to a new set of IFAs which are developing into the definer for A's processes in terms of their functional character. Since they are IFAs, not CFAs, there are no grounded PAT chains that are traceable as to their ultimate validity, and so the idea emerges of mental processes following the rules of logic (since it happened to be such a rule-type that got introduced into the whole dynamic, but the same applies to any type)—the rules being an inherent part of the processes are taken as given. The result is two-fold: whatever is not or cannot be articulated is coming under the influence of a denominative, introduced set of rules held to be authoritative, and what is being expressed may or may not concur with these rules. If it does, the argument is accepted as valid; if it does not, it is rejected—with the rules left untouched. Not only are they left alone, being IFAs and spread across a substantial segment of sub-TSs (after all, it is only the conscious part of ourselves which delimits notions into separate entities) they contribute to the fundamental character of one's ego-TPs. As such they emerge whenever the integrity of the self is seen to be challenged, without however providing grounded links to objectively justify the response.

And finally, Popper's response consists in fragmenting the original complex A for the purposes of applying V. To deconstruct something into smaller units is not trivial, especially not in the cognitive context. As a single entity the effects by any outside FAs or OAs exist in relation to its entirety. That is to say, the denominative influence of some content is measured not by the number of links it was able to establish with the complex, but through the result it has achieved in terms of the whole complex's character. Once turned into many parts however, a successful influence will have left its mark on each. To do that, it needs to negotiate its significance over and over again. In addition, in a separate and smaller unit the various denominative influences of its *own* structure will have a greater chance of being addressed in such an operation, and so the capacity of some V needs to be multifaceted. This is why a problem can seem insurmountable when confronted as a whole (ie, whatever knowledge one possesses can only be applied to the sum-total of the re-representation), yet when broken down into several parts the solution can begin with a small, easily understandable step—provided of course the essential information is within one's grasp in the first place.

If a decomposed statement is more vulnerable to attack compared to its greater momentum when left as a whole, then we can turn the issue around and say there is a higher degree of acceptance for the complete version. In other words, the cognitive elements which contributed to the formation of the complex and which made it into that current representation of data, meaning, and significance, all of which the mind's owner will eventually sensate, these elements will have a greater chance of evoking a positive response towards the statement if it stays together. In terms of the current situation, that is questioning a statement as to its likelihood or its probability, a positive response would be to say, "Yes, I think this is probable". Note that, as far as our cognitive perspective is concerned, the eventual verification through articulated and/or empirical analysis is only secondary; first and foremost comes the configuration of mental structures, and it is here that the ultimate fate of a statement is being decided. Given this interrelationship between the semantic probability of a notion and its chance of being falsified, it is interesting what Popper has to say about its outward manifestation:

> Whenever we can compare the degree of falsifiability of two statements, we can say that the one which is the less falsifiable is also the more probable, by virtue of its logical form. This probability I call '*logical probability*'; it must not be confused with that numerical probability which is employed in the theory of games of chance, and its statistics. *The logical probability of a statement is complementary to its degree of falsifiability:* it increases with decreasing degree of falsifiability. [39]

Amongst all these considerations about statements and what they can and should mean one theme occurs repeatedly: the quality of the statement itself. While there may be differing interpretations of the word 'quality' offered by a mathematician, an empirical scientist, or a philosopher (and I am not entering any arguments about how such groups may see each other), cognitively speaking the issue is one of accessibility. More precisely, the expression has to be such that its salient points can be recognised and related to by someone else. For example, a carpenter looking at a piece of wood about to be planed might say, "This piece is no good". Knowing nothing about timber the meaning is not clear, but someone else will be alerted about impurities in the material which can damage the blade. In terms of TSs the associative domain in the carpenter's mind and accessed via the sentence "This piece is no good" contains enough detail to be used productively in the present context; someone lacking such content needs further associations in order to make sense. We can say then that a statement is clear once its *semantic resolution* is sufficient for the purpose. The degree of semantic resolution can be applied to both structures, the one responsible for the statement and the other attempting to relate to it. For instance, pointing to a stack of boards in a timber yard and saying "They are no good" is not meaningful even for a carpenter since the expression lacks the context to provide the detail. As such the phenomenon spans two systems, and is a factor in the contextual association between TSs.

We can also identify a relationship between the semantic resolution and the autonomy of a structure. Recall that for a TS to be autonomous it has to have a sufficient number of abstracts to become independent of previous links; provided those abstracts are relevant, a greater number of abstracts means a higher semantic resolution and leads to greater autonomy. A greater autonomy in turn allows for more associations, hence more realised potential, and so for an increased applicability within the structure's resource space. To put it simply, the more I understand a concept through its detail, the more powerful it becomes in my hands.

Popper concerns himself with the issue in a way which at first seems contradictory. As a corollary to a statement's logical value he writes about the simplicity of a theory, and what it takes to attract such a description to begin with. There may be two versions of a mathematical proof, one simpler or more elegant than the other. This, he says, has nothing to do with their actual information value and is more a matter of aesthetics and pragmatism [40]. What is more important is the usefulness suggested by an expression which is simple compared to one which is not, simple statements attaining the aura of a law. This is still not satisfactory however because it does not explain the inherent nature of simplicity.

There is another angle which is also pertinent to our context, although in the end Popper rejects it as well: suppose we place twenty x,y coordinate values of some function on a graph paper where y depends on x, and they happen to produce a straight line. We are then driven to conclude that there must be some law at work which produces straight lines for the x,y values of that function. This however must be a conjecture since we only assume such a law because of the simplicity of the line and because it would be highly improbable for the chosen values to create the line by sheer coincidence. If we relate the above to what we have said about semantic resolution, it can be seen that the final appearance of some condition's articulation is mainly a result of that condition's initial definition.

Popper then offers to equate 'simplicity' with the degree of falsifiability, at the same time remarking that one should not end up arguing about the meaning of the word itself but rather for the purpose it is meant to uphold in the current context. Falsifiability is about opportunities for attack and so,

> Simple statements, if knowledge is our object, are to be prized more highly than less simple ones *because they tell us more; because their empirical content is greater; and because they are better testable*. [41]

Simplicity then, deprived of its superfluous connotations (according to Popper), may *seem* to offer less volume but in reality enables such content to be made applicable across a wider area of the cognitive network, thereby offering interpretative potential to a greater number of sub-systems within the complex. This of course is what generalisations are all about; one higher abstract that stands for many of its lower parents, and is intended for all those extra ones which have not been articulated yet. The appearance of a generalisation may be simple in the present sense, but what it effectively stands for is not. Connecting this to our concept of semantic resolution, we can say that the abstraction process of some content entails a decrease of structural detail, but without diminishing the degree of semantic resolution such content has had already. To point out the efficiency of abstractions during the course of thought processes is nothing new, except to remark that its benefits have been illustrated coming from a direction quite different to what we have considered so far. Such efficiency comes at a cost however, and we will re-visit the issue in greater detail—at a higher resolution—in Part II.

The matters of decomposition, detail, resolution, and extent bring us back to the question of volume in a resource space. Whether it is the number of products, the range of their realised potentials, the resultant complexity and thus the amount of associations and linkages that can be built upon an existing network of PAT chains, it all has to do with the dimensions of scale operating within a dynamic system. Each event, in so far as it is the result of a previous event, can be defined in terms of its source—or can it? The parameters surrounding probability can be expressed through the participating elements and their associations together with the degrees of resonance and autonomy, but all our descriptions are but a representative shadow of their real-world counterparts. For the purposes of explanation we have relayed the behaviour of certain entities and hopefully their interdependent behaviour has been made clear. They do not however, and never will, reflect the comprehensive multitude of occurrences which form our 'real' thoughts, and the immense detail accompanying even the simplest of musings. Yet behind any of those considerations waits the desire to somehow predict the outcome of a particular constellation, the significance of the initial condition, the ultimate accuracy of a description. We have referred to the inexpediency of insisting on a universal resonance for some statement, lest the expression be considered useless. When it comes to the type of scale inherent in mental structures, it should be enough to ascertain a probability value of greater than zero for an event to be taken seriously. Talking about chance in relation to the framing of laws, Popper writes,

> We encounter 'chance' in the objective sense, it may be said, when our probability estimates are corroborated; just as we encounter causal regularities when our predictions deduced from laws are corroborated. [42]

Although he refers to observable states and experiments, the significance of his words goes further. The result of an observation is a cognitive representation, and an experiment is the implementation of a set of TSs. Therefore the "probability estimates" originate with the affinity potential of abstracts, and what matters most is not what value can be estimated but how often an estimate can be made in the first place—in other words, the complexity of the notion. Such a change in point of view attains even greater importance when we consider that each present state in a system is the basis for the next; that is to say, the validity for a subsequent state rests with *its* antecedent and no-one else's, and probability values do not involve memories of the past. Hence even an attempt at a reasonably comprehensive description of some development is served better by identifying the current state of affairs, rather than some conjecture about an original intent. This is not to deny humans the ability to plan, but to assert the impossibility of predicting the future. If the future cannot be predicted then there cannot be an effective intent either, such a characteristic becoming the more acute the further one proceeds in the direction of increasing scale and more fundamental states.

The affinity potential in our nomenclature translates into "frequency statements" in Popper's version, his main area of concern being physics.

> Statistical estimates, or frequency statements, can never be derived from laws of a 'deterministic' kind, for the reason that in order to deduce any prediction from such laws, initial conditions are needed. In their place, assumptions about the statistical *distribution* of initial conditions—that is to say specific statistical assumptions—enter into every deduction in which statistical laws are obtained from micro assumptions of a deterministic or 'precise' character. [43]

Popper gives us an example of the manifested interplay between the associations within a domain and the probabilities of certain abstractions (ie generalisations) that can emerge [44]. Suppose some bones of a bird have been found in New Zealand and nowhere else, consequently identified as coming from an extinct species called 'moa' (so far the story is true). Let us assume, he says, that under ideal conditions such a bird could live to the age of sixty years or longer, but that in New Zealand they were less favourable and so no bird ever found had been older than fifty years. According to the logic of universal statements the declaration, "All moas die before reaching the age of fifty" is certainly true, but it does not represent a natural law: "..it is only due to *accidental or contingent* conditions—such as the co-presence of a certain virus—that in fact no moa did live longer". In the TS context this amounts to saying that although there exist structures whose content relate to particular abstractions, the source of the latter should not be taken as an exclusive or even necessary pre-condition for their type to occur.

Somewhat closer to the realm of every-day thought processes is the following: the residents living next to an expanded freeway have their backyard fences changed into noise-reducing walls, 2-metre-high structures snaking their way through suburbia. Some of them decide to use the solid boarding as a support for decorative plants. Assuming the initial condition (ie the expansion of the freeway) had not been of the residents' choosing, can one therefore declare that erecting hanging gardens in one's backyard is a lifestyle choice for certain urban dwellers? Of course not—had there been no noise disruption the gardens would not have eventuated in the first place. Yet in terms of mental processes it is quite possible to identify positive associations about flowering walls, shady corners, and the like.

To guard against misleading PAT chains that follow along a certain associative path because of some momentary preferential affinity, the options for more extraneous links have to be made available. In previous chapters we have encountered the role of grounded abstracts vs incongruent ones, and the ramifications of either. More than anything it is the *quality* of these connections which determine the efficacy of TSs, that is the utilisation of their resource space. Talking about science in general, Popper writes,

> The advance of science is not due to the fact that more and more perceptual experiences accumulate in the course of time. Nor is it due to the fact that we are making ever better use of our senses. Out of uninterrupted sense-experiences science cannot be distilled, no matter how industriously we gather and sort them. Bold ideas, unjustified anticipation, and speculative thought, are our only means for interpreting nature: our only organon, our only instrument, for grasping her. And we must hazard them to win our prize. [45]

The movement of content from one complex to another via the associative properties of its contingents and the ability of abstractions to enhance the autonomy of those parts balances any tendency for progression locks. It is a kind of escape mechanism preventing the calcification of existing domains, not just in the evolution of physics as Popper has remarked [46] but generally in any society which advanced during its history. Even the proliferation of symbolism, referred to by Popper as "a new expression of the old quest for certainty, a new symbolic ritual, a new substitute for religion" [47], can be seen in this light, so long as its articulations represent the creation of a separate resource space in which to erect once again pristine edifices that cannot be assailed by residents of more traditional systems; unfortunately, such flights sometimes also serve, Mayflower-like, to escape the challenges of sophistication and to revert back to more comfortable strictures.

For the owner of a mind which provides for the widest-possible access to his mind's content, not only via lateral connections but also through the enriched potential of multiple abstraction levels, the rewards translate into a cognitive environment that provides a vigorous arsenal for life's eventualities. Popper's advice to permit one's thoughts to be challenged can be compared to a society which is tolerant towards exterior standards; it is not an act of tolerance to accept what the law admits anyway, but rather to allow the unusual to be considered. It is a fundamental precept for any group of humans who wish to call themselves civilised. And so, as we began our journey through the philosophies with Socrates, who questioned the binding acceptance of doctrine, it is only fitting that we end it with someone who equally criticised the allure of mental walls.

As elaborate as the inner worlds of philosophers are, even as relevant as their portrayals of life may have been, they were never a true reflection of most other contemporary mentalities. Particularly before the explosive pervasion of data of every kind during the last few generations, the intellectual gradient between the few educated and articulate and the rest was quite steep. A Descartes may have mused about the similarities between knowledge and public infrastructure, but such points would have been utterly lost to members of the common drudgery. *Their* minds, born into precarious situations from the start, reared by rigid strictures and ever-present punishments, deprived of the experience of higher abstractions and continuously compacted into a solid mass of human utility, *their* minds followed the prescribed dictums, controlled by their own inner devils, doing their subconscious' bidding—at the most shame-facedly nudging the borders but without any inkling of anything that could lie beyond. Such brutishness extended across the demographics as we understand them today, sparing neither high nor low. For varying reasons the criteria for social success were hardly ever in tandem with levels of intelligence and wisdom, although naturally these two contributed to its achievement in an indirect way. The system of the mind, which can be laid out so beautifully by studying the architecture of an impressive intellect, can also be a tangled clutter of unseemly matter the actual nature of which even its bearer fails to grasp.

For a human being the world is as his mind sensates it to him; nothing more, nothing less. The fundamentals of the system contain enough flexibility for their owner to be a religious zealot; a rationalist; a wallflower; a dictator; an Anglican logician; a computer programming tarot reader; an existentialist palliative nurse; a poetic idealist soldier—*and* do their job so well their master (?) remains convinced it is all so natural. If the structures are sophisticated, wide-ranging, and elegant, they will be capable of relating to something of a similar nature when exposed to it. Conversely, if they are simple, beset with inconsistency and the resultant fear of the real, any experience, no matter how noble in itself, will be relegated to the personal standard. Apart from the exceptional mind which understands *and* hates with a definite purpose, the ordinary human likes and fears what makes sense in terms of his own mind's potential for affinity. As such the path of progress in any society is characterised by suspicion towards the unfamiliar, followed by a gradual acceptance as the benefits of the new slowly trickle into society's consciousness, but in a form its common structures are able to comprehend. If an idea cannot follow this path it is doomed, with sometimes murderous consequences for its holder. Just as the philosophical landscape featured ever more complex sceneries over the centuries, so did the mental surface of other people change, albeit with considerable offset towards the mediocre.

To import data and to process them are not one and the same. The existence of first-level abstracts does not ensure a comprehensive set of higher-level structures capable of transposing their contents along congruent PAT-chains. To feel pain therefore does not automatically provide the insight necessary for alleviating it; rallying one's emotions in a righteous manner does not make for a righteous person; the act of pointing out an error does not represent the knowledge how to fix it. Accompanying the destruction of visionaries then is the manifest commitment towards rectifying a perceived problem, conducted after a fashion. The proverb, "The road to hell is paved with good intentions" is as true in its immediate, direct form, as it is in the deeper and technical sense we have come to understand by now. The functionalities we have identified (articulated in the form of abstract symbols of this or that functionality) served to describe some of the dynamics the mind is able to perform. We could do this by examining the intellectual edifices of exceptional people, because they happened to take the time and show the courage to articulate what their minds managed to make of reality. We should be comfortable with the basic performance of these symbols, because we are now turning to the internal dynamics of such entities. By doing so we are leaving the world of perception, of objects and actions, and enter the realm where existence is not what we *think* it is but what our mental operating system puts together *before* our conscious self is ever made aware of it. From now on our journey will take us to an altogether different region.

As a form of technical summary let us have a brief look at a simplified illustration of a simple thought, to demonstrate the sheer complexity even at that level. The boy Calvin, he of the comic strip *Calvin & Hobbes* fame by Bill Watterson [48], thinks he's got monsters under his bed. In terms of our symbolism, part of his mind could look something like Fig. 12-1.

Fig. 12-1: "There are monsters under my bed!" — see text for details.

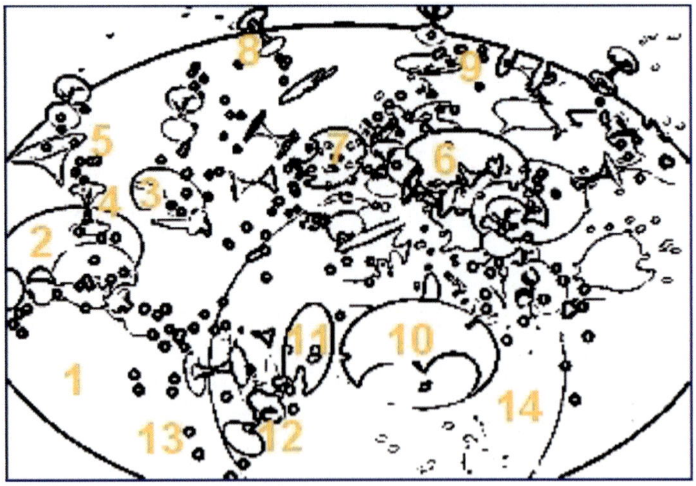

1 'ego' container
2 'outside' container
3 'strangers' container
4 illusion tree in 'outside' domain
5 abstracts in 'outside' domain
6 'have' container
7 'own room' container
8 illusion tree in 'have' domain
9 abstracts in 'have' domain
10 'fantasy' container
11 'monsters' container
12 illusion tree in 'fantasy' domain
13 abstracts in 'fantasy' domain
14 logic boundary around 'fantasy' domain

There is the 'ego' container (*1* in the legend), a vast structure representing everything which Calvin uses to identify himself—his feelings, his appearance (as perceived by him), his inclinations, and all the notions that comprise one's situatedness in the world. There is also a container holding the experiences that relate to the 'outside' (*2*), and it is fed by the processes inherent in sub-containers dealing with such things as spatial extensions, other people (*3*), and associations having to do with the concepts of 'unfamiliarity', 'strangeness', 'threat', but also 'inconsistency' and 'impermanence' (unless of course one happens to personally identify with such things). We also expect to find a comparable structure relating to 'possession' and consequently everything which can be linked to 'I have' (*6*), for example one's room (*7*). And, since Calvin is a child which believes in monsters, his mind features a 'fantasy' container, serving as an attractor as well as a placement for all that which in one way or another is make-belief (*10*), so 'monsters' would have a place here (*11*). Connecting those various containers are illusion trees (*4*, *8*, *12*) which channel the abstracts from one structure to another, processing them while doing so. In this case there may have been an imported perception of some strange creature (note that the labels 'strange' and 'creature' are already the results of abstraction processes) which now turns into something which lives under the bed, having established links with the structures about 'I have' and 'strange', but in such a way that the characteristics 'strangeness' and 'threat' are being maintained. Therefore, while the concept 'dragon' for instance would not be seen as disturbing (ie, it exists in its own sub-domain within the wider concept of 'strange'), once the thing identified as 'monster' gets connected with 'my own bed' the result is fear because it resides now inside 'one's own'. The abstracts placed and distributed in these manners (*5*, *9*, *13*) can be object or functional ones (not differentiated in our picture for simplicity's sake), and they carry their information value along the PAT chains at varying levels of abstraction. Surrounding most of the 'fantasy' domain is the logic boundary, seen here as a greenish sphere (*14*). Everything inside is removed from the congruence existing beyond its extension (assuming, again for simplicity's sake, that everything on the outside is indeed congruent), and the only way its content can be reached is through those structural entities which are allowed to transport data across the boundary. The criterion for being able to do so is a function of the type of incongruence which it protects; that is to say, it is a matter of *how* the information value inherent in certain abstracts had to be modified in order for the structure to be placed inside the bubble to begin with. The specific nature of functional abstracts is important here, and as usual their associative properties decide on the access rights one way or another. For example, to lower the degree of apprehension about monsters under the bed, a reference to something equally fantastic would be more effective than some recourse to pure logic (in this case, Calvin's stuffed tiger Hobbes is more valuable than turning the light on and demonstrating the emptiness of the space under the bed). Or, on a more serious note, to a religious person an exorcism is more meaningful than a course of antibiotics.

The types of shapes used in these descriptions have been chosen for good reason. Containers looking like vortices, illusion trees having the shape of wind funnels, and abstracts in the form of little bubbles riding along the currents—they are meant to suggest the dynamic fluidity of the system. How close that picture gets to the real we will see shortly.

Bibliography

On Internet material:

(1) Due to the changing nature of domain names and addresses and the complexity of some links, a source from the Internet is (a) listed under the title of the text, the name of the local site's proprietor and the date accessed, but not its URL; and (b), quoted from the then current web page.

(2) Because page numbers are irrelevant in browser-style displays, three words from the relevant tract have been quoted instead for use in text search (eg, <ab cd ef>).

1. The start of the journey

[1] Korzybski, A., 1941. *Science and Sanity*, International Non-aristotelian Publishing Co., Lancaster, Pennsylvania. (p. 90)

2. The mind-body myth

[1] Plato, new ed. 1977. *The Last Days of Socrates*, Penguin Books Ltd, Middlesex, England. (p. 19)

[2] *ibid*, (p. 99)

[3] Hutchinson, D.S., 1986. *The Virtues of Aristotle*, Routledge & Kegan Paul, London, New York. (p. 20)

[4] *ibid*, (p. 26)

[5] *ibid*, (p. 40)

[6] Cooper, J.M., 1975. *Reason and Human Good in Aristotle*, Harvard University Press, Cambridge, Mass. (p. 65)

3. The universe as a fractal

[1] Chakravartinayanar, A., Upadhye, A.N., 1975. *Pañcastikayasara, The Building of the Cosmos*, Bharatiya Jnanpith Publication, New Delhi.

[2] *ibid*, (p. 40)

[3] *ibid*, (p. 41)

[4] *ibid*, (p. 46)

[5] *ibid*, (p. 56)

[6] *ibid*, (p. 64)

[7] *ibid*, (p. 66)

[8] *ibid*, (p. 73)

[9] *ibid*, (p. 77)

[10] *ibid*, (p. 97)

[11] *ibid*, (p. 91)

[12] *ibid*, (p. 104)

[13] *ibid*, (p. 106)

[14] *ibid*, (p. 130)

[15] *ibid*, (p. lvi)

[16] *ibid*, (p. lxvi)

[17] Eco, U., 1989. *Foucault's Pendulum*, Secker & Warburg, London. (p. 362)

4. Building Europe's mind: the foundations

[1] Murray, D., 22 July 1998. *Practising pagans land in hot water*, The Courier Mail, Brisbane.

[2] Widengren, G., 1965. *Mani and Manichaeism*, Weidenfeld and Nicolson, London. (p. 43)

[3] Grant, K., 1976. *Cults of the Shadow*, Samuel Weiser, New York. (p. 75)

[4] Widengren, G., *op. cit.*, (p. 97)

[5] *ibid*, (p. 47)

[6] *ibid*, (p. 119)

[7] Wright, E., 1985. *History of the World, Prehistory to the Renaissance*, Newnes Books, Middlesex. (p. 270)

[8] Widengren, G., *op. cit.*, (p. 48)

[9] *ibid*, (p. 53)

[10] *ibid*, (pp. 55-58)

[11] *ibid*, (p. 68)

[12] Gallagher, D.A., Gallagher, I.J., (transl) 1966. *St. Augustine, The Catholic and Manichaean Ways of Life*, The Catholic University of America Press, Washington. (pp. xi-xv)

[13] *ibid*, (p. 4)

[14] *ibid*, (p.5)

[15] *ibid*, (inside cover)

[16] *ibid*, (p. 7)

[17] *ibid*, (p. 11)

[18] *ibid*, (p. 20)

[19] *ibid*, (p. 22)

[20] *ibid*, (p. 30)

[21] *ibid*, (p. 31)

[22] *ibid*, (p. 40)

[23] *ibid*, (p. 66)

[24] *ibid*, (p. 69)

[25] *ibid*, (p. 68)

[26] *ibid*, (p. 92)

[27] Benjamin, A.S., Hackstaff, L.H. (transl), 1964. *Saint Augustine, On the Choice of Free Will*, The Bobbs-Merrill Company, Inc, New York.

[28] *ibid*, (p. 13)

[29] *ibid*, (p. 15)

[30] *ibid*, (p. 15)

[31] *ibid*, (p. 32)

[32] *ibid*, (p. 36)

[33] *ibid*, (p. 42)

[34] *ibid*, (p. 44)

[35] *ibid*, (p. 61)

[36] *ibid*, (p. 66)

[37] *ibid*, (p. 71)

[38] *ibid*, (p. 75)

[39] *ibid*, (p. 80)

[40] *ibid*, (p. 83)

5. Rights and wrongs

[1] Reith, H., 1958. *The Metaphysics of St. Thomas Aquinas*, The Bruce Publishing Company, Milwaukee.

[2] *ibid*, (p. vi)

[3] *ibid*, (p. 2)

[4] *ibid*, (p. 3)

[5] *ibid*, (p. 3)

[6] *ibid*, (p. 27)

[7] *ibid*, (p. 30)

[8] *ibid*, (p. 32)

[9] *ibid*, (p. 49)

[10] *ibid*, (p. 51)

[11] *ibid*, (p. 52)

[12] *ibid*, (p. 53)

[13] *ibid*, (p. 57)

[14] *ibid*, (p. 59)

[15] *ibid*, (p. 70)

[16] *ibid*, (p. 79)

[17] *ibid*, (p. 75)

[18] *ibid*, (p. 112)

[19] *ibid*, (p. 73)

[20] *ibid*, (p. 22)

[21] *ibid*, (p. 24)

[22] *ibid*, (p. 25)

[23] *ibid*, (p. 43)

[24] *ibid*, (p. 44)

[25] *ibid*, (p. 46)

[26] *ibid*, (p. 92)

[27] *ibid*, (p. 135)

[28] *ibid*, (p. 27)

[29] *ibid*, (p. 178)

[30] *ibid*, (p. 183)

[31] *ibid*, (p. 185)

[32] *ibid*, (p. 195)

[33] *ibid*, (p. 197)

[34] Birks, W., Gilbert R.A., 1987. *The Treasure of Montségur, A Study of the Cathar Heresy and the Nature of the Cathar Sect*, Crucible, Great Britain.

[35] Madaule, J., 1967. *The Albigensian Crusade*, Burns & Oates Limited, London.

[36] Birks, W., Gilbert R.A., *op. cit.*, (p. 56)

[37] Madaule, J., *op. cit.*, (p. 48)

[38] Birks, W., Gilbert R.A., *op. cit.*, (p. 56)

[39] *ibid*, (p. 63)

[40] von Eschenbach, W., new ed. 1961. *Parzival*, Vintage Books, New York.

[41] Baigent, M., Leigh, R., Lincoln, H., 1982. *The Holy Blood and The Holy Grail*, Corgi Books, Great Britain.

[42] Ludowyk, F. (ed.), *Ozwords*, Oxford University Press, South Melbourne.

[43] Birks, W., Gilbert R.A., *op. cit.*, (p. 85)

[44] *ibid*, (p. 87)

[45] Madaule, J., *op. cit.*, (p. 45)

[46] *ibid*, (p. 31)

[47] *ibid*, (p. 39)

[48] Le Roy Ladurie, E., 1979. *Carnival, A People's Uprising at Romans 1579-1580*, Scolar Press, London.

[49] de Coster, C., year not given. *Uilenspiegel und Lamme Goedzak, Ein fröhliches Buch trotz Tod und Tränen*, Insel Verlag, Leipzig.

[50] Huxley, A., 1977. *The Devils of Loudun*, Triad/Panther Books, Frogmore, St Albans, Herts.

[51] Huysmans, J.K., 1923. *Saint Lydwine of Schiedam*, Tan Books and Publishers, Inc., Illinois.

[52] Hope, C. (ed.), new ed. 1983. *The Autobiography of Benvenuto Cellini*, Phaidon Press Limited, Oxford.

[53] Kamen, H., 1997. *The Spanish Inquisition, An Historical Revision*, Phoenix, London.

[54] Mathers, S.L.M., new ed. 1975. *The Book of the Sacred Magic of Abramelin the Mage*, Dover Publications, Inc., New York.

[55] Robb, P., 1998. *M*, Duffy & Snellgrove, Sydney.

[56] Kiple, K. (ed.), 1999. *Plague, Pox & Pestilence*, Phoenix Illustrated, London.

[57] Kohn, G.C. (ed.), 1995. *Encyclopedia of Plague and Pestilence*, Facts on File Inc., New York.

[58] Stafford, P., 1977. *Psychedelics Encyclopedia*, And/Or Press, Berkeley.

[59] Hurwood, B.J., 1974. *Torture Through The Ages*, Tandem, London.

[60] Villon, F., new ed. 1968. *Complete Poems of François Villon*, Everyman's Library, London.

6. Into the light

[1] de Rico, U., Solti, G., 1980. *The Ring of the Nibelung*, Thames and Hudson Ltd, London. (p. 14)

[2] Descartes, R., new ed. 2001. *Discourse On The Method Of Rightly Conducting The Reason, And Seeking Truth In The Sciences*, Project Gutenberg Association. Accessed on Internet 29 Jan 2001.

[3] *ibid*, (p. 9)

[4] *ibid*, (p. 35)

[5] *ibid*, (p. 40)

[6] *ibid*, (p. 18)

[7] *ibid*, (p. 9)

[8] *ibid*, (p. 10)

[9] *ibid*, (p. 12)

[10] *ibid*, (p. 12)

[11] *ibid*, (p. 6)

[12] *ibid*, (p. 16)

[13] *ibid*, (p. 13)

[14] *ibid*, (p. 15)

[15] *ibid*, (p. 20)

[16] *ibid*, (p. 31)

[17] *ibid*, (p. 6)

[18] *ibid*, (p. 20)

[19] *ibid*, (p. 21)

[20] *ibid*, (p. 22)

[21] *ibid*, (p. 29)

[22] *ibid*, (p. 36)

[23] *ibid*, (p. 37)

[24] de Spinoza, B., new ed. 1955. *On the Improvement of the Understanding, The Ethics, Correspondence*, Dover Publications, Inc, New York.

[25] *ibid*, (p. 7)

[26] *ibid*, (p. 8)

[27] *ibid*, (p. 9)

[28] *ibid*, (p. 12)

[29] *ibid*, (p. 13)

[30] *ibid*, (p. 22)

[31] *ibid*, (p. 29)

[32] *ibid*, (p. 33)

[33] *ibid*, (p. 35)

[34] *ibid*, (p. 36)

[35] *ibid*, (p. 189)

[36] *ibid*, (p. 39)

[37] *ibid*, (p. 45)

[38] *ibid*, (p. 46)

[39] *ibid*, (p. 59)

[40] *ibid*, (p. 66)

[41] *ibid*, (p. 70)

[42] *ibid*, (p. 75)

7. The mind as government

[1] Locke, J., new ed. 1975. *An Essay concerning Human Understanding*, Oxford University Press, Oxford.

[2] *ibid*, (p. 236)

[3] *ibid*, (p. 620)

[4] *ibid*, (p. 621)

[5] *ibid*, (p. 624)

[6] *ibid*, (p. 708)

[7] *ibid*, (p. 682)

[8] *ibid*, (p. 684)

[9] *ibid*, (p. 82)

[10] *ibid*, (p. 104)

[11] *ibid*, (p. 115)

[12] *ibid*, (p. 116)

[13] *ibid*, (p. 144)

[14] *ibid*, (p. 162)

[15] *ibid*, (p. 182)

[16] *ibid*, (p. 188)

[17] *ibid*, (p. 393)

[18] *ibid*, (p. 397)

[19] *ibid*, (p. 429)

[20] *ibid*, (p. 430)

[21] *ibid*, (p. 537)

[22] *ibid*, (p. 661)

[23] *ibid*, (p. 689)

[24] *ibid*, (p. 712)

[25] *ibid*, (p. 714)

[26] *ibid*, (p. 718)

[27] *ibid*, (p. 230)

[28] *ibid*, (p. 248)

[29] *ibid*, (p. 254)

[30] *ibid*, (p. 267)

[31] *ibid*, (p. 699)

[32] *ibid*, (p. 170)

[33] *ibid*, (p. 296)

[34] *ibid*, (p. 352)

[35] *ibid*, (p. 419)

[36] *ibid*, (p. 578)

[37] *ibid*, (p. 598)

[38] *ibid*, (p. 607)

[39] *ibid*, (p. 669)

[40] *ibid*, (p. 670)

[41] *ibid*, (p. 671)

[42] *ibid*, (p. 705)

[43] *ibid*, (p. 44)

[44] *ibid*, (p. 50)

[45] *ibid*, (p. 55)

[46] *ibid*, (p. 101)

[47] *ibid*, (p. 137)

[48] *ibid*, (p. 146)

[49] *ibid*, (p. 147)

[50] *ibid*, (p. 154)

[51] *ibid*, (p. 158)

[52] *ibid*, (p. 211)

[53] *ibid*, (p. 220)

[54] *ibid*, (p. 318)

8. God by any other name

[1] Rousseau, J-J., new ed. 1999. *Discourse on Political Economy*, *The Social Contract*, Oxford University Press, Oxford.

[2] *ibid*, (p. 6)

[3] *ibid*, (p. 7)

[4] *ibid*, (p. 8)

[5] *ibid*, (p. 11)

[6] *ibid*, (p. 17)

[7] *ibid*, (p. 31)

[8] *ibid*, (p. 76)

[9] *ibid*, (p. 14)

[10] *ibid*, (p. 25)

[11] *ibid*, (p. 45)

[12] *ibid*, (p. 46)

[13] *ibid*, (p. 51)

[14] *ibid*, (p. 78)

[15] *ibid*, (p. 84)

[16] *ibid*, (p. 97)

[17] *ibid*, (p. 143)

[18] *ibid*, (p. 162)

[19] *ibid*, (p. 165)

[20] *ibid*, (p. 167)

[21] *ibid*, (p. 54)

[22] *ibid*, (p. 60)

[23] *ibid*, (p. 66)

[24] *ibid*, (p. 73)

[25] *ibid*, (p. 83)

[26] *ibid*, (p. 112)

[27] *ibid*, (p. 113)

[28] *ibid*, (p. 134)

[29] *ibid*, (p. 89)

[30] Kant, I., new ed. 1966. *Critique of Pure Reason*, Anchor Books, New York.

[31] *ibid*, (p. xxii)

[32] *ibid*, (p. 245)

[33] *ibid*, (p. 330)

[34] *ibid*, (p. xxxii)

[35] *ibid*, (p. 14)

[36] *ibid*, (p. xlii)

[37] *ibid*, (p. 266)

[38] *ibid*, (p. 1)

[39] *ibid*, (p. 389)

[40] *ibid*, (p. 465)

[41] *ibid*, (p. 496)

[42] 7 July 2001. *say what?*, The Courier Mail, Brisbane.

[43] Kant, I., *op. cit.*, (p. 2)

[44] *ibid*, (p. 4)

[45] *ibid*, (p. 5)

[46] *ibid*, (p. 7)

[47] *ibid*, (p. 8)

[48] *ibid*, (p. 16)

[49] *ibid*, (p. 21)

[50] *ibid*, (p. 132)

[51] *ibid*, (p. 25)

[52] *ibid*, (p. 26)

[53] *ibid*, (p. 40)

[54] *ibid*, (p. 427)

[55] *ibid*, (p. 304)

[56] *ibid*, (p. 306)

[57] *ibid*, (p. 307)

9. The will for independence

[1] Hegel, W.F.G., new ed. 1991. *The Philosophy of History*, Prometheus Books, New York.

[2] *ibid*, (p. 1)

[3] *ibid*, (p. 4)

[4] *ibid*, (p. 6)

[5] *ibid*, (p. 8)

[6] *ibid*, (p. 9)

[7] *ibid*, (p. 103)

[8] *ibid*, (p. 10)

[9] *ibid*, (p. 15)

[10] *ibid*, (p. 109)

[11] *ibid*, (p. 278)

[12] *ibid*, (p. 321)

[13] *ibid*, (p. 324)

[14] *ibid*, (p. 442)

[15] *ibid*, (p. 331)

[16] *ibid*, (p. 16)

[17] *ibid*, (p. 17)

[18] *ibid*, (p. 23)

[19] *ibid*, (p. 24)

[20] *ibid*, (p. 27)

[21] *ibid*, (p. 29)

[22] *ibid*, (p. 74)

[23] Harris, M., 29 January 1999. Correspondence to the author, Queensland Parks and Wildlife Service.

[24] Hegel, W.F.G., *op. cit.*, (p. 35)

[25] *ibid*, (p. 39)

[26] *ibid*, (p. 40)

[27] *ibid*, (p. 86)

[28] Darwin, C., new ed. 1999. *On the Origin of Species by Means of Natural Selection, or the Preservation of Favoured Races in the Struggle for Life*, Knowledge Matters Ltd. Accessed on Internet 3 Feb 2001.

[29] *ibid*, (p. 35)

[30] *ibid*, (p. 42)

[31] *ibid*, (p. 76)

[32] *ibid*, (p. 159)

[33] *ibid*, (p. 232)

[34] *ibid*, (p. 236)

[35] *ibid*, (p. 237)

[36] *ibid*, (p. 240)

[37] *ibid*, (p. 241)

[38] Reitter, E., 1908. *Fauna Germanica, Die Käfer des Deutschen Reiches*, Vol I, K.G. Lutz' Verlag, Stuttgart. (p. vii)

[39] Nietzsche, F., 1978. (A) *Thus spoke Zarathustra, a book for all and none*, Penguin Books, New York.

[40] Nietzsche, F., 1990. (B) *Beyond Good and Evil, prelude to a philosophy of the future*, Penguin Books, London.

[41] Nietzsche, F., (A), *op. cit.*, (p. xiii)

[42] *ibid*, (p. xv)

[43] Drury, N., 1985. *Dictionary of Mysticism and the Occult*, Harper & Row, New York. (p. 279)

[44] Nietzsche, F., (A), *op. cit.*, (p. 14)

[45] *ibid*, (p. 15)

[46] *ibid*, (p. 52)

[47] *ibid*, (p. 119)

[48] *ibid*, (p. 81)

[49] *ibid*, (p. 135)

[50] *ibid*, (p. 212)

[51] *ibid*, (p. 221)

[52] *ibid*, (p. 151)

[53] *ibid*, (p. 93)

[54] Hurwood, B.J., *op. cit.*, (p. 73)

[55] Nietzsche, F., (B), *op. cit.*, (p. 7)

[56] *ibid*, (p. 75)

[57] *ibid*, (p. 76)

[58] *ibid*, (p. 88)

[59] *ibid*, (p. 203)

10. Anatomy of a universe

[1] Marx, K., new ed 1993. *The Communist Manifesto*, Project Gutenberg Association. Accessed on Internet 30 Jul 2001.

[2] *ibid*, (p. 5)

[3] *ibid*, (p. 6)

[4] *ibid*, (p. 7)

[5] *ibid*, (p. 9)

[6] *ibid*, (p. 10)

[7] *ibid*, (p. 11)

[8] *ibid*, (p. 12)

[9] *ibid*, (p. 13)

[10] 1983. *The Macquarie History of Ideas*, Macquarie Library, Sydney.

[11] Marx, K., *op. cit.*, (p. 14)

[12] Foucault, M., new ed. 1991. *Discipline and Punish, The Birth of the Prison*, Penguin Books, London. (fig. 9)

[13] Tannahill, R., 1981. *Sex in History*, Abacus, London. (after p. 244)

[14] Marx, K., *op. cit.*, (p. 15)

[15] *ibid*, (p. 16)
[16] *ibid*, (p. 17)

[17] *ibid*, (p. 18)

[18] *ibid*, (p. 18)

[19] *ibid*, (p. 19)

[20] *ibid*, (p. 20)

[21] *ibid*, (p. 20)

[22] *ibid*, (p. 22)

[23] Husserl, E., new ed. 1998. (A) *Ideas pertaining to a pure phenomenology and to a phenomenological philosophy, First Book - General introduction to a pure phenomenology*, Kluwer Academic Publishers, Dordrecht.

Husserl, E., new ed. 1989. (B) *Ideas pertaining to a pure phenomenology and to a phenomenological philosophy, Second Book - Studies in the phenomenology of constitution*, Kluwer Academic Publishers, Dordrecht.

[24] *ibid*, (p. A-xxii)

[25] *ibid*, (p. A-xxiii)

[26] *ibid*, (p. A-290)

[27] *ibid*, (p. A-15)

[28] *ibid*, (p. A-17)

[29] *ibid*, (p. A-92)

[30] *ibid*, (p. A-90)

[31] *ibid*, (p. A-100)

[32] *ibid*, (p. A-131)

[33] *ibid*, (p. A-132)

[34] *ibid*, (p. A-134)

[35] *ibid*, (p. A-187)

[36] *ibid*, (p. A-8)

[37] *ibid*, (p. A-11)

[38] *ibid*, (p. A-22)

[39] *ibid*, (p. A-42)

[40] *ibid*, (p. A-43)

[41] *ibid*, (p. A-139)

[42] *ibid*, (p. A-276)

[43] *ibid*, (p. A-277)

[44] *ibid*, (p. A-332)

[45] *ibid*, (p. B-9)

[46] *ibid*, (p. A-54)

[47] *ibid*, (p. A-65)

[48] *ibid*, (p. A-113)

[49] *ibid*, (p. A-116)

[50] *ibid*, (p. A-124)

[51] *ibid*, (p. A-125)

[52] *ibid*, (p. A-291)

[53] *ibid*, (p. B-96)

[54] *ibid*, (p. B-126)

[55] *ibid*, (p. B-84)

[56] *ibid*, (p. B-86)

[57] *ibid*, (p. B-90)

[58] Jung, C.G., new ed. 1995. *Memories, Dreams, Reflections*, Fontana Press, London.

[59] *ibid*, (p. 7)

[60] *ibid*, (p. 9)

[61] *ibid*, (p. 11)

[62] *ibid*, (p. 19)

[63] Boden, M., 1994. *Creativity and Computers*, from: Dartnall T. (ed). 1994. *Artificial Intelligence and Creativity, An Interdisciplinary Approach*, Kluwer Academic Publishers, Dortrecht. (p. 3)

[64] Jung, C.G., *op. cit.*, (p. 26)

[65] *ibid*, (p. 28)

[66] 18 February 1995. *Naughty Bits*, The Courier Mail, Brisbane.

[67] Jung, C.G., *op. cit.*, (p. 62)

[68] *ibid*, (p. 64)

[69] *ibid*, (p. 31)

[70] *ibid*, (p. 41)

[71] *ibid*, (p. 61)

[72] *ibid*, (p. 52)

[73] *ibid*, (p. 55)

[74] Lindow, W. (ed), 1997. *Ein kurzweiliges Lesen von Dil Ulenspiegel*, Philipp Reclam Jun., Stuttgart. (p. 30)

[75] *ibid*, (p. 243)

[76] Burroughs, W., 1974. *The Soft Machine*, Corgi Books, London.

[77] Jung, C.G., *op. cit.*, (p. 61)

[78] *ibid*, (p. 353)

[79] *ibid*, (p. 35)

[80] *ibid*, (p. 44)

[81] *ibid*, (p. 85)

[82] *ibid*, (p. 380)

[83] *ibid*, (p. 197)

[84] *ibid*, (p. 248)

[85] *ibid*, (p. 290)

[86] *ibid*, (p. 333)

[87] *ibid*, (p. 262)

[88] *ibid*, (p. 281)

[89] *ibid*, (p. 191)

[90] *ibid*, (p. 377)

11. In search of freedom?

[1] Marcuse, H., new ed. 1998. *Eros and Civilization, A Philosophical Inquiry into Freud*, Routledge, London.

[2] *ibid*, (p. 187)

[3] *ibid*, (p. 189)

[4] *ibid*, (p. 192)

[5] *ibid*, (p. xxi)

[6] *ibid*, (p. 3)

[7] *ibid*, (p. 4)

[8] *ibid*, (p. 12)

[9] *ibid*, (p. 14)

[10] *ibid*, (p. 19)

[11] *ibid*, (p. 34)

[12] *ibid*, (p. 35)

[13] *ibid*, (p. 90)

[14] *ibid*, (p. 96)

[15] *ibid*, (p. 98)

[16] *ibid*, (p. 100)

[17] *ibid*, (p. 102)

[18] *ibid*, (p. 151)

[19] *ibid*, (p. 169)

[20] Petronius; Seneca, new ed. 1981. *The Satyricon*; *The Apocolocyntosis*, Penguin Books, London.

[21] Marcuse, H., *op. cit.*, (p. 22)

[22] *ibid*, (p. 23)

[23] Thornton, E.M., 1986. *The Freudian Fallacy*, Paladin Grafton Books, London.

[24] *ibid*, (p. 10)

[25] *ibid*, (p. 250)

[26] Smith, D.C., Sparks, W., 1986. *The Naked Child, Growing up without Shame*, Elysium Growth Press, Los Angeles. (p. 19)

[27] Marcuse, H., *op. cit.*, (p. 49)

[28] *ibid*, (p. 159)

[29] Chick, J.T., 1975. *Exorcists*, Chick Publications, California.

[30] Marcuse, H., *op. cit.*, (p. 269)

[31] Ballard, J.G., 1974. *Salvador Dali (the innocent as paranoid)*, from: *Best SF Stories from New Worlds 8*, Granada Publishing Limited, Great Britain. (p. 97)

[32] de la Serna, R.G., 1979. *Dalí*, Park Lane, New York. (p. 15)

[33] Sartre, J.-P., 1998. *Being and Nothingness, An Essay on Phenomenological Ontology*, Routledge, London.

[34] *ibid*, (p. xxii)

[35] *ibid*, (p. xxiii)

[36] *ibid*, (p. xxix)

[37] *ibid*, (p. xxx)

[38] *ibid*, (p. xxxi)

[39] *ibid*, (p. xxxiii)

[40] *ibid*, (p. 4)

[41] *ibid*, (p. 8)

[42] *ibid*, (p. 9)

[43] *ibid*, (p. 25)

[44] *ibid*, (p. 29)

[45] *ibid*, (p. 47)

[46] *ibid*, (p. 48)

[47] *ibid*, (p. 50)

[48] *ibid*, (p. 54)

[49] *ibid*, (p. 433)

[50] *ibid*, (p. 435)

[51] *ibid*, (p. 436)

[52] *ibid*, (p. 479)

12. Back to earth

[1] Wittgenstein, L., 1998. *Philosophical Investigations, Second Edition*, Blackwell Publishers, Oxford.

[2] *ibid*, (p. 2e)

[3] *ibid*, (p. 8e)

[4] *ibid*, (p. 12e)

[5] *ibid*, (p. 31e)

[6] *ibid*, (p. 19e)

[7] *ibid*, (p. 26e)

[8] *ibid*, (p. 30e)

[9] *ibid*, (p. 34e)

[10] *ibid*, (p. 35e)

[11] *ibid*, (p. 72e)

[12] *ibid*, (p. 76e)

[13] *ibid*, (p. 77e)

[14] *ibid*, (p. 92e)

[15] *ibid*, (p. 95e)

[16] *ibid*, (p. 97e)

[17] *ibid*, (p. 98e)

[18] *ibid*, (p. 103e)

[19] *ibid*, (p. 108e)

[20] *ibid*, (p. 110e)

[21] *ibid*, (p. 124e)

[22] *ibid*, (p. 174e)

[23] February 2002. *What the Benq?*, Issue 278, Australian Personal Computer, Sydney. (p. 15)

[24] Wittgenstein, L., *op. cit.*, (p. 217e)

[25] *ibid*, (p. 218e)

[26] Popper, K., 2001. *The logic of scientific discovery*, Routledge, New York.

[27] *ibid*, (p. 16)

[28] *ibid*, (p. 29)

[29] *ibid*, (p. 32)

[30] *ibid*, (p. 34)

[31] *ibid*, (p. 40)

[32] *ibid*, (p. 63)

[33] *ibid*, (p. 65)

[34] *ibid*, (p. 82)

[35] *ibid*, (p. 93)

[36] *ibid*, (p. 96)

[37] *ibid*, (p. 98)

[38] *ibid*, (p. 99)

[39] *ibid*, (p. 118)

[40] *ibid*, (p. 137)

[41] *ibid*, (p. 142)

[42] *ibid*, (p. 206)

[43] *ibid*, (p. 208)

[44] *ibid*, (p. 427)

[45] *ibid*, (p. 280)

[46] *ibid*, (p. 276)

[47] *ibid*, (p. 394)

[48] Watterson, B., 1990. *Calvin & Hobbes*, Universal Press Syndicate, Kansas City.

Index

Made in United States
Orlando, FL
04 November 2026

90137848R00250